Psychology

Genetics of Man

GENETICS OF MAN

Second Edition

Paul Amos Moody

University of Vermont

W · W · NORTON & COMPANY · INC · *NEW YORK*

To the memory of my friend and predecessor
Henry Farnham Perkins, Ph.D.
Founder and Director of the
Eugenics Survey of Vermont

Second Edition

Library of Congress Cataloging in Publication Data
Moody, Paul Amos, 1903–
Genetics of man.
Includes bibliographies.
1. Human genetics. I. Title. [DNLM: 1. Genetics, Human. QH431 M816g]
QH431.M565 1975 573.2'1 75-2259
ISBN 0-393-09228-3

Published simultaneously in Canada
by George J. McLeod Limited, Toronto
Printed in the United States of America
2 3 4 5 6 7 8 9 0

Contents

Preface

The goal that I have had constantly before me in writing this book is to provide a meaningful understanding of the principles of human genetics without delving too far into the intricacies of mathematics and biochemistry, both of which are essential for research workers in the field. In music, there are subtleties and depths of enjoyment open only to listeners schooled, for example, in the piano or violin. Yet a listener with no musical training can gain much enjoyment from listening to Rubenstein or Heifetz. In somewhat the same way, there are profundities in the science of human genetics that can only be plumbed by the thoroughly trained, yet an understanding and appreciation of basic principles can be gained by any intelligent person who has the desire to do so. I have deliberately attempted to present a discussion of human genetics that will be comprehensible to readers with little or no formal background for the subject and that will enable them to understand how geneticists think, and what they are discovering about aspects of our lives that affect us all.

No attempt has been made to encompass the vast field of medical genetics. Human traits, normal and abnormal, have been selected for inclusion on the basis of their general interest, importance in our lives, and value in illustrating genetic principles. I have carefully documented discussions by including references to original sources and more comprehensive treatments. I particularly commend to attention the lists of references at the ends of the chapters, intended for readers who wish to delve more deeply.

Users of the first edition will find that the present one has been extensively rewritten to bring discussions up to date, to incorporate new material, and to mirror changing emphases in interpretation of older concepts. There is a new chapter on bodily traits that are of general interest and readily observed, even though the genetic bases are frequently not simple (e.g., finger and palm prints, hair color and form, handedness, polydactyly, PTC tasting, and so on).

The present edition gives added emphasis to the inborn errors of metabolism. The discussion of chromosomes has been expanded to include the newer means of identifying them, and more material on anomalies connected with changes in chromosome number and structure, a field in which knowledge increases steadily. The exciting new field of somatic cell genetics is introduced, with emphasis on cell hybridization as an important tool for mapping chromosomes.

The modern eugenics movement is discussed, with its changing concepts and goals as compared to those of earlier days. The discussion of genetic counseling has been greatly expanded to include newly developed means for increasing its precision (e.g., detection of the carrier state; fetal diagnosis), and human and ethical problems involved.

When mathematical analyses are introduced, they employ the simplest algebra, and I have attempted to explain them so thoroughly that even the least mathematically minded reader can follow the reasoning. Problems have been appended to certain chapters; these are the chapters for which problem solving provides a particularly appropriate means of checking one's understanding of subject matter.

The meanings of technical terms are indicated when the terms are first introduced. In addition, a glossary is appended for ready reference. The definitions given in the glossary are for use in understanding the text and are not intended as substitutes for complete dictionary definitions.

It is a pleasure to acknowledge the assistance of the many persons who have made the book possible. First of all, I gladly express my gratitude to the artists, Dr. Frances Ann McKittrick and Mr. Fred Haynes, whose skill has added so much to the value and attractiveness of the book. I am indebted to many authors and publishers for illustrations, and these are indicated in the captions of the figures. I wish to thank especially the people who have supplied original photographs: Mrs. Robert H. Erdmann and Mrs. Marvin G. Sheffield; Dr. Dick Hoefnagel; Dr. Victor A. McKusick; Dr. Barbara R. Migeon; Dr. James R. Miller; Dr. Orlando J. Miller; Dr. Raymond M. Mulcahy; Dr. Ronald C. Picoff and Mrs. Elizabeth M. Hier; Dr. Carl J. Witkop, Jr.; Dr. Ernest J. DuPraw; and Mr. Howard H. Hirschhorn.

I owe a special debt of gratitude to Dr. Sheldon C. Reed for reading the entire manuscript, and to Dr. Richard J. Albertini, Dr. V. Elving Anderson, Dr. William E. Hodgkin, Dr. Clyde V. Kiser, and Dr. Alexander S. Wiener for reading portions of it. The book has profited greatly from their advice, but its shortcomings are my own.

I am indebted to the Literary Executor of the late Sir Ronald A. Fisher, F.R.S., Cambridge, to Dr. Frank Yates, F.R.S., Rothamsted, and to Messrs. Oliver & Boyd Ltd., Edinburgh, for permission to reprint a portion of Table No. IV from their book *Statistical Tables for Biological, Agricultural, and Medical Research.*

I am grateful to my wife, Judith, for constant encouragement (and for supplying the thumb for Fig. 10.1), to Mrs. Anne E. B. Howland for assistance with the glossary, to my typists, Mrs. Jeannette Brown, Mrs. Madeline W. Cota, and Mrs. Susan Harkay, whose accuracy so lightened the task of revising, and to Mr. Kenneth B. Demaree and the many others responsible for the metamorphosis of a manuscript into a book.

It is my hope that this book will be useful in courses designed for students who desire a general knowledge of the subject. At the same time I shall be gratified if some readers are stimulated by the unsolved problems to equip themselves with the needed mathematical, biochemical, and cytological tools so that they, in turn, may be able to make constructive contributions to the science.

Burlington, Vermont

Paul A. Moody

Genetics of Man

1

Why Study the Genetics of Man?

As we approach the question of why we study the genetics of man, we note at the outset that men and women are in many respects rather unpromising subjects for genetic research. We cannot experiment with them so freely as we can with bacteria, corn, fruit flies, or mice. For this very reason, study of other organisms has provided absolutely essential knowledge of genetic principles applicable to ourselves. Yet it is natural that to the majority of human beings, other human beings are the most interesting organisms on earth. We find strong motivation for the study of any aspect of ourselves. One of the most interesting aspects is the manner in which we replace ourselves from generation to generation. The production of a baby is the most intricate and elaborate "manufacturing" process known. What are the controls that ensure the production of a perfect individual? We can learn something of them by studying instances in which a control failed to function. What determines the nature of the "product": eye color, skin color, hair color, blood group, intelligence, and countless other traits? This book is an attempt to supply some of the answers, although frequently we shall encounter limits to our knowledge—limits that are challenges to further investigation.

The desire to understand ourselves is, then, the strongest reason for studying human genetics. This subject combines with many others to form the intellectual field that we call science. Science, in turn, combines with philosophy, religion, literature, and art to form our cultural or social inheritance, which is our most distinctive attribute. No other organism on earth even approaches us in this respect. Hence a contribution to the knowledge of human genetics is a contribution to one facet of our crowning achievement, our greatest distinction.

Medical Applications Some readers will ask: "Are there not more *practical* reasons for studying human genetics?" Indeed there are. Let us consider first the application of human genetics to the theory and practice of medicine. When I was a young man, many of my medical colleagues looked on students of human genetics with thinly veiled amusement. What the latter were doing might be somewhat interesting in a dilettante sort of way, but it was of no real significance. Now many medical schools have their own departments of human genetics; even those that do not, treat human genetics with utmost seriousness. This revolution in thinking has come about partly because the diseases in which heredity was relatively unimportant have been largely conquered—as for example, the familiar childhood diseases. (Even here, however, heredity may play some role. If one twin contracts measles, why is the other twin more likely to do so if he is an identical twin than if he is a fraternal twin? See Chapter 23.)

This successful conquest left medical science relatively free to concentrate on (a) birth defects—abnormalities present at birth or that develop because of abnormal genes or chromosomes—and (b) the degenerative diseases of advancing years, in which heredity frequently is an important factor. Furthermore, increasing knowledge led to realization of the importance of heredity in diseases and disorders in which its importance had gone unrecognized.

As we shall see in Chapter 28, knowledge of inheritance and of the patient's ancestors can be of assistance to the physician in determining diagnosis and in deciding on the most suitable treatment. For example, perhaps a patient's blood does not clot normally. Is it hemophilia (pp. 227–230), afibrinogenemia (pp. 173–174), or some other condition? A knowledge of the patient's family and ancestry may aid in answering the question.

Genetic Counseling Although most of us are not going to be physicians, we all profit from advances in medical science. However, most of us will be parents, if we are not already, and many genetic problems arise in connection with marriage and parenthood.

At one time I received the following inquiry: "A friend of mine is desirous of marrying his half-sister's daughter. Would such a relationship be too close, relative to 'blood lines' for normal healthy children?" (This case is discussed further in Chapter 11.)

A couple whose first child was an albino consulted a geneticist to ask whether a second child was also likely to be an albino.

If a mother has schizophrenia, are her children likely to be similarly afflicted?

We do not have all of the answers by any means, but as we progress through the book we shall find that some useful answers are available for these and other questions. Furthermore, new "tools" are steadily being perfected to improve the precision of a counselor's advice. We may mention especially (a) improved methods of studying chromosomes, (b) an ever-increasing list of conditions for which the "carrier state" can be recognized (identification of parents who do not show an abnormal trait but have the defective gene that may cause it in their children), and (c) development of means of prenatal diagnosis—identification of defective fetuses before birth, early enough so that the pregnancy can be terminated if that seems wise (Chapter 28).

Eugenics Apparently from time immemorial, some people have been concerned that we are not managing wisely matters of marriage and parenthood. Thus the Greek poet Theognis, viewing his contemporaries in the sixth century B.C., wrote: "We look for rams and asses and stallions of good stock, and one believes that good will come from good; yet a good man minds not to wed an evil daughter to an evil sire, if he but give her much wealth. Wealth confounds our stock. Marvel not that the stock of our folk is tarnished, for the good is mingling with the base" (Popenoe and Johnson, 1933*). More than 250 years later Sir Francis Galton founded the eugenics movement, which survives today among people looking forward to possible genetic improvement of mankind and apprehensive of possible genetic deterioration (see Chap. 25).

For centuries, observers of the human scene have feared that the best-endowed segments of society were not producing their share of children, and they have speculated about the consequences. From Francis Bacon in the seventeenth century to Francis Galton in the nineteenth, we find concern over the fact that distinguished men were leaving relatively few descendants. Today, many people are less concerned with the make-up of the world's population than with its sheer size. One of the most pressing problems of our own time is control of population size. We use the term "population explosion" to express the speed with which the world is approaching overpopulation. The problem is so urgent that voluntary control of human reproduction is becoming more and more common and will surely become the rule in the not-distant future. Geneticists are nevertheless anxious to ensure that as quantity is brought under control, quality does not suffer. We are concerned that the genes of people who contribute most to human life and culture shall not decline in relative abundance as generation follows generation. This matter will claim our attention again in Chapter 26.

*Dates in parentheses following authors' names indicate references at the ends of the chapters.

As we learn more and more about human genetics, our hope increases that man may be able to exercise control over his own heredity. This control may have two aspects: prevention of deterioration and production of improvement.

We have hinted at the first aspect in our mention of the eugenics movement and of the problems arising from differences in fertility among different groups of people. Also involved are problems arising from increased exposure of our population to X-rays and other man-made radiations. Because such radiations are known to induce genetic changes (mutations) and many of these changes are harmful, geneticists are concerned that our descendants may inherit a "genetic load" of mutations that will give rise to a multitude of infirmities (see Chap. 21).

Parental Selection Many students of human genetics are interested in the possibility of employing new methods to prevent genetic deterioration and, if possible, bring about genetic improvement. One possibility is conscious selection of the sperm cells to be used in fertilizing the ovum. Sometimes if a husband is sterile or is highly likely to be carrying some defective gene, the couple may choose to have the wife fertilized by artificial insemination using sperm from a man lacking the defective gene. Semen from a healthy, fertile male is introduced by instrumental means into the uterus of the woman. The donor is selected on the basis of health and general racial and physical similarity to the parents, but the selection is usually made by the physician rather than by the parents. In fact, to avoid emotional involvements, the identity of the donor of the semen is kept secret.

Proponents of parental selection suggest that sometimes a couple might choose to have this done even when avoidance of genetic defects is not involved. If some man, for example, is noted for especially outstanding qualities, might not a couple choose to have his semen used to father a child instead of the semen of the husband? Because sperm cells can be preserved in the frozen state and subsequently revived, a couple might conceivably choose to have a child fathered by some eminent man of the past *if* during his lifetime he had contributed to a "sperm bank." To avoid the emotional involvements to which we have alluded, Muller, a leading advocate of the method, suggested that only semen from a man who had been dead for at least 20 years be used in the program. Moreover, the intervening years would not only help to confirm the judgment that the man did, indeed, possess qualities of eminence, but also might allow an evaluation of the quality of the children he had produced during his lifetime. This may be compared to the progeny test in the breeding of domestic animals, and it has relevance because, as everyone knows, some celebrated men fail to produce equally accomplished children. In the program we visualize, it would be important, if possible, to select as sperm donors those eminent men whose superiority had a genetic basis and could be transmitted to offspring.

In a way, such selection would be similar to adoption, and so it was called "pre-adoption" by Huxley, "prenatal adoption" by Glass.

This is not the place to argue the pros and cons of parental selection. Espousal of the program will be found in Muller, 1961, 1963, 1965a, 1965b, and in Glass, 1971. We introduce the subject to indicate one of the possible applications of human genetics. In passing, we may note that we cannot now store ova in the frozen state as we do sperm cells. Human tissues can be grown outside the body in tissue culture, however. Perhaps ovarian tissue can eventually be grown in this way, so that we can produce ova as desired. When this day comes, banks of living ovarian tissue might be built up to supplement the sperm bank. In theory, a couple might then decide that they wished a child to inherit not from themselves, but from an eminent man *and* an eminent woman of the past. A selected ovum would be fertilized by a selected sperm and then implanted into the uterus of the wife, where it would undergo normal development.

Success has already been achieved in removing from women ova in the oöcyte stage (p. 32), fertilizing these oöcytes, and growing embryos to the 16-celled stage (Edwards and Fowler, 1970). Such embryos could then be implanted in the uterus of the woman who supplied the oöcytes, or in the uterus of any woman in the proper stage of her menstrual cycle. We can easily imagine ways in which this technique could be applied to the selection of biological parentage for one's children. To what extent would people really wish to do this?

Genetic Engineering Still more visionary possibilities are found in the fertile minds of geneticists. These ideas may seem "far out" to the reader, but all are based on manipulations that have proven successful in microorganisms, plants, or lower animals.

For example, why not dispense with sperm cells and have our babies produced from *unfertilized* eggs? Reproduction in this way is called **parthenogenesis**, and is common in many kinds of lower animals. Experimenters have learned how to stimulate development of unfertilized ova in many species that are not naturally parthenogenetic. Fatherless rabbits have been produced in this way. It could conceivably be done with human ova. All offspring would be female (lacking a Y chromosome; Chap. 17). Thus, in theory, a race consisting entirely of women could be produced. Would this be desirable?

Another possibility arises from the fact that experimental embryologists working with such creatures as salamanders and frogs have learned how to remove the nucleus from an ovum and substitute for it a nucleus taken from some other cell. Suppose that a nucleus could be taken from a connective-tissue cell, for example, of a man or woman considered to have desirable traits, and transplanted into a human ovum from which the nucleus had been removed. The resulting baby would have all the desirable traits of the person from whom the nucleus was obtained (in so far as those traits were gene-determined and not the result of environment). Such a procedure would avoid the recombination of genes that occurs in the process of sexual reproduction, a subject we shall discuss in subsequent chapters.

Going one step further: Might not *both* sperm and ovum be dispensed with? Complete carrot plants can be produced from single cells. Might it not be possible eventually to produce babies from single somatic (body) cells—starting with a single connective-tissue cell, for example? A baby arising from such a cell would be the "identical twin" of the person from whom the starter cell was taken. In theory, large numbers of identical individuals could be produced in this way. Would this be desirable?

These ideas afford only a sample of possibilities envisioned for the future (see, for example, Lederberg, 1966; Glass, 1971). We may wonder: When these possibilities become actualities, will they be used wisely and for the highest good of all? Many people are concerned about the ethical issues involved (Peter, 1971; Kass, 1971; Hilton et al., 1973).

Seemingly, still further in the future than the possibilities we have mentioned is the possibility that the genes themselves can be altered by what Muller (1965b) called **genetic surgery.** Genes are immensely long molecules of DNA (Chap. 4). We may compare a gene to the tape of a magnetic tape recorder of the type now popular. Much as a defective section of such a tape can be removed and a perfect section substituted for it, in the future it may be possible to substitute a "good" gene for a "bad" one. Or it may be possible to rebuild the "bad" gene so that it will no longer produce an abnormality. Such things are already being done with bacteria and viruses, and the future application of these techniques to man is possible, even probable. For fuller discussion see Muller (1965b); Sonneborn (1965), Tatum (1965), Friedmann and Roblin (1972).

Our brief survey has given a hint of possible future applications of human genetics. But before we can apply it we must know more about it. Human genetics is the application to man of principles discovered largely by investigations on viruses, bacteria, plants, and lower animals. Investigations with these organisms offer the advantage that experimentation is possible in ways prohibited when one is dealing with people. Yet as we shall see in Chapter 19, much is now being learned by growing human cells in laboratory glassware and studying them somewhat as the microbiologist investigates his organisms.

On the other hand, human populations offer advantages, too. Large numbers are available if desired, and verbal communication is an inestimable boon to investigation. For no other organism do we have such detailed knowledge of biochemistry, anatomy, physiology, and psychology, including anomalies and aberrations that frequently give clues to factors at work in normal development. Hence the investigations on man complement and supplement those on lower animals and plants of which they are the outgrowth.

We consider Gregor Mendel the father of modern genetics. Because the genetics of man is included in this term, we most appropriately begin our discussion with a brief review of his experiments. In this we wish to emphasize the thinking he employed in deducing from them those principles on which our science is founded. Then we shall apply those principles to human investigations to see how they can be utilized, and how modified, to fit studies in which the investigator cannot arrange the matings of his subjects.

Science and art are more akin than most people realize. Both have their beauties. There is esthetic appeal in a logical progresssion of thought or in a simple principle that, at a stroke, reveals the explanation of diverse phenomena that had been mysterious. Although human genetics has its utilitarian aspects, it also has its full measure of beauty. I hope that this will become evident as we proceed, and that it will add to the enjoyment of the pages that follow.

References

Edwards, R. G., and R. E. Fowler, 1970. "Human embryos in the laboratory," *Scientific American,* **223**:45–54.

Friedmann, T., and R. Roblin, 1972. "Gene therapy for human genetic disease?," *Science,* **175**:949–955.

Glass, B., 1971. "Science: Endless horizons or golden age?," *Science,* **171**:23–29.

Hilton, B., D. Callahan, M. Harris, P. Condliffe, and B. Berkley, (eds.), 1973. *Ethical Issues in Human Genetics.* Fogarty International Proceedings No. 13. New York: Plenum Press.

Kass, L. R., 1971. "The new biology: What price relieving man's estate?," *Science,* **174**:779–788.

Lederberg, J., 1966. "Experimental genetics and human evolution," *American Naturalist,* **100**:519–531.

Muller, H. J., 1961. "Human evolution by voluntary choice of germ plasm," *Science,* **134**:643–649.

Muller, H. J., 1963. "Genetic progress by voluntarily conducted germinal choice." In Wolstenholme, G. (ed.), 1963. *Man and His Future.* Ciba Foundation Volume. Boston: Little, Brown & Co. Pp. 247–262.
Muller, H. J., 1965a. "Better genes for tomorrow." In L. K. Y. Ng and S. Mudd (eds.). *The Population Crisis.* Bloomington: Indiana University Press. Pp. 223–247.
Muller, H. J., 1965b. "Means and aims in human genetic development." In T. M. Sonneborn, 1965. Pp. 100–122.
Peter, W. G. III, 1971. "Ethical perspectives in the use of genetic knowledge," *BioScience,* **21**:1133–1137.
Popenoe, P., and R. H. Johnson, 1933. *Applied Eugenics,* rev. ed. New York: The Macmillan Company.
Sonneborn, T. M. (ed.), 1965. *The Control of Human Heredity and Evolution.* New York: The Macmillan Company.
Tatum, E. L., 1959. "A case history in biological research," *Science,* **129**: 1711–1714. (This is the lecture given when Dr. Tatum received the Nobel prize.)
Tatum, E. L., 1965. "Perspectives from physiological genetics." In T. M. Sonneborn, 1965. Pp. 20–34.

2

Essentials of Mendelian Thinking

Variability provides the raw materials for genetics, the science of heredity. If all individuals were brown-eyed, we could learn nothing of the hereditary basis of eye color. However, because some people are blue-eyed and marry brown-eyed people, we can gain some knowledge of the genetics involved. (As we shall see in Chap. 8, this particular example has a more complicated genetic basis than people generally realize.) So it is with all genetic study: We constantly analyze the genetic bases of the *differences* between individuals. Some of these differences occur normally, such as differences in eye color, blood groups, or ability to taste PTC (phenylthiocarbamide). Other differences are abnormal, involving physical abnormalities and abnormalities in the processes of metabolism. For example, although most of us can break down the amino acid phenylalanine, derived from protein in our diet, to simpler compounds, for excretion, some people have a gene-determined inability to do this. They suffer from what is called phenylketonuria, which, if unchecked, leads to severe mental deficiency. We shall find examples of this kind, involving gene control of enzyme systems, particularly instructive as we attempt to answer the question: How do genes work?

We might note in passing that the necessity of concentrating on differences imposes some limitations on genetics. There are definite limits to the variations that an individual can have and still survive. For example, we might wish to know the genetic basis for the formation of the liver, but because no human being could possibly exist without a liver, we are never likely to learn all the genetic complexities underlying liver formation. We can, however, analyze the genetic bases of differences in liver function that are not severe enough to be lethal.

Our knowledge of genetics is based on investigations of plants and animals, which can be experimented on in a way impossible with human beings. Once hereditary differences are discovered, it is of great advantage to be able to arrange matings between the differing parents and to raise large numbers of offspring.

The father of the modern science of genetics was Gregor Mendel, an Austrian monk whose classic experiments on garden peas revealed principles fundamental to all genetic thinking.

Mendel's Experiments

Mendel concentrated attention on seven differences in peas. Some were tall; some were dwarf. Some had white flowers; some had colored ones. Some had seeds that were smooth when dry; some had wrinkled seeds. Because the experiments performed and the conclusions reached were similar in all cases, we shall concentrate on only one typical series of experiments.

One of the seven pairs of contrasting characteristics Mendel studied concerned the color of the cotyledons (rudimentary leaves) in the seed. These were yellow in the seeds produced by some plants and green in the seeds produced by other plants. The yellow-seeded plants were pure-breeding, and so were the green-seeded ones.

Mendel started his experiment by cross-pollinating these two varieties: placing pollen from the yellow-seeded variety on the stigmas of flowers of the green-seeded variety (the stamens of these flowers having been removed), and vice versa. In all cases, the seeds produced had yellow cotyledons. We might suspect that the green color and its genetic basis had simply been destroyed; this did not prove to be the case, as we shall see.

It has become customary to refer to such first-generation hybrids as the "first filial generation," or **F_1** generation. Of the pair of contrasting traits been considered, the one that produced a visible effect in the F_1 generation Mendel called **dominant**. Conversely, he used the term **recessive** for the trait that did not produce a visible effect in this first generation. In genetics the term "dominant" lacks many of the connotations of the word in common speech. Dominant traits are not necessarily superior to or better than recessive ones. For example, one of the most severe types of mental defect in man (Huntington's chorea) is transmitted by a dominant gene.

Mendel next planted the hybrid seeds; 258 of them germinated. (In higher plants the seed is a miniature individual, enclosed in a seed coat and waiting suitable conditions to enlarge and develop. Hence the seed and the plant that develops from it both represent an F_1 individual in different stages of its life history.) These F_1 plants produced flowers, which were permitted to fertilize themselves (normal procedure in peas). The 258 plants produced a total of 8023 seeds. These seeds constituted a "second filial" or $\mathbf{F_2}$ generation. Of these seeds, 6022 had yellow cotyledons and 2001 had green cotyledons. Thus the ratio of yellow to green was 3.01:1.

In his experiments with the other six pairs of contrasting traits, Mendel also found that approximately three-fourths of the F_2 individuals showed the dominant trait, and one-fourth showed the recessive trait. Here was a regularity to be explained. Mendel explained it as an indication that each pair of traits depended on a single pair of hereditary factors, which we call genes today. This bring us to the point of our discussion: *Why is obtaining uniformity in the F_1 generation, followed by a 3:1 ratio in the F_2 generation, regarded as evidence that inheritance depends on a single pair of genes?*

In answering the question, we assume at the outset that with regard to the traits under consideration, male and female parents contribute equally (through pollen grain and ovum). Although we may think that this point is too self-evident to stress, in the early history of biology equal contribution by the parents was not recognized. There are also situations in which the contributions of the parents are not equal (e.g., sex linkage, Chap. 14).

If inheritance from the two parents is equal, then a plant of the pure-breeding, yellow-seeded variety must have inherited "yellowness" from both of its parents. Both the ovum from which it developed and the pollen grain that fertilized that ovum must have contained a gene for yellowness. If the initial *Y* represents the gene "for" yellow, then a pure-breeding, yellow-seeded plant would have the genetic formula *YY*. The genetic formula is called a **genotype.** The visible trait, yellow color in this case, is called the **phenotype.**

A pure-breeding, green-seeded plant would have the genotype *yy* (*y* representing the gene for green color). *Y* and *y* constitute a gene pair, or, as we say, a pair of **alleles.** *Y* is the allele of *y* (its opposite number, so to speak) and vice versa. The occurrence of genes in allelic pairs is a fundamental assumption of Mendelian thinking.

Mendel crossed pure-breeding yellow peas with pure-breeding green ones to produce the F_1 offspring. In terms of our assumed genotypes we may represent this as $YY \times yy$. Let us suppose that in this cross the yellow parent contributed the pollen grains (as was true part of the time; which parent contributed pollen and which contributed ova made no difference in the outcome). We assume, following Mendel's thinking, that when pollen grains are produced the members of a pair of alleles separate, so that each pollen grain receives only one member of the pair (only one *Y* gene, in this case). This assumed separation of genes is sometimes called the **law of segregation, the first Mendelian law.** It asserts that in the cells of individuals, genes occur in pairs, and that when those individuals produce germ cells each germ cell receives only one member of the pair. This law applies equally to pollen grains (or sperm) and to ova.

Hence we may represent the cross as follows:

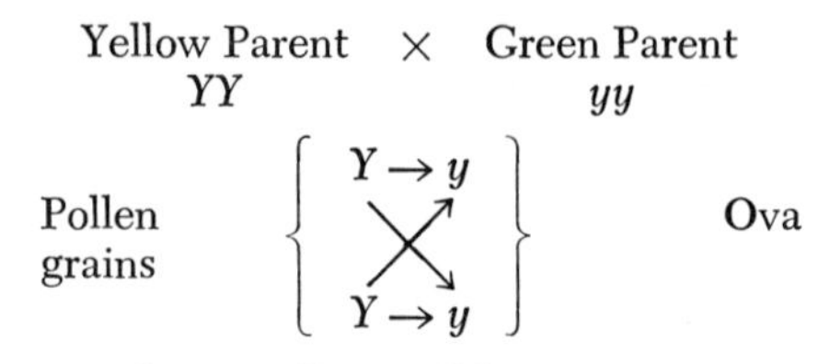

The arrows indicate all possible combinations of pollen grains and ova. We note that all combinations yield the genotype *Yy*. All these F_1 seeds had yellow cotyledons, yellow being dominant to green, or in genetic terms, *Y* being dominant to *y*. These F_1 individuals resembled the yellow parent in phenotype (being yellow) but not in genotype (*Yy* as opposed to *YY*). Both parents were ***homozygous.*** Both members of that pair of alleles were the same. The yellow parent was homozygous for *Y* and the green parent for *y*. The F_1 individuals were **heterozygous,** having one *Y* and one *y*.

Succeeding events demonstrated that the genes *Y* and *y* were not altered or modified in any way by being combined in the cells of a heterozygote. We note especially that the gene *y*, although it produced no phenotypic effect in the heterozygote, retained all of its properties so that later, in a homozygous descendant (*yy*), the phenotype "green" was produced.

The flowers on the F_1 plants were permitted to fertilize themselves, each flower producing both ova and pollen. To facilitate comparison with animals and man, however, we shall represent a cross between two F_1 individuals as if one parent produced all the pollen grains, the other all the ova. (This really does no violence to Mendel's experiments, because he would have obtained the same results if he had used a technique of artificial pollination, taking pollen from one F_1 flower and placing it on the stigma of another one.)

F_1 parents   Yellow × Yellow
Genotype   *Yy*   *Yy*

Pollen   ½*Y* → ½*Y*   Ova
   ½*y* → ½*y*

What happens when a heterozygote (*Yy*) produces germ cells? As in the preceding generation, each germ cell (**gamete**) receives only one member of a pair of alleles, either *Y* or *y*. We assume that the chances are equal that a given gamete will receive Y or y. Hence, on the average, half the pollen grains will contain Y and half will contain y; in the same way, half the ova will contain Y and the other half will contain y. On the face of it, this seems an eminently reasonable assumption to make. Mendel, however, was not familiar with the cellular mechanism that seems to underlie this production of two kinds of germ cells in equal numbers, which will be discussed in the next chapter.

The Role of Chance or Probability Now we are ready for the combinations of the two kinds of pollen grains with the two kinds of ova. Mendel assumed that this combining followed the laws of chance or probability. This means that a pollen grain containing Y is equally likely to fertilize an ovum containing Y or an ovum containing y (because these two types of ova occur in equal numbers). The same is true of a pollen grain containing y. Note that we specifically assume that there is no tendency to preferential pairing (e.g., no tendency for Y-bearing pollen grains to fertilize Y-bearing ova more frequently than do y-bearing ones). Thus the four combinations of pollen and ova (above diagram) are expected to occur as follows:

$$\left.\begin{array}{lcl} \tfrac{1}{4}\,YY & = & \text{Yellow} \\ \tfrac{1}{4}\,Yy & = & \text{Yellow} \\ \tfrac{1}{4}\,Yy & = & \text{Yellow} \end{array}\right\}\ \tfrac{3}{4}$$
$$\begin{array}{lcl} \tfrac{1}{4}\,yy & = & \text{Green} \quad \tfrac{1}{4} \end{array}$$

It is most important to realize that these fractions depend on the operation of the laws of probability. A model using coins will help to emphasize the point. Such a model consists of two coins—a dime and a penny, perhaps—tossed at the same time. The dime may represent the pollen: At any given toss the chances are equal that the dime will come up "heads" or that it will come up "tails." Similarly, at any given fertilization the chances are equal that a Y-bearing pollen grain or that a y-bearing one will be transmitted.

The penny represents the ovum. Again, the chances are equal for "heads" or "tails," just as the chances are equal that in any fertilization a Y-bearing or a y-bearing ovum will be involved. If we toss the two coins together and do it many times, we shall obtain an approximation of the following:

¼ dime heads; penny heads [$= YY$]
¼ dime heads; penny tails [$= Yy$]
¼ dime tails; penny heads [$= yY$]
¼ dime tails; penny tails [$= yy$]

If we regard "heads" as dominant, we find that three-fourths of the time there is at least one "head," one-fourth of the time no "heads" (both coins "tails"). Hence this gives us a model of the 3:1 ratio dependent on the laws of probability.

We can now see that obtaining a 3:1 F_2 ratio can be most simply explained by assuming that a single pair of genes is involved in the inheritance. It is further assumed (a) that the genes are paired in the cells of individuals, (b) that each germ cell (gamete) receives only one gene of the pair, (c) that heterozygotes produce in equal numbers gametes containing the dominant member of the pair and gametes containing the recessive member, and (d) that the laws of probability are followed when the two types of male gametes fertilize the two types of female gametes.

This is an appropriate place to point out that phenotypes are usually, if not always, the result of many genes working together. We speak, for example, of the gene "for" yellow. What we mean is that when this gene is absent, yellow color will not appear. However, the phenotype of the cotyledons was not determined solely by this gene. Mendel noted that the color of the yellow seeds might be pale yellow, bright yellow, or orange hued. All had the gene Y. What caused the difference in shade? Two additional influences were doubtless also at work: (a) environmental factors and (b) other genes, which affected the phenotype determined by the "main gene." These other genes may have been of the type called **modifying genes,** perhaps genes that determined the quantity of yellow pigment present after the gene Y had determined that yellow pigment, in whatever quantity, should be present. When genetic analysis is thorough enough, the presence of modifying genes of this sort is almost always revealed.

We note that the 3:1 ratio we have been stressing is a *phenotypic* ratio. The *genotypic* F_2 ratio is $\frac{1}{4}YY:\frac{2}{4}Yy:\frac{1}{4}yy$. By further experiments, Mendel determined that the green F_2 plants bred true, but that not all the yellow plants did so. He found that one out of three of the yellow plants bred true but that two-thirds of them behaved like their F_1 parents and gave 3:1 ratios in their turn (in an F_3 generation). These findings, agreeing with theoretical expectation, add further evidence that Mendel's explanation is correct—that single gene-pair inheritance is involved in the difference between yellow and green cotyledons in peas. The same conclusion applies to the other six differences he investigated.

Mendelian Inheritance in Animals

Dominance Present Since Mendel's time, examples almost without number of single gene-pair inheritance have been investigated in many kinds of plants and animals. Figure 2.1 illustrates a well-known example from the animal kingdom, black versus white color in guinea pigs. As shown, if homozygous black guinea pigs are mated to white ones, the F_1

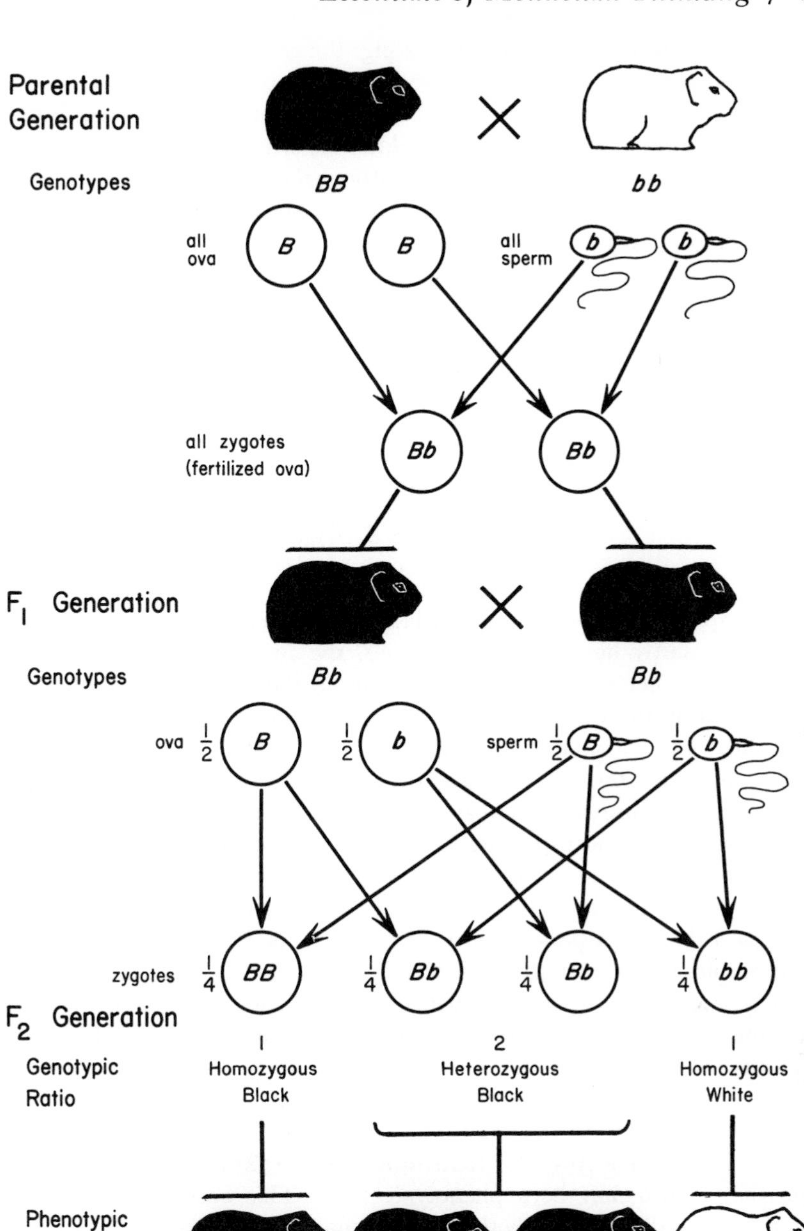

FIG. 2.1 Simple Mendelian inheritance of coat color (black versus white) in guinea pigs.

offspring are all black. Thus in guinea pigs the gene for black (B) is dominant, because these F_1 individuals are heterozygous.

When F_1 females are mated to F_1 males, the F_2 offspring display a phenotypic ratio of 3 blacks to 1 white. As with Mendel's peas, this is based on the laws of probability operating when two equally numerous kinds of sperm fertilize two equally numerous kinds of ova. The genotypic ratio in the F_2 generation is 1 BB:2 Bb:1 bb.

This is a good place to emphasize the fact that the most fundamental Mendelian ratio is 1:1, the ratio in which the two types of gametes are produced (½ B-containing:½ b-containing). The 1:2:1 ratio is simply two of these 1:1 ratios multiplied together:

$$\begin{array}{lll} 1B + 1b & & \\ 1B + 1b & & \\ \hline 1BB + 1Bb & & \\ & 1Bb + 1bb & \\ \hline 1BB + 2Bb + 1bb & & \end{array}$$

At times this 1:1 gametic ratio may also be the phenotypic ratio. For example, when heterozygous black guinea pigs are mated to white ones (necessarily homozygous recessive), the two kinds of gametes produced by the heterozygous parent determine the two kinds of offspring and determine the ratio in which they shall occur (for the white parent contributes only b-containing gametes).

Such a cross is diagramed in Figure 2.2. It is sometimes called a **back cross,** because the heterozygous individual is genetically similar to an F_1 individual, and the white individual is genetically similar to one of the parents of such an F_1 individual. It is also known as a **test cross** because it is a means by which a geneticist can test whether an individual showing a dominant phenotype is homozygous or heterozygous. Suppose, for example, that you wish to know whether a certain black guinea pig is homozygous or heterozygous. If you mate it to a white one and they produce one or more white offspring, you know at once that the black individual is heterozygous.

Dominance Absent Returning to our 1:2:1 genotypic ratio, we note that it also may be a phenotypic ratio. This occurs whenever the heterozygotes differ from both types of homozygotes, and this in turn occurs when dominance is lacking or incomplete. It so happened that Mendel worked with pairs of traits in which one trait did exhibit dominance over the other. Hence he did not observe a 1:2:1 phenotypic ratio.

A well-known and easily visualized case in point is provided by Blue Andalusian fowls (Fig. 2.3). The "blue" birds are heterozygotes, produced in the first instance by mating a certain type of black fowl to

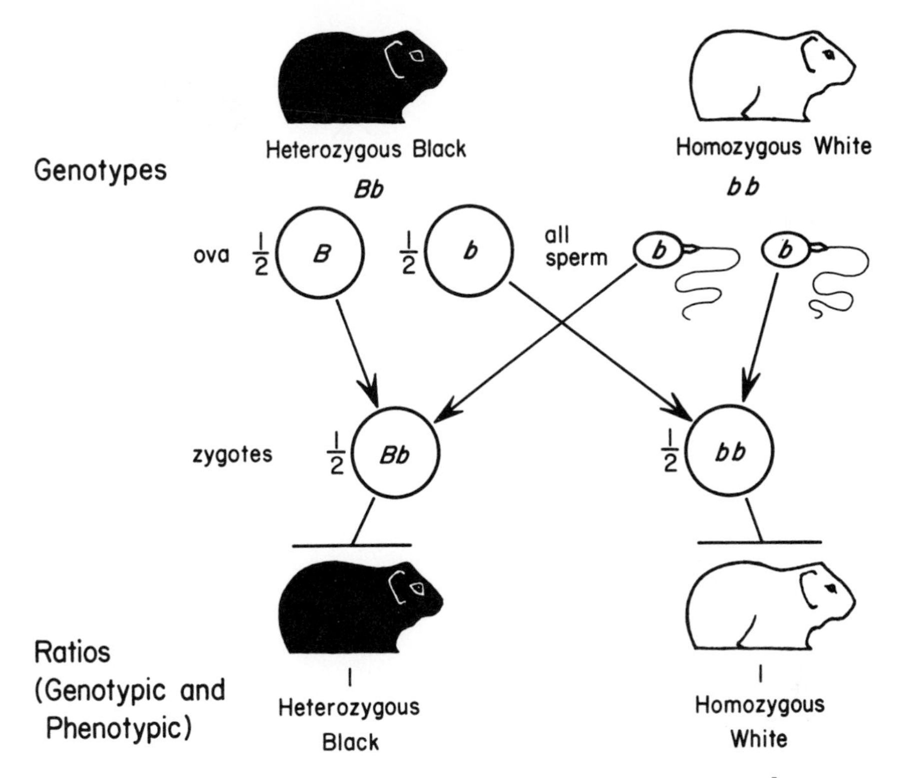

FIG. 2.2 A simple Mendelian back cross: heterozygote mated to homozygous recessive.

"splashed white" ones. These latter birds are white, but some of the feathers are margined with black. If the black parents are represented as having the genotype *BB* and the splashed white parents as having the genotype *bb,* all the F_1 offspring will have the genotype *Bb.* These F_1 individuals are blue in color, and so afford an example of the situation mentioned above in which heterozygotes differ from both types of homozygotes.

Very probably this is a "dosage effect": two *B* genes together produce much black pigment, and so the *BB* individual appears black. One *B* gene by itself produces less black pigment spread thinly throughout the feathers. The blue color seen by the observer is the combined effect of this pigment and the refraction of light by the surface layer of the feathers. It is not, therefore, a matter of blue pigment versus black pigment. The optical effect "blue" occurs when there is less pigment than the amount that gives the optical effect "black." The genetically determined difference between *BB* and *Bb* individuals is doubtless mainly a quantitative one. We shall encounter many differences that are quantitative rather than qualitative (Chap. 8).

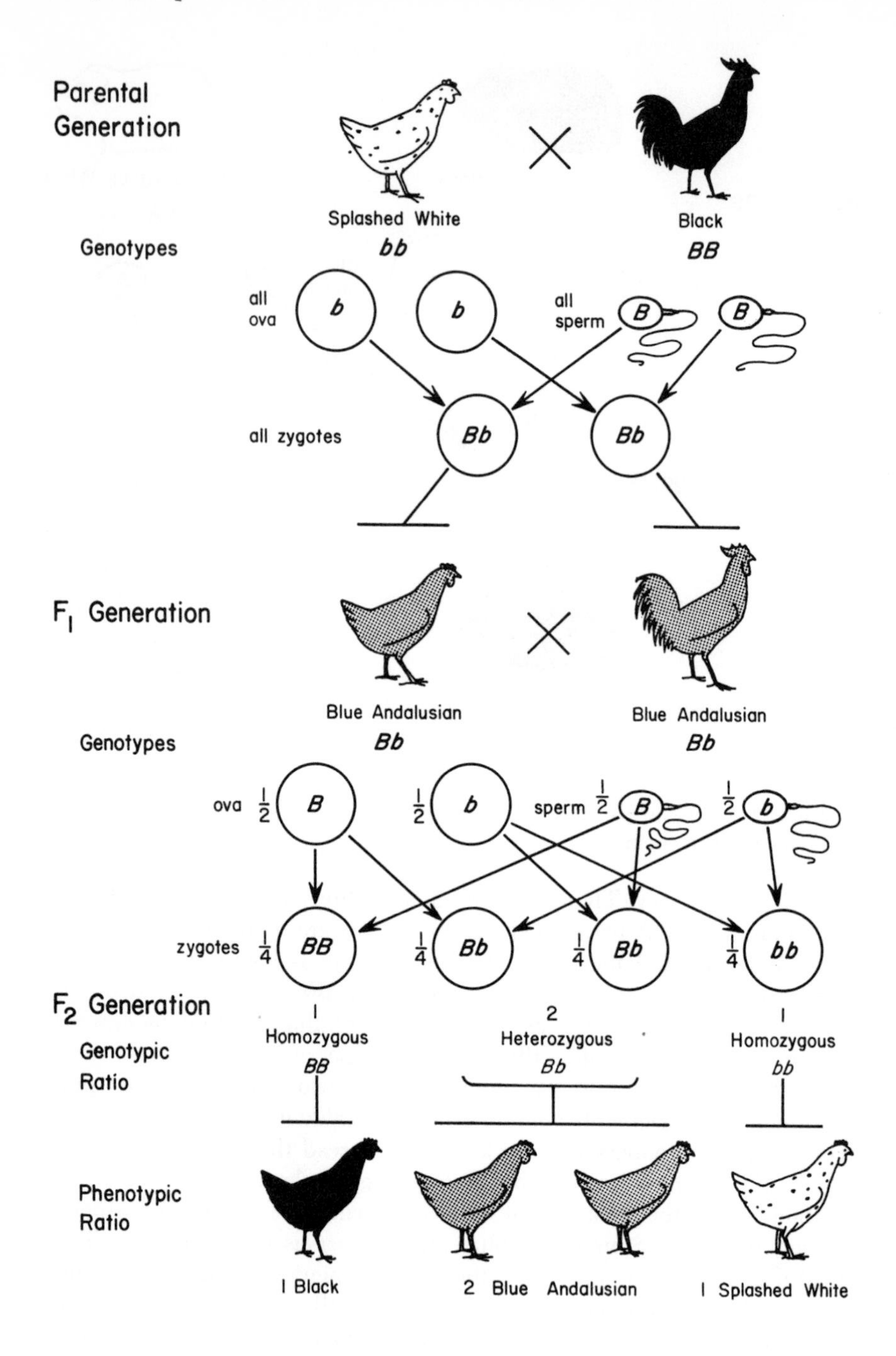

FIG. 2.3 Simple Mendelian inheritance with dominance absent or incomplete: Blue Andalusian fowls.

As shown in Figure 2.3, when Blue Andalusians are mated together they produce offspring in the ratio of 1 black:2 blue:1 splashed white, the genotypic and phenotypic ratios being the same.

Mendelian Ratios and the Number of Individuals

One aspect of Mendel's experiments is frequently overlooked although it has real significance for our thinking. Not all the F_1 plants produced yellow or green seeds in a 3:1 ratio. As we have seen, the *average* seed production of the 258 plants approximated the 3:1 ratio very closely, but individual plants varied widely. As extremes, Mendel recorded (a) one plant that had 32 yellow seeds and only 1 green seed, and (b) one plant that had 20 yellow seeds and 19 green seeds. In his paper he gave the seed counts for the first ten plants counted. The results were as follows:

Plant No.	Yellow Seeds	Green Seeds
1	25	11
2	32	7
3	14	5
4	70	27
5	24	13
6	20	6
7	32	13
8	44	9
9	50	14
10	44	18

Note how widely some of these departed from the 3:1 ratio. Yet plants having less than one-fourth of the green seeds were compensated for by plants having more than one-fourth of the green seeds, so that the average (based on 8023 seeds) closely approximated 3:1.

We emphasize this point because it is exactly what is expected when phenomena depend on the laws of probability. We have emphasized that these laws form the basis of all Mendelian thinking.

Because this is true, large numbers of offspring must be raised and studied if we are to attain close approximations to the theoretical ratios. Suppose, for example, that Mendel had raised only one F_1 plant and that the plant had been the one designated as No. 8 in the preceding table. It produced 53 seeds; 44 were yellow, 9 were green. This one plant by itself would never have suggested to Mendel that an underlying 3:1 ratio was present, for it gave an almost perfect 5:1 ratio.

Large numbers can be attained easily with plants and lower animals, and notably with bacteria, molds, and other microorganisms. In recent years, many geneticists have turned to investigations on these latter, in part because of the tremendous numbers that can be produced with ease.

However, we are presently concerned with human genetics where, save for certain national population studies, numbers are always relatively small. We must draw what conclusions we can from small numbers; means of doing this are discussed in Chapter 7. At present, we wish to emphasize the point that when numbers are small, close approximations to the theoretical ratios are not to be expected.

A human characteristic that behaves as a Mendelian recessive is albinism (lack of pigment in skin and eyes). Albinism is discussed more fully in Chapter 5. As explained there, we represent the gene by *c*, the allele for normal pigment by *C*. What will be expected when two normally pigmented but heterozygous individuals marry each other (the human equivalent of an $F_1 \times F_1$ mating)?

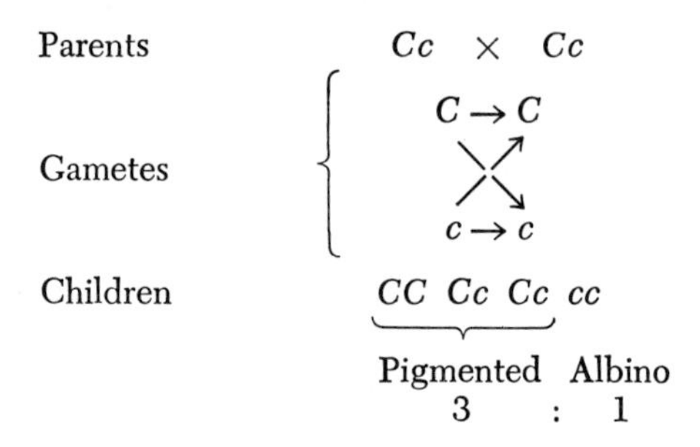

Thus three-fourths of the children can be expected to be pigmented, one-fourth to be albino.

Now suppose, as may well be the case, that this couple has only four children. What are the chances that three will be pigmented and one will be an albino? People who know just a little about Mendel's experiments would say that such an outcome is certain. Do not the Mendelian laws prescribe it? As we have seen, the Mendelian laws are special applications of the laws of probability. They prescribe that if, for example, a couple had 400 children (what a delight that would be to a human geneticist!) *about* 100 of them would be albinos, the rest pigmented.

But suppose we pursue our question a little further: What *are* the chances that a family (**sibship**) of four, born to heterozygous parents, will consist of three pigmented children and one albino child?

The albino child may be any one of the four. The chance that any one child will be pigmented is ¾, the chance that he will be albino is ¼. What are the chances that the first child will be an albino and the other three pigmented? The chance that the first child will be an albino is ¼. The chance that the second child will be pigmented is ¾. The chance that the third child will be pigmented is ¾. The chance that the fourth child will be pigmented is $3/4$. Now each child is an "independent event"; whether or not the first child is an albino has no influence on whether or not the second child will be, and so on. The total probability of several independent events occurring together is the product of the probabilities that they will occur singly (thus the chance that three coins tossed together will all fall "heads" is $1/2 \times 1/2 \times 1/2 = 1/8$). In the present instance, then, we multiply the four probabilities: $1/4 \times 3/4 \times 3/4 \times 3/4 = 27/256$.

Thus the chance that the first child will be an albino and the following three pigmented is $27/256$.

However, it may equally well be the second child who is an albino (the first, third, and fourth pigmented). The chance of this is also $27/256$. In the same way, the chance that the albino child is the third one (first, second, and fourth being pigmented) is $27/256$. And the chance that the albino child is the fourth one (first, second, and third being pigmented) is $27/256$. The total chance that the family of four will consist of one albino child and three pigmented ones is $4 \times 27/256 = 108/256$ or about $2/5$. (A less laborious way of calculating such expectations is presented in Chapter 7, p. 86). Thus we see that less than half the time are families of four, born to heterozygous parents, expected to exhibit the 3:1 ratio exactly.

Now we readily understand the error in thinking made by a young couple of whom I once read. They had one child, an albino (thereby demonstrating that they were both heterozygous). They came to a genetic counselor with the question as to the chances that future children would also be albinos. He explained Mendelian principles to them as best he could, and they grasped the idea that they could expect one child in four to be an albino. Then they had the happy thought: "We have had our albino child; now we can go ahead and have three more children and be sure that they will all be normal!" They did not grasp the idea that each child is an independent event.

Not only do individual small samples varying from the 3:1 ratio not disprove Mendelian principles, as has sometimes been contended, but their occurrence can be predicted, statistically, in accordance with Mendelian principles. To return to our family of four children, if two-fifths of such families are expected to consist of one albino and three normally pigmented children, then three-fifths of such families are expected *not* to fit the 3:1 ratio exactly (i.e., to consist of four pigmented, or of two pigmented and two albino, or of one pigmented and three albino, or of four albino children).

In the next chapter we shall discuss the chromosomal basis of Mendelian genetics, a basis unknown to Mendel himself. Then in Chapter 7 we shall discuss ways in which Mendelian thinking can be applied to human genetics, where breeding experiments of the type employed by Mendel and his successors are not possible.

Problems

1. One of the traits studied by Mendel in peas was height of stem. He found that tallness was dominant to dwarfness. If you had homozygous tall peas and crossed them with dwarf peas, what proportion of the F_1 offspring would be tall? What proportion would be dwarf? If you allowed the F_1 plants to self-fertilize and thus produced 1200 seeds, how many of these seeds would you expect to grow into tall F_2 plants? How many into dwarf F_2 plants? How many of the dwarf F_2 plants would you expect to be homozygous? How many of the tall F_2 plants would you expect to be homozygous?
2. In guinea pigs, short hair is dominant to long. A short-haired guinea pig was mated to a long-haired one. Five offspring were produced: three long-haired and two short-haired. Give the genotypes of the parents and of the offspring.
3. If homozygous short-haired male guinea pigs are mated to heterozygous short-haired females, what proportion of the offspring will be expected to be (a) homozygous short-haired, (b) homozygous long-haired, (c) heterozygous short-haired, (d) heterozygous long-haired?
4. Suppose that you experiment with a certain species of insect and discover that whereas most of the individuals are dark brown in color, an occasional individual is tan. When you mate dark brown individuals to tan ones, all of the F_1 offspring are dark brown. When these F_1 offspring are mated to each other, you obtain the following F_2 generation: 338 dark brown, 112 tan. What is the most probable explanation for these results? Assign symbols to the genes and give the genotypes of (a) the original parents, (b) the F_1 offspring, (c) the F_2 offspring.
5. How would your interpretation of the genetic situation in Problem 4 have differed if the F_1 offspring had been light brown in color, and if the F_2 offspring had been as follows: 110 dark brown, 228 light brown, 112 tan?
6. A husband and wife both have normal skin pigmentation. Their first child is an albino. Give the genotypes of the parents and of the albino child. What is the chance that if they have a second child, this child will be an albino? What is the chance that if they then have a third child, this child will be an albino?
7. An albino man marries a woman who has normal pigmentation. Their first child is an albino. Give the genotype of the mother.

Reference

Mendel, G. 1865. "Experiments in plant hybridization." (Original paper in *Verhandlungen naturforschender Verein in Brünn, Abhandlungen, iv,* [1865], which appeared in 1866.) Two easily available sources for the English translation are the following: Peters, J. A. (ed.), 1959. *Classic Papers in Genetics*. Englewood Cliffs, N.J.: Prentice-Hall, Inc. Pp. 1–20. Sinnott, E. W., L. C. Dunn, and T. Dobzhansky, 1958. *Principles of Genetics,* 5th ed. New York: McGraw-Hill Book Company, pp. 419–443.

3

Chromosomes, Mitosis, and Meiosis

Mendel's report on his research was published in 1866, and then completely ignored for about 30 years, only to be rediscovered in 1900 when three other geneticists, independently, came to conclusions that were substantially the same.

During this 30 years, great progress was made in the study of the microscopic structure of cells and of the changes involved when cells divide and when gametes are formed. Early in the present century, two authors pointed out that in the process of gamete formation the **chromosomes**, deeply staining bodies in the nuclei of cells, behave in such a manner as to provide the mechanism for the type of inheritance Mendel had postulated (see Sutton, 1903).

As was noted in the preceding chapter, the Mendelian explanation of inheritance involves two assumptions: (a) Units of heredity, which today are called genes, occur in pairs, and (b) in the formation of gametes the members of a pair of genes separate so that each gamete receives only one of the two. Sutton and Boveri both pointed out (a) that chromosomes occur in pairs, and (b) that in the formation of gametes the members of a pair separate so that each gamete receives only one member of the pair. (See Sturtevant, 1965, for historical background.) This striking parallel between the behavior of genes and chromosomes suggests that the chromosomes are indeed the bearers of the genes. Although this parallelism does not in itself prove that the chromosomes are bearers of the genes, the overwhelming amount of corroboratory evidence amassed in the last 60 years has firmly established the connection between gene and chromosome. (For an orderly survey of the evidence, see Snyder and David, 1957).

Chromosomes

Chromosomes are extremely small. In the condensed state usually studied, the largest human chromosome is not over 10 microns (abbreviated μ) in length ($1/100$ of a millimeter). Chromosomes are most easily seen with the light microscope when the cell is dividing. At this time they stain deeply with biological stains.

Mitosis

When a cell divides, the division of the cell body (cytoplasm) is preceded by a most precise distribution of the chromosomes in a process called mitosis.

Figure 3.1A shows a cell prior to mitosis. The material of the chromosome is comparable to an extremely slender thread. Careful studies demonstrate that each such attenuated chromosome has already duplicated itself. The only visible indication that mitosis is about to occur is the fact that a tiny body outside the nucleus, the **centriole**, has divided, the two centrioles thus formed have started to move apart, and a spindle-shaped arrangement of fibers has formed between them. The centrioles continue to migrate until they reach the poles of the cell (Figs. 3.1B and 3.1C), and the spindle becomes large. Although details of the mechanisms involved are still obscure despite much research, the spindle in some manner aids or guides the distribution of the chromosomes.

During the portion of mitosis called the **prophase** (Fig. 3.1B), each chromosome coils tightly and condenses. Because each chromosome has already duplicated itself, the result of this condensation is the formation of a double structure composed of two **chromatids** held together by an unseparated region called the **centromere** (represented in Fig. 3.1 by a tiny clear circle). Biochemically, each chromatid is a complete chromosome. The processes by which these results are achieved are described more fully in the next chapter.

Four chromosomes are shown in Figure 3.1, two long ones and two short ones; a human cell contains 46, but to include them all would make an unwieldly diagram.

The chromosomes line up at the equator of the spindle (Fig. 3.1C). The cell is then said to be in **metaphase.** This arrangement of chromosomes is called an **equatorial,** or **metaphase, plate.** At this time the chromosomes are in the best position for counting and study (Chap. 18).

As we have noted, each chromosome consists of two chromatids united at the centromere. Separation now occurs at the centromere. This frees the sister chromatids from each other, and one moves toward one

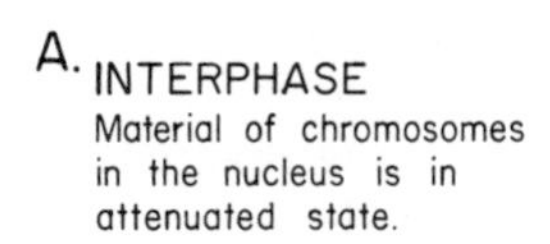

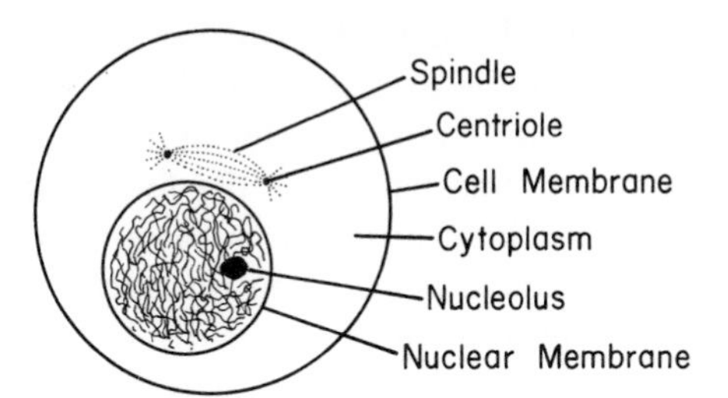

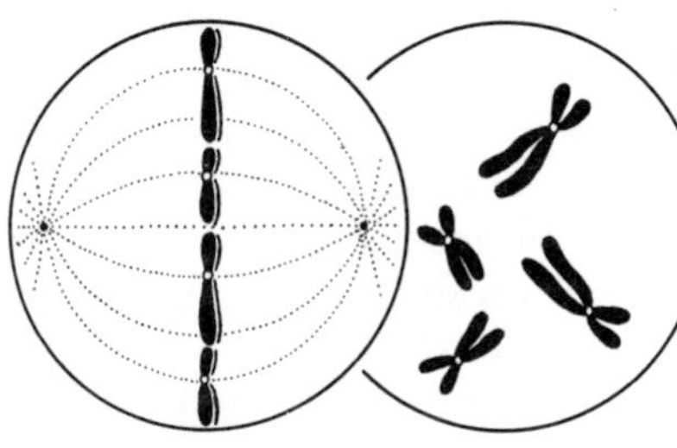

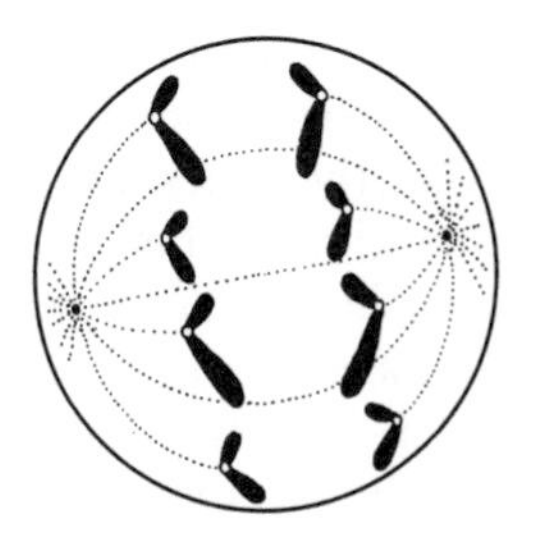

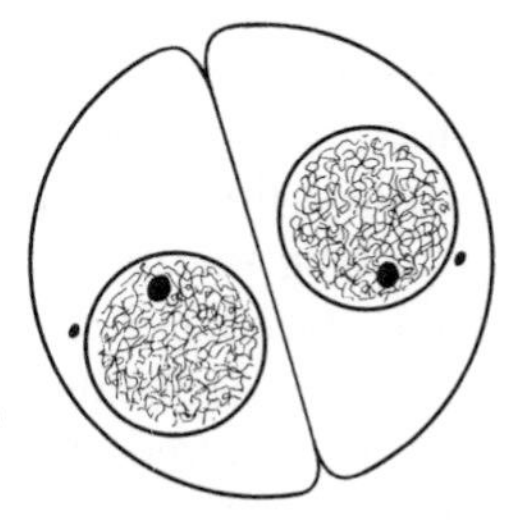

FIG. 3.1 The genetically significant highlights of mitosis.

pole of the spindle, the other toward the other pole (Fig. 3.1D). The cell is now said to be in **anaphase.** The motive force seems to be centered in the centromere, the remainder of the chromatid being dragged along rather passively. The nature of the forces that act on the centromere is still uncertain. The centromere is sometimes called the "spindle fiber attachment," or kinetochore. Some investigators conclude that the attached fiber exerts a pull on the centromere, but the evidence is not conclusive either for or against this idea. At any rate the chromatids, each of which can now be called a chromosome in its own right, move toward their respective poles until they cluster about the latter.

About this time the cytoplasm of the cell starts to divide; in animal and human cells this is usually by formation of a furrow which grows deeper until the cell is cleaved in two (Fig. 3.1E). At this stage, the **telophase,** a nuclear membrane is formed around each mass of chromosomes. The chromosomes elongate and uncoil until they regain the attenuated state they had exhibited before mitosis began. Each daughter cell has now become a cell in **interphase.**

During interphase, each chromosome duplicates itself in preparation for the next mitosis. This doubling of the amount of chromosomal material can be detected by chemical means even though the chromosomes themselves cannot be seen at this time.

What is the significance of mitosis? This rather elaborate process ensures that when a cell divides, each daughter cell will receive exactly the same amount and kind of chromosomal material received by the other one, and that the material received by both will be exactly the same in amount and kind as that possessed by the cell before division. Evidently the chromosomes are so important that nothing can be left to chance in their duplication and subsequent distribution. Evidence from plants and experimental animals indicates that the genes are arranged in a row (linear order) along the chromosome. Thus when a chromosome duplicates itself, all the genes throughout its length duplicate themselves. We may think of one as an original and the other as a carbon copy, but in this case the carbon copy is the equal of the original in every respect. Through mitosis one daughter cell receives the original, the other the carbon copy. As a result, both receive exactly the same genetic information.

Although the linear order of the genes cannot be seen, linearly arranged structures in chromosomes can be observed in favorable material. This is particularly true of the salivary gland cells of some of the flies (order Diptera), *Drosophila* being the most thoroughly analyzed example (Fig. 3.2). In these cells the attenuated chromosomes remain in the extended, uncondensed state. They duplicate themselves several or many times, forming a mass of threads referred to as a giant chromosome. Each thread is not uniform in diameter throughout its length but rather has irregularly spaced enlargements called **chromomeres.** The massing to-

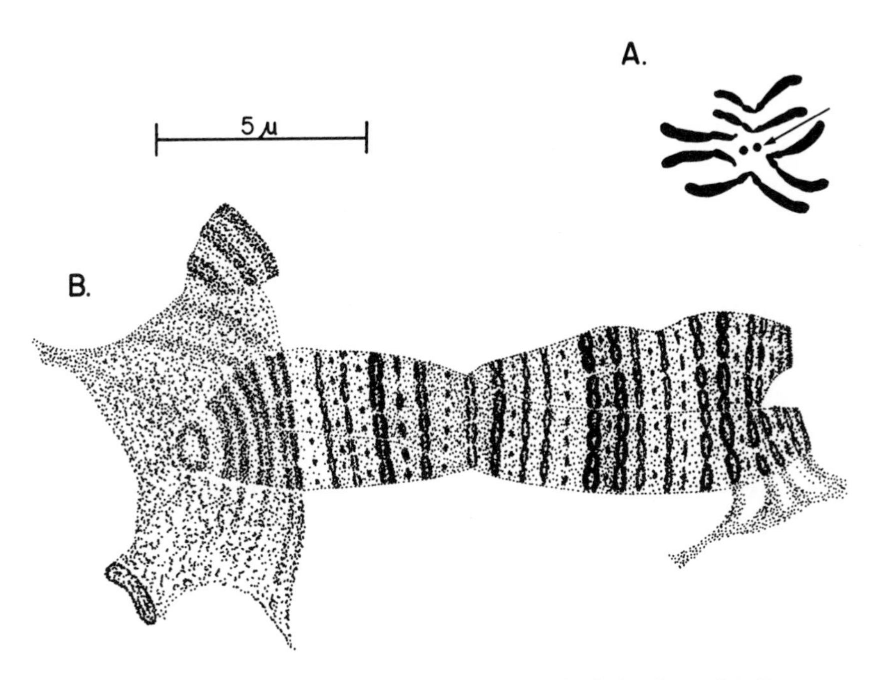

FIG. 3.2 Salivary gland chromosome No. IV of the fruit fly, *Drosophila,* compared with the same chromosome as seen at metaphase in a germ cell. A. Metaphase plate of *Drosophila* chromosomes in a germ cell. The No. IV chromosome is indicated by the arrow. B. The No. IV chromosome as seen in a cell of the salivary gland of larval *Drosophila*. Both drawings are on the same scale (5 microns = 5 thousandths of a millimeter). (Redrawn from Bridges, C. B., 1935. "Salivary chromosome maps," *Journal of Heredity,* **26**: 60–64.)

gether of many chromomeres results in the banded appearance evident in the figure. Because the bands have a great variety of patterns, chromosomes and portions of chromosomes can be identified by experts almost as accurately as other experts can identify fingerprints. Hence genetic changes can frequently be seen to be associated with visible chromosomal changes. In many cases the location of a gene on or near a given band can be pinpointed with accuracy.

Prior to 1970, chromomeres had been discerned in favorable human material (Fig. 3.3), although banding had not. But since that year, banding patterns have been demonstrated and are used increasingly in identification of human chromosomes and portions of chromosomes (see Chap. 18).

Meiosis

The role of chromosomes in Mendelian inheritance is most clearly

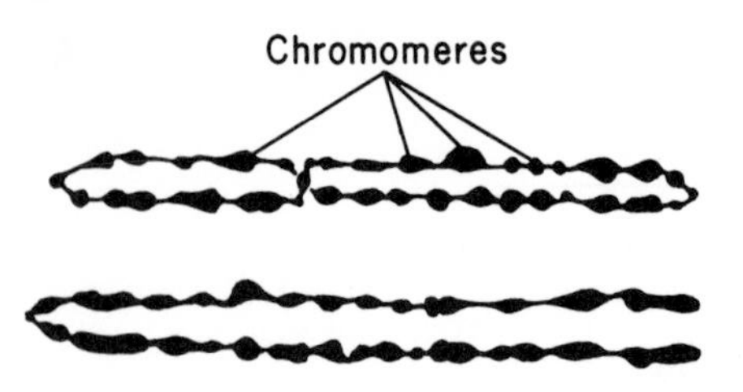

FIG. 3.3 Human chromosomes showing chromomeres. (Redrawn from Yerganian, G., 1957. "Cytologic maps of some isolated human pachytene chromosomes," *American Journal of Human Genetics,* 9:42–54.)

revealed in the process by which gametes are produced from precursor cells, the process of **meiosis.**

As mentioned earlier, chromosomes occur in pairs in most of the cells of the body. The number of pairs per cell varies from species to species; in man, each cell normally has 23 pairs, a total of 46 chromosomes. One member of each pair came from the mother of the individual (and hence is called maternal) and the other member came from the father (paternal). In human beings, the ovum contains 23 chromosomes and the sperm contains 23; hence a fertilized ovum (formed by fusion of a sperm with an ovum) contains 46. Because the whole body of the offspring that develops from that fertilized ovum arises from the latter by repeated divisions and subdivisions (mitosis), all the cells of the body contain 46 chromosomes. (Although there may be some variation in number from cell to cell, this is the rule.)

As an embryo develops from a fertilized ovum, certain cells are set aside to form the reproductive cells of the individual. These are called **primordial germ cells.** Like the other cells (of skin, muscle, liver, etc.), they contain chromosomes in pairs. Because a diagram showing 23 pairs would be unwieldy, Figure 3.4 is drawn with two pairs only, as in the diagram of mitosis (Fig. 3.1). Figure 3.4 represents meiosis in a male (**spermatogenesis**); the primordial germ cells are called **spermatogonia.** The figure summarizes the history of a single one of these, although an embryo produces many of them.

Figure 3.4 shows the spermatogonium in two of its several stages. The first diagram represents it as a cell in interphase, when the chromosomes are present as long, slender threads (p. 36) that have not yet duplicated themselves. Actually, each thread is longer and more twisted than the diagram indicates. A human spermatogonium at this stage has 46 threads so twisted and massed together that the individual ones would not be discernible. In the diagram, two long threads and two short ones are shown. As in mitosis, each chromosome duplicates itself and then coils in complex fashion. At the conclusion of the process, the resulting chro-

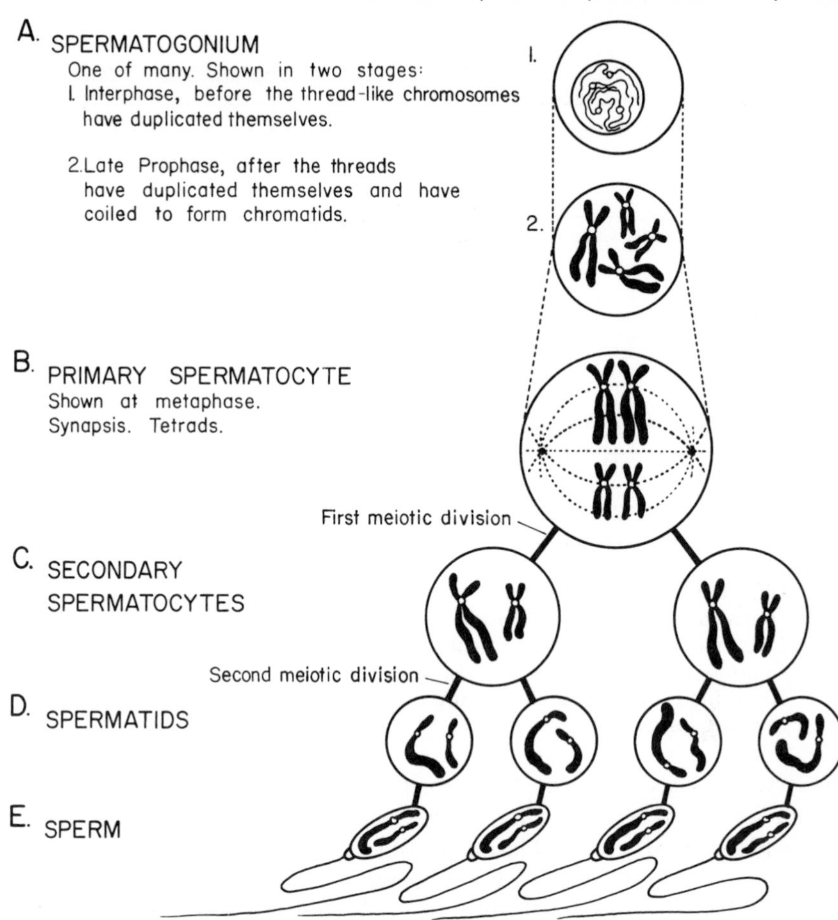

FIG. 3.4 The genetically significant highlights of meiosis in the male (spermatogenesis).

mosomes present the appearance shown in Figure 3.4A,2. Each chromosome has become two chromatids held together by a centromere.

The homologous chromosomes now pair together in a process called **synapsis.** Thus, in terms of our diagram, the two "long chromosomes" lie closely side by side, and so do the two short ones. Because each synapsed pair of chromosomes consists of four chromatids, the pair is now called a **tetrad.** The chromatids composing a tetrad are tightly bunched together (Fig. 3.5A) and frequently twisted about each other. At this time, portions of one chromatid may be exchanged with corresponding portions of another one in the same tetrad, in a process known as **crossing over.** Because each chromatid contains many genes, such crossing over increases the number and variety of combinations of genes. Soon the chromosomes comprising a tetrad start to separate, and at this time an

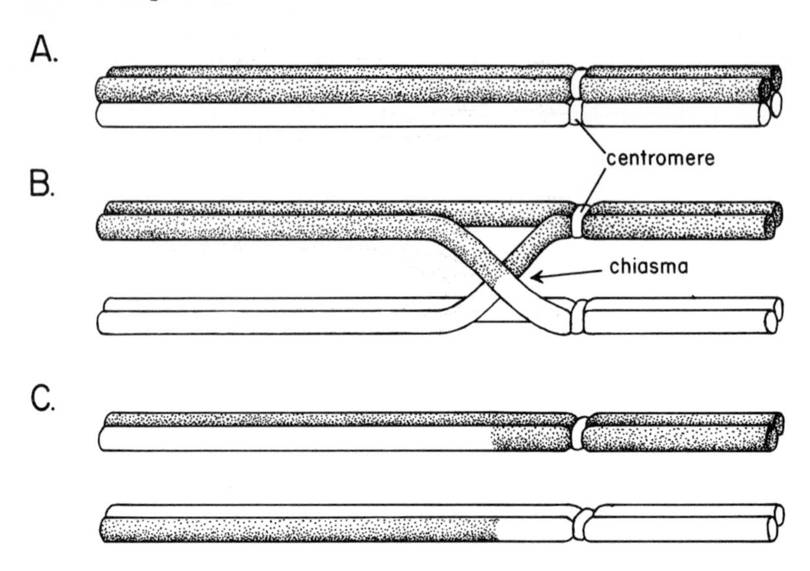

FIG. 3.5 Synapsis, crossing over, and chiasma formation. A. Tetrad consisting of four chromatids (two chromosomes) in synapsis. B. Chiasma formed as chromosomes separate following crossing over (represented by exchange of portions between two chromatids). C. Resulting chromosomes following crossing over and the disappearance of the chiasma.

observer can see evidence that crossing over has occurred because of a tendency of crossed-over chromatids to cling together, forming temporary cross-like patterns called **chiasmata** (singular: **chiasma;** Fig. 3.5B).

The name **primary spermatocyte** (rather than spermatogonium) is applied to a cell in which these events take place. The tetrads now line up on the spindle, which has been forming meanwhile (Fig. 3.4B). When the primary spermatocyte divides, the **first meiotic division,** the chromosomes composing a tetrad separate from each other, each daughter cell receiving one of each kind (Fig. 3.4C). The resulting cells are called **secondary spermatocytes.** Because each chromosome is composed of two chromatids, each chromosome in a secondary spermatocyte is sometimes called a **dyad** (cf. tetrad). (Usage varies here; some workers regard each chromatid as a separate chromosome and hence a dyad as being composed of two chromosomes.)

The **second meiotic division** now occurs. As a result of separation at the centromere, the sister chromatids (chromosomes) are freed, to be distributed to different daughter cells (Fig. 3.4D), much as in mitosis. These cells are called **spermatids.**

As a last step, each spermatid undergoes a sort of metamorphosis, developing a flagellum (tail), which enables it to swim. It is now a mature **sperm** (Fig. 3.4E).

In summary, we note that the spermatogonia contain chromosomes in pairs. The total number of chromosomes in such a cell (46 in man) is called the **diploid** number. Each sperm cell contains one of each pair of chromosomes present in the spermatogonium. The total number of chromosomes in a sperm cell is called the **haploid** number (23 in man). Thus meiosis is an orderly process by which haploid gametes are produced from diploid primordial cells.

We can readily understand how this process provides the mechanism for Mendelian inheritance. Suppose, for example, that the spermatogenesis diagramed in Figure 3.4 occurs in a heterozygous black guinea pig, and that the genes concerned are borne by the long chromosomes, one of the latter containing *B*, one containing *b*. Figure 3.6 shows clearly how the behavior of the chromosomes results in the separation of *B* from

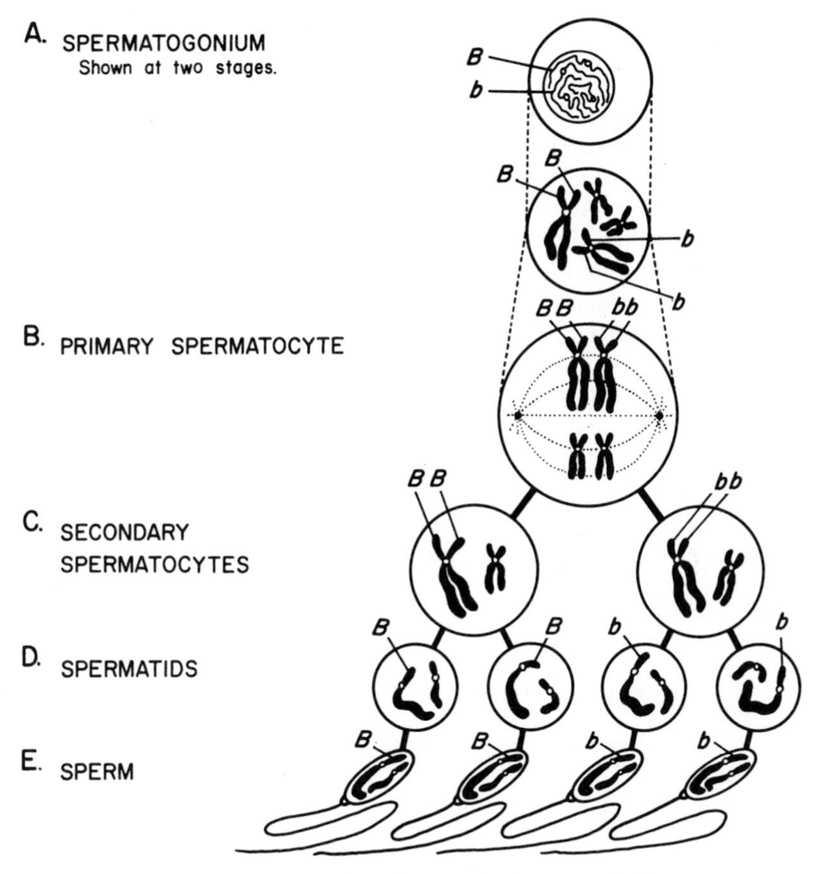

FIG. 3.6 How meiosis provides the mechanism by which a heterozygous male produces sperm of two kinds, half of them containing the dominant gene, half of them containing its recessive allele.

b, and in the production of *B*-containing and *b*-containing gametes in equal numbers, as postulated by Mendelian theory.

Meiosis in the female (**oögenesis**) is in principle much like that in the male (Fig. 3.7). The primordial germ cells are called **oögonia.** Tetrads are formed as in the male, the stage in which they occur being called the **primary oöcyte.** When the primary oöcyte divides, the chromosomes separate just as they do in the male, but the division of the cell body (cytoplasm) is extremely unequal, one daughter cell being just large enough to enclose the chromosomes. This tiny cell is called the **first polar body;** although it may divide, nothing ever comes from it. This is merely a method of discarding chromosomes while retaining most of the cytoplasm in the other cell formed by the division, the **secondary oöcyte.**

Similarly, when the secondary oöcyte divides, the freed chromatids (now separate chromosomes) behave as they do in the male. The division of cytoplasm is again highly unequal, the **second polar body** receiving only enough to envelop the chromosomes. The second polar body, like the first one, distintegrates, leaving a single functional cell, the **ovum.** Note that the ovum contains the haploid number of chromosomes with a relatively large amount of cytoplasm. The importance of this large amount of cytoplasm is evident when we recall that nourishment for the first stages of embryonic development must be provided by the ovum.

Figure 3.7 represents oögenesis in a heterozygous black female guinea pig. The fate of genes *B* and *b* may be followed as in the corresponding spermatogenesis (Fig. 3.6). In this case we have represented *b* as being discarded in the first polar body, the single ovum produced by that primary oöcyte containing the gene *B*. It is equally likely that the *B* gene will be discarded in the polar body; so other ova produced by other oöcytes will contain the gene *b*. Because the chances are equal that the *B* gene or the *b* gene will be retained in the ovum, a heterozygous female may be expected to produce during her reproductive life *B*-containing and *b*-containing ova in approximately equal numbers. Again, this accords with the postulates of Mendelian theory.

Independent Assortment

In the foregoing discussion of the genetical significance of meiosis, we have concentrated on one pair of genes in one pair of chromosomes. Frequently we wish to study the simultaneous distribution of two or more pairs of genes in different pairs of chromosomes. In such cases we find that the manner in which the members of one pair of genes are distributed to gametes does not influence the manner in which the members of another pair are distributed. This principle of independent assortment is called the

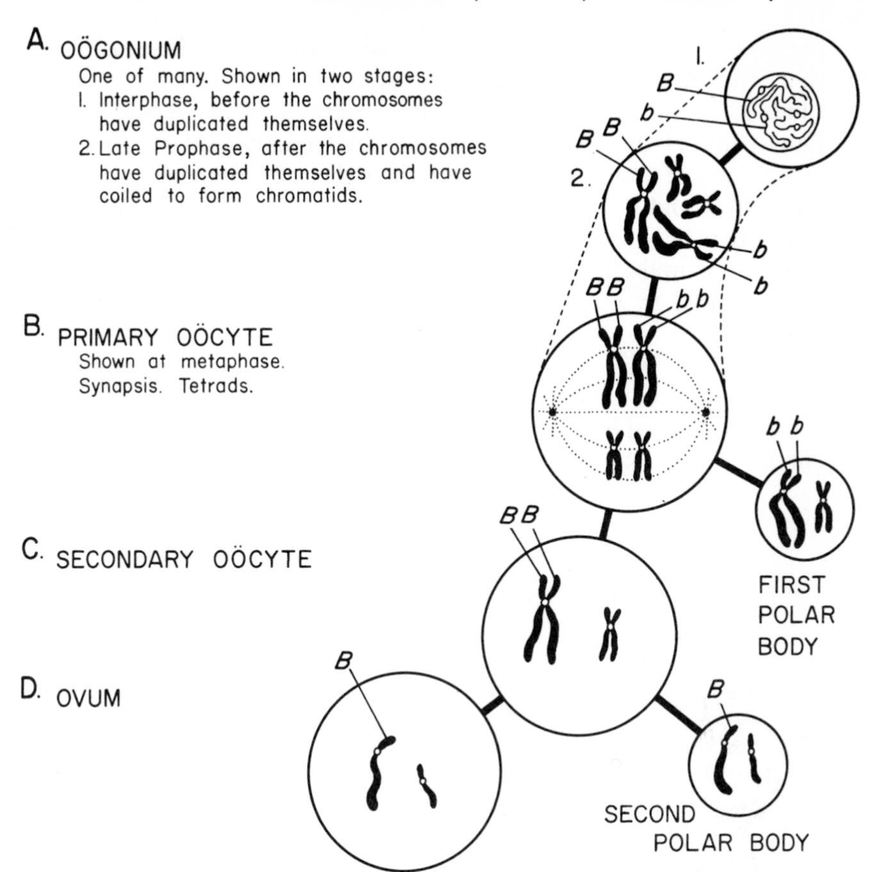

FIG. 3.7 The genetically significant highlights of meiosis in the female (oögenesis). The female is shown as heterozygous, *Bb*. The recessive gene, *b*, is shown discarded in the first polar body. Because the dominant gene, *B*, is equally likely to be so discarded, heterozygous females produce in equal numbers ova containing the dominant gene and ova containing the recessive gene.

second Mendelian law (the first Mendelian law, the law of segregation, was discussed on p. 12).

We may illustrate the principle by considering the gametes produced by an individual who has the genotype *AaBb*. Following meiosis, half of his gametes will contain gene *A*, half of them will contain gene *a*.

Similarly, half will contain *B*, half will contain *b*. However, an *A*-containing gamete is equally likely to contain *B* or *b*. Hence a person with the genotype *AaBa* produces four kinds of gametes in equal numbers: *AB, Ab, aB, ab*.

It is important to realize that independent assortment depends on independent distribution of chromosomes in meiosis. Chance determines the arrangement of tetrads on the spindle prior to the first meiotic division. Thus if the long chromosomes carry the genes *A* and *a,* respectively (Fig. 3.8) and the short chromosomes carry the genes *B* and *b,* two arrangements of the tetrads are possible and equally likely: (1) *A* and *B* may be on the left, with *a* and *b* on the right; (2) *A* and *b* may be on the left, with *a* and *B* on the right. As shown in the figure, the outcome is the production of the four kinds of gametes in equal numbers. A similar mechanism underlies independent assortment in the female; a series of ova produced by an *AaBb* female will include all four kinds with equal frequencies.

Thus, we see that independent assortment characterizes genes located on different chromosomes. The converse of independent assortment is **linkage,** caused by two or more genes (e.g., *A* and *B*) being in the same chromosome (Chap. 16).

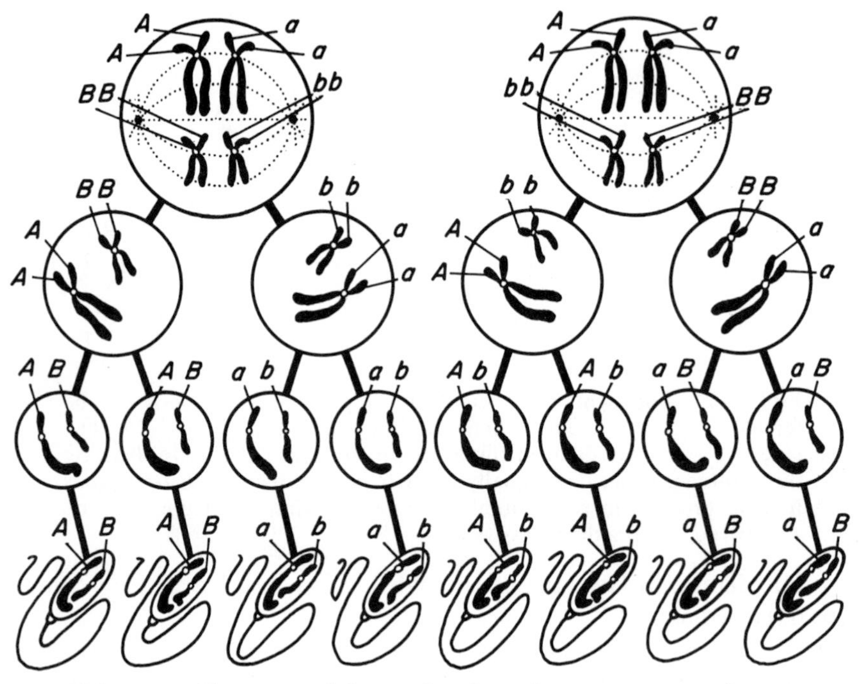

FIG. 3.8 Chromosomal basis of independent assortment of genes, starting with two primary spermatocytes in a doubly heterozygous individual: *AaBb.* Subsequent stages are those shown in Fig. 3.4.

Problems

1. Before mitosis begins, a human skin cell contains 46 chromosomes. When that cell is at the metaphase stage of mitosis, how many chromatids does it

contain? How many centromeres? At the conclusion of mitosis, how many chromosomes does each of the two daughter cells contain? How many centromeres?

2. When a human primary spermatocyte is at metaphase, (a) how many chromatids does it contain? (b) How many centromeres? (c) How many tetrads?
3. How many chromatids does a human secondary spermatocyte contain? How many centromeres?
4. How many chromosomes does a human spermatid contain? How many does a sperm contain?
5. How many chromatids does a human primary oöcyte contain? How many does a secondary oöcyte contain? A first polar body? A second polar body?
6. If a woman has the genotype *Aa,* what proportion of her primary oöcytes will contain gene *A?* In each primary oöcyte, how many *A* genes will be present? How many *a* genes? If a first polar body contains gene *A,* what does the corresponding secondary oöcyte contain? If an ovum contains gene *A,* what does the corresponding second polar body contain?
7. If a man has the genotype *Aa,* what proportion of his primary spermatocytes will contain gene *a?* What proportion of his secondary spermatocytes will contain *a?* What proportion of his spermatids? Of his sperms?
8. A man has the genotype *MmTt.* (a) What proportion of his primary spermatocytes will contain gene *M* and gene *T* (but not necessarily these only)? (b) What proportion of his spermatids will contain gene *M* and gene *t?* (c) What proportion of his sperms will contain gene *M* and gene *m?* (d) What proportion of the sperms that contain gene *M* will also contain gene *T?* Gene *t?*
9. During the lifetime of a woman having the genotype *MmTt,* what proportion of the ova she produces will contain (a) *M* and *T,* (b) *m* and *T,* (c) *T* and *T,* (d) *T* and *t?*
10. During the lifetime of a woman having the genotype *MmTT,* what proportion of the ova she produces will contain (a) *M* and *T,* (b) *m* and *T,* (c) *m* and *t?*
11. A man with the genotype *Mm* produces two kinds of sperms relative to these genes. A man with the genotype *MmTt* produces four kinds of sperms. How many kinds of sperms are produced by a man with the genotype *MmTtWw?* List the different kinds he produces.

References

Bridges, C. B., 1935. "Salivary chromosome maps," *Journal of Heredity,* **26**:60–64.

Snyder, L. H., and P. R. David, 1957. *The Principles of Heredity,* 5th ed. Boston: D. C. Heath & Company.

Sturtevant, A. H., 1965. *A History of Genetics.* New York: Harper & Row.

Sutton, W. S., 1903. "The chromosomes in heredity," *Biological Bulletin,* 4:231–251. Reprinted in J. A. Peters (ed.), 1961. *Classic Papers in Genetics.* Englewood Cliffs, N.J.: Prentice-Hall, Inc. Pp. 27–41.

Yerganian, G., 1957. "Cytologic maps of isolated human pachytene chromosomes," *American Journal of Human Genetics,* **9**:42–54.

4

Chromosomes and Genes

Fine Structure of Chromosomes

In the preceding chapter we stated that during interphase the chromosomes are in an attenuated (thread-like) form. These threads—sometimes called chromonemata—can be seen with a light microscope in favorable material but are best studied with an electron microscope because of the vastly greater magnification achievable. They are usually found to have a diameter of from 200 to 300 angstrom units. Since an angstrom unit, Å, is only one ten-millionth of a millimeter, the threads are almost unimaginably fine by our ordinary human scales of measurement.

Studies with the electron microscope reveal that a chromosome at metaphase of mitosis or meiosis is composed of a seemingly tangled mass of such fibers (Fig. 4.1). Although we cannot be certain, evidence accumulates that each chromatid of a metaphase chromosome consists of a single fiber intricately coiled and folded (DuPraw's "folded fiber model," DuPraw 1965, 1966, 1968). This would be in accord with what we stated in the preceding chapter about the formation of metaphase chromosomes by coiling and condensation of the attenuated interphase chromosomes.

Of what are the 200–300 Å fibers composed? Their structure cannot be observed directly, even with an electron microscope, but indirect evidence suggests that each such fiber consists of a supercoiled filament (a "coiled coil") that is 35–60 Å in diameter. Such filaments are found and are measurable when in the uncoiled state.

This leads to the question: Of what are these finer filaments composed? They are composed of **nucleoprotein,** complexes consisting of

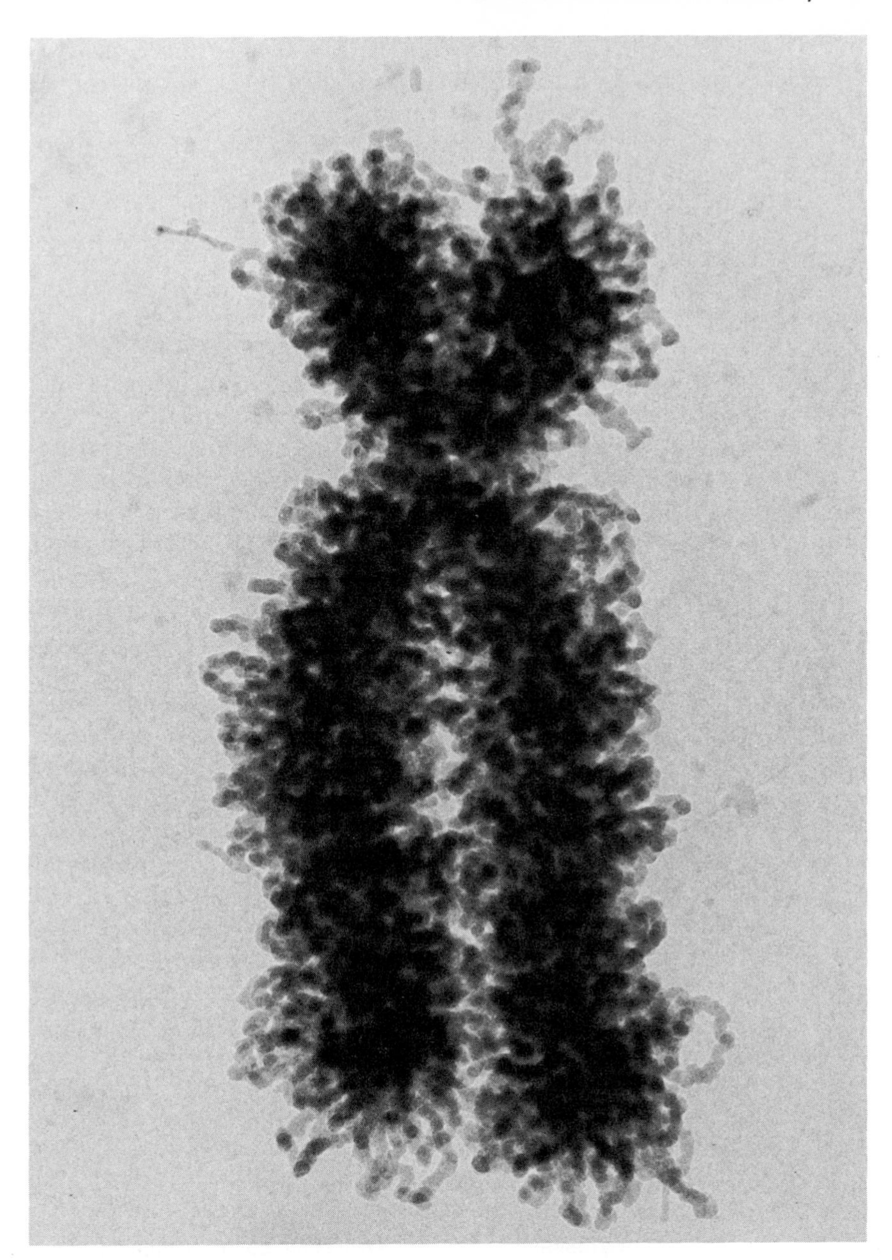

FIG. 4.1 Fibrous structure of a chromosome revealed by an electron microscope. (From *DNA and Chromosomes* by E. J. DuPraw. Copyright © 1970 by Holt, Rinehart and Winston, Inc. Reproduced by permission of Holt, Rinehart and Winston, Inc.)

nucleic acid joined to proteins. The nucleic acid in this case is **deoxyribonucleic acid (DNA).** It has been suggested that this forms the core of the filament, the proteins providing a sheath around the core. (For details of the ultrastructure of chromosomes, see Abuelo and Moore, 1969; DuPraw, 1968, 1970.)

Thus, by successively finer dissections of the chromosome we reach the material generally regarded as being of greatest genetic importance, the DNA.

DNA

The DNA molecule is very long, being really a macromolecule composed of smaller molecules in orderly array. The unit of structure is called a **nucleotide,** and consists of three molecules: a pentose sugar (deoxyribose), a phosphate, and an organic base (a purine or pyrimidine) (Fig. 4.2). The sugar of one nucleotide is joined to the phosphate of the next one, and in this way a long chain is formed (Fig. 4.3A). To each sugar

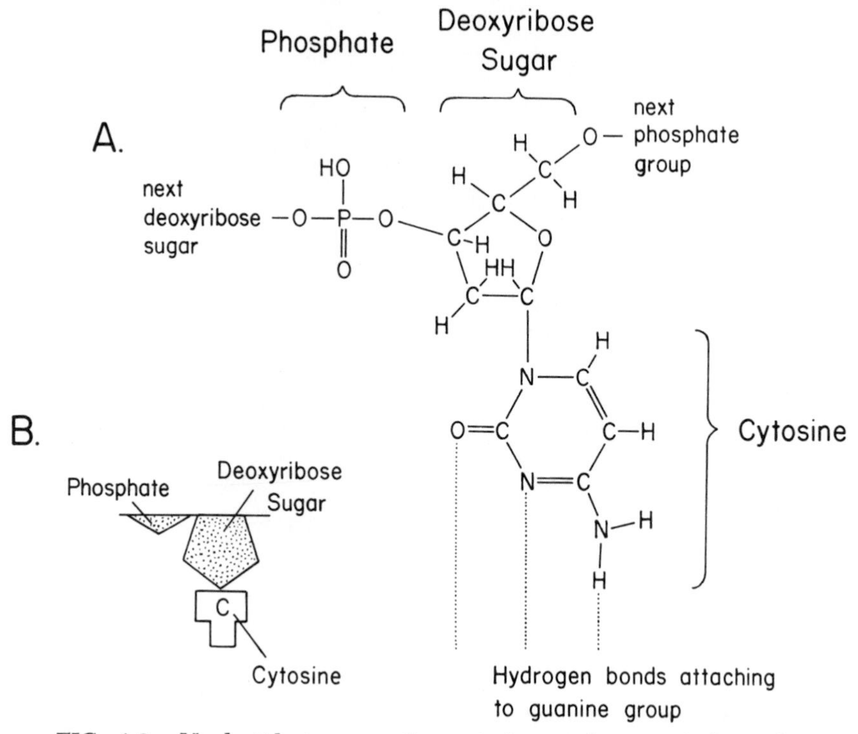

FIG. 4.2 Nucleotide incorporating cytosine as the organic base. A. Structural formula of the nucleotide. B. Stereotyped diagram used to represent this nucleotide in Fig. 4.3. The other three nucleotides in DNA differ only in the shape of the outline used to represent the organic base.

molecule is also attached an organic base. In DNA there are four of these: the purines, adenine (A) and guanine (G), and the pyrimidines, cytosine (C) and thymine (T). Thus DNA is composed of four kinds of nucleotides (distinguished by the four bases) arranged in a long chain. If current thinking is correct, these nucleotides are linked in all possible arrangements, and the arrangement constitutes a code conveying genetic information much as the arrangement of letters in a sentence conveys information to the reader. Figure 4.3 shows an arbitrary example: one possible arrangement of these nucleotides.

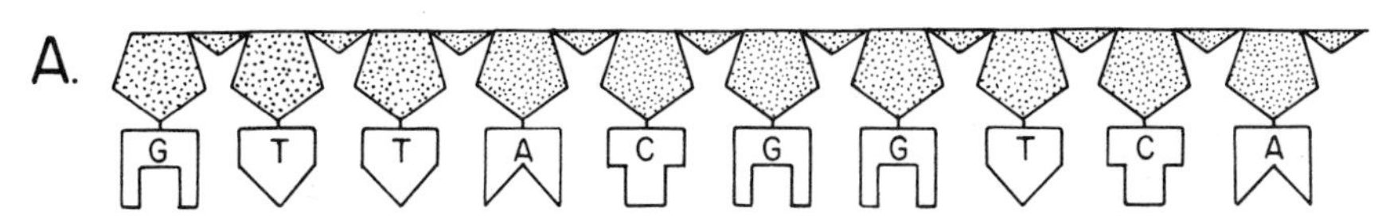

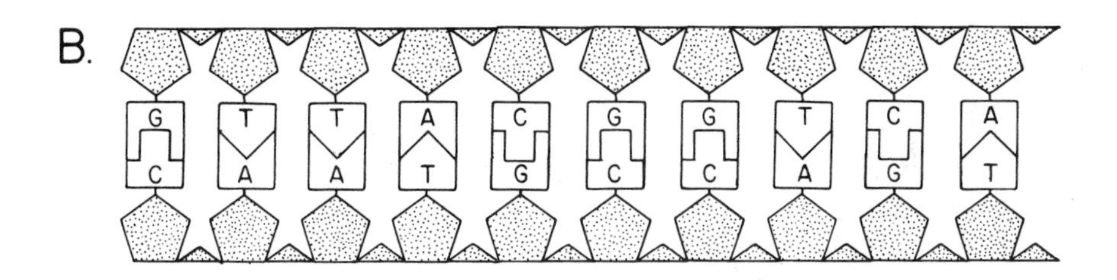

FIG. 4.3 Stereotyped diagram of a portion of the DNA molecule. A. Single strand composed of a chain of nucleotides (Fig. 4.2). B. Double strand composed of two chains of nucleotides joined by their complementary organic bases. C. A single strand duplicating itself by serving as a template or pattern on which nucleotides are assembled in a sequence complementary to the sequence of template strand itself.

The chain we have described forms only half of the DNA macromolecule. Running parallel to it is a second chain. The two chains are joined by hydrogen bonding of their organic bases. This occurs in such a manner that cytosine of one chain is always joined to guanine of the other, and thymine is always joined to adenine (Fig. 4.3B). Furthermore, X-ray diffraction studies, revealing the patterns in which atoms are arranged, suggest that this whole complex is twisted in the form of a double helix. This famous Watson-Crick model of DNA structure is shown in Figure 4.4, in which the paired bases are represented as forming the rungs of a

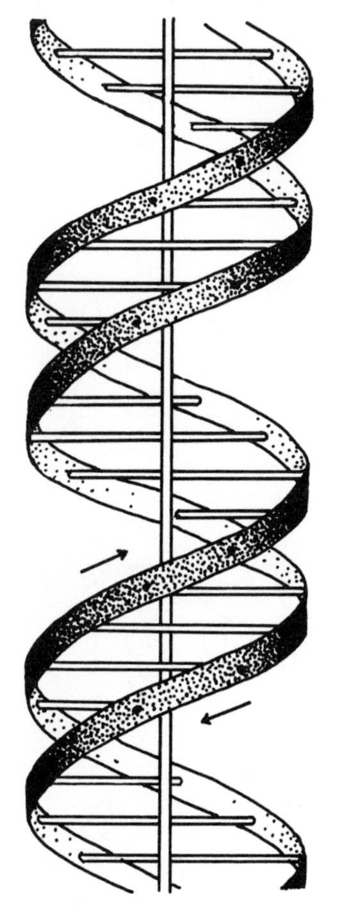

FIG. 4.4 The Watson-Crick model of DNA structure. The sides of the helical "ladder" are composed of sugar-phosphate chains (Fig. 4.2). Each "rung" is formed of two organic bases held together by hydrogen bonds (Fig. 4.2B). The vertical lines show the axis of the helix. (Redrawn from Watson, J. D., and F. H. C. Crick, 1953. "The structure of DNA," *Cold Spring Harbor Symposia on Quantitative Biology,* 18:123–131.)

ladder. The diameter of this helix is about 20 Å; it is believed to form the central core of the 35–60 Å filaments shown by the electron microscope (see above), the remainder of the filament being composed of associated proteins.

We have seen that exact duplication of chromosomes is of fundamental importance in mitosis and meiosis (Chap. 3.) The DNA macromolecule provides a precise mechanism for this. When the two chains (Fig. 4.3B) separate (by breaking of the hydrogen bonds between organic bases), each chain becomes a template or pattern on which a new chain like the other can be reconstructed from material present in the nucleus. For example, in Figure 4.3C, we show the upper chain serving as a template for construction of a duplicate of the lower one, nucleotides represented as being "loose" in the surrounding medium being utilized. The original lower chain would serve as a pattern for construction of a new upper chain in like manner. In this way, two identical double chains would arise where only one had existed at first: precisely what is needed in chromosome duplication. (This process is not nearly so simple as it looks; many other molecules, acting as enzymes, sources of energy, etc., are involved, but consideration of them would lead us far afield. Many good accounts are available, e.g., Watson, 1970.)

Function of DNA

What does DNA *do?* It is generally believed that the principal, if not the only, function of DNA is to control the production of **proteins.** Proteins are the most complex compounds known; without them there would be no life. Proteins enter into the *structure* of all cells, tissues, and organs of the body. Other proteins serve as *enzymes,* organic catalysts without which the chemical processes involved in living would not occur, at least at rates making life possible. All aspects of metabolism and bodily functioning, including embryonic development and growth, depend upon enzymes. Hence the importance of proteins cannot be overemphasized.

The huge protein molecules are composed of smaller units called **amino acids.** There are 20 of these (Table 4.1), and they are linked together in a seemingly endless variety of combinations to form long chains. The linkage between two neighboring amino acids is called a pep-

TABLE 4.1 The 20 amino acids commonly found in organisms.

Amino acid	Abbreviation	Amino acid	Abbreviation
Alanine	Ala	Leucine	Leu
Arginine	Arg	Lysine	Lys
Asparagine	Asn	Methionine	Met
Aspartic acid	Asp	Phenylalanine	Phe
Cysteine	Cys	Proline	Pro
Glutamine	Gln	Serine	Ser
Glutamic acid	Glu	Threonine	Thr
Glycine	Gly	Tryptophan	Trp
Histidine	His	Tyrosine	Tyr
Isoleucine	Ile	Valine	Val

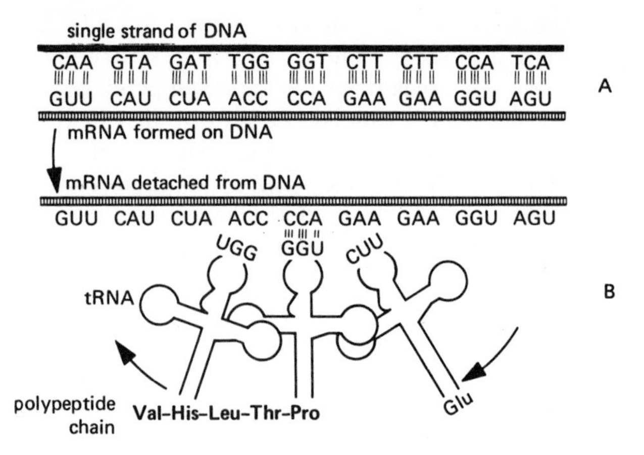

FIG. 4.5 Highlights of processes by which DNA structure is translated into protein structure. At A, the sugar-phosphate chain of DNA is represented by a solid line, that of mRNA by a hatched line. For clarity, spaces are shown between the triplets of organic bases although such spaces are not found in nature. See text.

tide linkage, and so a chain of amino acids is called a **polypeptide chain.** A protein molecule is composed of one or more of these polypeptide chains.

The sequence of amino acids in a polypeptide chain is called its *primary structure*. What is sometimes called the "central dogma of molecular biology" is the idea that the sequence of nucleotides in the DNA molecule determines the sequence of amino acids in the polypeptide chain. Much evidence supports the idea.

How is the DNA structure translated into protein (polypeptide) structure? Investigations reveal a process so complex as to stagger the imagination (see, for example, Watson, 1970). Here we must content ourselves with sketching the highlights.

At the outset we note that while there are only four different types of nucleotides—containing adenine, guanine, cytosine, and thymine; respectively—there are 20 amino acids to be coded for. Thus there cannot be a one-to-one correspondence. If a *pair* of nucleotides coded for each amino acid, only 16 amino acids would be provided for. But if each amino acid is represented by a triplet of nucleotides, more than enough such triplets (64) are available. Various lines of evidence combine to indicate that each amino acid is in fact coded for by a triplet of nucleotides.

DNA does not serve directly as a template for the lining up of amino acids to form a polypeptide chain. There are intermediaries in the process, composed of a different type of nucleic acid: **ribonucleic acid (RNA).** RNA differs from DNA in two respects: (1) the sugar molecule in each nucleotide is ribose rather than deoxyribose; (2) uracil (U) replaces thymine as a pairing partner for adenine (A).

When a cell is preparing to synthesize a polypeptide chain, one strand of the DNA serves as a template for the synthesis of a strand of RNA (Fig. 4.5A). This strand is called **messenger RNA** (mRNA) because it detaches from the DNA and moves out into the cytoplasm of the cell. There it serves as a template for the lining up of amino acids. At the outset each of the amino acids is attached to a small RNA molecule called **transfer RNA** (tRNA). Each of these molecules contains 70–80 nucleotides arranged in such a manner that the strand is folded, with the nucleotides in some regions pairing by hydrogen bonds, in other regions remaining unpaired. The result is thought to be a "cloverleaf pattern" somewhat resembling the outlines in Figure 4.5B, although it undoubtedly is not really flat. As shown in the figure, an amino acid is attached to one end of the strand, and in one of the loops there are three nucleotides that will pair with the three appropriate nucleotides in the mRNA.

In Figure 4.5B, a tRNA molecule carrying the amino acid proline is shown as attached to the mRNA strand at a point where the nucleotide triplet CCA is found. CCA is said to be a "code word" or **codon** for proline. The tRNA is shown with the corresponding triplet GGU; this is called an **anticodon.** Pairing between codon and anticodon is very brief, just long enough for a peptide bond to be formed uniting the amino acid—proline in this case—to the growing polypeptide chain. Then the "empty" tRNA molecule moves out of position and a new one moves up to the next codon. (In Fig. 4.5B, the tRNA on the left is shown moving away after having contributed threonine to the chain. The tRNA on the right, bearing gluctamic acid, is moving into place so that its anticodon CUU will pair with the codon GAA in the mRNA.) So step by step, but in fact with great rapidity, a polypeptide chain is formed.

What brings the correct tRNA molecules into position? What causes the formation of peptide bonds between two amino acids brought together in this way? Light on these and many kindred questions is being shed by many investigations. Of vital importance in the process are tiny bodies called ribosomes, which contain a third type of RNA. Much is being learned about the roles of ribosomes, and of many enzymes, in this complex process (we recommend Watson, 1970, to interested readers).

The Genetic Code

There are 64 of the mRNA codons we have just mentioned. Extremely ingenious experimentation has made possible identification of the amino acid for which each codes. This information is summarized in a "dictionary," Table 4.2. Although this genetic code is given in terms of mRNA codons, we recall that each codon is determined by a nucleotide triplet in the DNA itself.

The genetic code is said to be "degenerate." This is cryptographers' jargon for the fact that it contains synonyms: Some amino acids are coded for by more than one codon. Phenylalanine, for example, is coded for by UUU and UUC, while valine has four code words: GUU, GUC, GUA, and GUG (Table 4.2).

TABLE 4.2 Dictionary of the messenger RNA codons[a].

First Nucleo-tide	Second Nucleotide				Third Nucleo-tide
	U	C	A	G	
U	UUU Phe UUC Phe UUA Leu UUG Leu	UCU Ser UCC Ser UCA Ser UCG Ser	UAU Tyr UAC Tyr UAA Ochre UAG Amber	UGU Cys UGC Cys UGA Cys[b] UGG Trp	U C A G
C	CUU Leu CUC Leu CUA Leu CUG Leu	CCU Pro CCC Pro CCA Pro CCG Pro	CAU His CAC His CAA Gln CAG Gln	CGU Arg CGC Arg CGA Arg CGG Arg	U C A G
A	AUU Ile AUC Ile AUA Ile AUG Met	ACU Thr ACC Thr ACA Thr ACG Thr	AAU Asn AAC Asn AAA Lys AAG Lys	AGU Ser AGC Ser AGA Arg AGG Arg	U C A G
G	GUU Val GUC Val GUA Val GUG Val	GCU Ala GCC Ala GCA Ala GCG Ala	GAU Asp GAC Asp GAA Glu GAG Glu	GGU Gly GGC Gly GGA Gly GGG Gly	U C A G

[a]Abbreviations for names of amino acids are those given in Table 4.1. ("Ochre" and "amber" signal the termination of a polypeptide chain; they are somewhat like punctuation marks in a sentence.) Based on the researches of Nirenberg, Ochoa, Khorana, and many others.

[b]UGA codes for cysteine in some organisms but not in the bacterium *Escherichia coli,* where it is a chain terminator.

(From Moody, P. A., 1970. *Introduction to Evolution,* 3rd ed. New York: Harper & Row.)

We should mention that this genetic code was worked out for the common colon bacillus, *Escherichia coli,* an organism most intensively investigated. To what extent is this code also utilized by other organisms, including man? In other words, is the code universal? A beginning has been made in answering this question. For example, Marshall, Caskey, and Nirenberg (1967) investigated the question for the African clawed toad (*Xenopus*) and the guinea pig (an amphibian and a mammal). They found that these vertebrates utilize the same genetic code as does *E. coli,* but that different organisms differ in what we might call their "preference" for certain codons. Thus GCU, GCC, GCA, and GCG code for alanine in all three organisms, but GCG, the "preferred" codon for *E. coli,* is least "preferred" by the two vertebrates. Other investigations have yielded comparable results. In the words of Fraenkel-Conrat (1964): "It is as if the genetic code, although universal in principle, contained varying dialects, the cells of different species using different versions of the general language."

What Is a Gene?

If we were so-minded, we might devote an entire chapter to answering this question. The word "gene" is defined in various ways depending upon one's point of view and immediate objective. For our purposes, however, a definition emphasizing *function* will be most useful. The gene is an amount of genetic material having a unitary function in determining the phenotype, or, to extract from a definition by Watson (1970), "a discrete chromosomal region which is responsible for a specific cellular product." Benzer's term **cistron** (see Benzer, 1957) is frequently used for a gene defined in this manner. In accordance with this concept, a gene is frequently defined as *that portion of DNA that codes for one polypeptide chain.* As we shall see in the next chapter, change in a polypeptide chain may alter or eliminate the enzymatic activity of the protein of which the chain is a part, and hence have a great effect on the phenotype, or even the viability, of the organism. We mention this to emphasize the importance of polypeptide chains as units of function.

Because polypeptide chains differ in length, genes also differ in length correspondingly. Variations of from 500 to 1500 nucleotides perhaps cover the usual range (Watson, 1970).

Problems

1. Using Table 4.2, translate the following sequence of messenger RNA nucleotides into a sequence of amino acids forming part of a polypeptide chain:

   C-U-G-U-U-U-U-G-C-A-U-A-G-U-G-G-U-U-G-A-U-G-G-C

2. Write the sequence of nucleotides for the portion of a DNA molecule that served as a pattern or template for the formation of the messenger RNA in Problem 1.
3. A certain protein contains the following sequence of amino acids: isoleucine-serine-arginine-glutamic acid-serine-proline-valine-glutamic acid. Using Table 4.2, write the sequence of RNA triplets that would code for this sequence

of amino acids. Then write the sequence of DNA triplets that would code for the formation of the RNA.

References

Abuelo, J. G., and D. E. Moore, 1969. "The human chromosome. Electron microscopic observations on chromatin fiber organization," *Journal of Cell Biology,* **41**:73–90.

Benzer, S., 1957. "The elementary units of heredity." In W. D. McElroy and B. Glass (eds.). *A Symposium on the Chemical Basis of Heredity.* Baltimore: The Johns Hopkins Press. Pp. 70–93.

DuPraw, E. J., 1965. "Macromolecular organization of nuclei and chromosomes: A folded fibre model based on whole-mount electron microscopy," *Nature,* **206**:338–343.

DuPraw, E. J., 1966. "Evidence for a 'folded-fibre' organization in human chromosomes," *Nature,* **209**:577–581.

DuPraw, E. J., 1968. *Cell and Molecular Biology.* New York: Academic Press.

DuPraw, E. J., 1970. *DNA and Chromosomes.* New York: Holt, Rinehart and Winston, Inc.

Fraenkel-Conrat, H., 1964. "The genetic code of a virus," *Scientific American,* **211**:46–54.

Marshall, R. E., C. T. Caskey, and M. Nirenberg, 1967. "Fine structure of RNA codewords, recognized by bacterial, amphibian, and mammalian transfer RNA," *Science,* **155**:820–826.

Moody, P. A., 1970. *Introduction to Evolution,* 3rd ed. New York: Harper & Row.

Watson, J. D., 1970. *Molecular Biology of the Gene,* 2nd ed. New York: W. A. Benjamin, Inc. A lucid presentation by one of the originators of the Watson-Crick model.

Watson, J. D., and F. H. C. Crick, 1953. "The structure of DNA," *Cold Spring Harbor Symposia on Quantitative Biology,* **18**:123–131.

5

Mutations and the "Inborn Errors"

When a gene undergoes chemical change, we say that a mutation has occurred. Such changes are called **gene mutations** to distinguish them from changes connected with microscopically visible alterations of chromosomes: **chromosomal aberrations** (pp. 311–325). Gene mutations, then, arise as changes in the DNA macromolecule. The most completely analyzed examples we have are the abnormal hemoglobins, of which the type of hemoglobin found in sickle-cell anemia is best known.

Hemoglobin S: Sickle-Cell Hemoglobin

Our red blood cells contain hemoglobin as a means of transporting oxygen. The molecule consists of two portions: an iron-containing heme group and an associated protein (globin). It is this protein that concerns us here. The protein is composed of *four* polypeptide chains—two called **alpha** (α) **chains** and two called **beta** (β) **chains.** The two α chains are identical with each other; the two β chains are also identical.

The amino acid sequences in these chains have been analyzed. The α chains consist of 141 amino acids each, the β chains of 146 amino acids each. Hence the total protein is comprised of 574 amino acids (Ingram, 1963).

Most people have normal hemoglobin, called **Hb A,** but occasional individuals have an abnormal variant called **Hb S.** Red blood cells, normally round, assume bizarre shapes when deprived of oxygen if **Hb S** is present (Fig. 5.1). This change of shape is called "sickling" and arises

FIG. 5.1 Red blood cells exhibiting the phenomenon of sickling. A. Cells of a heterozygote showing the sickle trait. B. Cells of a homozygote (sickle-cell anemia patient). (Reprinted from *Human Heredity* by J. V. Neel and W. J. Schull by permission of The University of Chicago Press. Copyright © 1954 by The University of Chicago.)

from the fact that **Hb S,** when deprived of oxygen (deoxygenated), is much less soluble than is **Hb A** under the same conditions. Thus it comes out of free solution in the cell and deforms the shape of the latter (Harris, 1970).

Most persons exhibiting this "sickle trait" suffer few, if any, unfavorable effects from it, although under some conditions the trait may be harmful. Their red cells do not sickle inside the body, although they will do so when tested experimentally. But in occasional individuals sickling is accompanied by severe symptoms characterized as **sickle-cell anemia.** The cells may sickle within the veins of such individuals because of the low oxygen content of venous blood. The life expectancy of children stricken with this anemia is low.

Genetic analyses have shown that sufferers from sickle-cell anemia are homozygous for a gene for which possessors of the simple sickle trait are heterozygous (Neel, 1949, 1951). The gene for the α chain is designated by the symbol Hb_{α}, that for the beta chain by Hb_{β}. In the following discussion we shall be concerned with the β chain only and accordingly shall simplify our symbolism by omitting the Greek subscripts. The gene for the normal β chain will be designated by Hb^A, that for the abnormal β chain found in sickle-cell anemia by Hb^S (the α chains being normal in both cases). Using these symbols, we designate individual genotypes as follows:

Normal	$= Hb^A\ Hb^A$
Sickle trait	$= Hb^A\ Hb^S$
Sickle-cell anemia	$= Hb^S\ Hb^S$

People with the genotype $Hb^A\ Hb^A$ have **Hb A** only, those with the genotype $Hb^S\ Hb^S$ have **Hb S** only. But heterozygotes, $Hb^A\ Hb^S$, have *both* types of hemoglobin. Hence neither gene is dominant to the other.

The two types of hemoglobin differ somewhat in properties, including the electrical charge carried by the molecules. In an electrical field, dissolved **Hb A** moves toward the negative pole, whereas **Hb S** moves toward the positive pole (Pauling et al., 1949). Advantage has been taken of this fact to separate the two types and to demonstrate that heterozygotes possess both types (Fig. 5.2). Such technique is called electrophoresis.

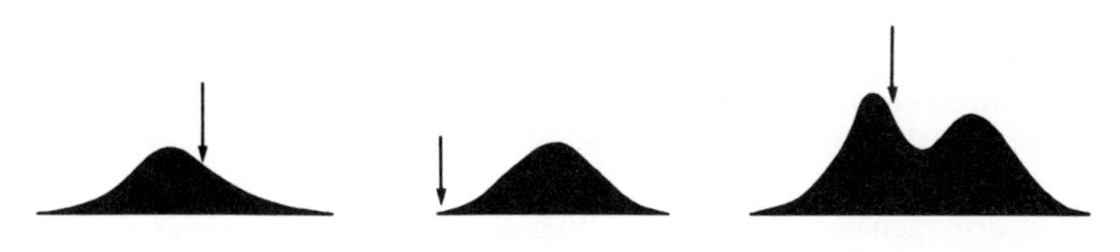

FIG. 5.2 Movement of normal hemoglobin (**Hb A**) and of sickle-cell hemoglobin (**Hb S**) in an electrical field. (a) **Hb A** moves toward the negative pole (to the left of the vertical arrow). (b) **Hb S** moves toward the positive pole. (c) Heterozygotes, Hb^A Hb^S, possess both types of hemoglobin. (Redrawn from Pauling, L., et al., 1949. "Sickle cell anemia, a molecular disease," *Science,* **110**:543–548. Copyright © 1949 by the American Association for the Advancement of Science.)

Intensive investigation of the amino acid sequences in the two hemoglobins revealed that their α chains are identical, but that there is one difference, and only one, in their β chains (Ingram, 1956, 1958, 1963). We recall that the β chains consist of 146 amino acids each. Listed in the accepted manner, the first six of these in **Hb A** are as follows: valine-histidine-leucine-threonine-proline-*glutamic acid.*
In **Hb S** the sequence is valine-histidine-leucine-threonine-proline-*valine.*

Thus we see that a difference in one amino acid in one position in a large molecule may make an important genetic difference—virtually a life-and-death difference for homozygotes, in fact.

On the likely assumption that **Hb S** arose from **Hb A** by mutation, we may now ask: What sort of genetic change would lead to the substitution af valine for glutamic acid in position No. 6 of the β chain? From Table 4.2 (p. 38) we note that the mRNA codons for glutamic acid are GAA and GAG (corresponding to DNA triplets of CTT and CTC, respectively). The mRNA codons for valine are GUU, GUC, GUA, and GUG (the corresponding DNA triplets being CAA, CAG, CAT, and CAC). Hence, if the human codon for glutamic acid were GAA, substitution of U for A (uracil for adenine) in the second position would change the codon to one for valine (GUA). If the glutamic acid codon were GAG, it would be changed to the valine codon GUG by the same means.

Figure 4.5 (p. 42) was designed to show the first six amino

acids in the β chain of hemoglobin. Note that the tRNA molecule shown at the right as just moving up to the mRNA strand is carrying glutamic acid into position No. 6. The tRNA anticodon is about to pair with the mRNA codon GAA. From the upper part of the diagram, we see that the corresponding DNA triplet is CTT. Evidently, then, the genetic change involved is a change in a DNA triplet from CTT to CAT. The point we wish to emphasize is that a significant mutation can arise as a change in a single DNA triplet involving, as an end product, substitution of one nucleotide for another. This would seem to constitute mutation reduced to its lowest common denominator.

In this case the nucleotide substitution results in an amino acid substitution that alters the protein in such a way as to change its solubility when deoxygenated. And the resulting changed solubility causes the sickling. Here we have one of the most complete series of events connecting genetic change with phenotypic change known in human genetics.

Other Abnormal Hemoglobins

Hemoglobin C, Hb C This variant of ***Hb A*** was discovered soon after the discovery of sickle-cell hemoglobin. Like the latter, it is inherited in simple Mendelian fashion. Heterozygotes are normal, and homozygotes may have a moderate amount of anemia, called "hemoglobin C disease" (Harris, 1970).

Analysis of amino acid sequences has shown that the α chains are identical to those in **Hb A,** and that the β chains differ only at one point: position No. 6, the same position at which **Hb A** and **Hb S** differ. **Hb C,** however, has lysine in this position. So the substitution was lysine for glutamic acid. The gene concerned is designated as Hb^C.

As we noted, the codons for glutamic acid are GAA and GAG. Codons for lysine are AAA and AAG (Table 4.2, p. 44). Thus, exchanging A for G in the first position will accomplish the change. What is the change in DNA here? The DNA triplet for formation of the mRNA codon GAA is CTT. That for mRNA codon AAA is TTT. Thus the change involved is substitution of thymine for cytosine in the first position of the triplet.

Multiple Alleles It is most likely that genes Hb^S and Hb^C arose by mutations of Hb^A. Thus all three are genetic variations of the same gene, and all three are concerned with the same polypeptide chain (β). This means that all three are alleles of each other and so constitute a series of **multiple alleles.**

The position in a chromosome occupied by a gene is called that gene's locus. Thus some one of the human chromosomes contains a position that we may call the *Hb* locus. This locus may be occupied by Hb^A, or by the Hb^S gene, or by the Hb^C gene. Because the chromosomes occur in pairs, any one individual can have only two of these three alleles, however. This is true no matter how long the series of multiple alleles may be, and series longer than three are common.

In Chapter 12 we shall discuss the chromosomal basis further; here we list the possible genotypes for the three under discussion:

Hb^A Hb^A = **Hb A**; normal
Hb^A Hb^S = **Hb A, Hb S**; sickle-cell trait
Hb^A Hb^C = **Hb A, Hb C**; hemoglobin C trait
Hb^S Hb^S = **Hb S**; sickle-cell anemia
Hb^C Hb^C = **Hb C**; hemoglobin C disease
Hb^S Hb^C = **Hb S, Hb C**; sickle-cell-hemoglobin C disease

The abnormal hemoglobins afford us our best human example of the molecular basis of multiple allelism—a gene undergoing different mutations.

Other Variant Hemoglobins ***Hb S*** and ***Hb C*** are but two of many abnormal hemoglobins discovered. The number increases constantly as investigations continue. For the most part, variants are discovered when the hemoglobins from large numbers of people are tested by electrophoresis, as noted for **Hb S** above. By this sort of screening, variant hemoglobins can be discovered even when, as is usually the case, no injury to health is present. In many cases a variant is known only from heterozygotes, homozygotes not having been found. But homozygotes for some variants seem to suffer no ill effects. Why substitution of an amino acid in some cases is harmful while other substitutions are not, presents an interesting problem. (See Harris, 1970, for further discussion.)

Some idea of the variant hemoglobins known up to a given date may be gained from a study such as that of Rucknagel and Laros, 1969. In a table, these authors listed 43 variants of the α chain and 54 variants of the β chain. In most cases a variant differs from normal by substitution of one amino acid at one position in a chain. There are as many positions as there are amino acid molecules in the chain (i.e., 141 in the α chain, 146 in the β chain), and substitutions have been found at many of these positions. Some of the known variants are duplicates, in the sense of being the same change, at the same point in the same chain, discovered separately in different populations that can hardly be assumed to have the same mutation because of inheritance from common ancestry. For example, the hemoglobin labeled **D**, which involves substitution of glutamine for glutamic acid at position No. 121 of the β chain, has been found six times.

It is customary to include in the name of an aberrant hemoglobin the name of the locality where it was found. Thus we may list the six **D** variants as **D**-Punjab, **D**-Cyprus, **D**-Chicago, **D**-Los Angeles, **D**-Portugal, **D**-N. Carolina.

Taking into account this duplication, we note that the table given by Rucknagel and Laros lists 21 *different* α-chain variants and 36 *different* β-chain ones. How many may be discovered in the future? It has been estimated that over 2200 single amino acid substitutions would be possible in the hemoglobin molecule, and that 700 of these would result in changing the electrical charge of the molecule and so be detectable by electrophoresis. Undoubtedly, therefore, there will be many more such discoveries.

For our purposes the most significant point to be made about these variant hemoglobins is that almost all of them differ from **Hb A** by substitution of *one amino acid molecule,* and that in almost all cases this substitution could result from change of *a single nucleotide* in the mRNA codon (and hence in the DNA). We have illustrated this for **Hb S** and **Hb C.** In the case of **Hb D,** above, glutamine substituted for glutamic acid, the nucleotide exchange is of C (cytosine) for G (guanine) in the first position of the triplet (i.e., change of GAA to CAA, or of GAG to CAG). If space permitted we might give many more examples, all illustrating the point we have made (for complete data and fuller discussion see Lehmann and Carrell, 1969.)

Before leaving the subject of hemoglobin, we should mention that there are other abnormalities of hemoglobin, especially the thalassemias, that may lead to varying degrees of anemia. In the thalassemias we see genetically caused defects in the mechanisms controlling production of one or another of the chains comprising the molecule. The genetics has been studied and interesting interrelations found with genes we have discussed above (see Rucknagel and Laros, 1969; Harris, 1970).

Inborn Errors of Metabolism

We have used the hemoglobins to illustrate the point that genes act by controlling the production of proteins. Hemoglobins provide an excellent example because of our detailed knowledge of their molecular structure, and the precision with which that structure can be correlated with DNA structure. But in a sense hemoglobin is a rather unusual molecule in that the protein portion, globin, serves a structural function, supporting the heme group, which is the oxygen-carrying part of the molecule. Most of the other proteins that interest us, on the other hand, are *enzymes,* proteins that in themselves are active in controlling all manner of life processes (p. 41). As a rule, genes cause their phenotypic effects by controlling the nature of enzymes and the production of these enzymes.

The pioneer investigator in this field was the distinguished British physician, Sir Archibald E. Garrod, who early in this century recognized the importance of chemical deficiencies in causing such disorders as alkaptonuria (see below). In 1902 he published a paper on alkaptonuria in which, in the light of Bateson's writings on Mendelian principles, he recognized that his observations of patients and their families indicated recessive inheritance. Recall that in 1902 Mendelian principles had only recently been rediscovered (p. 23). Hence Garrod's paper "gives the first substantial account of recessive inheritance in man" (Harris, 1963). In 1909 Garrod's book, *Inborn Errors of Metabolism,* was published. This is the classic in the field, although recognition of its true significance awaited further development of the sciences of genetics and biochemistry. Garrod was ahead of his time.

Garrod discussed four inborn errors: albinism, alkaptonuria, cystinuria, and pentosuria. The number has been greatly extended since then. (Harris, 1970, listed 63, and the number is surely larger by now.) Whenever a new discovery is made of an enzyme deficiency based on a genetic change, that discovery adds a new inborn error to the list.

We select four of the inborn errors for discussion: phenylketonuria (PKU), alkaptonuria, albinism, and galactosemia. These are selected because of their intrinsic interest and illustrative value, and because they are ones that people generally are somewhat concerned about.

Phenylketonuria (PKU)

Three of the inborn errors in which we are interested are concerned at one point or another in the long chain of reactions by which the amino acids phenylalanine and tyrosine are metabolized to carbon dioxide and water (Fig. 5.3). These amino acids, like others, enter our bodies when we eat protein food. The proteins are broken down into their constituent amino acids, which are then used as building blocks from which our bodies manufacture their own proteins, or, when present in excess of such need, are metabolized to CO_2 and H_2O for excretion. Unneeded amino acids from the breakdown of our own "worn-out" proteins receive the same treatment.

As shown in Figure 5.3, one step in the process is the conversion, in the liver, of phenylalanine to the slightly simpler amino acid tyrosine. This is made possible by an enzyme called phenylalanine hydroxylase. Like all enzymes, this one is a protein under the control of genes in the manner discussed for the hemoglobins. Although genetic analysis is not so complete as it is for the latter, we cannot doubt that comparable mechanisms are involved.

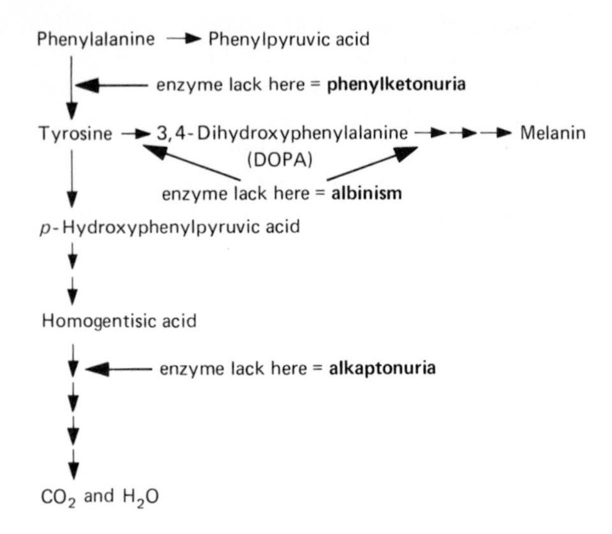

FIG. 5.3 Three inborn errors of metabolism connected with the metabolism of phenylalanine and tyrosine.

What will happen if a mutation occurs in the gene responsible for one of the polypeptide chains comprising the enzyme molecule? The enzyme may be so altered that it will no longer be an effective catalyst, if, indeed, it is produced at all. As a result, phenylalanine will be accumulated in the body in undesirable amounts, and so will phenylpyruvic acid, an alternative breakdown product of phenylalanine (Fig. 5.3). That such a mutation has occurred is evidenced by the fact that some people are found to have the defective gene, and that the gene behaves as a simple Mendelian recessive in inheritance. Criteria for recognizing such a mode of inheritance will be discussed in the next chapter. Here we note that people who are homozygous for this gene suffer from a condition called **phenylketonuria, PKU** for short. If we use *p* for the abnormal gene, the genotypes are as follows:

PP = homozygous normal
Pp = heterozygous normal; "carriers"
pp = phenylketonuric

In the blood of children with **PKU**, the content of phenylalanine may be more than 30 times the normal level (Harris, 1970). As a result of the blocking of the normal route for the breakdown of phenylalanine, various other abnormal compounds also accumulate in the body, their presence being detectable in blood and urine. The most important, and unfortunate, consequence of these biochemical disturbances is the fact that mental development is so inhibited that a child is likely to suffer severe *mental retardation* unless corrective steps are taken early. These consist of removing milk from the baby's diet and substituting a diet that contains little phenylalanine (*some* must be included, because the amino acid is essential for the building of proteins). To be most effective, the change in diet must be made early, within the first month of life if possible, so many hospitals now test all newborn babies for the defect. Estimates vary as to the frequency of occurrence of PKU, but one baby in 15,000 births seems a reasonable approximation (see Levy, 1973). Evidence indicates that the diet can be discontinued after three years or more with little danger of mental deterioration.

So far we have discussed typical or "classical" phenylketonuria. But there are many questions yet to be answered. For instance, from 20 to 25 percent of children having elevated levels of phenylalaline have intelligence within the normal range. Why is this? Seemingly, mental retardation results only if the concentration of phenylalanine and substances derived from it reaches a critically high level at a time when brain tissue is being actively formed. Furthermore, we do not yet know what actually causes the mental deterioration in those children with elevated phenylalanine level who *do* show such deterioration. Also, many investigations indicate that there are varied causes for a raised level of phenylalanine (hyperphenylalanemia). It seems that genes other than the **PKU** gene (*p*) may be involved, and we are uncertain at present whether these other genes are alleles of *p* (another case of multiple alleles, p. 50) or are nonallelic modifying genes. The enzyme activity involved is being investigated. Results so far seem to indicate that different genetic constitutions result in different levels of enzyme (phenylalanine hydroxylase) activity. Gaps in our knowledge concerning all these matters are sure to be filled by further investigation (for more detailed discussion see Hsia, 1970).

Phenylketonuric Mothers What happens when women who are diagnosed as having **PKU** become mothers? Their genotype is *pp*. If they marry normal men, the genotype of the father is likely to be *PP*. If so, all children will be heterozygous, *Pp,* that is, nonphenylketonuric. But will such children be normal? Howell and Stevenson (1971) reported findings on 33 women with **PKU** who were known to have children. The number of children totaled 121, and of these only three were considered normal. Most of the affected children do not have **PKU**, but they suffer from a variety of physical deformities, and virtually all of them are mentally retarded (Hsia, 1970). Presumably these physical and mental abnormalities arose because of the high phenylalanine levels in the mother. Phenylalanine and its derivatives can be passed from mother to fetus through the placenta to harm the developing brain and other organs. Clearly, women who have or have had **PKU** should be warned not to produce children.

Detection of Heterozygotes In almost all cases, children who have **PKU** are born to parents both of whom are normal but are heterozygous (*Pp*). For purposes of genetic counseling, can such persons be identified before they have proven their genotype by having a **PKU** child? Sometimes heterozygotes have a somewhat elevated phenylalanine level. Various phenylalanine-tolerance tests have been developed. In such tests a person is given a rather large amount of phenylalanine and then tested to determine how efficiently his body converts it to tyrosine. Hsia, a worker in this field, stated (1970): "Despite the enthusiasm of certain proponents, it would appear that all of these techniques are useful in clearly distinguishing a population of heterozygotes from a population of normals, but the discriminants are not sufficiently sensitive to classify a specific individual as being definitely a heterozygote for phenylketonuria." But "sharper" tests will surely be developed. A married couple will be vitally interested in knowing their chances of having a **PKU** child, particularly if they know of relatives who have the defect, and so have reason to wonder whether or not they are carriers themselves.

Alkaptonuria

There are many steps in the process of the metabolic breakdown (catabolism) of tyrosine. Figure 5.3 omits many of the intermediate products. The intermediate product at one point in the chain is homogentisic acid (alkapton). Most people have an enzyme (homogentisic acid oxidase) that metabolizes this further. But occasional individuals lack the enzyme, or have it in inactive form, with the result that homogentisic acid accumulates and is excreted in the urine. Such people are said to have **alkaptonuria.** The most striking feature of this is the fact that homogentisic acid, and hence urine containing it, turns black on exposure to the air (oxidation). Diapers stained in this manner call a mother's attention to the fact that an infant is alkaptonuric. The trait usually has no seriously detrimental effects, although arthritis later in life sometimes seems to accompany it.

As we noted previously, this was the first inborn error Garrod studied extensively. He was led to postulate a hereditary basis for it by observing that, while the condition is a rare one, it is common in some families (kindreds). Furthermore, he noticed early that of four pairs of parents having alkaptonuric children, three pairs consisted of first cousins married to each other. His further investigation increased the number of examples of consanguinity in parentage of alkaptonurics. Influenced by the early geneticist Bateson, he recognized such consanguinity in the parentage as evidence for recessive inheritance.

Increased consanguinity in the parentage for a trait is now generally recognized as an important indication that the trait depends on a recessive gene. Why is this? If we represent the gene for alkaptonuria by *a*, genotypes of individuals are as follows:

AA = homozygous normal
Aa = heterozygous normal (carrier)
aa = alkaptonuric

As we have said, alkaptonurics are rare in the general population. So are carriers (*Aa*), but these are not so rare as are the *aa* individuals (Chap. 10). In point of fact, most alkaptonurics are the offspring of normal parents. Hence the normal parents of an alkaptonuric child must be carriers, and the mating must be *Aa* × *Aa*. Because carriers are rare in the population, one carrier will seldom happen to marry another carrier *unless* he marries a relative. Relatives of a carrier are more likely to have the same recessive gene than are unrelated members of the population. As a result, surveys of persons who have a rare recessive trait show that an unusually large proportion of them are the offspring of consanguineous matings. We will discuss this further in Chapter 11. Here we give as an example a large kindred in which there are 15 cases of alkaptonuria (Hall, Hawkins, and Child, 1950).

Figure 5.4 presents a portion of this kindred including 12 of the 15 cases. In pedigree charts of this kind, circles represent females; squares represent males. Solid or shaded squares and circles represent individuals who show the trait in question. A horizontal line connecting a circle to a square represents a mating, and a vertical line extending

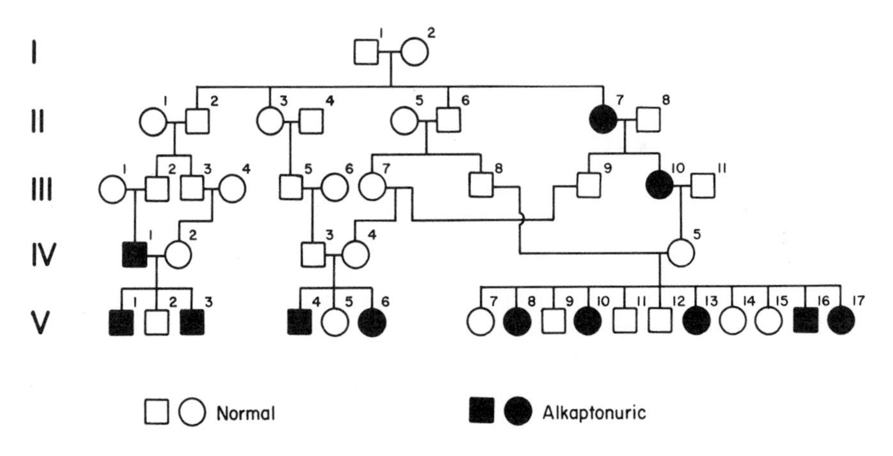

FIG. 5.4 Inheritance of alkaptonuria in a kindred characterized by consanguinity. (Based on a chart in Hall, W. K., K. R. Hawkins, and G. P. Child, 1950. "The inheritance of alkaptonuria in a large human family," *Journal of Heredity*, **41**:23–25.)

downward from the horizontal one connects to a child or children produced by that mating. If there are several children, the vertical line ends in a horizontal one, to which are attached the squares and circles representing the children. The generations are numbered at the left with Roman numerals, and each individual in a generation is designated by an Arabic numeral. Thus each individual can be referred to without the use of names. For example, the earliest ancestors shown in Figure 5.4 are I-1 and I-2.

The diagram is interesting for the varying degrees of consanguinity present. The parents of V-1, V-2, and V-3 were first cousins, the father being alkaptonuric himself (thus the mating was *aa* × *Aa*).

Individuals IV-3 and IV-4, both carriers, were double second cousins. The father of IV-3 (III-5) was the first cousin of *both* of his wife's parents (III-7 and III-9). Such an arrangement is well calculated to concentrate recessive genes.

The parents of the largest family in the last generation (III-8 × IV-5) were first cousins once removed. The husband married his first cousin's daughter. Evidently both were carriers.

Albinism

Albinism, more or less complete absence of pigment from skin, hair, and eyes, occurs widely throughout the animal kingdom. There are various types of albinism, but except as otherwise indicated we shall use the word to mean the commonest form: oculocutaneous albinism.

In man, and in those other species that have been investigated, it is often caused by lack or inactivity of the enzyme **tyrosinase.** As indicated in Figure 5.3, tyrosine is the precursor of the dark pigment, melanin, found to varying extents in the skin and eyes of persons who are not albinos. The synthesis of melanin involves a long chain of reactions, with intermediate products and enzymes. The first step in the process is the conversion of tyrosine to 3,4-dihydroxyphenylalanine (DOPA). Tyrosinase is involved in this conversion and also in the oxidation of DOPA, which forms the next step in the series (Duchon, Fitzpatrick, and Seiji, 1968). So if the enzyme is absent or inactive, melanin formation is stopped at its source, and skin, hair, and eyes remain devoid of pigment (Fig. 5.5). The principal disadvantage of this lack is the fact that eyes without normal pigment do not function as well as do normal eyes. In some cases the eyes are pink from the color of the blood, in other cases blue [but as we shall see (p. 107), blue in eyes is an optical effect, not caused by blue pigment]. Suppression of melanin formation is not always complete.

Evidence both from human and animal studies indicates that tyrosinase deficiency depends upon a recessive gene, commonly represented by the initial *c*. Hence genotypes are as follows:

CC = homozygous normal
Cc = heterozygous normal (carrier)
cc = albino

An interesting corollary of this mode of inheritance is the fact that when albinos marry each other ($cc \times cc$), we should expect *all* chil-

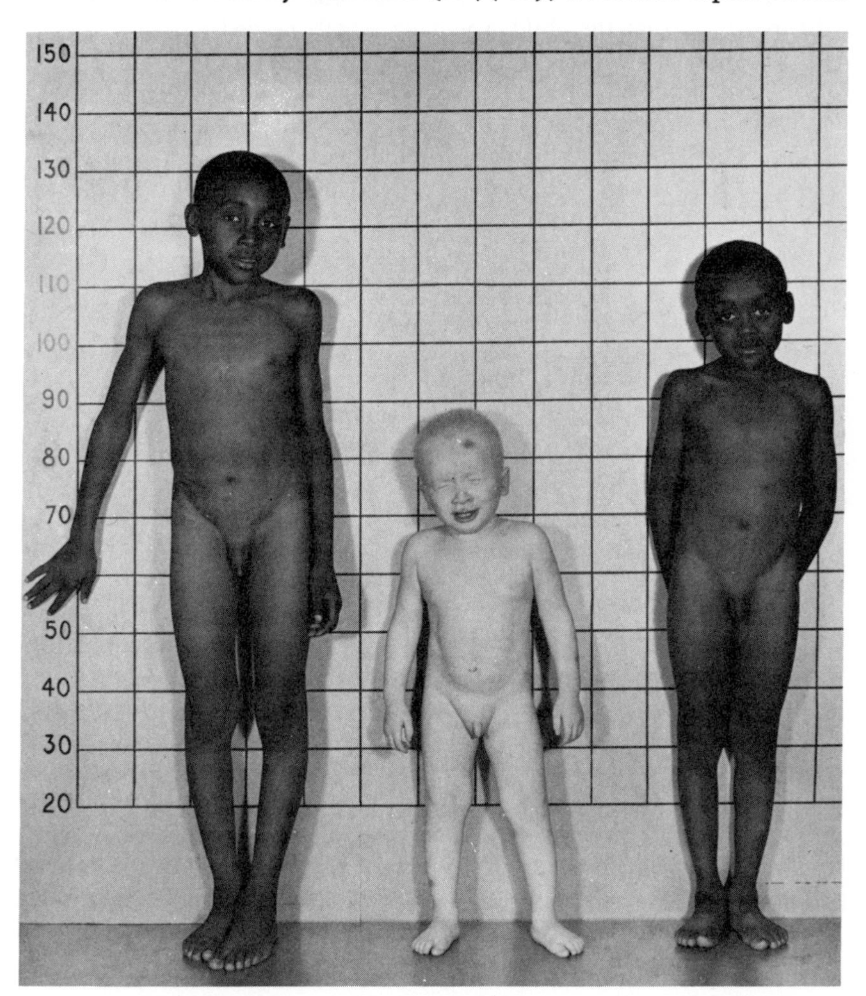

FIG. 5.5 An albino boy, 26 months of age, and his normally pigmented brothers. The albino brother probably closed his eyes because the photographer's bright lights hurt them. Recall that albino eyes lack pigment. (Courtesy of Victor A. McKusick, M.D.)

dren to be albinos. Commonly this is found to be the case, but occasionally two albino parents will have a normally pigmented child. How can this be explained, in cases where we are confident that the two albinos are the actual biological parents? We might anticipate that the reason lies in the fact that the parents are albinos for different reasons—because of lack of different enzymes. If this were the case, the father would have the enzyme lacked by the mother, who, in turn, would have the enzyme lacked by the father. Then the father could pass on to a child the gene for the enzyme he had, and the mother could pass on the gene for the enzyme she had, with the result that the child would have both needed enzymes, and be normally pigmented.

Witkop et al. (1970) cited two families in which this explanation is clearly correct. That the parents in each case had different types of albinism was indicated in part by differences in physical appearance, but most clearly by a hair-bulb incubation test. Single hairs with bulb intact were plucked from the head of individuals concerned and incubated in a solution of tyrosine. If the hair bulb contained tyrosinase, pigment developed and the albino was said to be tyrosinase-positive. But if the hair bulb did not contain tyrosinase, no pigment developed and the albino was called tyrosinase-negative (Fig. 5.6). Such an albino would be homozygous for gene *c*. In both families studied, one parent's hair bulbs showed melanin production when tested, the other parent's hair bulbs did not.

What is the deficiency in tyrosinase-positive albinos? The answer is not known with certainty, but the authors suggested some deficiency in the mechanism by which tyrosine enters the pigment-forming cells—perhaps a deficiency in an enzyme (permease) affecting the permeability of the cell membrane so that tyrosine cannot enter in the quantity needed.

If we represent the gene for the defective permease by *p*, the genetic situation can be represented as follows:

tyrosinase-positive parent × tyrosinase-negative parent
CCpp *ccPP*
Gametes = *Cp* Gametes = *cP*
normal child
CcPp

(The child has the gene C for normal tyrosinase, and the gene P for normal permease, and they complement each other to produce normal pigmentation.) For additional information on albinism see Witkop et al., 1970, and Witkop, 1971.

Galactosemia

The fourth inborn error we shall consider is not directly concerned with the metabolism of phenylalanine and tyrosine. Instead it is

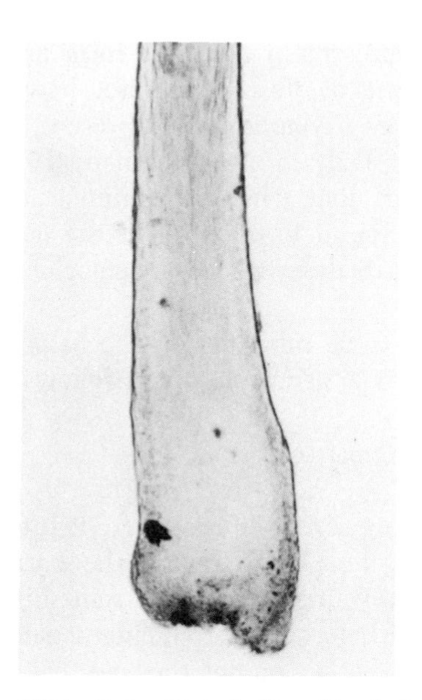

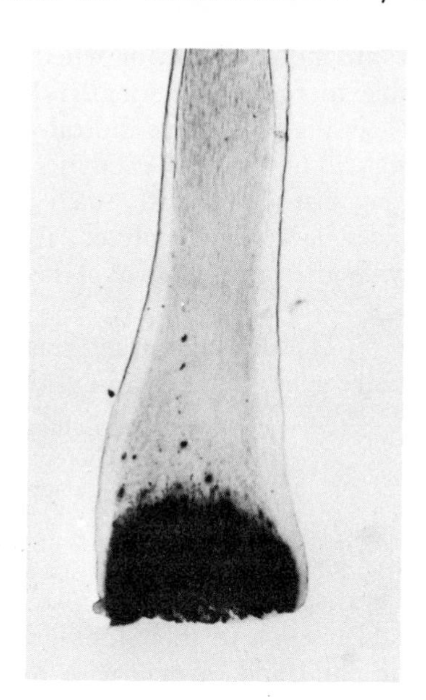

FIG. 5.6 **Hair bulbs after being incubated in tyrosine.** No pigment was formed in the bulb on the left: "tyrosinase-negative." The bulb on the right formed pigment: "tyrosinase-positive." The hair bulb on the left came from the albino father, that on the right from the albino mother, of a normally pigmented child. (Courtesy of Carl J. Witkop, Jr., D.D.S. From Witkop, C. J., Jr., et al., 1970. "Autosomal recessive oculocutaneous albinism in man: Evidence for genetic heterogeneity," *American Journal of Human Genetics,* **22**:55–74.)

an error in the metabolism of carbohydrates, those substances (starches, sugars) on which we depend largely for energy.

The main carbohydrate present in milk is lactose. Our body splits (hydrolyzes) this into the two simpler sugars glucose and galactose. Then these two simple sugars are broken down further to release the energy stored in their molecules. An occasional individual cannot metabolize **galactose** in this manner, with the result that this sugar and some harmful compounds derived from it accumulate in the body with severe consequences. Such individuals have **galactosemia.** An infant so afflicted "fails to thrive, weight gain is slow, mental development is retarded, the liver becomes enlarged and eventually cirrhotic, and cataracts develop in the lens" (Harris, 1970). Death in infancy frequently results. Fortunately, if the condition is recognized early, the baby can be placed on a diet free of milk and other sources of galactose, with marked improvement, indeed a return to normal growth and development.

The defective enzyme in galactosemia has been identified; it is named galactose-1-phosphate uridyl tranferase (we shall call it simply transferase). As noted previously, we are often unsure whether an abnormal enzyme is absent, or whether it is present in defective form and so unable to function normally. In the case of this transferase, however. serological studies have indicated that the enzyme is indeed present, even though it has lost its catalytic activity (Tedesco and Mellman, 1971). We may prophesy with some confidence that when its amino acid sequences have been analyzed, it will be found to differ from the normal enzyme after the manner of hemoglobin S differing from hemoglobin A (pp. 47–50).

The defective transferase seems to be inherited on the basis of a recessive gene. If we represent the gene by g, genotypes are as follows:

GG = homozygous normal
Gg = heterozygous normal (carrier)
gg = galactosemic

As with other inborn errors having severe effects, there is interest in the possibility of detecting heterozygotes. Even though the harmful effects can be prevented by a strict diet, a young couple with some history of galactosemia in their families may wish to avoid producing a galactosemic child and so may wish to know whether or not they are carriers. Although the transferase functions in the liver primarily, it is present in other bodily tissues, including the red blood cells. Tests indicate that blood cells of galactosemics have practically no transferase activity, while cells of heterozygotes have about half as much activity of the enzyme as do cells of homozygous normals (Fig. 5.7). The heterozygotes in this case were identified by the fact that they *had* produced a galactosemic child. We note from the chart that a few of the heterozygotes had as much transferase activity as did the normals with the least activity. Thus an occasional carrier parent might not be identified as such. Tests will undoubtedly be sharpened. All such tests improve the accuracy of genetic counseling as compared to advice given without them (Chap. 28).

Résumé

In this chapter we have given some of the best human examples of genes controlling production of proteins and of mutations of these genes resulting in production of altered proteins. Knowledge of change-in-gene producing change-in-protein is most complete for the hemoglobins, but the same sort of relationship will undoubtedly be found when proteins that are enzymes, and their DNA basis, become as completely known as are the hemoglobins and their genes.

The inborn errors are inherited exactly as we should expect them to be according to the one-gene, one-polypeptide chain concept. Genes probably always act by determining the nature and production of proteins. A large proportion of these proteins are enzymes. All life processes, including embryonic development, are controlled by enzymes. Thus the type of

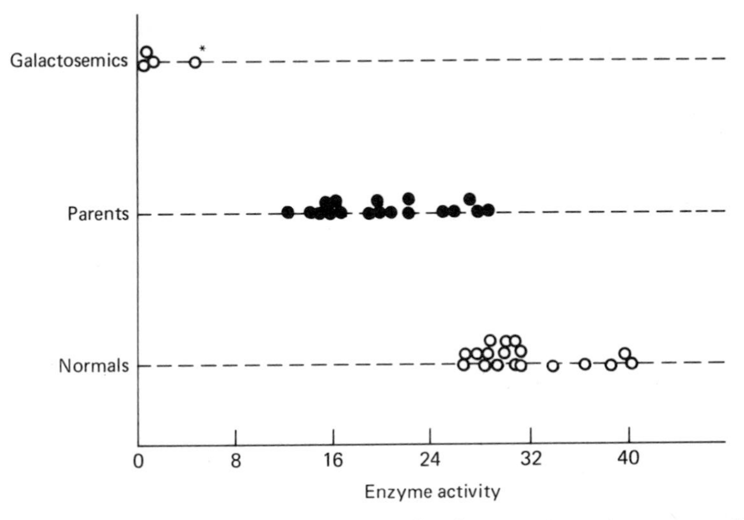

FIG. 5.7 Identification of carriers of galactosemia, using a test of red blood cells for activity of the enzyme Gal-l-P uridyl transferase. See text. *This patient had had a blood transfusion. (Redrawn from Kirkman, H. N., and E. Bynum, 1959. "Enzymic evidence of a galactosemic trait in parents of galactosemic children," *Annals of Human Genetics,* **23**:117–126.)

mechanism we see here operates in the more complex chains of events that lead to the production of many human traits in which we are interested. The inborn errors are good examples with which to begin, because the chain of events between changed gene and changed phenotype is relatively short and direct.

Problems

1. Two normal parents produce a child suffering from sickle-cell anemia. What is the chance that the next child will also have this anemia?
2. A man with the sickle-cell trait marries a woman with the hemoglobin C trait (p. 51). What types of children may they expect to have?
3. Two normal parents have a child with phenylketonuria (PKU). Give the genotypes of the child and the parents.
4. If both a husband and his wife were recovered phenylketonurics, what proportion of their children would be expected to have PKU?
5. Referring to Figure 5.4, what is the genotype of V-2? If he were to marry his distant cousin V-6, what proportion of their children would be expected to be alkaptonuric?
6. Referring to information about the two types of albinism, would it be possible for an albino to have the genotype *ccpp*? If such a person married one having the genotype *CcPp,* what types of children would they be

expected to have, and in what proportions? (Recall our discussion of independent assortment in Chap. 3.)

7. A child born to normal parents has tyrosinase-negative albinism and galactosemia. Give the genotypes of parents and child.
8. A normal woman has a sister with galactosemia. If the parents of the two sisters are normal, what is the chance that the woman is heterozygous (a carrier)? How can this be determined more accurately?
9. Two albinos are married to each other. When hair bulbs from the wife are incubated in tyrosine, they produce melanin. When hair bulbs from the husband are similarly treated, they do not produce melanin. Give the most probable phenotype or phenotypes of children they may produce.

References

Duchon, J., T. B. Fitzpatrick, and M. Seiji, 1968. "Melanin 1968: Some definitions and problems." In Kopf, A. W., and R. Andrade (eds.), *The Year Book of Dermatology, 1967–1968 Series.* Chicago: Year Book Medical Publishers. Pp. 6–33.

Garrod, A. E., 1909. *Inborn Errors of Metabolism.* London: Oxford University Press. Reprinted with a supplement by H. Harris, 1963.

Hall, W. K., K. R. Hawkins, and G. P. Child, 1950. "The inheritance of alkaptonuria in a large human family," *Journal of Heredity,* **41**:23–25.

Harris, H., 1963. See supplemental material added in the reprinting of Garrod (1909).

Harris, H., 1970. *The Principles of Human Biochemical Genetics.* New York: American Elsevier Publishing Company, Inc.

Howell, R. R., and R. E. Stevenson, 1971. "The offspring of phenylketonuric women," *Social Biology,* **18** (Supplement):19–29.

Hsia, D. Y. Y., 1970. "Phenylketonuria and its variants." In Steinberg, A. G., and A. G. Bearn (eds.), *Progress in Medical Genetics. Vol VII.* New York: Grune & Stratton. Pp. 29–68.

Ingram, V. M., 1956. "A specific chemical difference between the globins of normal human and sickle-cell anaemia haemoglobin," *Nature,* **178**:792–794.

Ingram, V. M., 1958. "How do genes act?" *Scientific American,* **198**:68–76.

Ingram, V. M., 1963. *The Hemoglobins in Genetics and Evolution.* New York: Columbia University Press.

Kirkman, H. N., and E. Bynum, 1959, "Enzymic evidence of a galactosemic trait in parents of galactosemic children," *Annals of Human Genetics,* **23**:117–126.

Lehmann, H., and R. W. Carrell, 1969. "Variations in the structure of human haemoglobin," *British Medical Bulletin,* 25:14–23.

Levy, H. L., 1973. "Genetic screening." In Harris, H., and K. Hirschhorn (eds.), *Advances in Human Genetics, Vol. 4.* New York: Plenum Press. Pp. 1–104; 389–394.

Neel, J. V., 1949. "The inheritance of sickle cell anemia," *Science,* **110**:64–66.

Neel, J. V., 1951. "The inheritance of the sickling phenomenon, with particular reference to sickle cell disease," *Blood,* **6**:389–412.

Neel, J. V., and W. J. Schull, 1954. *Human Heredity.* Chicago: University of Chicago Press.

Pauling, L., H. A. Itano, S. J. Singer, and I. C. Wells, 1949. "Sickle cell anemia, a molecular disease," *Science,* **110**:543–548.

Rucknagel, D. L., and R. K. Laros, Jr., 1969. "Hemoglobinopathies: Genetics and implications for studies of human reproduction," *Clinical Obstetrics and Gynecology,* **12**:49–75.

Tedesco, T. A., and W. J. Mellman, 1971. "Galactosemia: Evidence for a structural gene mutation," *Science,* **172**:727–728.

Witkop, C. J., Jr., 1971. "Albinism." In Harris, H., and K. Hirschhorn (eds.), *Advances in Human Genetics, Vol.* 2. New York: Plenum Press. Pp. 61–142.

Witkop, C. J., Jr., W. E. Nance, R. F. Rawls, and J. G. White, 1970. "Autosomal recessive oculocutaneous albinism in man: Evidence for genetic heterogeneity," *American Journal of Human Genetics,* **22**:55–74.

6

Studying Human Pedigrees

In the preceding chapter we mentioned that the abnormal hemoglobins and the inborn errors of metabolism are inherited on the basis of recessive genes. We now ask: In human genetics, where breeding experiments are impossible, how do we decide the question of dominance and recessiveness in inheritance? The time-honored method consists of studying as many generations as possible in kindreds ("families" in the inclusive and continuing sense of the word). Data concerning such kinships are conveniently summarized in pedigree charts ("family trees").

Dominant Inheritance

If a characteristic depends on a dominant gene, how will this fact manifest itself in a pedigree?

As a first approximation, we may make this statement: *A trait determined by a dominant gene will not appear in an offspring unless it also appears in one or both parents.* If the offspring shows the dominant phenotype A, his genotype must be *AA* or *Aa*. Therefore, he must have received an *A* from at least one parent, and that parent (being *AA* or *Aa*) will show the trait, too. This fact is sometimes stated by saying that a dominant trait never "skips."

Huntington's Chorea (The term **Huntington's disease** is preferred by some.) Figure 6.1 is an abbreviated version of a large pedigree chart of a kindred in which Huntington's chorea has been inherited for at

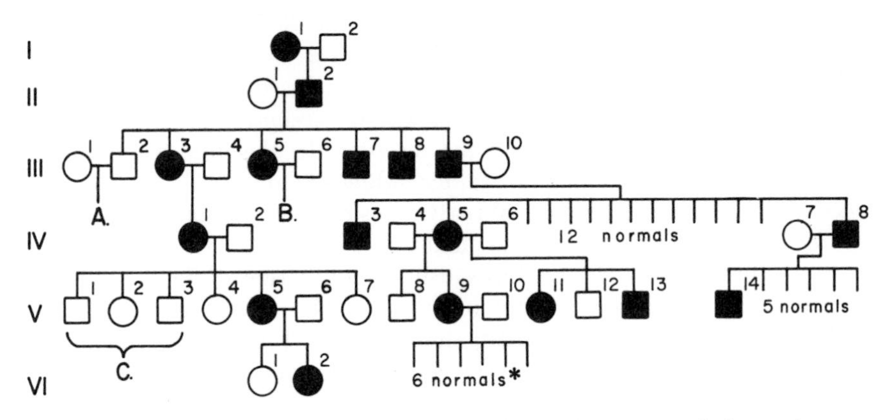

FIG. 6.1 Inheritance of Huntington's chorea (Huntington's disease). Black squares and circles represent males and females, respectively, who had the disease. A. This couple had 9 children and 53 grandchildren, all normal. B. This couple had 11 children, 3 with Huntington's chorea. These three married but did not pass on the trait, although they had families of 6, 8, and 8 children, respectively. C. These three married, and produced a total of 19 children, all normal. *It is possible that some members were young at the time of investigation, and that symptoms may have developed later in life. (Abbreviated from a pedigree chart prepared by the Eugenics Survey of Vermont.)

least six generations. This is a mental disorder in which mental deterioration is accompanied by uncontrollable, involuntary movements. At the present time, afflicted persons do not recover from it.

The symbols employed in this chart and subsequent ones are those described on pp. 57–58.

I-1 was a woman in whom Huntington's chorea developed. Her husband, I-2, did not have the disorder. They had one son, II-2, who showed the trait. He married a woman, II-1, who did not. They, in turn, had 6 children: a boy, III-2, who did not have the trait, and 2 daughters and 3 sons who did (III-3; III-5; III-7; III-8; III-9).

Let us consider III-2, because he and his normal wife, III-1, illustrate a most important corollary of the general statement we have made about dominant inheritance. *If the trait depends on a dominant gene, a marriage of two people who do not show the trait will produce only normal descendants,* barring the rare occurrence of a new mutation. If we let H represent the gene for Huntington's chorea and h the gene for normality, then III-1 and III-2 must have had the genotypes $hh \times hh$. In that case, all descendants would be hh, too, unless the gene H were brought in by marriage from "outside the family." As Figure 6.1 shows, III-1 and III-2 had 9 children, all normal, and these in turn married and produced 53 grandchildren, all normal. In no case did two normal parents produce a child who later had Huntington's chorea.

In contrast to the descendants of III-2, consider those of his sister III-3. She married a normal man and they had one daughter, IV-1, in whom the trait developed. She married and produced a family of 6, only one of whom, V-5, became choreic. V-5 married and produced 2 daughters, one of whom (VI-2) developed the disorder. In this line we have six generations of Huntington's chorea, extending without a skip from I-1 to VI-2. Note again that the normal individuals of V generation who married (V-1, V-2, V-3) had only normal children.

Huntington's chorea thus satisfies the criterion of never skipping, which is indicative of dominant inheritance. Many pedigrees have been published giving the same results; the fact that Huntington's chorea depends on a dominant gene may be regarded as well established. Unlike some other types of mental disorder, little if any environmental influence is involved. The trait will develop in persons who have the gene (if they live long enough), regardless of good or bad environment.

Another expectation for simple dominant-gene inheritance is satisfied by the data in Figure 6.1. *The trait shows up with about equal frequency in males and in females.*

However, the pedigree chart does not show the ratio of affected to normal children that we might expect. We note that in no case did an affected individual marry another affected individual; all had normal spouses. (This is almost invariably the case when a dominant gene is rare—as that for Huntington's chorea fortunately is, outside of certain families). Thus the matings were $Hh \times hh$. In such families, we should expect affected (Hh) and normal (hh) children in equal numbers—the 1:1 ratio. However, inspection of Figure 6.1 shows that the number of offspring, from $Hh \times hh$ matings, classified as normal far exceeds the number of affected offspring. For example, the family of 15 born to III-9 and III-10 included only 3 choreics. And III-5 × III-6 produced 11 children, of whom 3 were affected; these three married and produced a total of 22 children, all classified as normal.

How can we explain this discrepancy between expectation and actuality? In the first place, Huntington's chorea has one complicating aspect: Although people may develop the symptoms at any age, they normally do not do so until at least middle age (the average age of onset is in the 40's). This means, then, that an investigator often cannot accurately classify children and young people. Hence, too many young people are likely to be classified as normal.

Similarly, the family of III-9 × III-10, for example, lived about four generations ago, a long time in terms of human memory and of the history of medicine. It seems possible that some of the 12 children classified as normal may have died young (of childhood diseases, tuberculosis, cancer), before they were likely to have manifested choreic symptoms. Therefore, some people may have been classified as normal when in fact they had the *Hh* genotype and would have become choreic had they lived long enough.

This is a complication that arises when hereditary trait has a late onset. Fortunately, many traits are detectable at birth or soon after. In such cases, the children of $Aa \times aa$ matings usually approximate the expected 1:1 ratio very closely (when data from several or many such matings are pooled, of course).

Retinoblastoma Steinberg (1959) cited an example that is applicable here. The trait is called **retinoblastoma**, a cancer of the retina of the eye in children, and it is quickly fatal unless the affected eye is removed. If the eye is removed in time, the child may live to adulthood, marry, and have children of his own. Much evidence indicates that retinoblastoma depends on a dominant gene. If we represent the gene by *R* and the normal allele by *r*, then persons who have retinoblastoma have the genotype *Rr*. If they survive the disease and marry a normal mate, we have an example of an $Rr \times rr$ mating. (In theory, of course, a person with retinoblastoma could also be homozygous, *RR*. But an *RR* genotype could arise only if two persons who survived retinoblastoma married each other. To date, there is no record of such a marriage.)

Reese (1954) gathered data concerning 15 families, in each of which one parent had survived retinoblastoma. Figure 6.2 presents two of the pedigrees studied by Reese (1954). Note that there is no skipping.

In summary, the following are the hallmarks of dominant, autosomal inheritance. (1) Children do not show the trait unless one parent does. As a corollary, the trait never develops in the descendants of two normal parents unless a new mutation occurs or unless the gene is introduced from an outside source by marriage. (2) The trait appears with equal frequency in males and females. (3) If the trait is rare, most, if not.

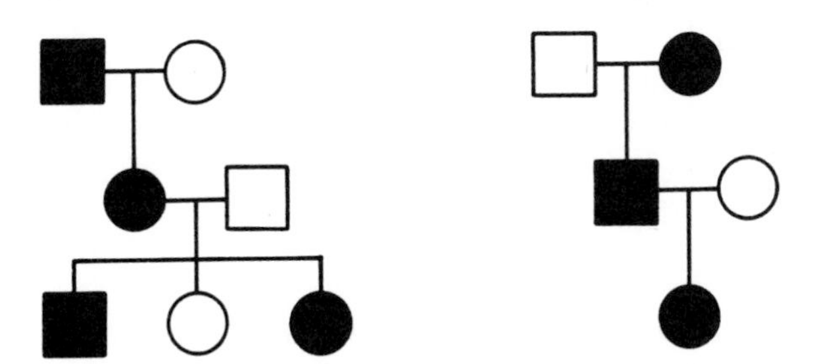

FIG. 6.2 Two pedigrees of inheritance of retinoblastoma. Black squares and circles represent males and females, respectively, in whom retinoblastoma developed. (Redrawn from Reese, A. B., 1954. "Frequency of retinoblastoma in the progeny of parents who have survived the disease," *Archives of Ophthalmology*, **52**:815–818.

all, persons who have it will be heterozygotes. (4) When such heterozygotes marry unaffected mates, half the children, on the average, can be expected to show the trait and half can be expected not to do so.

Mention should also be made of another hallmark of a *completely* dominant gene: Homozygotes for the gene (*AA*) have the same phenotype as do heterozygotes (*Aa*). The emphasis here is on the word "completely." In fact, in plants and experimental animals all degrees of dominance are found. Sometimes homozygotes seem to be exactly like heterozygotes. Sometimes heterozygotes differ from homozygotes—slightly or more distinctly. If the difference is great enough, we say that dominance is lacking (as in Blue Andalusian fowls, p. 16).

In the case of man, we can apply this criterion only with dominant genes that are common. People who belong to blood group A, for example, may be homozygous or heterozygous (I^AI^A or I^AI^O, p. 182). Ordinary testing serums do not distinguish between the red blood cells of homozygotes and heterozygotes. Therefore, we may say that I^A is completely dominant to I^O. In the future, however, more refined testing methods may enable us to distinguish between the two. When that becomes possible, we shall no longer regard I^A as completely dominant to I^O. This illustrates the point that the distinction between dominant and recessive genes is somewhat arbitrary and varies with the ability to distinguish homozygotes from heterozygotes.

Because most of the dominant genes studied in man are rare, the criterion of homozygotes being indistinguishable from heterozygotes cannot be applied. This is because all, or almost all, persons showing the dominant trait are heterozygotes. As noted in the case of retinoblastoma, homozygotes can occur only when heterozygotes marry each other (e.g., $Aa \times Aa$). When such a mating occurs, only one-fourth of the children, on the average, are expected to be homozygous *AA;* the small size of human families adds to the unlikelihood of actually finding *AA* individuals. Thus, many dominant genes are known only from their effects in heterozygotes, and we do not know what the phenotype of homozygosity would be. In some instances we have reason to suspect that homozygosity for harmful genes would cause more severe symptoms than does having the gene in single dose (heterozygosity).

Partial Penetrance Let us return to the first criterion, that of a dominant trait not skipping. Suppose that we collected information about a certain trait, inherited in a kindred for several generations, and then constructed a pedigree chart resembling Figure 6.1. In this study a child usually does not show the trait unless one parent does, but suppose that there are occasional exceptions. What conclusions should be drawn?

Although in most respects the criteria for a dominant gene are satisfied, must we conclude because of these exceptions that the gene is not dominant at all? How can we account for these exceptions?

We have already mentioned the possibility that a mutation may have occurred. Human geneticists must always be alert to two other possibilities. There is always the chance, even probability, that a small proportion of the children studied in any investigation may not be the biological offspring of one or both of the people who appear to be their parents. When adoption and illegitimacy are kept secret, as they frequently are, they introduce sources of error into the calculations of a geneticist.

In the present instance, however, let us suppose that the exceptions in our pedigree cannot be accounted for by adoption, illegitimacy, or the rare occurrence of new mutations. We are left with the fact that at times a gene does not produce its usual phenotypic effect. Gene *A* usually results in phenotype A, but sometimes the phenotype fails to appear although gene *A* is present. We then say that gene *A* has incomplete or partial penetrance. When a dominant gene is completely penetrant, every possessor of the gene shows the trait. The gene for Huntington's chorea seems to be completely penetrant, or very nearly so; the symptoms develop in every possessor of the gene, if he lives long enough. In contrast, the gene for retinoblastoma is calculated to have about 90 percent penetrance (Neel and Schull, 1954); malignant tumors of the retina develop in about 90 percent of the children who have the gene.

Why is it that a dominant gene does not always produce the effect it usually does? A phenotype is the result of complicated processes conditioned by both heredity and environment. In environment we include all the factors that may impinge on an embryo during its development from ovum to birth. At times, environmental factors may override genetic factors, rendering impossible the usual outcome of a genetic endowment. Undoubtedly tumors of the retina have requirements in addition to the mere presence of gene *R*. Perhaps when tumors do not develop in possessors of gene *R*, the failure (a happy one) is due to lack of some essential constituent for tumor formation, or to the development of some inhibitor to tumor formation. Future research will undoubtedly yield much more information than we have at present on this point.

We have mentioned absence of essential constituents and presence of inhibitors as examples of environmental factors that can modify the action of genes. In fact, however, so inextricably are heredity and environment interwoven that essential constituents and inhibitors may themselves be under genetic control. Here we are referring to what are usually called modifying genes (p. 112). The phenotype is the result of the action of all genes present (plus environmental factors); no gene operates all by itself. If a gene is to produce its usual effect, complicated systems of enzymes must operate during development. Some of these enzymes are under the control of other genes. Hence, changes in these other genes may lead to failure of the main gene to produce its usual phenotype.

For example, consider some of the genes for color in mice. Genes *B* and *b* are alleles. Normally, mice having gene *B* are black, and homozygotes for gene *b* are brown. The color depends largely upon the amount of melanin present. As we have seen (pp. 58–60), the enzyme tyrosinase, dependent upon the dominant gene *C*, is essential for the production of melanin (this is as true of mice as it is of humans). In the absence of *C* (i.e., in *cc* mice), no melanin is produced, and therefore neither gene *B* nor gene *b* can have any effect at all. As a result, mice with the genotype *ccbb, ccBb,* or *ccBB* are all albinos. Thus we may say that the absence of *C* reduces the penetrance of *B* and *b* to zero.

Modifying genes are so common that probably we should always find them if our investigation were intensive enough. To return to the subject of Huntington's chorea, there is evidence that modifying genes control the age of onset of the symptoms (Neel and Schull, 1954). It seems to be genetically determined in some families that the symptoms will appear in early life, whereas the genes present in other families determine that the symptoms will appear later in life. In this case, we are not dealing with lack of penetrance, although we might use our imaginations to picture how an apparent lack of penetrance might arise from the situation. Suppose that in the modifying genes in a certain kindred there should be a change (a mutation perhaps), causing the symptoms to appear in possessors of the mutation only after age 90. In that kindred we should then have examples of people with the *Hh* genotype who married, had children, and died of heart disease or cancer *before* they reached the age of 90 and developed choreic symptoms. However, if some of their children became choreic, we should have an example of apparent skipping, in other words, an apparent lack of penetrance. Although this example is imaginary, it is entirely in line with known genetic principles.

We shall have occasion to refer to partial penetrance from time to time in subsequent discussions. Its existence complicates the work of the human geneticist. How much simpler it would be if a certain genotype produced its corresponding phenotype in all the individuals possessing it.

Variable Expressivity To parallel the last statement: How much simpler if a certain genotype always produced the *same* phenotype. As a matter of fact, even when the genes does produce a detectable effect, the expression of the gene may vary. (Of course, if the gene does not express itself at all, we have lack of penetrance.) Thus a gene may produce somewhat different effects in one individual than a gene of the same kind produces in another individual.

An apparent example is afforded by the study of Lutman and Neel (1945) on juvenile cataract of the eyes. When cataracts develop, the crystalline lens of the eye gradually becomes more and more opaque, with the result that vision is progressively impaired. Removal of the cloudy lens restores vision. We usually think of cataracts as an accompaniment of old age, but occasionally cataracts develop in the eyes of young children. Figure 6.3 represents a portion of a large kindred of this kind studied by Lutman and Neel. The presence of juvenile cataract seems to be determined by a dominant gene. Analysis of the entire pedigree (of which Fig. 6.3 is only a portion) shows that the trait never skips: no child has cataracts unless one of the parents has. There were 24 marriages in which one parent was cataractous. Because no two cataractous persons married each other, all persons with cataract must have been heterozygotes. Hence, if we represent the gene for the trait by *C* (not to be confused with the gene for tyrosinase we have been discussing), their marriages were *Cc* × *cc* matings. From such matings we expect a 1:1 ratio. The 24 marriages produced 79 children. The nature of 2 of these was undeterminable, but of the remaining 77, 34 were normal and 43 cataractous. This agrees fairly well with the expectation, which would be 38:39. Of the 43 affected individuals, 23 were males, 20 females.

When an ophthalmologist examines a cataract, the opacity is seen to be caused by more-or-less densely packed white flakes and granules, which form a pattern. The patterns may differ greatly from individual to

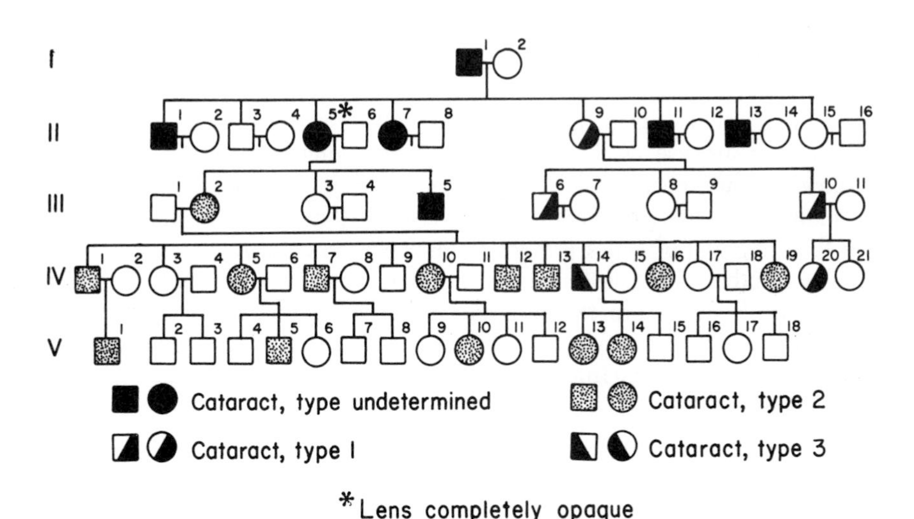

FIG. 6.3 Variable expressivity in inherited cataract. (Based on a chart in Lutman, F. C., and J. V. Neel, 1945. "Inherited cataract in the B. genealogy," *Archives of Ophthalmology*, **33**:341–357.)

individual.* Despite the diversity of these patterns, Lutman and Neel were able to classify the cataracts into three types. In type 1 the opacities were flake-like and fenestrated (like tiny irregular doughnuts) and were concentrated toward the front of the lens. As you will see from Figure 6.3, this type was confined to II-9 and her descendants. Type 2, characterized by more solid flakes massed in the center of the lens, was more common; it was possessed by III-2 and most of her numerous cataractous descendants. Yet one of her sons was the only individual investigated who had a type 3 cataract. In him the opacity was feather-like and fibrous, rather than flake-like as in the other two types. Nevertheless, his cataractous daughters (V-13 and 14) displayed the more usual type 2.

What caused the variability? Probably we should look for the cause in modifying genes. It seems likely that II-9 and her descendants were characterized by certain modifiers of the main gene, III-2 and her descendants by different modifiers. The authors pointed out that the simplest explanation would be that II-9 had a dominant modifying gene (by mutation?) not present in her brothers and sisters. If we represent this gene by *M*, then she and her cataractous descendants would have the genotype *CcMm* (whereas III-2 and her cataractous descendants, lacking *M*, would be *Ccmm*).

There is, of course, the additional possibility of the influence of environmental factors operating during embryonic development. Perhaps such factors might account for the strange case of IV-14, who differed so markedly from his mother, sibs, and daughters. However, further conclusions on this point must await additional research.

Oepen's extensive investigation of patients with Huntington's chorea has emphasized the great diversity of symptoms displayed by different individuals (Oepen, 1961). Actually, such variable expressivity should not surprise us. Without fear of contradiction, we can state that individual differences will be found in any trait whatsoever if that trait is analyzed intensively enough. We have stated that the phenotype depends on the sum total of genes present (plus environmental factors), all interacting together. No two human individuals (with the possible exception of identical twins) are exactly alike in all genes (let alone in environmental factors). Therefore, we should expect variations in phenotype to be the rule, even among individuals who share certain main genes.

Recessive Inheritance

To a large extent as we study human pedigrees, we identify recessive traits by virtue of the fact that they do *not* follow the rules that

*The original paper contains excellent drawings of these varied patterns.

we have laid down for dominant traits. Dominant traits do not skip (children do not show the trait unless a parent did); recessive traits commonly skip (although they may not do so).

Figure 6.4 presents a pedigree of the inheritance of albinism based on a study by Pearson, Nettleship, and Usher (1911–1913). The two types of albinism had not been distinguished at the time. Generation II was the first one about which the authors obtained definite information. In this generation, we find a family consisting of two normally pigmented brothers (II-2 and II-4) and their albino sister (II-3). Evidently, then, if neither parent (I-1 and I-2) was an albino, both must have been heterozygous $Cc \times Cc$, assuming that the albinism was the tyrosinase-negative form. Both brothers married, and each marriage started a line of descent in which no albinos appeared until the fifth generation. A great-granddaughter (V-1) of II-2 was an albino, and three of the great-grandchildren of II-4 were also albinos. This is a fine example of skipping generations. Moreover, if the families had not known that there was an albino great-aunt, the appearance of albino children would have been a complete surprise, for there was no other indication that the gene was in the family. Because not only two, but three, four, or more generations may be skipped, albino children often are born to parents who had no idea that they or their relatives and ancestors possessed the gene.

As far as the records indicated, none of the fathers (IV-2, IV-6, IV-8) of the fifth-generation albino children were related to each other or to their wives. Thus it may have been coincidence that the three women, IV-1, IV-5, and IV-7, carriers of albinism, happened to marry men who were

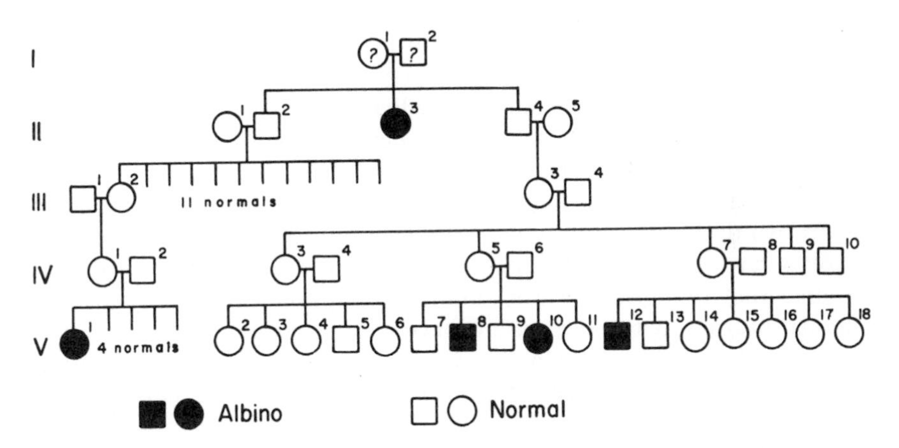

FIG. 6.4 Inheritance of albinism. (Based on a chart in Pearson, K., E. Nettleship, and C. H. Usher, 1911–1913. "A monograph on albinism in man," *Drapers' Company Research Memoirs, Biometric Series*, **6**, **8**, **9**. London: Dulau & Company Ltd. Used by permission of Galton Laboratory, University College, London.)

also heterozygous. As we noted in our discussion of alkaptonuria, chances are small that one carrier of a rare gene will happen to marry another carrier *unless* he marries a relative (pp. 56–58; see also Chap. 11).

Another way in which pedigrees showing recessive inheritance usually differ from those showing dominant inheritance is in the smaller number of affected individuals. Commonly, homozygous recessives are the children of two heterozygotes. In Figure 6.4, the matings of IV-1 × IV-2, IV-5 × IV-6, and IV-7 × IV-8 were surely of this kind, and the mating of I-1 × I-2 probably was also. From the four matings, the numbers of normal and of albino children are, respectively, 4 + 1; 3 + 2; 6 + 1; 2 + 1. Thus, among 20 children, 5 are albinos. How many albino children would be expected in these four families, which were selected because they had at least one albino child? This is a significant question, and the way in which an answer can be determined will be discussed in the next chapter.

Another criterion of a recessive trait is the fact that when two persons showing the phenotype marry each other, *all* their children will show this phenotype. We must immediately qualify this statement by pointing out that it is true *if* the parents with the same phenotype have the same *genotype*. As we noted in the preceding chapter (pp. 59–60), all children of albinos married to each other will be albinos if both parents are tyrosinase-positive, or if both parents are tyrosinase-negative. But if one parent is of one type, one of the other type, normally pigmented children will be produced. In this case, one parent has the normal gene that the other parent lacks, and vice versa. The gene (*C*) for tyrosinase and the gene (*P*) for permease in this example are called **complementary genes,** because both must be present to produce melanin. Because most, if not all, synthetic processes are complex chain reactions, like this one, we may expect to find more and more individuals that are similar in phenotype while differing in genotype as our knowledge of the basic biochemical processes underlying phenotypes increases.

Some types of hereditary deafness form another example. Sometimes two deaf parents have only deaf children. In other cases, two deaf parents have children with normal hearing. We can explain this if we assume that two genes, usually designated *D* and *E,* are necessary for normal hearing. If two deaf people having the genotype *DDee* marry each other, all the children (lacking *E*) will be deaf. However, two deaf parents having the genotypes *DDee* and *ddEE,* respectively, will have children with normal hearing (*DdEe*).

In summary, we note three criteria of recessive inheritance to be looked for in studying pedigrees: (1) Traits dependent on recessive genes frequently skip one or more generations in a direct ancestral line. (2) When two parents showing a phenotype dependent on recessive genes have the same genotype, all children show the phenotype of the parents.

(3) In a pedigree showing inheritance of an otherwise rare recessive trait, the amount of consanguinity is frequently greater than it is in the population at large (see Chap. 11). (4) In pooled data of families in which both parents show the normal phenotype (i.e., that produced by the dominant allele), only a minority of the children will be expected to show the phenotype caused by homozygosity for the recessive allele—a subject we shall explore further in the next chapter.

Problems

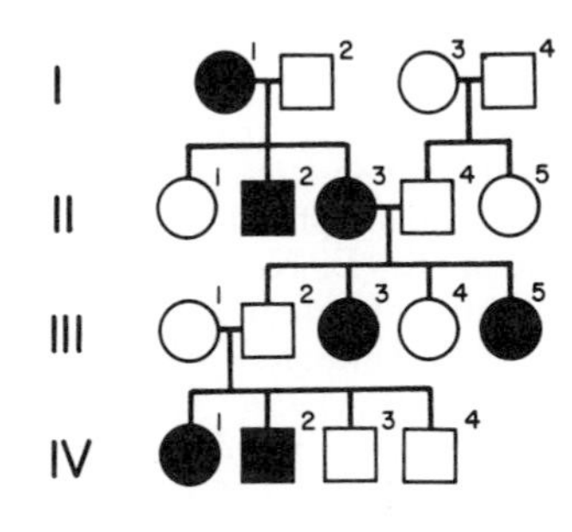

1. The individuals represented by solid squares and circles in the accompanying pedigree chart show a certain trait.
   (a) Could the trait be inherited on the basis of a dominant gene with complete penetrance? A dominant gene with partial penetrance? A recessive gene?
   (b) Assuming that the trait depends on a dominant gene, *A*, having partial penetrance, give the genotypes of III-2, IV-1, IV-2, IV-3, IV-4. (Assume that III-1 does not have the gene.)
   (c) Assuming that the trait depends on a recessive gene, *a*, give the genotypes of III-1, III-2, IV-1, IV-2, IV-3, IV-4. (If more than one genotype is possible, give all genotypes that the individual may have.)
2. If black mice having the genotype *CcBb* are mated to white (pigmentless) mice having the genotype *ccbb*, what colors will be expected in the offspring, and in what proportions will the different colors be expected to occur?
3. If the suggestion of Lutman and Neel (p. 74) is correct, would it have been possible for II-9 (Fig. 6.3) to have had a child with type 2 cataract? (Assume that II-10 had the genotype *ccmm* and that independent assortment occurred.)
4. Following the hypothesis concerning hereditary deafness stated on p. 76, explain the following cases:
   (a) Two deaf parents whose children are all deaf. Give possible genotypes of parents and children.
   (b) Two deaf parents whose children all have normal hearing. Give possible genotypes of parents and children.
   (c) Father with normal hearing, mother deaf; half of the children deaf, half normal. Give possible genotypes of parents and children.
5. Referring to Figure 6.1, if V-7 had married V-8, what would be the chance that their first child would develop Huntington's chorea in later years? If V-7 had married V-13, what would be the chance that their first child would develop Huntington's chorea in later years?
6. Referring to Figure 6.2, if the son in the third generation of the family shown on the left married the daughter in the third generation of the family shown on the right, what would be the chance that their first child

would have the gene for retinoblastoma? If, as stated, the penetrance of the gene is 90 percent, what would be the chance that their first child would develop the symptoms of retinoblastoma?

7. In Figure 6.4, what is the chance that V-13 is heterozygous for albinism? If he married V-3, what would be the chance that their first child would be an albino?

References

Lutman, F. C., and J. V. Neel, 1945. "Inherited cataract in the B. genealogy," *Archives of Ophthalmology,* **33**:341–357.

Neel, J. V., and W. J. Schull, 1954. *Human Heredity.* Chicago: University of Chicago Press.

Oepen, H., 1961. "The specificity and expressivity of the Huntington gene exemplified by paroxysmal disturbances," *Second International Congress of Human Genetics.* International Congress Series No. 32, p. E 167. New York: Excerpta Medica Foundation.

Pearson, K., E. Nettleship, and C. H. Usher, 1911–1913. "A monograph on albinism in man," *Drapers' Company Research Memoirs, Biometric Series,* **6, 8, 9.** London: Dulau & Company Ltd.

Reese, A. B., 1954. "Frequency of retinoblastoma in the progeny of parents who have survived the disease," *Archives of Ophthalmology,* **52**:815–818.

Steinberg, A. G., 1959. "Methodology in human genetics," *Journal of Medical Education,* **34**:315–334.

7

Mendelian Thinking Applied to Human Families

In Chapter 2 we discussed the fundamentals of thinking employed by Mendel and his successors in drawing conclusions from experimental data. We saw that obtaining a 3:1 (or a 1:2:1) ratio among F_2 offspring can be taken as evidence of single gene-pair inheritance if we assume the following: (1) Individual organisms possess genes in pairs. (2) In the production of germ cells, members of a pair of genes separate from each other so that each germ cell receives only one member of a pair. (3) A heterozygote produces two kinds of germ cells (for each gene pair) in equal numbers. (4) The laws of probability operate when the two kinds of sperm (or pollen grains) fertilize the two kinds of ova.

This interpretation arose from a need to explain data obtained in experiments with plants and animals. We must now transfer the interpretation to man himself.

Differences from Genetical Studies of Lower Organisms

In the first place, investigators of human genetics cannot arrange the matings of the persons they are studying. They must work with whatever the matrimonial whims of people provide in the way of "crosses."

Second, there is no such thing in human reproduction as an F_2 generation, as this term was employed in Chapter 2, and as it appears, for example, in Figure 2.1 (p. 15). In the experiments represented by this figure, homozygous black guinea pigs were mated to homozygous white ones. From the matings, F_1 males and females were produced and were mated among themselves (brother-sister matings) to produce the F_2 generation. Although brother-sister marriage has been practiced under exceptional conditions (e.g., the Pharaohs of an ancient Egyptian dynasty), it is not a social institution on which the modern human geneticist can rely for data.

What is the closest human approximation to an F_2 generation? We recall that the parents of an F_2 generation are both heterozygous. Therefore, when two individuals heterozygous for the same gene pair (e.g., both *Bb*) marry each other, we have the genetic equivalent of an $F_1 \times F_1$ mating, even though the wife is not the sister of her husband. The advantage of brother-sister matings lies in the certainty that each is indeed heterozygous for the same gene pair. When husband and wife are not brother and sister, an element of uncertainty is introduced; the genes under study may not in fact be identical even though their phenotypic effects may be the same. We have seen that there is more than one gene for albinism in man, for example. We shall see other examples of similar phenotypes caused by dissimilar genotypes. The element of uncertainty, however, is not so serious as to make impossible the obtaining of valuable data. Following Mendelian thinking, we should expect the children in such families to approximate a 3:1 ratio if the families are large (at least a score of children, preferably a hundred or more!).

That last statement points up another handicap for the human geneticist. In point of fact, human families are small. Families (at least those born to one wife) that a geneticist would consider large are practically nonexistent. Geneticists can overcome this handicap by pooling data—by finding as many families as possible in which both parents seem to be heterozygous for the gene pair under study and then adding together all the children. Pooling data from many families does present some pitfalls, however. For example, not all individuals who seem to be heterozygous for the same gene pair are so in fact. This is a real problem of genetic identification, and it undoubtedly has produced errors.

We can now return to the question discussed in Chapter 2, but this time we shall phrase it in terms of human genetics. *How can we determine whether or not a pair of alternative characteristics depends on the action of a single pair of alternative genes (alleles)?* There are several answers to this question, but the one of immediate interest at this point parallels Mendel's thinking: If a single pair of alleles is involved, pooled data of children in families in which both parents are of the presumed heterozygous type should approximate a 3:1 ratio (1:2:1 ratio if dominance is lacking).

An example will be helpful at this point. Blood types form a genetically simple example, although the phenotypes cannot be seen (like the color of cotyledons or hair) but can be discovered only by serological means.

The M-N Blood Types

The genetics of human blood groups and types receives much attention for many reasons. Human differences in blood type are important in clinical medicine (in blood transfusion and in the causation of erythroblastosis as a result of Rh-incompatibility of the parents, to mention two of the best-known instances). Of even more interest to the geneticist is the fact that the relationship between these genes and their products can be traced more readily than can the relationships between many other genes and their products. The products in the case of these genes are complex chemical compounds called **antigens.** Many typical antigens are proteins. Those that are not, undoubtedly depend on enzymes (proteins) for their production. So, as in the cases discussed in Chapter 5, the road to be traversed between gene and phenotype is a relatively short one.

An antigen is a substance that, under suitable circumstances, will induce the formation of **antibodies**, and then will react with these antibodies if brought into contact with them (in a test tube, for example). The antigens with which we are concerned here are complex substances in (or more probably on the surface of) red blood cells (erythrocytes). The most widely known of these antigens are those designated A, B, and Rh; they will be discussed later (pp. 179–212). Two other antigens, called M and N, were discovered by Landsteiner and Levine in 1927. They found that an individual's red blood cells may contain one or both of these antigens. If, for example, a suspension of red blood cells containing M is inoculated into a rabbit, it will form antibodies against the foreign cells. The antibodies are found in the serum (fluid portion) of the rabbit's blood. When some of the serum from this rabbit (called an **antiserum** because it contains antibodies) is mixed with human red blood cells containing antigen M, these cells will clump or agglutinate. An antigen such as M, which causes the formation of antibodies that induce agglutination, is called an **agglutinogen.** The antibodies formed by stimulus of an agglutinogen are called **agglutinins.**

In a similar manner, red blood cells containing agglutinogen N will cause a rabbit to form anti-N agglutinins. This serum is used to test for the presence of antigen N in unknown red blood cells, just as the serum from a rabbit inoculated with red blood cells containing agglutinogen M is used to test for the presence of antigen M. Using these two serums, a serologist can determine the blood type to which any individual belongs, on the basis of whether that individual's red cells react with

(1) anti-M serum only, (2) anti-N serum only, (3) both anti-M and anti-N serums. On this basis, as shown in Table 7.1, people may be divided into three categories or blood types: M, N, and MN.

TABLE 7.1 M-N blood types—Reactions of cells with antiserums.

If red blood cells contain—	Reactions with antiserums Anti-M	Anti-N	Blood type of the donor of the cells
M only	+	−	M
M and N	+	+	MN
N only	−	+	N

+ = agglutination of the cells occurs.
− = agglutination of the cells does not occur.

With this much serological background, let us inquire into the genetic basis of these blood types. Landsteiner and Levine advanced the hypothesis that antigens M and N depend on a single pair of alleles (and that, of course, is the reason we chose to use the example here). Thus, following custom, we may designate the gene that results in M being present as L^M; the gene that results in N being present as L^N. If the hypothesis is correct, therefore, the genotypes are as follows:

L^ML^M = type M; antigen M only
L^ML^N = type MN; antigens M and N
L^NL^N = type N; antigen N only

This hypothesis was tested in various ways, some of which we shall refer to later. At this time we are interested in studying the human equivalent of an F_2 generation. *If the hypothesis is correct, children in families of which both parents belong to type MN should approximate a 1:2:1 ratio.*

	Father	Mother	
Genotype	L^ML^N	L^ML^N	
Germ cells	$L^M \rightarrow L^M$ (crossed with) $L^N \rightarrow L^N$		

Children's genotypes	L^ML^M	L^ML^N — L^ML^N	L^NL^N
Phenotypes	M	MN	N
Ratio	1 :	2 :	1
Percentages	25	50	25

At the outset, Landsteiner and Levine (1928) had data from only 11 families having both parents of type MN. The children in these families were as follows: type M = 17; type MN = 31; type N = 7.

Taken by itself, this would hardly suggest a 1 : 2 : 1 ratio. However, these data, plus those concerning children from other parental combinations (e.g., MN × N; MN × M; M × N, and so on), encouraged Landsteiner and Levine to conclude that they had support for their hypothesis.

As the years passed, many investigators in various countries accumulated additional data. By 1953, Wiener and his colleagues were able to amass information concerning 377 MN × MN matings (see Wiener and Wexler, 1958). The children of these matings was as follows.

Blood types	M	MN	N
Number	199	405	196
Percentages	24.7	50.8	24.5

We can see that this is indeed a close approximation to the ideal 1 : 2 : 1 ratio. Again we must emphasize that a close approximation to the Mendelian ratios is to be expected only when the numbers of offspring are large.

So far, we have discussed MN × MN matings because they directly parallel Mendel's experiments. Landsteiner and Levine, and their successors, also analyzed children from other parental combinations (see table on p. 43 of Wiener and Wexler, 1958). If single gene-pair inheritance is involved, other criteria, in addition to the one stressed above, should indicate it. Thus when both parents are of type M ($L^ML^M \times L^ML^M$), all of the children should be of type M. When both parents are of type N, all children should be of type N. When one parent is type M and the other type MN, half the children should be type M and half type MN.

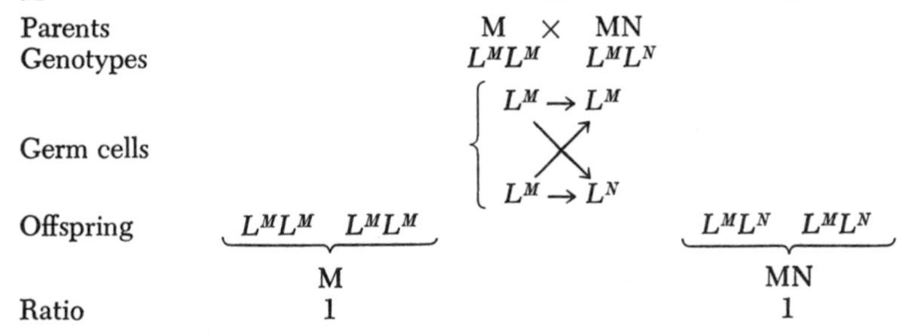

In such families, no type N children are to be expected. Similarly, in families in which one parent is of type N, the other of type MN, no type M children are to be expected.

Do the cumulative data indicate any exceptions to the expectations stated? The data recorded in the literature contain results for thousands of individuals; among them only six exceptions to the rules have been found. Why should there be any exceptions at all?

In the first place, there is always the possibility of human error in the serological tests. The testing procedures are somewhat complicated, and error is not impossible, particularly for antigen N. Still, the tests were performed by experts, and the recent results, at least, are not likely to be in error.

Another possibility is that a rare mutation was encountered in some particular case. For example, suppose that two type M parents have a type MN child. What was the source of the L^N gene in the child? Perhaps during meiosis in one of the parents, one of the normally occurring L^M genes underwent chemical alteration, causing it to mutate to L^N. (Until we know more about the chemical difference beween antigens M and N we cannot say for sure, but such an alteration of the genes might conceivably involve only one or two nucleotide pairs in the DNA molecule.) Mutations are relatively rare events, but they do occur.

Finally we come to the explanation favored by the geneticists engaged in this particular study: that in these six cases one or both of the people who seemed to be the parents were not the actual biological parents (see p. 71).

Investigations of MN blood types are a fine example of the type of thinking involved in determining the mode of inheritance of human characteristics. And they have one great advantage: the presumed heterozygous type ($L^M L^N$) can be clearly identified by serological tests. Consequently, when we say we have an MN × MN mating, we are sure that we do have one. The basis of this advantage is the fact that one allele is not dominant to the other. However, in a great many cases in which we are interested, one allele is dominant to the other.

Dominance and the Problem of Ascertainment

Dominance complicates the problem. Letting *A* and *a* represent a pair of alleles, we say that from pooled *Aa* × *Aa* matings we expect a 3:1 ratio (our criterion for single gene-pair inheritance). Now, when one of the genes is dominant, how are we going to distingish *Aa* individuals from *AA* individuals? They may just be alike in phenotype. If so, how can we distinguish *Aa* × *Aa* matings from *AA* × *AA* matings, or from *Aa* × *AA* matings? In fact, there may be no way of doing so by examining the phenotypes of the parents. In many cases the only way to distinguish an *Aa* × *Aa* mating from the other two is by means of what in domestic animals is called a progeny test. *Aa* × *Aa* matings can produce *aa* children; the other two matings cannot. If a marriage between two people, both of whom possess the dominant phenotype, results in at least one child showing the recessive phenotype, we know that both parents are heterozygous (that it is in fact an *Aa* × *Aa* mating).

In such families, what ratio of dominant-phenotype to recessive-phenotype children can be expected? We must not be too quick to answer: 3:1. We can expect a 3:1 ratio in pooled data only when we have complete information about *Aa* × *Aa* matings among the people in the study. In the case of M-N blood types, we can identify all MN × MN matings regardless of the type of children produced or indeed of whether there are any children at all). Now, on the other hand, when we study genes that show dominance, such as our hypothetical *A* and *a*, we count only matings that have produced at least one *aa* child. We leave out all $Aa \times Aa$ matings in which all the children are *AA* or *Aa*. Thus we do not have an unbiased sample of all $Aa \times Aa$ matings; we have a truncate distribution. We must therefore rephrase our question: If inheritance is based on a single pair of alleles, what proportion of *A*-phenotype to *a*-phenotype children can be expected in $Aa \times Aa$ matings *identified by having produced at least one a-phenotype child?*

We can attack the problem most directly by asking: What proportion of $Aa \times Aa$ matings do in fact produce *no aa* children? Because the answer depends on the size of the sibship (number of children in the family), we inquire first: What proportion of one-child sibships will be expected to contain no *aa* child?

Parents	$Aa \times Aa$
Germ cells	$A \rightarrow A$, $a \rightarrow a$ (crossed arrows)
Children	*AA* *Aa* *Aa* (¾) *aa* (¼)

For each child produced, the chance that he or she will show the dominant phenotype is ¾; we therefore expect that in three-fourths of the one-child sibships the one child will show that dominant phenotype. In other words, three-fourths of one-child sibships (of $Aa \times Aa$ matings) will contain no *aa* child.

What proportion of two-child sibships will be expected to contain no *aa* child? The chance that the older child will show the dominant phenotype is ¾; the chance that the younger child will show the dominant phenotype is also ¾. Hence the chance that both will show the dominant phenotype is $3/4 \times 3/4 = (3/4)^2 = 9/16$. In other words, $9/16$ of two-child sibships of $Aa \times Aa$ matings will contain no *aa* child.

Application of the Binomial Formula What is the total expectation with regard to two-child sibships born to $Aa \times Aa$ matings?

Let c = chances for one dominant-phenotype child = ¾
d = chances for one recessive-phenotype child = ¼*

*The initials p and q are traditionally used here, but in this book p and q will always refer to gene frequencies in the Hardy-Weinberg formula.

Then expectation concerning a two-child family can be expressed by squaring the binomial $(c + d)$ as follows:

$$(c + d)^2 = c^2 + 2cd + d^2$$

In this expansion:

c^2 = sibship of two dominant-phenotype children
$2cd$ = sibship of one dominant-phenotype and one recessive-phenotype child.
The coefficient 2 indicates that the sibship can arise in two ways: (1) the older child with dominant phenotype, the younger child with recessive, or (2) the older child with recessive phenotype, the younger child with dominant.
d^2 = sibship of two recessive-phenotype children.

Substituting the values of c and d, we obtain:

$c^2 = (3/4)^2 = 9/16$ (two dominant-phenotype children)
$2cd = 2 \times 3/4 \times 1/4 = 6/16$ (one dominant, one recessive)
$d^2 = (1/4)^2 = 1/16$ (two recessive-phenotype children)

Similarly, the proportion of three-child sibships expected to contain no *aa* child can be found by raising the binomial $(c + d)$ to the third power:

$$(c + d)^3 = c^3 + 3c^2d + 3cd^2 + d^3$$

Here we have all possible kinds of sibships of three:

c^3 = all three children with dominant phenotype = $(3/4)^3 = 27/64$
$3c^2d$ = two dominant-phenotype children and one recessive-phenotype child = $3 \times (3/4)^2 \times 1/4 = 3 \times 9/16 \times 1/4 = 27/64$†
And so on.

†The coefficient 3 indicates that this sibship may arise in three ways (note that the one *aa* child may be the oldest child, or the middle child, or the youngest child).

To find the proportion of four-child sibships that will be expected to contain no *aa* child, we raise $(c + d)$ to the fourth power:

$$(c \times d)^4 = c^4 + 4c^3d + 6c^2d^2 + 4cd^3 + d^4$$

In view of our discussion in Chapter 2, we may note especially the term $4c^3d$: three dominant-phenotype children and one recessive-phenotype child, the "perfect fit" in a family of four to the 3:1 ratio (see p. 21). In Chapter 2 we calculated this expectation; now we note that the binomial expansion states the same expectation: $4 \times (3/4)^3 \times 1/4 = 108/256$ or about $2/5$. The binomial expansion has two advantages: It is simple to calculate, and it makes a complete statement about expectations concerning all the possibilities in families of a given size.

The proportion of sibships of four that will contain no *aa* child is shown by the term $c^4 = (3/4)^4 = 81/256$.

Now we are able to generalize a little concerning the proportion of sibships expected to contain no *aa* child.

1-child sibships, $c = 3/4$
2-child sibships, $c^2 = 9/16$
3-child sibships, $c^3 = 27/64$
4-child sibships, $c^4 = 81/256$

In general, then, the proportion of sibships expected to contain no *aa* child is c^n, where n is the size of the sibship.

What we are really interested in is the proportion of $Aa \times Aa$ matings that are expected to produce at least one *aa* child. This is all $Aa \times Aa$ matings minus those that produce no *aa* child. Because c^n is a fraction, we can represent the total of $Aa \times Aa$ matings as 1 or unity. Then the proportion of matings expected to produce at least one *aa* child is $1 - c^n$. Notice how this works out in the case of the two-child sibships. Disregarding the $9/16$ of such sibships expected to consist of two dominant-phenotype children, we note that $6/16$ are expected to have one dominant-phenotype and one recessive-phenotype child, $1/16$ to have two recessive-phenotype children. Hence $6/16 + 1/16 = 7/16$ of the sibships will be expected to have at least one *aa* child. Note that $1 - c^n = 1 - (3/4)^2 = 1 - 9/16 = 7/16$ gives this value directly and easily.

Now we may return to the question: In $Aa \times Aa$ matings identified by producing at least one *aa* child, what proportion of the children will be expected to have the *aa* phenotype? If we had information concerning all $Aa \times Aa$ matings (regardless of children produced), this fraction would be $1/4$, or d. We may use the symbol d' to represent the fraction of recessive-phenotype children expected in sibships from $Aa \times Aa$ matings identified by having at least one such child.

If we knew of all $Aa \times Aa$ matings, d' would equal d (or $d/1$), but we can learn of only $(1 - c^n)$ of these matings. Hence

$$d' = \frac{d}{1 - c^n} \qquad (1)$$

For example, we apply this formula to a sibship of 4:

$$d' = \frac{d}{1 - c^n} = \frac{1/4}{1 - (3/4)^4} = \frac{1/4}{1 - 81/256} = \frac{1/4}{175/256} = 1/4 \times 256/175 =$$

$$64/175 = 0.366 = 36.6\%$$

Thus, in sibships of four from $Aa \times Aa$ matings identified because they produced at least one *aa* child, 36.6 percent of the children will be expected to be *aa* (show the recessive phenotype). If we could include $Aa \times Aa$ matings that produced no *aa* children, 25 percent would be expected to be *aa*. The observed value, d', is always greater than d because of the fact that the dominant-phenotype children in sibships consisting only of such children cannot be counted in computing the ratio of dominant-phenotype to recessive-phenotype children in the pooled data. As a result, families that have more *aa* children than expected are not offset by families having more *AA* or *Aa* children than expected when these families have no *aa* children at all. Hence, in the observable families, the proportion of recessive-phenotype children to dominant-phenotype ones will be higher than it would be if knowledge of all $Aa \times Aa$ matings could be included.

Correcting Pooled Data: *A priori* Method

As an example we shall use the tyrosinase-negative form of albinism (p. 60). Most albino children are born to normally pigmented parents. When two such parents produce an albino child, we know at once that the mating is $Cc \times Cc$.

Unlike the MN blood type situation, identifying *Cc* people (carriers of albinism) by other means is not possible at present. Tests, probably of a biochemical nature, may be developed eventually to make such identification possible. *Suppose* that we had a chemical test now and could identify all $Cc \times Cc$ matings in a given community or nation.

For the sake of simplicity, let us also suppose that all the sibships consist of four children, and that we learn of 256 such sibships. How many of the 256 would be expected to have no albino children, one albino child, two albino children, three albino children, four albino children? We can answer this question by raising the binomial $(c + d)$ to the fourth power.

$$(c+d)^4 = c^4 + 4c^3d + 6c^2d^2 + 4cd^3 + d^4$$
$$= (3/4)^4 + 4(3/4)^3\, 1/4 + 6(3/4)^2(1/4)^2 + 4(3/4)(1/4)^3 + (1/4)^4$$
$$= \frac{81}{256} + \frac{108}{256} + \frac{54}{256} + \frac{12}{256} + \frac{1}{256}$$

The resulting expectations for 256 sibships of four children each can be tabulated as follows:

Children per sibship

Normal	Albino	No. of sibships	Total No. of children	Normal children	Albino children
4	0	81	324	324	0
3	1	108	432	324	108
2	2	54	216	108	108
1	3	12′	48	12	36
0	4	1	4	0	4
		256	1024	768	256

3:1

∴ albinos = 25%

This model presents an ideal situation, closely realized in the study of MN blood types. Actually, we cannot attain such complete data about albinism until we have some means of detecting heterozygotes (carriers) other than by production of albino children. Without such a test we must omit the first line of the table: families that have *no* albino children. Our data, then, would be as follows:

Children per sibship

Normal	Albino	No. of sibships	Total No. of children	Normal children	Albino children
3	1	108	432	324	108
2	2	54	216	108	108
1	3	12	48	12	36
0	4	1	4	0	4
		175	700	444	256

1.73:1
∴ albinos = 36.6%

The only difference between the two tables is that the total number of normal sibs has been reduced by 324, whereas the number of albino sibs has not changed at all. As a result, the albinos now constitute 36.6 percent (instead of 25) of the total number of sibs. We recall that 36.6 percent was the percentage we obtained when we applied formula (1) to families of four (p. 87). Thus whether we use the formula or the model, we reach the conclusion that single gene-pair inheritance of albinism would be indicated by finding that 36.6 percent of the children in four-child sibships born to normal parents are albinos, when the sibships have been identified in the first place by having at least one albino child.

This percentage applies specifically to sibships of four. As indicated previously, the expected percentage varies with the size of the family and can be calculated by using formula (1). Values of d' for families of varying size are given in Table 7.2. Note that as the size of the sibship increases, the expected proportion of recessive-phenotype children approaches more and more closely to 25 percent. Thus, with sibships of 12, only a fraction of a percentage more than 25 percent of the children is expected to consist of children having the recessive phenotype. This close approach to the ideal percentage reflects the fact that few $Aa \times Aa$ matings that produce 12 children will go undetected. The chance is very small that no one of the 12 children will be *aa*. In the third column of Table 7.2, percentages are translated into the number of *aa* children expected per sibship.

TABLE 7.2 Expected occurrence of recessive-phenotype children in sibships identified by having at least one child with the recessive phenotype, *a priori* method.

Size of sibship n	Percentage of children expected to show recessive phenotype d'	No. of recessive-phenotype children expected per sibship $d' \cdot n$
1	100.00	1.000
2	57.14	1.143
3	43.24	1.297
4	36.57	1.463
5	32.78	1.639
6	30.41	1.825
7	28.85	2.020
8	27.78	2.222
9	27.03	2.433
10	26.49	2.649
11	26.10	2.871
12	25.82	3.098

This method of correcting pooled data is commonly called the **a priori method** because it starts with the assumption that if we had complete information concerning all relevant families, a 3:1 ratio would be approximated. To recapitulate, it is a way of compensating for the fact that when families of two heterozygous parents are identified only by virtue of having at least one child with the presumed recessive trait, *more* than one-fourth of the children will be expected to show the recessive trait.*

Now let us return to the question we asked when we were discussing the pedigree of albinism (Fig. 6.4, p. 75). We noted four matings of normal parents that resulted in a total of 20 normally pigmented children and 5 albinos. We then asked: How many albino children would be expected in these four families, which were selected because they had at least one albino child?

From Figure 6.4 we note that the four families may be classified as follows: one family of three children, two families of five children, one family of seven children. How many children in each family would be expected to be albinos? This expectation is expressed by $d' \cdot n$ (Table 7.2). We may organize the material as follows:

*Readers interested in further elaboration of the *a priori* method are referred to Stern, 1973 (Chap. 10); Levitan and Montagu, 1971 (Chap. 11); Li, 1961 (Chap. 5); Steinberg, 1959; Neel and Schull, 1954 (Chap. 14), and to the original paper by Bernstein, 1929.

Size of sibships	No. of sibships	$d' \cdot n$	$d' \cdot n \cdot$ No. of sibships
3	1	1.297	$1.297 \times 1 = 1.297$
5	2	1.639	$1.639 \times 2 = 3.278$
7	1	2.020	$2.020 \times 1 = 2.020$
Total number of albino children expected			$= 6.595$

Thus in the four families selected, we should expect 6 or 7 children to be albinos. The number actually observed, 5, is below expectation, but not significantly so considering the small number of children involved.

Why Actual Populations May Differ from Models We should not leave this subject without noting that the correction of data given by formula (1), or by our model, is not a perfect method applicable to all cases. As many students of the subject have emphasized, the efficacy of the method depends on obtaining the data in "correct" proportions. We can illustrate this from our model. We note that in the model, 108 of the 175 sibships are expected to have one albino child only, 54 are expected to have two albino children, 12 are expected to have three, 1 is expected to have four. Suppose that the proportions of the different types actually differ markedly from this. We can readily see that the results will be affected.

Why should the proportions observed differ markedly from the expected ones? To answer this we must bring up the subject of **ascertainment,** a thorn in the side of the human geneticist. There would be no problem if it were equally likely that a sibship having only one albino child, a sibship having two albino children, a sibship having three albino children, and a sibship having four would come to the attention of an investigator—"be ascertained." Such an equal chance of ascertainment could be realized if a complete canvass could be made (complete ascertainment). However, in most studies some method of sampling is employed. This may consist of asking people if they know of families in which there are albino children, or, in the case of pathological traits, it may consist of a study of medical records. In any method of sampling there is always the question of whether the sample is a true representation of the larger population. Is a sibship having only one albino child as likely to be brought to an investigator's attention as is a family in which all four sibs are albinos? Probably not. Some theorists postulate a mathematical regularity here: A sibship of four affected children is four times as likely to be ascertained as is a sibship in which only one of the four children is affected. At this point, all sorts of complications may enter the picture. Some inherited defects cause children to die young; in such cases, a sibship in which four children were affected is more likely to have at least one affected child still living at the time of an investigation than is a sibship in which only one child was affected (Roberts, 1963). We are touching here on a large and complicated problem; mathematical methods are available for correcting for bias in, or incompleteness of, ascertainment (see references listed on p. 96).

Correcting Pooled Data: Simple Sib Method

Although there are various means of ascertainment, the commonest is by looking for affected individuals (single ascertainment, Li, 1961). An investigator learns about a family by being told of an albino child, or in a survey of school children he may see an albino child and thus learn about the family of which the child is a member. An affected individual who attracts the investigator's attention to the family is called a **propositus** or **proband.** In our frame of reference, discovering a propositus is the way in which an investigator learns of the existence of a certain $Cc \times Cc$ mating.

In the preceding section, we applied a correction to pooled data to compensate for the fact that only $Cc \times Cc$ matings having a propositus could be included in the study. Another, and simpler, method of compensating for this fact is by counting the sibs of each propositus, omitting the propositus himself. Let us see how this would work in the case of sibships of four (p. 88). These observable families are of four kinds: (a) 1 albino, 3 normal; (b) 2 albino, 2 normal; (c) 3 albino, 1 normal; (d) 4 albino, 0 normal. Now let us regard one of the albino children in each family as a propositus. The four types can then be classified as follows:

(a) [1 albino propositus]; 3 normal sibs, 0 albino sibs.
(b) [1 albino propositus]; 2 normal sibs, 1 albino sib.
(c) [1 albino propositus]; 1 normal sib, 2 albino sibs.
(d) [1 albino propositus]; 0 normal sibs, 3 albino sibs.

Among the sibs, then, we have all possible combinations of three, from three normals only, to three albinos only. It is as if we had started out to study families of three in the first place and could in some way include sibships having only three normal children. We are able to do so because the one albino propositus has identified for us the fact that the mating is $Cc \times Cc$; that is the only purpose he serves in our computations this time. Among the other three children in each sibship, all possible combinations can occur, and among them single gene-pair inheritance should reveal itself by the occurrence of an approximation to a 3:1 ratio.

One other point should be mentioned in applying this sib method of correcting pooled data. When there are, for example, two albino children in a family, each one is regarded as a propositus in turn. Such a family may in fact be ascertained twice. One albino child may be in the fifth grade at a school, the other in the eighth. When the investigator studies the fifth grade, he sees one child and so learns of the family. If he also studies the eighth grade, he sees the other child and learns of the family again. One child is just as much a propositus as is the other child, so in computations each is used as a propositus. This has the effect of counting one family twice (or three times if there are three albino children).

Perhaps we can visualize it as follows. Suppose the albino children are Mary and John, and the normally pigmented ones are Ted and Susan. Mary tells the investigator that among her brothers and sisters there are 1 albino (John) and 2 normals (Ted and Susan). John, in his turn, tells the investigator that among his brothers and sisters there are 1 albino (Mary) and 2 normals (Ted and Susan). So the investigator tabulates the sibs of the propositi as follows:

	Normal	Albino
(1)	2	1
(2)	2	1

Let us see how this method can be applied to the model we set up earlier (p. 89). We may reorganize the data in the table as follows:

Children per sibship: Albino	Normal	No. of sibships	Total no. of children: Albino	Normal	In each sibship:	Sibs of propositi: Albino	Normal
1	3	108	108	324	1 propositus with 3 normal sibs (× 108)	0	324
2	2	54	108	108	2 propositi, each with 1 albino and 2 normal sibs (× 54)	108	216
3	1	12	36	12	3 propositi, each with 2 albino and 1 normal sibs (× 12)	72	36
4	0	1	4	0	4 propositi, each with 3 albino sibs (× 1)	12	0
		175	256	444		192	576

Thus we see that among the sibs of the propositi the 3:1 ratio is exhibited (576 normal:192 albino), even though the total number of children including propositi (444 normal:256 albino) does not demonstrate this ratio. We must not be confused by the fact that the total number of sibs of propositi (192 + 576 = 768) is greater than the total number of individual children (444 + 256 = 700). This situation arises because families having two albino children are counted twice; those having three albinos are counted three times, and so on (each albino child is regarded as a propositus).

In the ideal model of families of four children exhibiting single gene-pair inheritance, the underlying 3:1 ratio is laid bare by the simple sib method of correcting pooled data. Of course, actual data do not give exact 3:1 ratios with the precision revealed by our model, but the principle is the same.

This method of correcting pooled data, like the *a priori* method discussed earlier, has its sources of error. For example, it will work well only if complete information can be obtained about each sibship; sometimes this is difficult to achieve. Both methods will serve as examples of means by which human geneticists strive to overcome the handicap of being unable to carry out breeding experiments as do geneticists working with plants and lower animals. Despite this handicap, Mendelian thinking can be applied to human families, and reasonably secure conclusions concerning modes of inheritance can be drawn.

In conclusion, it would be appropriate to note that a variation of the simple sib method was applied by Steinberg (1959) to the data of Reese on the children in 15 families of which one parent had survived retinoblastoma (pp. 69–70). There were 30 children in these families; 23 were affected, 7 were not affected. One of the families was discovered by accident, but the other 14 were discovered because an affected child had been brought for treatment. We have here an example of the sort of biased ascertainment we are discussing; we do not have proper representation of families in which there are no affected children. Interestingly, the sibship discovered by accident consisted of 3 unaffected children. Steinberg applied the sib method of correction. He stated: "It is clear that no family selected via an affected child can have fewer than one affected child. Hence, families of one child each will have 100 percent of the children affected when only 50 percent were expected." Subtracting the 14 propositi from the 23 affected children leaves 9 affected. The ratio of 9 affected to 7 unaffected is as close to 8:8 as we can expect when numbers are so small. The evidence is that survivors of retinoblastoma married to normal mates produced affected and normal children in a 1:1 ratio as expected (because such matings are $Rr \times rr$).

Problems

1. What blood types will the children belong to when the father is type M and the mother is type N? When the father is type MN, the mother type N?

2. The ability to taste phenylthiocarbamide (PTC) depends on a dominant gene, T, inability to taste the substance on its recessive allele, t. Persons with genotypes TT or Tt are called tasters, persons with genotype tt, nontasters.

   (a) If a $Tt \times Tt$ mating results in four children, what is the chance that all four will be tasters? That all four will be nontasters? That three will be tasters, one a nontaster? That two will be tasters, two nontasters?

   (b) If a $Tt \times tt$ mating results in four children, what is the chance that all four will be tasters? That all four will be nontasters? That three will be tasters, one a nontaster? That two will be tasters, two nontasters?

   (c) If a $Tt \times tt$ mating results in eight children, what is the chance that all eight will be tasters? That seven will be tasters and one a nontaster? That four will be tasters and four nontasters?

3. What proportion of sibships of six born to $Tt \times Tt$ matings (Problem 2) will be expected to include no nontaster children? What proportion will be expected to include at least one nontaster child?

4. If II-3 (Fig. 6.4, p. 75) had married a heterozygous man and they had produced four children, what would have been the chances that *at least one* child would have been an albino? (Chap. 6; but recall that in this case the chance for a normal child is ½ rather than ¾.)

5. Suppose that you are making a study of a newly discovered human trait. You locate families in which the parents are normal but at least one child shows the trait. If the trait depends on a recessive gene, affected children being homozygous for it, what proportion of families in which both parents are heterozygous and so *could* have an affected child will you fail to find because they do not, in fact, *have* an affected child?

6. Let us imagine a trait called "forked eyelashes" and suppose that we wish to determine whether or not the trait depends on a recessive gene. We learn of 156 families that have at least one child with forked eyelashes. For simplicity, let us suppose that these are all families of five children each. Thus we learn of a total of 780 children; how many of them would you expect to have forked eyelashes if the trait is indeed recessive? (Work this without using Table 7.2, but check the correctness of your answer with that table.)

7. We shall suppose that you discover a "new" inborn error of metabolism. We shall call it lack of enzyme X. You make a study of hospital records and learn of 12 patients who lack the enzyme. The 12 belong to 12 different sibships that (by happy coincidence) all consist of four children each. You investigate these sibships, finding that all parents are normal (they have enzyme X). If children lacking enzyme X are homozygous for a recessive gene, *how many* of the total of 48 children in the 12 sibships shall we *expect* to find lacking the enzyme?

8. Suppose that you discover a rare defect and wish to determine its mode of inheritance. You make as extensive a survey as you can and obtain the following data. In every case the parents are normal.

Size of sibship	No. of sibships	No. of affected children per sibship	Total number of affected children
1	2	1	2
2	4	1	4
2	2	2	4
3	8	1	8
4	6	1	6
4	6	2	12
			36

Use the *a priori* method of correcting pooled data to determine whether the incidence of affected children in these families is evidence for or against the hypothesis that the trait is inherited as a simple Mendelian recessive (Table 7.2). What do you conclude, and why?

9. An investigator studying a certain trait learned of the following families: (a) 4 families of 2 children each, with 1 affected child in each family. (b) 8 families of 3 children each, with 1 affected child in each family. (c) 2 families of 3 children each, with 2 affected children in each family. (d) 1 family of 6 children, 3 of whom were affected. Use the simple sib method of correcting pooled data to determine whether or not these findings give evidence that the trait is inherited on the basis of a recessive autosomal gene. What is your conclusion, and why?

References

Bernstein, F., 1929. "Variations- und Erblichkeitsstatistik." In E. Baur and M. Hartmann (eds.). *Handbuch der Vererbungswissenschaft,* Lieferung 8 (1, C), Bd. 1, pp. 1–96. Berlin: Gebrüder Borntraeger.

Landsteiner, K., and P. Levine, 1928. "On the inheritance of agglutinogens of human blood demonstrable by immune agglutinins," *Journal of Experimental Medicine,* **48**:731–749.

Levitan, M., and A. Montagu, 1971. *Textbook of Human Genetics.* New York: Oxford University Press.

Li, C. C., 1961. *Human Genetics.* New York: McGraw-Hill Book Company, Inc.

Neel, J. V., and W. J. Schull, 1954. *Human Heredity.* Chicago: University of Chicago Press.

Roberts, J. A. F., 1963. *An Introduction to Medical Genetics,* 3rd. ed. London: Oxford University Press.

Steinberg, A. G., 1959. "Methodology in human genetics," *Journal of Medical Education,* **34**:315–334.

Stern, C., 1973. *Principles of Human Genetics,* 3rd. ed. San Francisco: W. H. Freeman & Company.

Wiener, A. S., and I. B. Wexler, 1958. *Heredity of the Human Blood Groups.* New York: Grune & Stratton, Inc.

8

Polygenes: Multiple Gene Inheritance

So far we have been considering traits that depend largely upon single pairs of alleles, and in which phenotypes present an alternative or "either-or" situation. One is either an albino or one is not; one has hemoglobin S or one does not; and so on. Such traits are most comparable to the ones with which Mendel worked in his classic experiments with peas. And they are the traits in which the genetic basis can be analyzed most easily. As we have seen, genetic analysis is most positive when it can be linked to presence or absence of clearly identifiable enzymes or antigens.

But most of the commonly observed human traits in which people are interested have a more complex genetic basis. We have already introduced the subject of modifying genes (pp. 71–72). In these cases there is typically a "main gene" whose phenotypic expression is altered by other genes. Now we turn our attention to situations in which there is no "main" gene, but in which several or many genes contribute more or less equally. Most easily discussed are cases in which the genes have a cumulative or additive effect. Such genes are called **polygenes** or **multiple factors,** and the type of inheritance is frequently termed **polygenic** or **multifactorial.** The traits they determine are of a quantitative nature (e.g., height, breadth, size, weight, degree of pigmentation).

Let us illustrate with a hypothetical but true-to-life example. Suppose that we have two varieties of a certain plant, one tall, the other dwarf. The tall variety has an average height of 32 inches, the dwarf variety an average height of 8 inches. When the experimenter crosses the two varieties, he finds that the F_1 hybrids have an average height of 20 inches. He then interbreeds the F_1 hybrids to produce an F_2 generation.

These F_2 offspring show great variability, ranging in height all the way from 32 inches to 8 inches, but with an average of 20 inches. Morover, they show an approximation to a normal distribution; a graph shows an approximation to the familiar, bell-shaped, normal frequency curve.

Here we note something quite unlike Mendel's experiments with tall and dwarf peas. The F_1 offspring in his experiment were all tall, and the F_2 generation showed a 3:1 ratio of tall to dwarf. Evidently in that case the difference in height between the tall and dwarf varieties depended on only a single pair of genes, with the gene for tallness dominant.

On the other hand, how can we explain our present example, in which (a) the F_1 offspring are intermediate between the parents, and (b) the F_2 offspring show wide variation with a tendency to a normal frequency distribution? We might note that in the early days of genetics this was called "blending inheritance" and was thought not to depend on pairs of genes at all. Investigators soon realized, however, that the results can be explained if we assume that the quantitative difference (e.g., in height) is dependent on two or more pairs of genes having cumulative effect.

Let us suppose that in our present example, *two* pairs of genes are involved, the tall variety having the genotype *AABB*, the dwarf variety the genotype *aabb*. Evidently, then, the *aabb* genotype produces the initial 8 inches of height, and the dominant alleles add to this. We may suppose that gene *A* and gene *B* have the same effect in increasing the height above the basic 8 inches. This need not be the case, but it forms a simplifying assumption for the present example. The difference in height between the two varieties is 24 inches. There are four "capital-letter genes" in the genotype of the tall variety. These capital letter genes are preferably termed **effective alleles.** Evidently, then, each effective allele produces a height increase of 6 inches. A genotype of *aabb* gives a height of 8 inches; substituting an effective allele for the corresponding "small-letter gene" increases the height by 6 inches. Thus a plant with the genotype *Aabb* is 8 + 6 or 14 inches high, a plant with the genotype *AAbb* is 8 + 12 or 20 inches high, and so on to the limiting genotype of *AABB*, which gives a plant 8 + 24 or 32 inches high. For the sake of simplicity, we speak of the contributions of the genes in terms of absolute units, inches in this case. Frequently, perhaps usually, the contribution may be a percentage increase, each effective allele causing the height to increase by a certain percentage.

In Figure 8.1 we present the results of the experiment in terms of two pairs of cumulative genes. All gametes produced by the 32-inch parent contain genes *A* and *B;* all gametes of the 8-inch parent contain *a* and *b*. Hence the F_1 hybrids have the genotype *AaBb* and are 20 inches tall (8 + 6 + 6). In accordance with the principle of independent assortment (pp. 32–34), the F_1 individuals produce four kinds of gametes

PARENTAL GENERATION

Phenotypes:	32 inches		8 inches
Genotypes:	*AABB*	X	*aabb*
Gametes:	*AB*		*ab*

F_1 GENERATION

Phenotypes:	20 inches		20 inches
Genotypes:	*AaBb*	X	*AaBb*
Gametes:	*AB, Ab, aB, ab*		*AB, Ab, aB, ab*

Combinations of gametes producing an F_2 generation:

	AB	*Ab*	*aB*	*ab*
AB	*AABB* 32"	*AABb* 26"	*AaBB* 26"	*AaBb* 20"
Ab	*AABb* 26"	*AAbb* 20"	*AaBb* 20"	*Aabb* 14"
aB	*AaBB* 26"	*AaBb* 20"	*aaBB* 20"	*aaBb* 14"
ab	*AaBb* 20"	*Aabb* 14"	*aaBb* 14"	*aabb* 8"

F_2 GENERATION

Heights:	32"	26"	20"	14"	8"
Ratio:	1 :	4 :	6 :	4 :	1

GRAPH OF F_2 GENERATION

Length of columns represents the relative number of individuals having each of the heights listed along the base line.

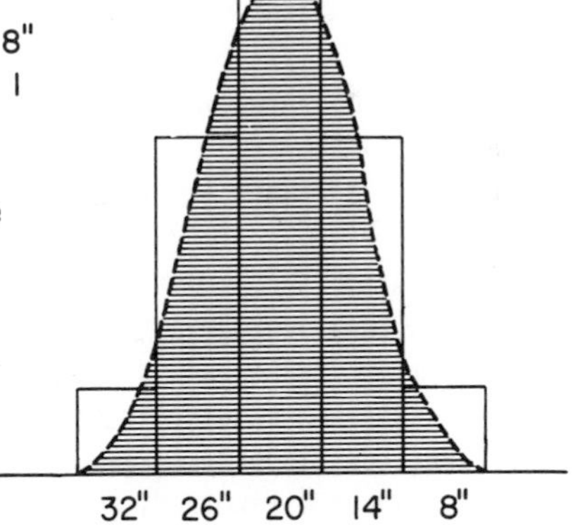

FIG. 8.1 Polygenic inheritance of difference in height when two imaginary plant varieties are crossed. It is assumed that two pairs of effective alleles are concerned with producing the 24-inch difference in height, each *A* and *B* increasing the height by 6 inches.

in equal numbers. All combinations of the four kinds of ova and the four kinds of pollen grains are shown in the checkerboard diagram. In each square of this diagram we have placed the phenotype, in inches. Assembling these data, we find that $\frac{1}{16}$ of the F_2 offspring are 32 inches tall; $\frac{4}{16}$ are 26 inches; $\frac{6}{16}$ are 20 inches; $\frac{4}{16}$ are 14 inches; $\frac{1}{16}$ are 8 inches. When we plot these results at the bottom of the figure, we note their (rather rough) approximation to a normal frequency curve.

Thus we see that the observed facts of this type of inheritance are explained by assuming the existence of pairs of polygenes having additive effect. We readily understand that more than two pairs of genes may be involved in a quantitative difference. Three pairs might be involved in our present example. In that case the 32-inch variety would have the genotype *AABBCC,* the 8-inch variety the genotype *aabbcc.* This would mean that each effective allele would produce $\frac{1}{6} \times 24 = 4$ inches of increase in height (over the basic 8 inches). The F_1 plants would have the genotype *AaBbCc,* and would produce eight kinds of gametes: *ABC, ABc, AbC, aBC, Abc, aBc, abC, abc.* A checkerboard diagram of eight squares on a side would be needed to compute the combinations. When we work this out, we find that the F_2 offspring occur as follows: $\frac{1}{64}$ = 32 inches; $\frac{6}{64}$ = 28 inches; $\frac{15}{64}$ = 24 inches; $\frac{20}{64}$ = 20 inches; $\frac{15}{64}$ = 16 inches; $\frac{6}{64}$ = 12 inches; $\frac{1}{64}$ = 8 inches. These data are plotted in Figure 8.2. Note that there are seven phenotypic (size) classes, instead of the five present when we assume only two pairs of genes. The result is a closer approximation to a normal frequency curve than we obtained originally.

A still closer approximation would be obtained if we assumed that the difference in height depended on four pairs of polygenes, the tall variety having the genotype *AABBCCDD.* In this case, each effective allele would produce $\frac{1}{8} \times 24 = 3$ inches of increase in height (over the basic 8 inches). The F_1 hybrids would produce 16 kinds of gametes, and we should require a checkerboard diagram, 16 squares on a side, to compute the combinations. There would be 9 size classes (32 inches, 29 inches, 26 inches, 23 inches, 20 inches, 17 inches, 14 inches, 11 inches, 8 inches) in the F_2 offspring, and accordingly, a closer approximation to a normal frequency distribution.

We may wonder whether or not there is some way of telling, in a given instance, how many pairs of genes are involved. The simplest way is to raise large numbers of F_2 offspring and observe how frequently one extreme or the other appears. We note from Figure 8.1 that if two pairs of genes are involved, $\frac{1}{16}$ of the F_2 plants are expected to be 32 inches high (and $\frac{1}{16}$ to be 8 inches high). If, however, three pairs of genes are involved, only $\frac{1}{64}$ of the F_2 plants will be 32 inches high (and $\frac{1}{64}$ will be 8 inches high), as shown in Figure 8.2. If four pairs of genes are involved, only $\frac{1}{256}$ of the F_2 offspring will be expected to be 32

FIG. 8.2 Graph of the F_2 generation of the cross diagramed in Fig. 8.1, but assuming that the 24-inch difference in height depends on three pairs of effective alleles instead of two.

inches high (and $1/256$ to be 8 inches high). Therefore, if we raise large numbers of F_2 individuals and find that only 1 in 256 shows one extreme or the other, we have evidence that the quantitative difference we are studying depends on four pairs of equal and cumulative alleles. Statistical methods of estimating the number of pairs of alleles are also available, and, of course, are particularly pertinent in human genetics, where experiments of the kind we have been discussing are impossible.

Human Skin Color

Probably the most famous investigation of polygenic inheritance in man is that of Dr. Charles B. Davenport on the heredity of skin color in Negro-white crosses (1913). Davenport and his co-workers carried on their investigation in Jamaica and Bermuda. The investigators employed a color top as a means of measuring skin pigments (Fig. 8.3). Colored paper disks were overlapped in such a way that varying proportions of each color were exposed. When the top was spun, the colors seemed to blend together. By varying the proportionate amounts of the

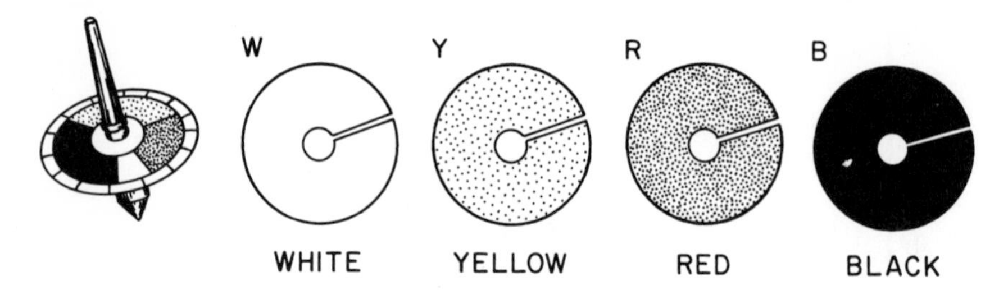

FIG. 8.3 Color top of the type used in Davenport's investigation of human skin color, with the four disks employed. (Redrawn from Davenport, C. B., 1926. "The skin colors of the races of mankind," *Natural History,* **26**:44–49.)

colors black, yellow, red, and white, the investigators could match the skin color of the persons studied. In these days of photoelectric instruments employing spectral light (see below), this may seem a crude method of measuring color, yet it sufficed to reveal the fundamentals of skin-color inheritance in the families studied. Figure 8.4 shows the setting of the disks that resulted in a match to the skin color of a "white" person and of a Negro. Needless to say, both whites and Negroes showed variation from person to person—the Negroes more than the whites, the investigators found.

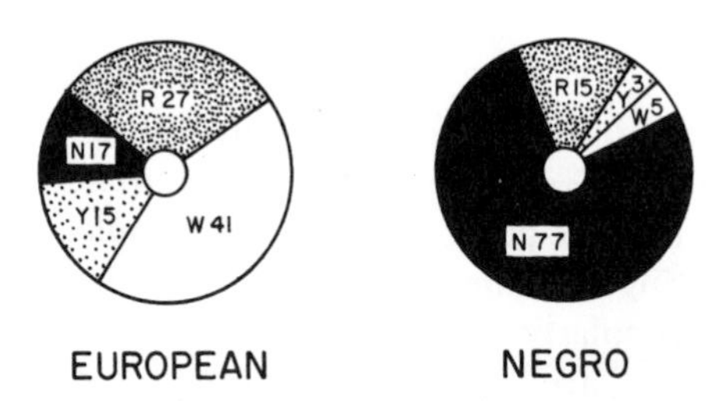

FIG. 8.4 Examples of color disks set to match the skin of a European and of a Negro. (Redrawn from Davenport, C. B., 1926. "The skin colors of the races of mankind," *Natural History,* **26**:44–49.)

At the present time we are interested in the inheritance of the black component measured by the investigators. This afforded a means of estimating the relative proportion of melanin in the skin (this is brown pigment, which may appear black when densely massed). With the exception of albinos, all persons possess melanin, but in varying amounts.

For comparison with the type of experimentation we have been discussing, we should like data on kindreds in which (a) Negroes marry whites, and (b) the offspring of such marriages (called mulattoes) marry each other:

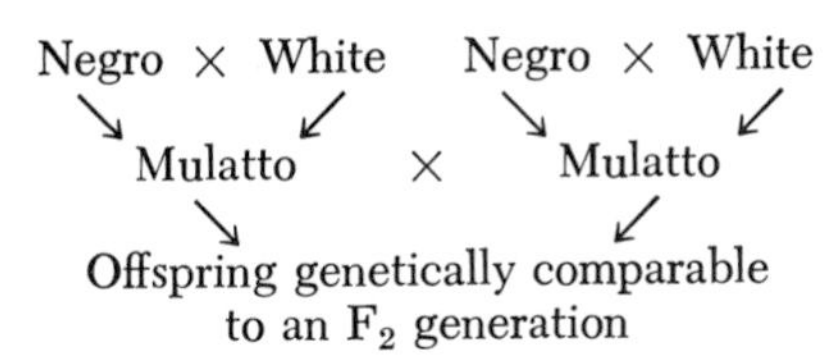

Davenport and his colleagues were able to obtain data on only six such families, although they did obtain information on a larger number of first-generation mulattoes (having one Negro and one white parent). The mulattoes were always intermediate between the parents in such cases. A typical example was a family in which the "black" reading for the father was 5 percent (5 percent of the black disk on the color top was exposed in the matching blend), and the comparable reading for the mother was 71 percent. The percentages for the seven children were 37, 35, 35, 43, 37, 35, and 35, respectively. These figures illustrate the fact that the readings for first-generation mulattoes were fairly uniform, although if the Negro parent was lighter than the one in our example, the children also tended to be somewhat lighter (averaging around 26 percent).

In the matter of an intermediate and fairly uniform F_1 generation, then, the findings agree with expectation for polygenic inheritance. What about the "F_2" generation? Do the children of first-generation mulattoes show great variability (as in the F_2 of our experiment with height in plants)? The six families available to the investigators produced 32 children. Their "black" percentages ranged from 10 to 56. In Figure 8.5 we have indicated the distribution of the readings for the 32 individuals. Although the numbers are small, there is some resemblance to a normal frequency curve, a bit skewed to the left.

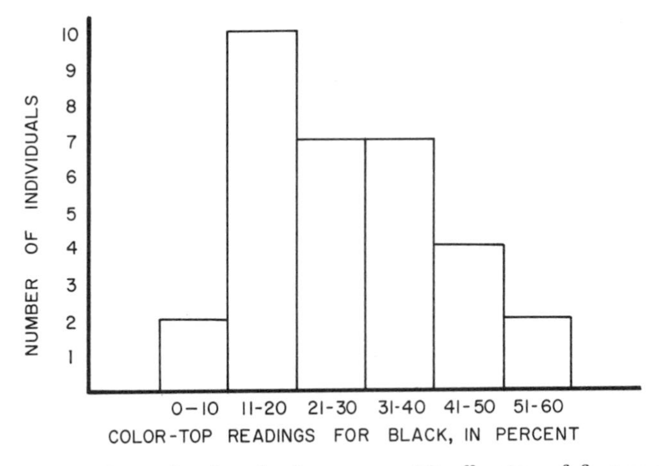

FIG. 8.5 Skin color distribution among 32 offspring of first-generation mulattoes married to each other. (Based on the data of Davenport, C. B., 1913. "Heredity of skin color in Negro-white crosses," *Carnegie Institution of Washington,* Publication No. 188, 1–106.)

Because Davenport found in his studies that many "white" persons showed a "black" reading of 10 percent or more, the two children having a reading of 10 percent were classified "white" (i.e., close to the skin color of the "ordinary brunet Caucasian") and thus represented one extreme of F_2 variability. Hence one extreme occurred twice in 32 times, or once in 16 times. We noted above that in the F_2 generation of experiments involving multiple genes, obtaining one extreme or the other 1/16 of the time is evidence that two pairs of genes are involved. Davenport advanced the hypothesis that two pairs of genes, which he designated *A* and *B*, are involved in this difference in skin color.

From our discussion of the importance of basing conclusions on large numbers, we can see that Davenport based his hypothesis on numbers so small that we might well question the validity of the hypothesis. However, Davenport tested the hypothesis in many parental combinations, and demonstrated good agreement between actual findings concerning the offspring and expected findings based on the hypothesis.

He assigned the following ranges of percentage values of blackness to the genotypes:

Genotype	Percentage of Blackness	
aabb	0–11	white
Aabb or *aaBb*	12–25	light colored
AaBb, or *AAbb*, or *aaBB*	26–40	medium colored
AABb or *AaBB*	41–55	dark colored
AABB	56–78	black

He then analyzed various parental combinations to determine whether actual findings agreed with expectation. Suppose, for example, that one parent has the genotype *aabb*, the other the genotype *Aabb*. What types of offspring will be expected, and in what proportions will they be expected to occur?

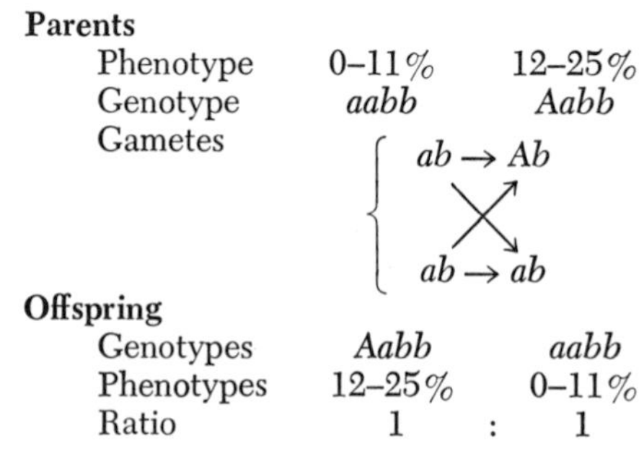

Half the children will be expected to be white, half to be light-colored. Davenport had data for 24 families of this kind, with a total of 99 children. Of the children, 42 were in the 0–11% category, 56 in the 12–25% category, 1 in the 26–40% category. Because the expectation was 49.5 in each of the first two categories mentioned and none in the third, it is evident that there was fairly good agreement between expectation and actual findings. Many other parental combinations were similarly analyzed; good agreement with expectation based on the hypothesis was usually demonstrated.

For sociological reasons, there is particular interest in families of which both parents are "white," or of which one or both are so light that they "pass for white." According to popular ideas, such people may have a "black" child. Davenport found three families in which both parents were in the 0–9% category. They had 9 children, none of whom graded higher than 9 percent. This is typical of other findings obtained since then. Students of the subject insist that no case has ever been proved of two "white" or very light parents having a "black" child. Proof in such a case would require removal of all doubt as to adoption or to the paternity of the child in question. It should be borne in mind that we refer only to the amount of pigment in the skin, not to other negroid physical characteristics.

As a general rule, in "mixed marriages" no child is darker than the darker parent. When darker children are produced, they are usually not much darker than the darker parent. For example, two light parents boh having the genotype *Aabb* could have a child with the genotype *AAbb,* who would probably be somewhat darker than either parent (but not "black").

This pioneer work by Davenport has stood the test of time in the sense that no one now questions the fact that skin color inheritance depends on multiple, cumulative genes. There is less agreement as to the actual number of pairs of these genes involved in the difference between Negro and white pigmentations. As we have seen, Davenport postulated two pairs, but with each resulting genotype represented by a considerable range of pigmentation. He recognized that superimposed on the effects of the genes were the effects of the environment. Exposure to the sun, for example, has a marked effect on skin pigmentation, as everyone knows. To minimize this effect on their results, the investigators matched, with their color tops, areas of skin normally covered by clothing—the upper arm, usually. Nevertheless, environmental effects were undoubtedly present. In addition, however, we should not be surprised to find that some of the variation was caused by genes other than the two pairs postulated by Davenport.

Thus, following a study of intermarriage between Negroes and whites in the United States, Gates (1949) suggested that three pairs of genes may be involved. Using a statistical approach and certain assumptions concerning the American Negro population, Stern (1953, 1954) has concluded that the range of variation in color exhibited by this group is best explained by postulating that the number of gene pairs is either four, five, or six, rather than fewer or more.

The simple color top of earlier investigators was replaced by a reflectance spectrophotometer in a study by Harrison and Owen (1964). This instrument measures the amount of light of different wavelengths reflected from a test surface:—the skin in this case. Using it, the investigators measured the reflectance of the skin of the following residents of Liverpool: 105 Europeans; 106 people from various parts of West Africa; 94 "F_1 hybrids" (offspring of intermarriage between Europeans and West Africans); 30 "backcross European hybrids" (offspring of marriages between Europeans and F_1 hybrids); 26 "backcross African hybrids" (offspring of marriages between West Africans and F_1 hybrids); 14 "F_2 hybrids" (offspring of F_1 hybrids married to each other*).

The results are shown in Figure 8.6 (the number of F_2 hybrids was considered too small to warrant inclusion). Along the base line of the graph the shorter wavelengths (violet) are at the left, the longer wavelengths (red) at the right. We note that for all wavelengths the African skin reflects much less light than does the European. The pigmentation of F_1 hybrids is intermediate, and the pigmentation of the respective back-

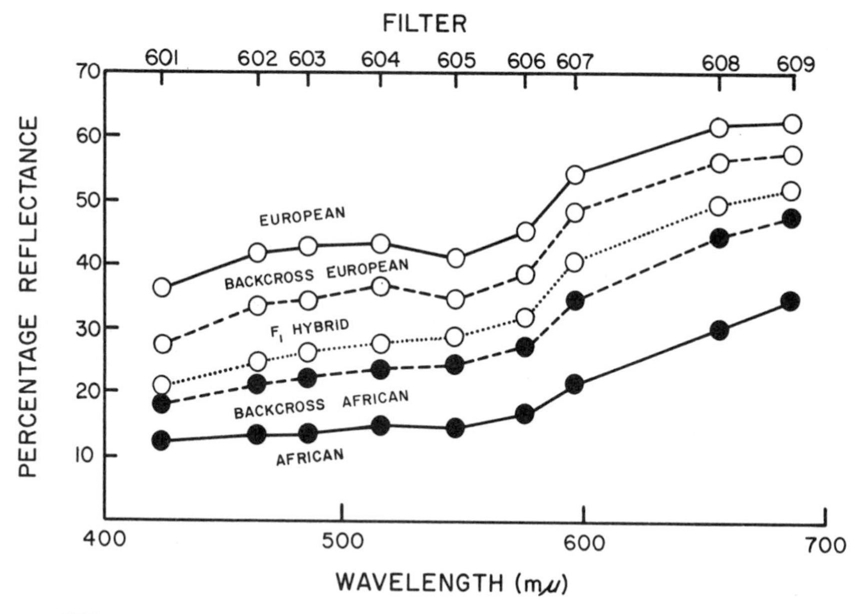

FIG. 8.6 Mean reflectance from the skins of Europeans, Africans, and various hybrid groups, as measured with a reflectance spectrophotometer. (Redrawn from Harrison, G. A., and J. J. T. Owen, 1964. "Studies on the inheritance of human skin color," *Annals of Human Genetics,* **28**:27–37.)

*This was not a literal F_2 generation as in animal experimentation, because these were not brother-sister matings.

cross hybrids is intermediate between the pigmentations of the parental groups concerned. Toward the violet end of the spectrum the F_1 curve approaches the African more closely, and toward the red end the F_1 curve approaches the European more closely. This reminds us that skin color is complex, contributed to not only by the amount of melanin present but also by other factors, such as the color of the blood in the skin capillaries, which is more evident in lighter skins than in darker ones.

These more accurate measurements reinforce the interpretation of the polygenic nature of skin color inheritance. Statistical analysis led Harrison and Owen to conclude that the number of "effective factors" responsible for the difference in skin color between Europeans and Africans is "between 3 and 4."

Human Eye Color

Some readers may be surprised to find the subject of eye color included in a discussion of polygenic inheritance. If there is one thing the average person thinks he knows about human inheritance, it is that "brown eyes are dominant to blue." Traditionally this has been regarded as the best human example of Mendelian inheritance based on a single pair of genes, brown-eyed people having the genotypes *BB* or *Bb*, blue-eyed people the genotype *bb*. Such simplicity is appealing, yet a little thought will convince us that we should not really expect it to be true.

Varied Hues and Shades In the first place, people cannot really be classified so easily as brown-eyed or blue-eyed. As we know from observation, eyes come in all sorts of colors between brown and blue—ranging through the various shades of brown to a brown so dark that we usually call it black, and including gray and green and intermediate shades and hues. This range of variation in itself suggests polygenic inheritance.

When we examine the eye itself, we find that in large measure the color depends on the presence in the iris, surrounding the pupil, of that brown pigment that colors the skin: melanin. If no observable amount of melanin is present, the eye appears blue, not because of blue pigment but because of the same optical effect that causes the sky to appear blue on a clear day. The scattering of light by particles in the atmosphere causes blue light to be reflected back to the eye of an observer. Similarly, colorless cells in the more superficial layers of the iris (viewed against the dark background of the deeper layers) scatter the light so that blue is reflected back to the eye of an observer. A blue eye, then, is one in which the superficial layers of the iris have no observable amount of brown pigment.

At this point we may well mention that studies with a microscope indicate that no iris, except that of some albinos, is completely without melanin. But the amount may be so small and the individual particles so minute that an observer who does not use a microscope is unaware of them.

If, instead of a completely colorless iris, we have an iris with just a little pigment present, what color does the eye appear to an observer? If the melanin is in fine particles evenly distributed but is so thinly spread out that it does not completely prevent blue from "shining through," the result may be a color that is a mixture of light brown and blue (some people might call it "hazel," although there does not seem to be general agreement on the meaning of that term).

How about yellow and green eyes? In addition to melanin, the iris may have specks or patches of a yellowish, fatty pigment material, a lipochrome. According to Gladstone (1969), it is the combination of this yellow pigment with the brown that makes the iris look green. The smaller the amount of melanin in proportion to the yellow, the yellower the iris would appear, of course.

Some people have gray ones. Gray has been variously accounted for. It may be simply the color of an iris that has little or no pigment yet is sufficiently opaque (from bundles of connective tissue, perhaps) so that little or no blue "shines through." On the other hand, Gladstone (1969) stated: "The combination of green eye color and blue eye color makes the iris appear gray." These explanations are not necessarily mutually exclusive, of course.

Heterochromia iridis Sometimes an individual's eyes differ in color—one brown, one blue, for example. Evidently, such an individual has genes for melanin formation but for some reason they are not expressed in one eye. Various pathological syndromes include this trait, but many normal people also show it. In some cases, at least, abnormalities of innervation from the sympathetic nervous system seem to underlie failure of the genes to be expressed in one eye.

Heterochromia is not always of a hereditary nature, but the fact that it is observed to "run in" some families indicates that genetic factors may be involved. Traditionally the gene for the trait is considered to be dominant, but because the trait sometimes skips a generation in a direct line of descent (Chap. 6), suspended judgment seems wise. As with so many human traits, we need many more data and much more complete knowledge of the underlying causes of the trait before the exact genetic basis can be determined with certainty.

Varied Shades of Brown Focusing our attention on melanin, we note that it may vary in amount all the way from complete absence (for an observer not using a microscope, at least) to such a large amount that the eye appears black. This is the sort of continuous variation in a quantitative characteristic that is usually found to depend on multiple, cumulative genes. Do we have evidence of this in the inheritance of eye color?

Hughes (1944) classified eye color into seven different grades. To avoid subjective differences between observers, he made all the observations himself, and in good natural light. He did not use magnification or photoelectric devices for measuring color. Recognizing that blue is an optical effect, and wishing to concentrate on the presence and absence of brown, he used the term "nonbrown" in place of "blue."

Hughes' seven eye-color grades were as follows: "1, no brown pigment observable; 2, a trace of brown; 3, one-fourth brown; 4, one-half brown; 5, three-fourths brown; 6, a trace of nonbrown; and 7, completely brown." He gathered data from 107 families with a total of 212 children, analyzing his results statistically. He classified the families by the eye-color grades of the parents: 1 × 1, 1 × 2, 1 × 3, and so on to 7 × 7 (actually he had no family that fitted this latter category). Although numbers were small in many categories, a general trend was indicated—the amount of pigment in the children's eyes increased with increased amounts in the eyes of their parents. Of more interest was the fact that the greatest variation in eye color among the children was found in families in which the parents had the "middle" grades. For example, there were 9 families of which the parentage was "2 × 4" (one parent had a trace of brown, the other one-half brown). The children of these families were as follows: 4 of grade 1; 2 of grade 2; 3 of grade 3; 4 of grade 4; 3 of grade 5; 2 of grade 6; 1 of grade 7. Thus they ranged all the way from completely nonbrown (blue) to completely brown. Such finding seems to indicate that the parents were heterozygous for several pairs of genes concerned with melanin production. There is nothing here, or elsewhere in the data, to suggest that dominance of brown over nonbrown is involved.

Can two blue-eyed parents have a brown-eyed child? They could not if blue-eyedness depended upon a recessive gene, blue-eyed people being homozygous recessive. But two people regarded by their acquaintances as blue-eyed sometimes *do* have a brown-eyed child. Certainly people of Hughes' grade 1 and probably people of grade 2 were regarded by family and friends as blue-eyed. While most of the children in families with parents of these grades were of the same grades, a few of them had darker eyes. No doubt, they inherited genes for pigment from both parents and so were darker-eyed than either parent—the same likelihood we mentioned for skin color inheritance.

The suggestion just made introduces the concept that a **threshold** may be present in polygenic inheritance. In this example we suggest that a certain level of melanin concentration must be reached before it becomes noticeable to observers and they classify the eye as brown rather than blue. As we shall note presently, the threshold concept may be important in other traits in which inheritance seems to be essentially polygenic but the trait is not found to vary continuously from one extreme to the other after the manner of the F_2 offspring in Figure 8.1.

Because Hughes investigated a population having northern European ancestry for the most part, "nonbrown" and lighter eye colors predominated. In connection with his investigation of skin color inheritance, Davenport (1913) studied inheritance of eye color in a population in which darker shades predominated. He found evidence suggesting polygenic inheritance and tentatively postulated that two pairs of genes are involved. Hughes made no postulation about number of genes.

Brues (1946) included along with eye color, structure of the iris, variations in which affect the color seen by an observer. The phenotypic complexity observed suggested genotypic complexity with genes in teracting in various ways. The data suggested that some of the genes are sex-linked (see Chap. 14), a possibility also suggested by earlier investigators who noted that women tend to have darker eyes than do men. (No evidence of sex-linkage was found by Hughes, 1944.) It would be well to note at this point that there is no *a priori* reason why some genes in a series of polygenes may not be sex-linked. Polygenes are in various chromosomes, and the X chromosome may well be one of those included in any given series.

Polygenes and Modifying Genes

In discussing polygenes, we tend to visualize situations in which the various genes of a cumulative series contribute equally. Such an assumption makes for ease of computation, but in many, if not most, instances it may be an oversimplification. There is no reason why one gene of a series may not contribute more than does another one.

To illustrate such a difference in genetic effect, let us return to our example of height inheritance in a hypothetical plant variety (Fig. 8.1). In our earlier discussion we assumed that *A* and *B* contributed equal increments of increase in height (above the basic 8 inches). Now let us suppose that *A* contributes 10 inches, whereas *B* contributes 2 inches. How will such differential contribution affect the results?

The experiment is diagramed in Figure 8.7. Note that, as before, when 32-inch plants are crossed with 8-inch ones the F_1 offspring are intermediate, 20 inches tall. Then the mating together of the F_1 individuals is represented by the same 16-square, checkerboard diagram we used previously. The difference lies in the interpretation of this diagram.

PARENTAL GENERATION

Phenotypes:	32 inches		8 inches
Genotypes:	*AABB*	X	*aabb*
Gametes:	*AB*		*ab*

F_1 GENERATION

Phenotypes:	20 inches		20 inches
Genotypes:	*AaBb*	X	*AaBb*
Gametes:	*AB, Ab, aB, ab*		*AB, Ab, aB, ab*

Combinations of gametes producing an F_2 generation:

	AB	*Ab*	*aB*	*ab*
AB	*AABB* 8+10+10+2 +2=32	*AABb* 8+10+10+2 =30	*AaBB* 8+10+2+2 =22	*AaBb* 8+10+2 =20
Ab	*AABb* 8+10+10+2 =30	*AAbb* 8+10+10 =28	*AaBb* 8+10+2 =20	*Aabb* 8+10 =18
aB	*AaBB* 8+10+2+2 =22	*AaBb* 8+10+2 =20	*aaBB* 8+2+2 =12	*aaBb* 8+2 =10
ab	*AaBb* 8+10+2 =20	*Aabb* 8+10 =18	*aaBb* 8+2 =10	*aabb* =8

F_2 GENERATION

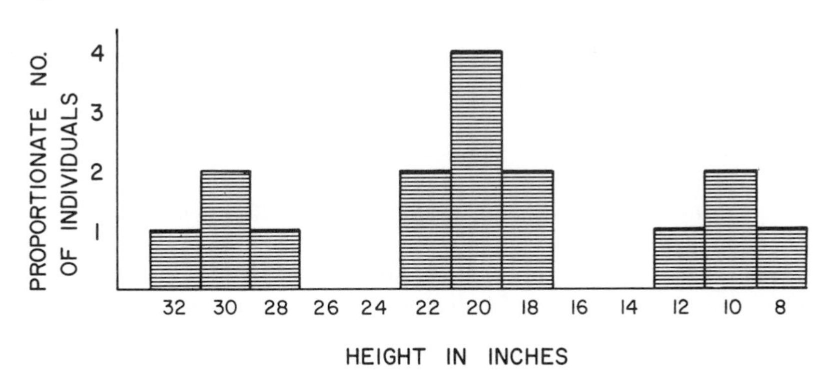

FIG. 8.7 Cross between two imaginary plant varieties differing in height by 24 inches (cf. Fig. 8.1). It is assumed that each *A* increases the height by 10 inches, each *B* increases it by 2 inches.

In each square, representing a zygote, we add, to the basic 8 inches, 10 inches for each *A*, 2 inches for each *B*.

The outcome, tabulated beneath the diagram, indicates that the F_2 will fall into three nonoverlapping groups: (1) plants from 28 to 32 inches tall, (2) plants from 18 to 22 inches tall, (3) plants from 8 to 12 inches tall. If we designate these as tall, medium, and short, respectively, we note that we have a 4:8:4, or 1:2:1 ratio.

Note the genotypes of these three groups. The tall individuals are all *AA;* the medium ones are *Aa;* the short ones are *aa* (neglecting the *B*'s for the moment).

There are two, equally valid, ways of interpreting these results. According to the first interpretation, we may state, as we did at the outset, that *A* and *B* are multiple genes, each contributing to increase in height but doing so unequally.

Or we may interpret the results as indicating that *A* is a *main* gene, and *B* is a *modifying* gene. According to this view, *A* is a gene for tallness, *a* its recessive allele for shortness, and dominance is lacking. Hence we obtain in the F_2 generation the ratio of ¼ *AA* (long):²⁄₄ *Aa* (medium):¼ *aa* (short).

Then, according to this view, the gene *B* modifies the action of the *A* genes. On the basis of the distribution of *B*'s we find, for example, that ¼ of the *AA* plants will be 32 inches tall, ²⁄₄ of the *AA* plants will be 30 inches tall, ¼ of the *AA* plants will be 28 inches tall. The *Aa* and *aa* plants vary similarly (note that each group has a secondary 1:2:1 ratio determined by the distribution of the *B*'s).

This example illustrates the fact that there is no clear-cut distinction between polygenes and modifying genes. In general usage, the term "polygenes" would probably be employed in cases in which the relative contributions of the genes in the series were of more or less the same order of magnitude, the term "modifying genes" being reserved for cases in which one gene contributed much more than did the genes called modifiers. Where to draw the line is more or less a matter of taste on the part of the investigator.

Not all modifying genes are merely quantitative, adding "more of the same." Modifying genes that are comparable to polygenes do have this quantitative similarity, but in other cases modifying genes may modify the action of the main gene *qualitatively*. For example, in color production the main gene may produce yellow, and a modifier may cause the yellow to have an orangish shade. (A distinction between quantitative and qualitative is rather artificial. In this case, the modifier gene would presumably act by producing red pigment to mix with the yellow. If the distinction is between "red pigment present" versus "red pigment absent," then we should be inclined to say that the action of the modifier is qualitative. However, also involved may be a question of the *amount* of red pigment present—a small amount producing no visible effect on the shade of yellow, a larger amount producing an effect. In this case, the modifier is also acting quantitatively.)

Other Polygenic Traits in Man

Thus far, our examples of human polygenic inheritance have involved differences in the amount of pigment produced. In point of fact, however, we may well suspect that this type of inheritance is the basis of all traits that vary quantitatively in a relatively continuous manner. Human stature is a familiar example. In view of the enormous volume of data on stature collected by anthropologists, we may wonder that so little of a definite nature can be stated concerning its inheritance. In any given population of men (or women), stature varies quite continuously from shortest to tallest, with an average near the midpoint between the extremes. As a consequence, bell-shaped, normal frequency curves are commonly approximated whenever data on stature are graphed. This suggests that inheritance is based on multiple genes acting cumulatively. As a matter of fact, stature afforded the first example of continuous variation to be analyzed—by two pioneers in genetic research, Galton and Pearson.

Unfortunately for genetic analysis, stature is a complex human trait. It is the total of the lengths of body segments that do not always vary correspondingly: head, neck, trunk, legs. (For example, we recall among our acquaintances tall people who have tall "sitting height," and others who have fairly short "sitting height" but are of equal total stature due to the length of their legs.)

Stature is influenced by hormones. We see evidence of this in differences in height between men and women. The effects of secretions of endocrine glands are most striking in cases of extreme over- or under-secretion. One of the secretions of the pituitary gland is a growth-controlling hormone. Overproduction of this hormone during infancy and childhood results in giantism, whereas midgets are produced if the hormone is in short supply during these critical years. These giants and midgets, respectively, are of fairly normal proportions (other types of dwarfism have different causes). Presumably, variations in stature between these extremes are also connected with varying amounts of this hormone, acting during the years of growth. In fact, some of the genes controlling stature may well act by regulating the production of the hormone.

Genes do not act alone, however. Environment plays a significant role in determining how tall a person will become. Without doubt, the most important environmental factor is diet. Many statistics show (a) that people are becoming taller as the generations pass, and (b) that when people migrate to the United States their children, on an average, are taller than their contemporaries living in "the old country." Both of these trends are well illustrated for Japanese boys by Figure 8.8. Although other factors may contribute, the principal cause of the increase is probably improvement in diet.

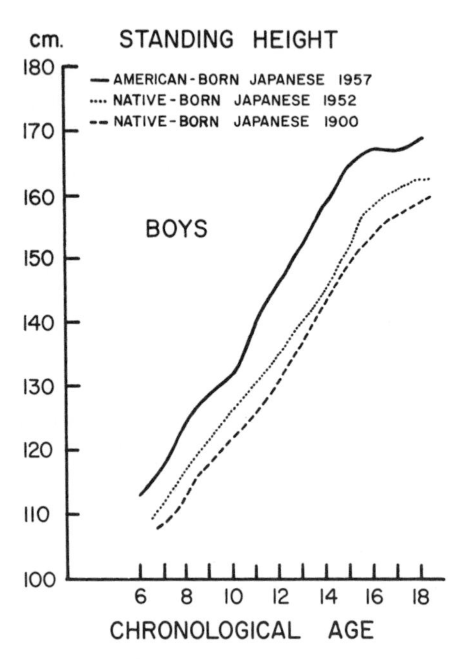

FIG. 8.8 Average standing height of American-born Japanese boys compared with that of boys in Japan, in 1900 and 1952. (Redrawn from Greulich, W. W., 1958. "Growth of children of the same race under different environmental conditions," *Science,* **127**:515–516. Copyright © 1958 by the American Association for the Advancement of Science.)

In view of these and other complexities, we can readily see that although there is general agreement that multiple, cumulative genes are involved in the determination of stature, knowledge of the number and mode of action of these genes must await future research. In Chapter 24 we shall discuss evidence derived from study of twins concerning inheritance of stature.

Weight is also a quantitative characteristic that varies continuously between extremes. Variations in weight show the hallmarks of polygenic inheritance. Yet here, even more than with stature, differences in diet have such large effects that we readily understand the difficulties of genetic analysis. Evidence from experiments with lower animals (e.g., rabbits), with careful control of diet, confirms our suspicions that weight depends on multiple genes.

Every schoolchild knows that when grades or the results of intelligence tests are graphed, a bell-shaped curve is approximated. We may, then, predict that when the time comes that the genetic basis of intelligence is analyzed, the variations of this trait will be found to depend on multiple genes. At the present time we are a long way from such an analysis. A first requirement will be the development of an accurate means of measuring inherent intellectual ability—a means of measurement independent of cultural background and educational experience. Then with this perfected yardstick, we must measure large groups of people and especially successive generations in many kindreds. Because human generations are so long, this means a continuing program of research stretching over many years. Until this utopian day arrives, we must draw as many conclusions as possible from data obtained with less perfect measuring instruments. Evidence as to the inheritance of intelligence will be presented in connection with a discussion of twins (Chap. 24).

Threshold Effect

In our discussion of the inheritance of eye color, we noted the probability that the amount of melanin in the iris must reach a certain level before observers will classify the eye as brown, rather than "blue." We find other traits in which there are two contrasting phenotypes—one either has or does not have a certain malformation, for example—and yet the genetic basis seems more complex than that of a single pair of genes. It is probable that the basis of many such traits is polygenic (multifactorial), but that there is a threshold in number of "defective genes" (i.e., those not contributing to a normal phenotype) that must be present if the trait is to appear in the phenotype.

The situation is pictured in Figure 8.9, which shows the number of genes distributed according to a normal frequency curve, as was characteristic of previous examples discussed in this chapter. But in these earlier examples we showed each differing number of genes as translated into a specific phenotype (e.g., a certain height). In the present instance, by contrast, possessors of all gene combinations represented by the portion of the curve to the left of the vertical line are likely to be normal. Only individuals having the higher proportions of defective genes represented by the shaded portion of the curve have the defect in question. As an oversimplified model we may picture a situation in which people having one, two, or three defective genes (e.g., *AaBBCCDD, AaBbCCDD,* or *AaBbCcDD*) are normal, while those having four or more defective genes (e.g., *AabbCcDD, AaBbCcdd*) are abnormal in some specific way.

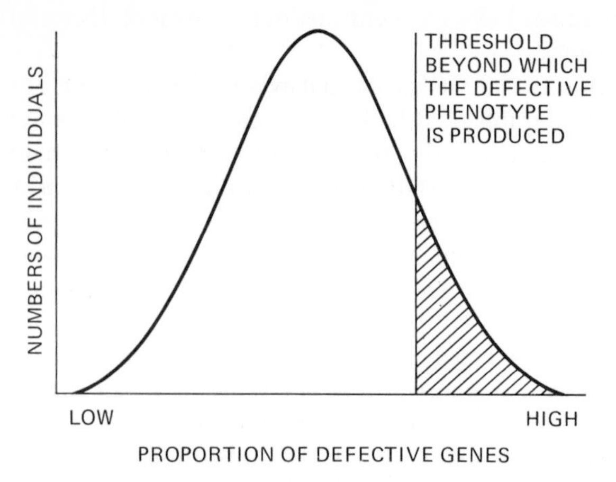

FIG. 8.9 Polygenic inheritance with a threshold effect, resulting in production of two contrasting phenotypes: normal versus defective.

An important consequence of this type of inheritance is the fact that the risk of having a defective child varies from family to family (Carter, 1969). This is because some normal parents will be carriers of a larger number of the defective genes than will other parents. To continue our simplified example: One pair of parents may be *AaBbCCDD* × *AaBBCcDD,* the mother being given first. They can produce a defective child only if the unfertilized ovum contains *a* and *b,* and the sperm contains *a* and *c* (thereby raising the number of defective genes in the fertilized ovum to four). Another pair of parents is *AaBbCcDD* × *AaBbCcDD.* In this case there are many combinations of ovum and sperm that will yield zygotes having four or more defective genes (e.g., ovum *abCD,* sperm *abCD;* ovum *abCD,* sperm *aBcD;* ovum *abcD,* sperm *aBcD,* and so on). Obviously, the risk of producing a defective child is much greater for the second parents than it is for the first.

How can we recognize those parents who have the greater risk? One criterion is the number of defective children already produced. For example, from an analysis of families in England, Carter and Roberts (1967) found that after the birth of one child having spina bifida or anencephaly (see Glossary), the risk of a second birth of the kind is 1 in 25. But in families having two children with the defect the risk of another birth of the same kind rises to 1 in 10. Hence for spina bifida and anencephaly, "the risk to later offspring is about twice as great for mothers who have already had two affected children, compared with those who already have had one such child. In contrast, with a recessive condition such as cystic fibrosis of the pancreas the risk remains 1 in 4 whether the mother has already had one, two or even three affected children" (Carter, 1969).

Carter noted that a similar situation is found with cleft lip, because if a parent showing the trait has a child with it, the risk of another affected child rises from 3 or 4 percent to over 10 percent.

In the case of pyloric stenosis, we find a sex difference. The ratio of affected males to affected females is 5:1. Carter (1961) postulated that the treshold for production of the defect is higher for females than it is for males. (If, for example, Fig. 8.9 represents the situation for males, a similar diagram having the threshold line moved further to the right would represent the situation for females, who must have a higher proportion of defective genes in order to show the trait.) This being the case, having affected daughters demonstrates that the parents have larger numbers of defective genes than is demonstrated merely by having defective sons. Thus it is found that parents who have a daughter with pyloric stenosis are about three times as likely to have another child with the condition as are parents of a son with pyloric stenosis (Carter, 1969). This reference should be consulted for additional examples and discussion.

Here we should remind ourselves again that most traits that we call inherited have an important environmental component in their causation. In many cases a given genetic constitution expresses itself under certain environmental conditions, not under others. In the present connection this general principle may mean that presence of a certain environmental influence in one individual may be so important that a trait is produced even when the number of defective genes is below the usual threshold (Fig. 8.9), that is, below the number required to produce the trait in some other individual who lacks the environmental influence in question.

We still know almost nothing of the number, type, and mode of interaction of the constitutent genes in such cases. But clearly this type of inheritance is highly significant for genetic counseling (Chap. 28).

Problems

1. Referring to Fig. 8.1, what types of progeny and in what proportions will be expected from each of the following crosses: (a) *AaBb* × *aabb;* (b) *AaBB* × *aaBb;* (c) *AaBb* × *Aabb?*
2. Assume that the difference between a variety of oats yielding 4 grams per plant and a variety yielding 10 grams is caused by two equal and cumulative pairs of multiple genes. A 4-gram variety having the genotype *sstt* is crossed with a 10-gram variety having the genotype *SSTT*. Give the genotype and phenotype of the F_1 offspring. When these F_1 plants are mated together, what types of F_2 offspring will be expected, and in what proportions?
3. One variety of a certain speces of plant is 48 inches high, another variety is 16 inches high. When the two varieties were crossed, the F_1 hybrids

averaged 32 inches high and varied little in height. The F_1 plants were mated together to produce an F_2 generation. These F_2 offspring averaged 32 inches in height but varied widely. Of 2560 F_2 plants, 10 were 48 inches high, and 8 were 16 inches high. How many pairs of genes were probably involved in producing the difference in height between the two varieties? How much increase in height did each effective allele contribute?

4. In skin color inheritance, if the genotypes postulated by Davenport are correct (p. **104**), what amounts of pigmentation will be expected in children produced by the following matings: (a) *AAbb* × *aaBB;* (b) *Aabb* × *Aabb;* (c) *AaBb* × *aabb;* (d) *Aabb* × *AaBb?*
5. Let us imagine that the amount of brown pigmentation of the eye is controlled by three pairs of polygenes (the number is almost certainly greater than this). Let us further imagine that there is an observational *threshold,* so that one effective allele by itself produces so little brown pigment that a microscope is needed to see it. People who have the genotype *aabbcc* are nonbrown-eyed (blue), and we may assume that people with the genotypes *Aabbcc, aaBbcc,* or *aabbCc* have so little pigment that their acquaintances regard them as blue-eyed. On the other hand, people with two effective alleles (*AAbbcc, aaBBcc, aabbCC, AaBbcc,* and so on) have enough iris pigmentation so that acquaintances regard them as brown-eyed. The greater the number of effective alleles above two, the darker the shade of brown. On the basis of this hypothesis, what eye colors would be expected from each of the following matings and in what proportions would the different colors be expected? In each cash give the phenotypes of the parents as well as those of the children. (a) *Aabbcc* × *Aabbcc;* (b) *AaBbcc* × *Aabbcc;* (c) *AaBbcc* × *AaBbcc;* (d) *AaBbcc* × *AAbbcc;* (e) *aaBbcc* × *aabbCc;* (f) *AaBbCc* × *aabbcc.*
6. Referring to Figure 8.7, what plant heights would be expected from the following mating: *AaBb* × *aabb*? What proportion of the offspring would be expected to have each of these heights?
7. Imagine a trait dependent on two pairs of genes: *A* and *a, B* and *b.* Only individuals having three or more "small-letter genes" show the trait. What is the chance that the first child will show the trait in each of the following matings of two normal parents?
   (a) *AaBb* × *AABb*
   (b) *AaBb* × *AaBb*
   (c) *AaBb* × *AABB*

References

Brues, A. M., 1946. "A genetic analysis of human eye color," *American Journal of Physical Anthropology,* New series **4**:1–36.

Carter, C. O., 1961. "The inheritance of congenital pyloric stenosis," *British Medical Bulletin,* **17**:251–254.

Carter, C. O., 1969. "Genetics of common disorders," *British Medical Bulletin,* **25**:52–57.

Carter, C. O., and J. A. F. Roberts, 1967. "The risk of recurrence after two children with central-nervous-system malformations," *Lancet,* **1967**:**1**: 306–308.

Davenport, C. B., 1913. "Heredity of skin color in Negro-white crosses," *Carnegie Institution of Washington,* Publication No. **188**, 1–106.

Davenport, C. B., 1926. "The Skin Colors of the Races of Mankind," *Natural History,* **26**:44–49.
Gates, R. R., 1949. *Pedigrees of Negro Families.* Philadelphia: The Blakiston Company.
Gladstone, R. M., 1969. "Development and significance of heterochromia of the iris," *Archives of Neurology,* **21**:184–192.
Greulich, W. W., 1958. "Growth of children of the same race under different environmental conditions," *Science,* **27**:515–516.
Harrison, G. A., and J. J. T. Owens, 1964. "Studies on the inheritance of human skin color," *Annals of Human Genetics,* **28**:27–37.
Hughes, B. O., 1944. "The inheritance of eye color in man—brown and nonbrown," *Contributions from the Laboratory of Vertebrate Biology, University of Michigan,* No. **27**:1–10.
Stern, C., 1953. "Model estimates of the frequency of white and near-white segregants in the American Negro," *Acta Genetica et Statistica Medica,* **4**:281–298.
Stern, C., 1954. "The biology of the Negro," *Scientific American,* **191**:81–85.

9

Some Other Readily Observable Traits

Of necessity, many of the traits included in the study of human genetics are unusual occurrences and abnormalities. Or they are discernible only by such technical means as serological testing or electrophoresis. But many people are interested in the heritability of normal, easily observed characteristics. So in this chapter we shall devote most of our attention to some of these. In doing so we shall make use of our knowledge of single-gene-pair inheritance, and of polygenes and modifying genes.

In the preceding chapter we discussed inheritance of some normal, commonly observed traits; skin pigmentation, eye color, height, and weight. In Chapter 24 we shall include discussion of inheritance of intelligence.

Hair Color

Varied Shades of Brown Because hair color may vary all the way from light blond to brown so dark that we usually call it black, we are not surprised that we have here another example of polygenic inheritance, as in skin color inheritance. Indeed, as we commonly observe, the two usually parallel each other. Although there are exceptions, people with light skin usually have light hair, and brunettes usually have dark hair. As with skin color, hair color varies with the amount of melanin in the hair.

No one knows how many genes are involved or how they interact. As with skin and eye color, the simplest way to picture them is as cumulative genes each contributing to melanin production. As a hypothetical model, we may imagine that four pair of genes are concerned (almost certainly the number is greater): B^1, B^2, B^3, and B^4 contributing to production of much melanin, the corresponding alleles, b^1, b^2, b^3, and b^4 contributing little. In this model an individual with the genotype $b^1b^1b^2b^2b^3b^3b^4b^4$ would have light blond hair. And various combinations of "capital-letter genes" (better termed: **effective alleles**) and "small-letter genes" would yield the varied shades of brown in between.

The statement is sometimes made that when parents differ in hair color, no child can have darker hair than the darker-haired parent. Perhaps it is commonly observed that children do not have darker hair than does this parent, but there is no theoretical reason why this *must* be true. In terms of our model, suppose the darker-haired parent had the genotype $B^1b^1B^2b^2B^3b^3b^4b^4$, the lighter-haired parent the genotype $B^1b^1b^2b^2B^3b^3b^4b^4$. Among the children there might be one with the genotype $B^1B^1B^2b^2B^3B^3$-b^4b^4. Having five effective alleles instead of the darker-haired parent's three, this child would have darker hair than the latter parent. At the same time, the model suggests that if the lighter-haired parent is quite blond, children are not likely to have hair much darker than that of the darker-haired parent. For example, the mating $B^1b^1B^2b^2B^3b^3b^4b^4 \times B^1b^1$-$b^2b^2b^3b^3b^4b^4$ could produce a child with one more effective allele than the darker-haired parent has, but not with a large number more.

Red Hair The general public interest in red hair seems to have been shared by geneticists, for there have been many investigations of it through the years. Despite this fact, considerable uncertainty remains about the precise mode of inheritance.

We may state as a first approximation that many family studies give results consistent with the idea that red-haired people are homozygous for a recessive gene. For example, Singleton and Ellis (1964) published a pedigree extending over six generations and showing this type of inheritance. Figure 9.1 is a portion of their chart. Note the hallmarks of recessive inheritance, especially the families in which red hair reappears in children even though their parents, and in one instance, their grandparents, do not have red hair. According to this hypothesis, genotypes would be as follows:

RR = homozygous nonred
Rr = heterozygous nonred
rr = red

The exact hair color of the "nonred" people would depend upon what other genes, mainly genes for melanin, were present.

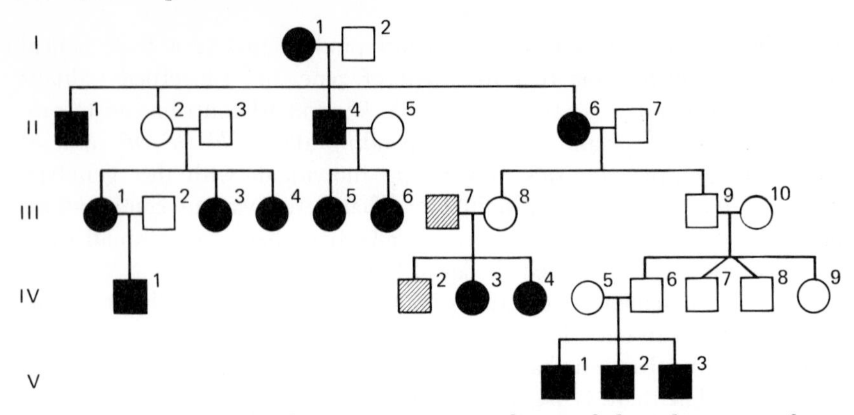

FIG. 9.1 Portion of a six-generation pedigree of the inheritance of red hair. Black symbols represent individuals with red hair, shaded symbols represent individuals with red beard and body hair. I-1 and I-2 had a total of 12 children, only four of whom are shown. (Redrawn from Singleton, W. R., and B. Ellis, 1964. "Inheritance of red hair for six generations," *Journal of Heredity,* **55**:261, 266.)

As Neel (1943) pointed out, special interest in the testing of this hypothesis centers in families in which both parents have red hair. According to this hypothesis they should have only red-haired children. Many such families have been studied through the years, and in almost all cases a pair of red-haired parents is found to have only red-haired children. The exceptions may reasonably be explained on the basis of lack of penetrance (pp. 70–72) of gene *R*. This means that one red-haired parent may have the genotype *Rr,* but for some reason gene *R* is inactive, gene *r* producing its phenotypic effect in this parent. Such a parent could pass *R* to a child in whom it might be expressed normally, the child having nonred hair in consequence.

Genes for red hair are evidently independent of those for brown hair discussed in the preceding section. If a person has genes for production of a large amount of melanin and also the genotype for red hair, the hair is likely to be so dark that the redness will not show (in terms of the model used above, such an individual's genotype might be $B^1B^1B^2B^2B^3b^3$ B^4B^4rr). If, however, there were fewer genes for melanin production, the result might be brownish-red or reddish-brown hair—auburn, perhaps (in terms of the model: $B^1b^1B^2b^2B^3b^3b^4b^4rr$, for example). In this way some of the varied hair colors actually observed may arise through mixtures of the red and brown pigments. People with bright red hair presumably are lacking most or all of the genes for melanin production in the hair, or have these genes "turned off" in some manner.

So far our discussion of red hair has made its inheritance sound simple. The appearance of simplicity is deceptive. Reed (1952) analyzed samples of hair with a reflectance spectrophotometer (much after the manner of Harrison and Owen in their analysis of skin color, pp. 106–107). The curves obtained with bright red hair differed somewhat from those for brown hair, but among the people tested, he found a continuous distribution from bright red to various shades of nonred. In other words, there was no clear cutoff point separating red from nonred. This might be caused in part by the genetic mechanism suggested in the preceding paragraph.

Nicholls (1968) utilized microspectrophotometry so that through a microscope he could measure the color of individual hairs (rather than of samples composed of many hairs). By this means he obtained quantitative measurements of the fact that individual hairs classified as red may differ greatly in actual hue. (As we should expect, he found the same to be true of brown hair.) Does this suggest that red hair, like brown hair, is polygenic or involves a "main gene" plus modifying genes? Nicholls also found that many brown-haired people have some red hairs, and that many red-haired people have some brown hairs. Evidently such individuals have the genes for both colors, but in some hair follicles the genes for melanin production are not penetrant (i.e., they are "turned off"). What is the control mechanism?

This question is one of many unanswered ones. Why do some people who were red haired as children become brown haired as they grow older? Does this mean that the penetrance of genes for melanin production varies with age? If so, why, and how is this controlled? Why do some men with brown hair on their heads have red beards and mustaches, or red hair on other portions of their bodies?

Perhaps such unanswered questions emphasize more than anything else the fact that no gene works alone, that the phenotype—the finished product—depends upon the whole genotype with many genes interacting in complex ways. And to varying extent the phenotype is also dependent upon environmental influences—including in that term all the forces acting upon an individual from the time that individual is a newly conceived, fertilized ovum (and even before that in some cases).

In summary, what can we conclude about the inheritance of red hair? Present evidence suggests that the gene or genes for red hair are recessive to the gene for nonred, but that many other genes, known and unknown, contribute to determination of the color produced.

Is blond hair a distinct entity, differing from very light brown in its genetic basis? Rife (1967) considered it so, and on the basis of statistical analysis of 22 families concluded that the gene for red hair is dominant to the gene for blond hair.

Other Characteristics of the Hair

Early Graying Observation leads us to conclude that in some family lines, hair turns gray earlier than it does in others. This suggests that the time in life when hair turns gray has a hereditary basis. In gray and white hair the pigment granules (melanin) are no longer present, and air bubbles are frequently found. Little is known of the mode of inheritance, but occasionally a family will exhibit an extreme type of premature graying that seems to follow a definite genetic pattern.

Such a lineage was described by Hare (1929). Six generations of this family are shown in Figure 9.2, in which the black squares and circles represent individuals whose hair started turning white when they were 17 or 18 years of age. In most cases, the hair was completely white by age 25. Inspection of this diagram seems to indicate that the gene is dominant, because the trait does not "skip" (p. 66). In other words, no child showed the trait unless a parent did. As Hare pointed out, it is possible that the gene is recessive, but in that case we would have to assume that all the normal parents of affected offspring were heterozygous. That seems highly improbable; none of the parents were close relatives.

Hair Form Genes for very curly ("kinky," "woolly," "frizzy") hair seem to be dominant to genes for less curly hair and straight hair. Straight hair is circular in cross section; wavy hair is oval, and as hair increases in curliness, the cross sections increase in the flatness of the oval.

The number of genes involved is uncertain. In the early days of human genetics, the Davenports suggested a single pair, the allele for curly being the dominant member, the allele for straight being recessive. Accord-

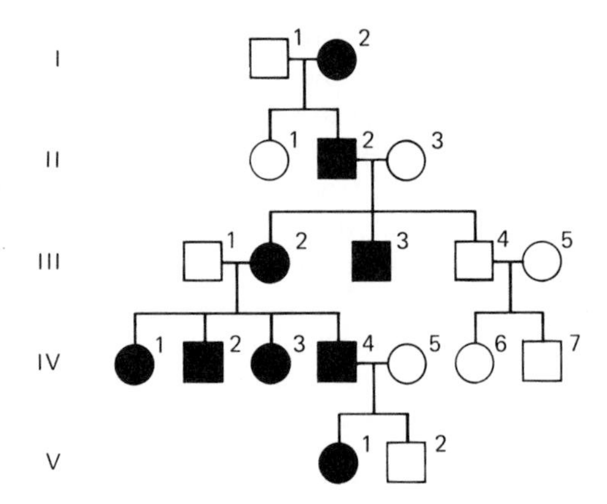

FIG. 9.2 Premature whitening of the hair. Black symbols represent individuals whose hair was white by age 25, except for IV-1 and V-1 whose hair started to whiten when they were in their later teens but "by exceptional care of the hair the development of the abnormal condition was checked." (Redrawn from Hare, H. J. H., 1929. "Premature whitening of the hair," *Journal of Heredity,* **20**:31–32.)

ing to this scheme, wavy-haired individuals are the heterozygotes. Gates (1946) suggested a set of three multiple alleles (p. 50), which we may symbolize as follows:

$$C^1 = \text{curly}$$
$$C^2 = \text{wavy}$$
$$c = \text{straight}$$

In this scheme each allele is regarded as dominant to the one or ones listed below it. This hypothesis helps to account for some of the variety actually found in families. In view of the varying degrees of curliness we observe, we may reasonably expect that several or many genes interacting in various ways may be involved. In the present state of our knowledge, we are probably wisest not to attempt a more definite formulation than the general statement made at the beginning of this section on hair form.

In actual experience, what is the basis for thinking that genes for very curly hair are dominant? The principal evidence is the fact that while two kinky-haired or curly-haired parents may have a straight-haired child, two straight-haired parents are very unlikely to have a curly-haired child.

The degree of curliness of a person's hair may change with age. When this occurs, usually the hair becomes straighter than it was in childhood, although rarely the reverse may be true.

White Forelock In occasional families, some individuals may have an unpigmented spot of skin above the forehead with white hair growing from it. Occasionally such people also have white areas in other parts of the body. The trait is analogous to the white spotting found in various varieties of animals. The gene concerned is said to be dominant, meaning that a child is unlikely to have a white forelock unless a parent does.

Widow's Peak Sometimes the hairline over the forehead dips downward in the midline, forming a rather definite point. The gene concerned is regarded as dominant, because children seldom show the trait unless a parent shows it.

Pattern baldness (See pp. 220–221.)

Mid-digital (Midphalangeal) Hair Some people have hair on the middle segments of their fingers (aside from the thumb), some people do not. This may seem a trifling matter to mention, but, as must be clearly evident from our discussions so far, human traits that are both easily observable and have a simple genetic basis are few. For that reason mid-digital hair has received much attention from both geneticists and anthropologists.

Although the trait is easily observable, it is best seen with a simple magnifier. The unaided eye may fail to see fine hairs, or hair follicles, in cases where the hairs themselves have been worn away by employment of the hands in work.

Since the pioneering research of Danforth (1921), there has been general agreement that people who have *no* hair on the middle segments of any of their fingers are homozygous for a *recessive* gene. But there is less agreement on the question of the number of genes allelic to this recessive gene. The problem is complex because people who have mid-digital hair may have it on one, or two, or three, or all four of their fingers. The simplest concept is that there is one "main gene" for the presence of the hair, and in addition modifying genes that determine which fingers will have hair. On the other hand, Bernstein and Burks (1942) presented statistical evidence in support of the hypothesis that we have here another case of multiple alleles. Simply stated, this is the idea that allelic to the gene for no mid-digital hair is a series of four genes; one gene determines that hair will be present on one finger, another gene that two fingers will have hair, another that three fingers will be hairy, and another that all four fingers will have hair.

Because of the uncertainty concerning the mode of inheritance of mid-digital hair when present, anthropologists have paid attention principally to presence versus absence of such hair, without regard for how many, or which, fingers are hairy. As we have stated, absence of all mid-digital hair has a simple genetic basis. Hence it may be useful in comparative studies of different populations of people. Differences have been found among ethnic groups. For example, Saldanha and Guinsburg (1961) stated that the proportion of people *lacking* mid-digital hair in northern European populations is 20 to 30 percent, while among Mediterranean peoples it is 30 to 50 percent, and among the Japanese, and American Indians, it is 60 to 90 percent. Many studies of the ethnic groups include data on the proportion of people lacking mid-digital hair.

Handedness

In this world of predominantly right-handed people, left-handedness has always aroused interest. Why do some people prefer to use their left hands for actions most people do with their right hands? Is left-handedness hereditary? If so, what is the mode of inheritance?

In attempting to answer such questions we encounter difficulties. In the first place, classification is not as simple as we might expect. While some people are completely right-handed or left-handed in such activities as writing, throwing, using a needle in sewing, using tools and utensils, and so on, other people are mixtures in the sense that they prefer one hand for some actions, the other hand for others. The present writer, for example, is strongly right-handed in most activities, yet he swings an axe or a baseball bat like a left-hander and would find doing otherwise extremely awkward. As a result, investigators must draw rather arbitrary lines in separating left-handers from right-handers.

Second, there is the problem of the extent to which handedness is influenced by environment, including early training of children. Sometimes parents consciously train children to use the right hand, even when a child shows a tendency to prefer to use the left one. But in very young children the training may be more or less unconscious—handing the child objects in such a manner that they are more easily grasped by the right hand than by the left one, for example. And to what extent do children learn which hand to use by imitating their parents?

There is no doubt that environment and training do influence handedness. But it would seem that in addition there are innate tendencies to prefer one hand to the other (a preference that can be observed in lower animals, incidentally). Innate, genetically determined tendencies have been denied by some investigators, however. One reason for this is that identical twins are as likely to differ from each other in handedness as are fraternal twins. Because identical twins have the same genes, shouldn't they always agree in hand preference, if handedness has a genetic basis? As we shall see in Chapter 23, however, there are difficulties in drawing correct conclusions from data on twins.

Study of the relatives of left-handed people suggests, although it does not prove, that the tendency may be of a hereditary nature. Rife (1950) stated that when both parents are right-handed only about 6 percent of the children, on the average, are left-handed. But when one parent is left-handed the percentage rises to 17, and when both parents are left-handed 50 percent of the children are left-handed. It seems unlikely that this trend is all ascribable to effects of training or to imitation of parents.

If there is a genetic basis, what is its nature? The most useful hypothesis seems to be that of Rife (1950), reaffirmed and expanded by Annett (1964). A single pair of genes is postulated; we represent them as *R* for right-handedness and *r* for left-handedness. People with the genotype *RR* are thought to be strongly right-handed, and not easily influenced to change preferences. Similarly, people with the *rr* genotype are thought to be strongly left-handed and not easily influenced to prefer the right hand. In contrast, heterozygotes, *Rr*, are thought to be more variable. They are potentially ambidextrous, and may be ambidextrous in fact. But they are thought to be easily influenced by environment or training, so that many in fact are either right-handed or left-handed. Thus a left-handed person may have gene *R* and transmit it to a child, who may then be right-handed. In this way we can explain how two left-handed parents can have a right-handed child. Also, this hypothesis can account for one identical twin being right-handed, the other left-handed. If both have the genotype *Rr,* gene *R* may produce its phenotypic effect in one twin, gene *r* in the other. It is even possible that the relative positions of the twin fetuses before birth may influence the handedness of each twin.

According to this hypothesis, the gene for left-handedness is not a typical recessive, because under some circumstances it may determine the phenotype of heterozygotes.

Finger and Palm Prints

Everyone is familiar with the importance of finger prints for identifying people, but there is not general knowledge of the value of finger printing to human genetics and anthropology.

The patterns of epidermal ridges on our hands and feet are formed early in embryonic development. By the end of the fourth month of fetal life, we have already acquired the patterns that will stay with us throughout our lives. As we know, no two individuals are exactly alike in the details of these patterns. This applies even to identical (monozygotic) twins. Such twins have strikingly similar patterns, however, and this fact is used in helping investigators to decide whether twins are identical or fraternal (dizygotic). Penrose (1968) told of the value of hand prints in a case where the probable success of a kidney transplant depended on whether or not the proposed donor was the identical twin of the recipient.

Patterns Figure 9.3 shows the principal finger-print patterns and some of the main lines of epidermal ridges on the palm. (Note that these lines are not the flexion creases where the skin folds when we close our hands.) In the figure the thumb (digit I) is shown as having a **whorl,** as is the ring finger (digit IV). The middle finger (digit III) is shown with an **arch.** Arches may be much higher than the one shown. The index and little fingers (digits II and V) are shown as having loops. The index finger has a **radial loop,** meaning that it opens toward the thumb side of the hand, which is also the side on which the wrist at-

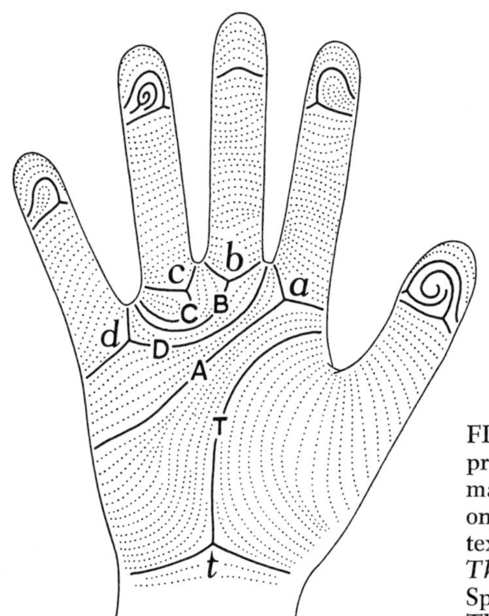

FIG. 9.3 The principal fingerprint patterns, and some of the main lines of epidermal ridges on the palm of the hand. See text. (From Holt, S. B., 1968. *The Genetics of Dermal Ridges.* Springfield, Ill.: Charles C Thomas.)

taches to the radius bone of the forearm. The little finger has an **ulnar loop,** meaning that it opens on the side on which the wrist attaches to the ulna bone. While these are the main types of patterns, numerous combinations and modifications of them occur.

Note that at the base of each loop there is a three-pronged configuration much like a broadly expanded Y. Such a marking is called a **triradius,** and forms an important landmark used in analysis of finger and palm prints. Note that there are two triradii at the base of the whorl on the thumb, and that there is a triradius at the base of each finger (marked *a, b, c,* and *d*) and another just above the wrist (*t*).

Finger and palm prints are frequently included among the traits studied by anthropologists. Peoples differ in the relative frequencies of the different patterns. For example, Chinese and other Eastern peoples have more whorls on their fingers than do Western Europeans, while Africans have fewer whorls (Penrose, 1969). When the relationships of an ethnic group to other groups is unclear, hand prints may help to solve the riddle.

Much effort by many investigators has been expended on attempts to analyze the mode of inheritance of these patterns. There is little doubt that heredity is in fact concerned, although chance and physical forces operating on the fingers as they form in fetal development play large roles. Evidence is afforded by the great similarity of the patterns in identical twins, and by many family studies. Particularly striking was a family studied by Wilder, one of the pioneers in the science of dermatoglyphics, as this type of study is called. Like the hands, the feet also have patterns of epidermal ridges, but usually the lines on the heel run smoothly from side to side; it is unusual to find a loop or whorl on the heel. Wilder discovered a family in which the father and mother had a heel pattern. Their three children showed the pattern, and some of the wife's relatives had vestiges of it (Cummins and Midlo, 1943). Other examples might be cited, all pointing to the influence of heredity. But more exact statements as to the manner of pattern inheritance, in terms of Mendelian genetics, would seem to be unjustified at present.

Ridge Counts The statement just made does not apply to the number of ridges characteristic of a person's finger prints. This number is counted in the manner shown in Figure 9.4. A line is drawn between the center of a loop, for example, and the corresponding triradius. The count is the number of ridges this line crosses. Certain rules are followed in counting more complex patterns than loops (see Holt, 1968). The counts for all ten fingers are totaled to give the **total finger ridge count.**

Statistical evidence indicates that total ridge count is based on simple polygenic inheritance in which the effects of the genes are additive (Holt, 1968). As we have seen (Chap. 8), in this type of inheritance the degree of development of the trait—number of ridges in this case—directly reflects the number of genes contributing. A child receives half of his genes from one parent (half from the other). This is expressed statistically by saying that the correlation coefficient between a parent and a child is one-half or 0.5. Analysis of total ridge-count data indicates that the correlation coefficient between parent and child is almost exactly 0.5. Similarly, the correlation coefficient between two sibs is 0.5; here again, findings for total ridge counts agree with this theoretical expectation.

Special interest attaches to twins in this connection. Because identical (monozygotic) twins have the same genes, the correlation coefficient should be 1.0. Analysis of total ridge-count data yielded a coefficient of 0.95. While this is very close to 1.0, the difference from 1.0 reflects the fact that other factors besides heredity enter into determination of number of ridges. As we expect on theoretical grounds, the correlation coefficient for fraternal (dizygotic) twins is about the same as that for pairs of sibs who are not twins: 0.5 (p. 376).

Thus, the similarity in number of ridges reflects faithfully the proportion of genes shared in common, a fact indicating "that a number of perfectly additive genes are concerned" (Holt, 1968, which should be consulted for further details and elaboration).

FIG. 9.4 Method of counting the ridges comprising a finger print, as applied to a loop. See text. (From Holt, S. B., 1968. *The Genetics of Dermal Ridges.* Springfield, Ill.: Charles C Thomas.)

Hand Patterns in Chromosomal Abnormalities

Many abnormalities with which one may be born—congenital defects—arise from abnormalities in the chromosomes. We shall discuss this in later chapters, but we mention it here because increasing evidence indicates that such chromosomal aberrations can sometimes be detected by studying finger and palm prints, a much simpler method than analyzing the chromosomes (Chap. 20). A newborn infant may show signs of being defective and yet not present sufficiently clear symptoms so that an exact diagnosis can be made. Early diagnosis is desirable, especially if it points to some appropriate treatment that should be started early. As an example, we shall use Down's syndrome.

Down's Syndrome This common type of mental retardation (also called mongolism) is usually caused by having three of the small chromosomes designated as No. 21, instead of the usual two (Fig. 20.7, p. 320). Symptoms are discussed in Chapter 20.

Persons with Down's syndrome tend to have a higher proportion of ulnar loops than do other people. While only 4 percent of the general population have ulnar loops on all ten fingers, 32 percent of Down's syndrome individuals do (Penrose, 1969). In Down's syndrome a radial loop is frequently found on the fourth (ring) finger; it seldom is on normal hands (Penrose, 1969). As shown in Figure 9.3, the base of the palm on the "little-finger" side—the hypothenar area—usually has only smoothly curving epidermal ridges. Some 12 percent of normal people have a large ulnar loop in this area, but 85 percent of individuals with Down's syndrome have such a hypothenar pattern (Walker, 1958).

Frequently in Down's syndrome the triradius labeled *t* in Figure 9.3 is much higher than shown in the figure, and may even be in the center of the palm. Quantitative expression of this fact is obtained by drawing a line from triradius *t* to triradius *a,* and another line from *t* to triradius *d.* The lines form an angle, called the *atd* angle. Penrose studied this trait and found that in some 90 percent of persons with Down's syndrome the sum of the *atd* angles of the two hands is significantly greater than it is in normal individuals (see Holt, 1968; Penrose, 1966).

Other details of ridge patterns have been analyzed, but the above examples suffice to demonstrate that, while no one trait indicates Down's syndrome with certainty, a combination of traits is likely to do so. Penrose (1969) stated: "A correct diagnosis of this syndrome can almost always be made from a good set of prints of the hands and feet."

Although they are not epidermal ridges, the flexion creases in our hands deserve mention in this connection. If most of us examine our palms, we shall see lines (where the skin folds when we close our hands) more or less similar to those in Figure 9.5A. Many persons with Down's syndrome have a crease extending completely across the palm (Fig. 9.5B), the so-called **simian crease.** Because some normal people have such a crease, and because it is common in some other chromosomal anomalies, its presence does not give clear-cut diagnosis of Down's syndrome, but it is one more bit of evidence to be taken into account in making a diagnosis.

Number of X Chromosomes and Number of Ridges

Evidence accumulates that for some reason yet undetermined there is an inverse relationship between the number of X chromosomes and the number of epidermal ridges on the fingers. Females with Turner's syndrome (pp. 268–274) have only one X chromosome in their cells; their total ridge count (p. 130) is found to average about 169, whereas that for normal females, having two X chromosomes, averages about 127 (Holt, 1968). Figure 9.6 illustrates the point in terms of single thumb prints, and includes a female who had *three* X chromosomes, with a further reduction in ridge count. We should emphasize, however, that there is broad overlapping in ridge counts: Some normal women have high ridge counts and some women with Turner's syndrome have low counts. So, as with our previous examples, this trait is not diagnostic

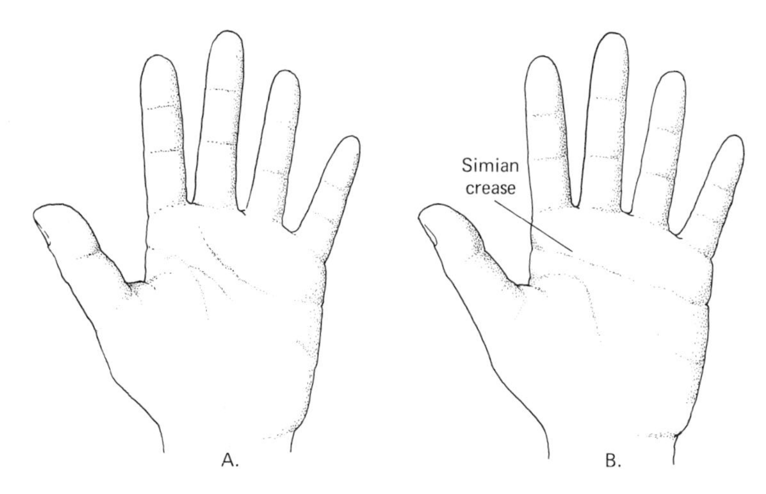

FIG. 9.5 Flexion creases in the palm, of (A) an average normal person, and (B) a person with the simian crease.

of Turner's syndrome by itself but forms evidence to be taken into account in making a diagnosis.

A similar trend is found in males. Normal males have one X chromosome in each cell; their total ridge count averages about 147. Males with Klinefelter's syndrome (pp. 274–276) have two X chromosomes; their count averages only about 118 (Holt, 1968).

Our brief excursion into dermatoglyphics has been most incomplete. Interested readers will find the books by Cummins and Midlo (1943), and Holt (1968) particularly informative.

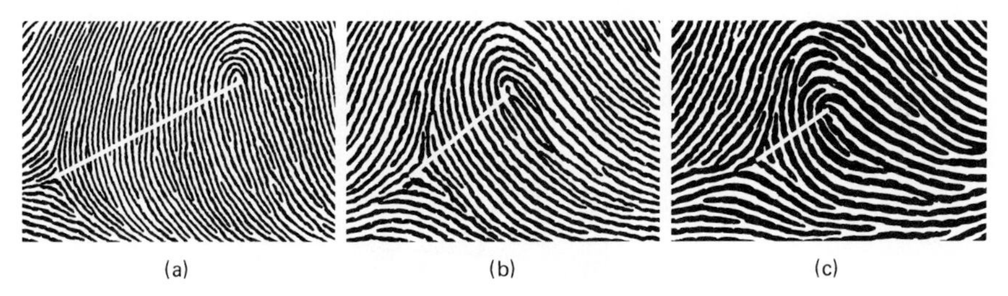

FIG. 9.6 Inverse relationship between number of X chromosomes and number of ridges comprising finger print patterns. (a) Thumb print of an XO individual (Turner's syndrome), showing an abnormally large number of ridges. (b) Thumb print of a normal woman with two X chromosomes. (c) Thumb print of a woman having three X chromosomes (note that the loop is composed of only eight ridges). (From Penrose, L. S., 1969. "Dermatoglyphics," *Scientific American,* **221**:72–84. Copyright © 1969 by Scientific American, Inc. All rights reserved.)

Anomalies of Fingers and Toes

Brachydactyly People with brachydactyly have all digits abnormally shortened. The middle joint (phalanx) of fingers and toes is short or rudimentary and is frequently fused to the terminal joint. When fusion occurs, the result is that the finger bends in only one place. This is especially likely to be true of the little finger (or toe)—digit V. Figure 9.7 shows the hands of a brachydactylous ten-year-old boy.

We have described what we may call classic brachydactyly (technically known as type A1 or Farabee type); it was the first human trait recognized as dependent on a Mendelian dominant gene. (Recall that alkaptonuria was the first human trait shown to be inherited on the basis of a recessive gene, p. 53.) Figure 9.8 shows a pedigree of brachydactyly. Note the hallmarks of dominant inheritance, especially the fact that the affected children are numerous and always have an affected parent. Evidently the gene has high penetrance (p. 70).

There are other types of brachydactyly. The fingers may be shortened in various ways, and in some types only certain of the digits are shortened. McKusick (1971) listed them all as being dependent on dominant genes.

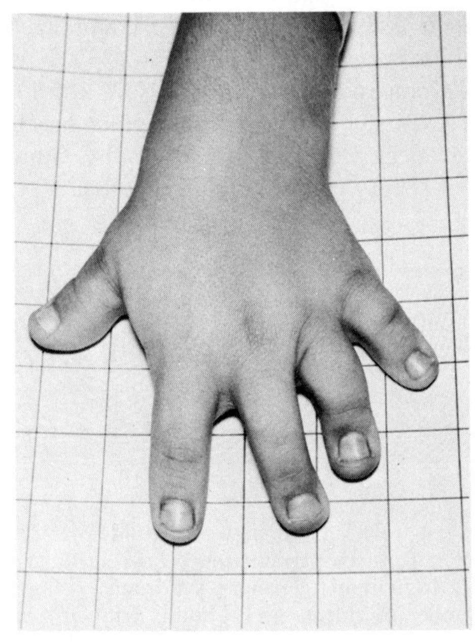

FIG. 9.7 Hand of a ten-year-old boy showing brachydactyly. (Courtesy of Dick Hoefnagel, M. D. From Hoefnagel, D., and P. S. Gerald, 1966. "Hereditary brachydactyly," *Annals of Human Genetics,* **29**: 377–382.)

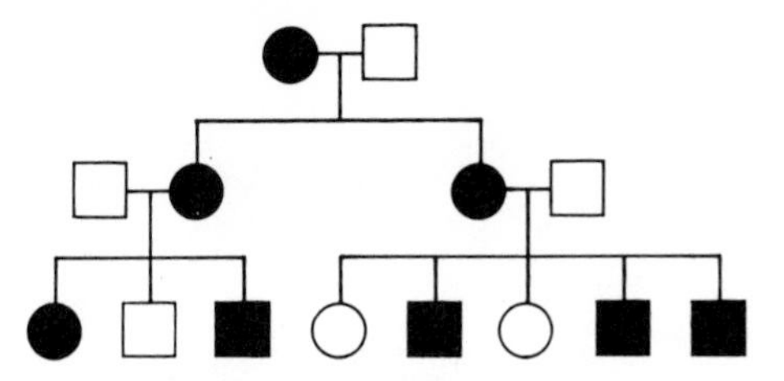

FIG. 9.8 Three generations of a family having hereditary brachydactyly. Black symbols represent brachydactylous individuals. Note the hallmarks of dominant, autosomal inheritance. (Redrawn from Hoefnagel, D., and P. S. Gerald, 1966. "Hereditary brachydactyly," *Annals of Human Genetics,* **29**:377–382.)

Polydactyly People with this anomaly usually have six digits on each of the hands and feet, although in some cases the trait may be confined to either the hands or the feet. In the most common type (postaxial), the extra finger is on the little-finger side of the hand. The extra digit may be well formed, with its own articulation to the hand, or it may be vestigial and more or less joined to the little finger. Study of families indicates that the trait depends upon a dominant gene (or genes), but with so much variation in expression of the trait that we may expect the inheritance is not simple.

There are also various types of *preaxial* polydactyly, in which the duplication is on the thumb side of the hand. These types show indications of being dependent on dominant genes (McKusick, 1971).

Syndactyly Under the term syndactyly we include a variety of conditions characterized by varying degrees of joining of the fingers, and sometimes toes. At times only two fingers are joined together, while at the other extreme are hands in which most of the fingers are joined so that the hand presents a "lobster-claw" appearance. The joining may involve actual fusion of the bones of two or more fingers. McKusick (1971) stated: "All are inherited as autosomal dominant traits and within any pedigree there is uniformity of type of syndactyly, allowing for the variation characteristic for dominant traits."

Distal Hyperextensibility of the Thumb (See pp. 157–159.)

Tongue Rolling and Folding

The ability to roll up the edges of the tongue to form a trough (Fig. 9.9) has long been regarded as dependent upon a dominant gene.

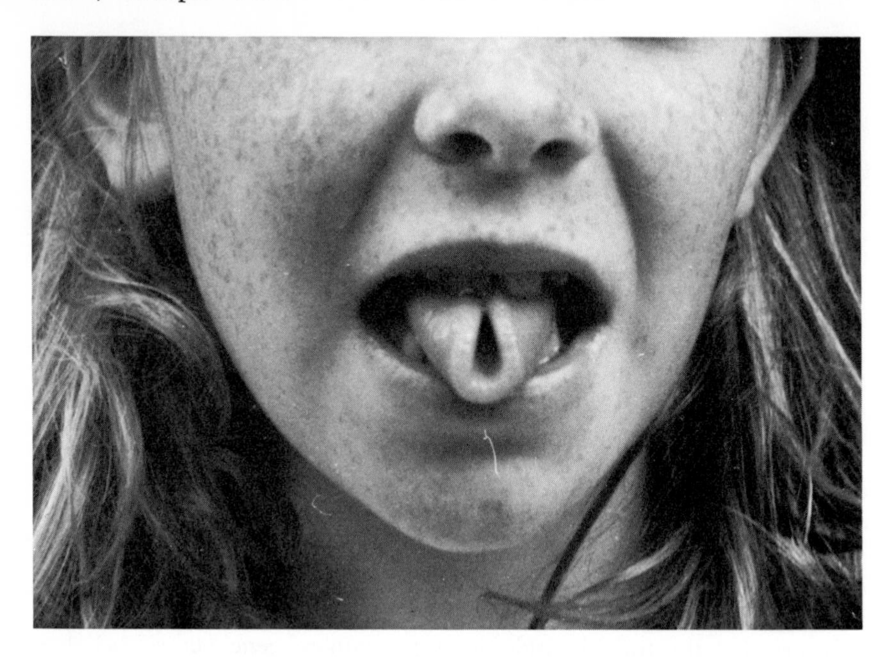

FIG. 9.9 Ability to roll the tongue lengthwise without using the lips. (Courtesy of Mr. Howard H. Hirschhorn. From Hirschhorn, H. H., 1970. "Transmission and learning of tongue gymnastic ability," *American Journal of Physical Anthropology,* **32**:451–454.)

Matlock (1952) studied 33 pairs of identical (monozygotic) twins and found that in seven pairs (21.2 percent) one twin had the ability but the other did not. Because identical twins have the same genotype, this finding suggested that other factors may be involved in addition to genetic constitution. It is still unclear to what extent imitation and practice may be concerned. On the basis of experience, Hirschhorn (1970) emphasized that "ample time and example must be provided for learning tongue tricks, even if they are genetically determined." Among the other such tricks are (a) ability to fold the tongue back on itself, (b) ability to twist the tongue so that the apex forms a 90° angle with the rear portion, (c) ability to fold the tongue with the tip up and the sides curled up (an extreme form of this results in formation of a "cloverleaf" form). Hirschhorn found that in some cases practice enabled a person to do a trick he was at first unable to do. Nevertheless, some individuals seem unable to do some or all of these things no matter how hard they try. This in itself suggests genetic differences. Hirschhorn's study warns that data collected quickly—by single contacts in many cases—may not be adequate to reveal the full genetically determined potentialities of individuals.

Earlier studies have yielded conflicting answers to the question of whether rolling and folding abilities are dependent upon separate genes. Hirschhorn's study suggested that they may be.

Ear Lobes

When we examine people's ears, we find great variety in the details of structure. Some of the variations are characteristic of certain families, such as the "Mozart ear" possessed by the composer and his father. Many of the variations have interested geneticists, but the trait most studied has related to the nature of the ear lobe. Many people have a free-hanging lobe extending below the lowest point at which the ear is attached to the head—like the girl in Figure 9.9. In other people, the lowest point of the ear is the point at which it attaches to the cheek. Such people are said to have attached or adherent lobes.

Some, though not all, early studies suggested that there is a dominant gene for free lobes, its recessive allele causing the lobes to be attached. As early as 1937, however, Wiener pointed out that we cannot divide people into the two categories so neatly, that a continuous series of gradations is found between the two extremes. He recognized two intermediate types. His limited data indicated that when both parents have attached lobes, not all children have attached lobes—as one would expect with simple recessive inheritance—but some have intermediate lobes. He postulated polygenic inheritance to explain the varying degrees of attachment, although he recognized the possibility that a "main" pair of genes with additional modifying genes might be involved.

Although Lai and Walsh (1966) classified people into only the two types, "attached" and "free," their results were similar to Wiener's in that they found that when both parents have attached lobes, some of the children have lobes that they classified as free. Some other workers, for example, Dutta and Ganguly (1965), have classified lobes into "free," "intermediate," and "attached," although they encountered gradations within the intermediate group.

Most recent workers recognize that the mode of inheritance is not yet well established, but favor the polygenic hypothesis.

Despite the uncertainty concerning its exact genetic basis, ear lobe type is a trait much used by physical anthropologists as they study the various peoples of the world. Racial differences are found in the proportions of the types. For example, Negroes and Mongolian peoples have a higher frequency of attached lobes than do Caucasians.

Hairy Ears (See pp. 230–232.)

PTC Tasting

People differ in the sensitivity with which they taste a variety of substances. Such individual differences have been tested for many chemicals and drugs, but the substance that has received most attention is phenylthiocarbamide (phenylthiourea), conveniently termed PTC. (The related compound propylthiouracil—PROP—is preferred by some workers.) Since the accidental discovery that PTC has a bitter taste for some people but is relatively tasteless for others, it has been the subject of extensive research.

First-Approximation Testing The differing sensitivities of people to the taste of PTC is most easily demonstrated with the use of "tasting paper." This is prepared by dipping filter paper (of the type used in chemistry laboratories) into a solution of PTC (e.g., a 0.5 percent solution in alcohol or acetone). The paper is dried and cut into small squares. Each square retains a bit of the chemical, and if put in the mouth, and perhaps chewed a little, will enable an individual to determine whether or not he is a "taster" of PTC.

Many investigations have indicated that tasting versus nontasting of PTC form a pair of alternative traits dependent upon a pair of alleles, the gene for tasting being dominant to the gene for nontasting. Accordingly, genotypes are as follows:

TT = taster, homozygous
Tt = taster, heterozygous
tt = nontaster

When testing is done in this manner, a clear demonstration of simple Mendelian inheritance is obtained. In fact, it is probably the most easily demonstrated case of simple Mendelian inheritance we have in human genetics. That is one reason it has received so much attention from geneticists and anthropologists. The latter find that peoples differ in the relative proportions of tasters and nontasters, and that fact gives some insight into the genetic basis of racial differences. Among people of European ancestry, for example, some 25 to 35 percent of the population are nontasters, while the percentage of nontasters is much lower (less than 15 percent) among Negroid and Mongolian peoples (Boobphanirojana, Chetanasilpin, and Saengudom, 1970).

What is the physiological difference between tasters and nontasters? Investigation indicates that the difference is in the enzymatic activity of the saliva. The saliva of nontasters oxidizes the chemical (PROP in this investigation) twice as fast as does the saliva of tasters (Fischer,

Griffin, and Pasamanik, 1963). The result would seem to be that in the mouths of nontasters, the substance is destroyed before it stimulates the taste buds.

More Discriminating Tests Whenever we test people with PTC paper, we find great differences among tasters in how bitter the substance seems. For some individuals it is extremely bitter, for others only slightly so. Such individual differences merit investigation, but we have no means of measuring the degrees of bitterness different people experience. We can, however, measure *thresholds:* the least amount of PTC (or PROP) that a person can taste. For this reason, much of the modern research in the field has been done with solutions of the substances, rather than with paper.

Most such investigations employ the method of Harris and Kalmus (1949) or some modification of it. This involves preparing a graded series of dilutions ranging from very concentrated to extremely dilute. Then each individual is tested with one dilution after another to see which one is the most dilute he can taste—this is his threshold. (As the test is usually run, the individual is presented with eight small glasses, four containing water, four the dilution being tested at the moment. They are arranged in random order, and he is asked to separate them correctly into the two groups of four. If he can do so, he is then tested with a more dilute solution. This is continued until a dilution is reached that he cannot distinguish accurately from plain water.)

Some of the results of the first investigation using this method are shown in Figure 9.10. The curves are typical of many obtained in subsequent investigations. The numbers along the base line represent the successive dilutions, the most concentrated solution (least dilution) being at the left. The vertical axis shows the percentages of tested individuals for whom each dilution was the threshold dilution.

As we look at the graph, we note that no one tested was completely unable to taste PTC when the concentration was strong enough. Then we note that the curve is bimodal—it has two peaks, one peak much smaller than the other. People for whom the thresholds occurred with the first five or six dilutions were relatively insensitive, and probably would be classified as nontasters by the tasting paper technique. People with lower thresholds (from dilution 6 or 7 through dilution 14) are the "tasters." Because no one was completely lacking in ability to taste PTC, however, the terms "insensitive taster" (rather than "nontaster") and "sensitive taster" are perhaps more appropriate (e.g., Fischer et al., 1963). But the terms "taster" and "nontaster" are so generally used that they will probably not be supplanted.

The fact that the curve is bimodal is usually regarded as evidence

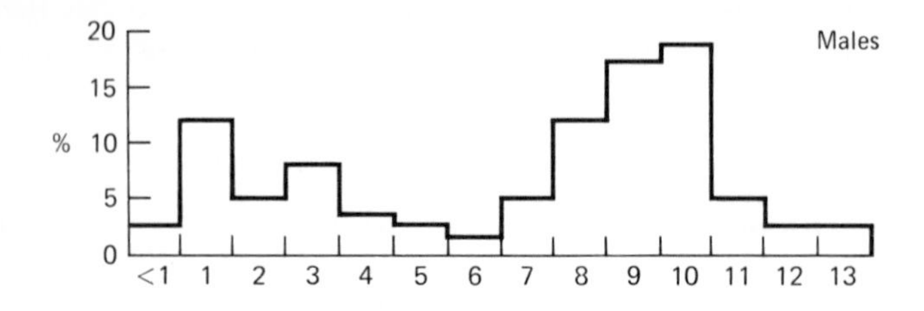

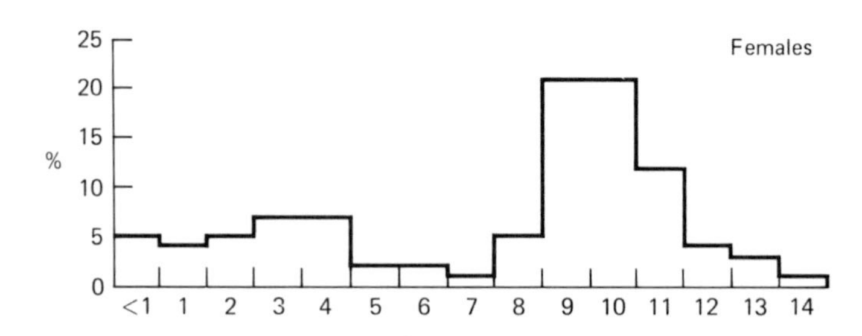

FIG. 9.10 PTC tasting thresholds for 100 females and 114 males between the ages of 20 and 59. Numbers along the base line represent the successive dilutions (see text). The vertical axis shows the percentages of tested individuals for whom each dilution was the threshold dilution. (Redrawn from Harris, H., and H. Kalmus, 1949. "The measurement of taste sensitivity to phenylthiourea (P.T.C.)," *Annals of Eugenics [Human Genetics]*, **15**:24–31.)

in support of the single-gene-pair mode of inheritance outlined above. According to this view, *T* and *t* would be the "main genes," and variations in sensitivity within the taster and nontaster categories could be explained by the influence of modifying genes, insofar as such variations have a genetic basis. By contrast, we might note that the curve obtained by plotting variations in sensitivity to the taste of quinine is unimodal, approaching a normal frequency curve. This is interpreted to indicate polygenic inheritance (Glanville and Kaplan, 1965).

This method of testing, while not so simple as the paper technique, is being employed by anthropologists. Some investigators are finding among certain groups "extremely sensitive tasters" whose thresholds are substantially lower than any usually found. For such situations, more complex genetic explanations may be necessary (Boobphanirojana et al., 1970).

Factors other than one's genes affect taste sensitivity. Some studies indicate that females have greater sensitivity than do males, but the investigations of Kaplan, Glanville, and Fischer (1965) led them to conclude that the principal difference lay in the fact that men on the average are heavier smokers than are women. They found that heavy smoking reduced taste sensitivity, particularly from age 20 onward. Among nonsmokers, they did not find difference in sensitivity between the sexes, or decrease in sensitivity with increasing age, as reported by some previous workers. In some women, taste sensitivity increases at the time of menstruation (Glanville and Kaplan, 1965). Differences in tasting sensitivity have been found to be associated with various pathological conditions (Fischer et al., 1963). As with most hereditary traits, the actual phenotype in taste sensitivity is determined in part by environmental influences.

A Final Comment

Thoughtful readers will have noted a recurrent theme throughout this chapter. In case after case the traits discussed were at one time thought to be inherited in simple Mendelian manner, and were classified as "dominant" or "recessive." Then further research revealed that the traits were actually complex, and that the genetic basis was correspondingly complicated. No wonder students of human genetics prefer to work with enzymes and antigens, in which the path from a gene to its product is much shorter, more direct, and less influenced by other genes and by environment!

Additional Traits Many readers will have in mind some particular trait about which they are curious—some trait not discussed in this chapter or elsewhere in the book. Where can they look up such traits? The first place to look is McKusick's *Mendelian Inheritance in Man* (1971). The only book of its kind, it is an annotated catalog of 1876 human traits, a large proportion of them pathological. A brief summary of current ideas as to the inheritance of each trait is accompanied by references to research publications, especially recent ones, concerning it. Gates' *Human Genetics* (1946) offers an almost encyclopedic coverage of the older literature. And we must not overlook the indexing and abstracting services: *Biological Abstracts* and, for those to whom a medical library is accessible, *Index Medicus* and *Excerpta Medica.* The latter has a section devoted to human genetics.

Problems

1. In terms of the hypothetical model of brown hair-color inheritance (p. 121), how many effective alleles would be possessed by the darkest-haired child to be expected from the mating:

$$B^1B^1B^2B^2b^3b^3b^4b^4 \times B^1B^1B^2b^2B^3b^3b^4b^4$$

How many effective alleles would be possessed by the lightest-haired child? Would the darkest-haired child have darker hair than either parent?

2. On p. 122 we included the presumed recessive gene for red hair in our model, and suggested a hypothetical genotype for auburn-haired people. Imagine two such auburn-haired people married to each other. Give the hypothetical genotype of the reddest-haired child they could produce; of the darkest-haired child they could produce.
3. According to the multiple-allele hypothesis of hair-form inheritance (p. 125), the genotypes of people with the various hair forms would be as follows:

   curly: C^1C^1, C^1C^2, C^1c
   wavy: C^2C^2, C^2c
   straight: cc

   Employing this hypothesis, give the hair forms children from the following mating would be expected to have: $C^1c \times C^2c$.
4. Following the hypothesis of Rife concerning the inheritance of handedness (p. 127), give the probable genotypes of two left-handed parents who have produced a right-handed child. Could they also produce a left-handed child?
5. Two parents who are brachydactylous produced a son without the trait. Give the genotypes of the parents (use the gene symbols B and b). Suppose that the nonbrachydactylous son married a normal woman who had a brachydactylous sister. What is the probability that their first child will be brachydactylous?
6. Following the suggestion (p. 140) that the genes T and t concerned with tasting PTC are subject to the action of modifying genes, we may imagine that there are four pairs of these: M^1 and m^1, M^2 and m^2, M^3 and m^3, M^4 and m^4, and that the tasting threshold is directly proportional to the number of effective alleles ("capital-letter genes").

   A taster with a low threshold may have the genotype $TtM^1m^1m^2m^2m^3m^3m^4m^4$, while a taster with a higher threshold may have the genotype $TtM^1m^1M^2m^2M^3m^3M^4m^4$. Suppose that these two people are married to each other. Give the genotypes of the following children of theirs:
   (a) a nontaster child having the highest threshold possible;
   (b) a nontaster child having the lowest threshold possible;
   (c) a taster child having the highest threshold possible;
   (d) a taster child having the lowest threshold possible.

References

Annett, M., 1964. "A model of the inheritance of handedness and cerebral dominance," *Nature,* **204**: 59–60.

Bernstein, M. M., and B. S. Burks, 1942. "The incidence and Mendelian transmission of mid-digital hair in man," *Journal of Heredity,* **33**:45–53.

Boobphanirojana, P., M. Chetanasilpin, and C. Saengudom, 1970. "Phenylthiocarbamide taste thresholds in the population of Thailand," *Humangenetik,* **10**:329–334.

Cummins, H., and C. Midlo, 1943. *Finger Prints, Palms and Soles.* Philadelphia: The Blakiston Company. (Reprinted 1962, New York: Dover.)

Danforth, C. H., 1921. "Distribution of hair on the digits in man," *American Journal of Physical Anthropology,* **4**: 189–204.
Dutta, P., and P. Ganguly, 1965. "Further observations on ear lobe attachment," *Acta Genetica et Statistica Medica,* **15**: 77–86.
Fischer, R., F. Griffin, and B. Pasamanik, 1963. "The perception of taste: Some psychophysiological, pathophysiological, pharmacological and clinical aspects." In Hoch, P. H., and J. Zubin (eds.). *Psychopathology of Perception* [Proceedings, 53rd annual meeting, American Psychopathological Association], pp. 129–163. New York: Grune & Stratton.
Gates, R. R., 1946. *Human Genetics,* Vols. I and II. New York: The Macmillan Company.
Glanville, E. V., and A. R. Kaplan, 1965. "The menstrual cycle and sensitivity of taste perception," *American Journal of Obstetrics and Gynecology,* **92**: 189–194.
Hare, H. J. H., 1929. "Premature whitening of the hair," *Journal of Heredity,* **20**: 31–32.
Harris, H., and H. Kalmus, 1949. "The measurement of taste sensitivity to phenylthiourea (P.T.C.)," *Annals of Eugenics* [*Human Genetics*], **15**: 24–31.
Hirschhorn, H. H., 1970. "Transmission and learning of tongue gymnastic ability," *American Journal of Physical Anthropology,* **32**: 451–454.
Hoefnagel, D., and P. S. Gerald, 1966. "Hereditary brachydactyly," *Annals of Human Genetics,* **29**: 377–382, 5 plates.
Holt, S. B., 1968. *The Genetics of Dermal Ridges.* Springfield, Ill.: Charles C Thomas.
Kaplan, A. R., E. V. Glanville, and R. Fischer, 1965. "Cumulative effect of age and smoking on taste sensitivity in males and females," *Journal of Gerontology,* **20**: 334–337.
Lai, L. Y. C., and R. J. Walsh, 1966. "Observations on ear lobe types," *Acta Genetica et Statistica Medica,* **16**: 250–257.
Matlock, P., 1952. "Identical twins discordant in tongue-rolling." *Journal of Heredity,* **43**: 24.
McKusick, V. A., 1971. *Mendelian Inheritance in Man,* 3rd ed. Baltimore: The Johns Hopkins Press. (Note added in proof: a 4th edition was published in 1975.)
Neel, J. V., 1943. "Concerning the inheritance of red hair," *Journal of Heredity,* **34**: 93–96.
Nicholls, E. M., 1968. "Microspectrophotometry in the study of red hair," *Annals of Human Genetics,* **32**: 15–26.
Penrose, L. S., 1966. *Down's Anomaly.* Boston: Little Brown & Company.
Penrose, L. S., 1968. "Medical significance of finger-prints and related phenomena," *British Medical Journal,* **2**: (No. 5601), 321–325.
Penrose, L. S., 1969. "Dermatoglyphics," *Scientific American,* **221**: 72–84.
Reed, T. E., 1952. "Red hair colour as a genetical character," *Annals of Eugenics,* **17**: 115–139.
Rife, D. C., 1950. "An application of gene frequency analysis to the interpretation of data from twins," *Human Biology,* **22**: 136–145.
Rife, D. C., 1967. "The inheritance of red hair," *Acta Geneticae Medicae et Gemellologiae,* **16**: 342–349.
Saldanha, P. H., and S. Guinsburg, 1961. "Distribution and inheritance of middle phalangeal hair in a white population of São Paulo, Brazil," *Human Biology,* **33**: 237–249.

Singleton, W. R., and B. Ellis, 1964. "Inheritance of red hair for six generations," *Journal of Heredity,* **55**:261, 266.

Walker, N. F., 1958. "The use of dermal configurations in the diagnosis of mongolism," *Pediatric Clinics of North America,* May, pp. 531–543.

Whitney, D. D., 1942. *Family Treasures.* Lancaster, Pa.: The Jaques Cattell Press. (Contains many fine photographs of human hereditary traits.)

Wiener, A. S., 1937. "Complications in ear genetics," *Journal of Heredity,* **28**:425–426.

10

Mendelian Thinking Applied to Populations

Some people may object to use of the term "Mendelian thinking" as a title for the subject matter to be presented in this chapter. Admittedly, Mendel did not think, or at least write, about the subject we shall discuss. However, what we mean by "Mendelian thinking" is the operation of the laws of probability in genetic phenomena. In Chapter 2 we emphasized the point that they operate (a) in the production by heterozygotes of two equally numerous types of germ cells, and (b) in the production of fertilized ova, when two equally numerous kinds of sperm combine with two equally numerous kinds of ova. We saw that application of the laws of probability at these two points results in the classic 3:1 (or 1:2:1) ratio.

Random Mating

Now we wish to apply the laws of probability to a third point: the arrangement of matings. Mendel arranged his matings of peas very carefully, leaving nothing to chance. Suppose that matings *are* left to chance; what may we then expect?

In the interest of clarity, let us talk in terms of a specific example. In Figure 2.1, p. 15, we showed a cross between homozygous black guinea pigs and white ones. The F_1 offspring are black and heterozygous. When they are mated among themselves, the F_2 offspring are ¼ *BB* (homozygous black), ²⁄₄ *Bb* (heterozygous black), and ¼ *bb* (homozygous white).

Suppose that the experimenter does not arrange the matings of these F_2 offspring, but lets the guinea pigs themselves do the choosing. What will be the result?

As we approach this question, we shall make two simplifying assumptions at the outset. The first is that all three genotypes (*BB, Bb,* and *bb*) are equally healthy and fertile. This is probably true in this case, but it is not true of all examples we might cite.

Our second assumption is that in their matings the guinea pigs have no preference based on color. If black guinea pigs preferred to mate with black ones, and white guinea pigs with white ones, this would be an example of **assortative mating.** Assortative mating frequently occurs, as we shall see later, but let us assume that it is not present in the case under consideration. The opposite of assortative mating is **random mating,** or **panmixis.** Let us assume that guinea pigs are panmictic as far as color is concerned, no preference being shown.

Suppose that we have a large group of these F_2 guinea pigs in the proportions stated above—¼ *BB,* ²⁄₄ *Bb,* ¼ *bb* (and that these proportions apply to both sexes). Now we let nature take its course. How will the matings occur if breeding is truly random? If it is, the chance that a given male will mate with a black female or a white female is determined by the relative frequencies of these two kinds of females in the population. There are three times as many black females as there are white ones. Hence the chance that a given male will mate with a black female is ¾, and the chance that he will mate with a white female is ¼. In terms of genotypes, the chance that a given male will mate with a *BB* female is ¼, that he will mate with a *Bb* female is ²⁄₄, that he will mate with a *bb* female is ¼. And, of course, we can turn the whole thing around and state that the chance that a given female will mate with a *BB* male is ¼, that she will mate with a *Bb* male is ²⁄₄, that she will mate with a *bb* male is ¼.

The various types of matings and their expected relative frequencies can be shown by a checkerboard diagram:

		Females		
		¼ *BB*	²⁄₄ *Bb*	¼ *bb*
	¼ *BB*	¹⁄₁₆ *BB* × *BB*	²⁄₁₆ *BB* × *Bb*	¹⁄₁₆ *BB* × *bb*
Males	²⁄₄ *Bb*	²⁄₁₆ *Bb* × *BB*	⁴⁄₁₆ *Bb* × *Bb*	²⁄₁₆ *Bb* × *bb*
	¼ *bb*	¹⁄₁₆ *bb* × *BB*	²⁄₁₆ *bb* × *Bb*	¹⁄₁₆ *bb* × *bb*

(The fraction in each space is obtained by multiplying the fraction at the head of that column by the fraction at the left of that horizontal row.)

These different matings and their relative frequencies are extracted from the checkerboard and listed as the first column of Table 10.1. This table gives the expectation with regard to the offspring, assuming as we do that all matings are equally fertile. If they are, the *BB* × *BB* matings, for example (first horizontal row in the table), will be expected to produce 1/16 of the offspring, and these will all be *BB*, as shown. Again, 4/16 of the matings will be *BB* × *Bb* (second horizontal row), but in this case half of the offspring (2/16 of the total) will be *BB*, half will be *Bb*. And so on for the remainder of the table (note the 1:2:1 ratio from the *Bb* × *Bb* matings).

From the totals at the bottom of Table 10.1 we note that the offspring, like their parents, will be expected to be 1/4 *BB*, 2/4 *Bb*, 1/4 *bb*. When random mating occurs in a population and the relative frequencies of the different genotypes remain unchanged from generation to generation, we say that the population is in **genetic equilibrium.** In this case, equilibrium depends on the laws of probability operating in the arrangement of matings as well as in the production of germ cells and the union of germ cells to produce fertilized ova (zygotes). In other cases, equilibrium may result from the operation of other forces, as we shall see later.

Let us now simplify our methods of computation somewhat. We started with females occurring in the proportions 1/4 *BB*, 2/4 *Bb*, 1/4 *bb*. Let us concentrate on the ova they produce. The *BB* females will produce only ova containing *B*, and these will constitute one-fourth of the total ova produced. The *Bb* females will produce *B*-containing and *b*-containing ova in equal numbers; because they produce two-fourths of the total number of ova, they will contribute one-fourth of the total as *B*-containing ova and one-fourth as *b*-containing ones. Finally, the *bb* females will produce only *b*-containing ova, which will constitute one-fourth of the whole number of ova. In summary, in this population, half of the ova contain *B* (one-fourth from *BB* females plus one-fourth from *Bb* females)

TABLE 10.1 Random mating in a population composed of 1/4 *BB*, 2/4 *Bb*, 1/4 *bb*.

Matings	Offspring		
	BB	*Bb*	*bb*
1/16 *BB* × *BB*	1/16	0	0
4/16 *BB* × *Bb*	2/16	2/16	0
2/16 *BB* × *bb*	0	2/16	0
4/16 *Bb* × *Bb*	1/16	2/16	1/16
4/16 *Bb* × *bb*	0	2/16	2/16
1/16 *bb* × *bb*	0	0	1/16
Totals	4/16 = 1/4	8/16 = 2/4	4/16 = 1/4

and half of the ova contain *b* (one-fourth from *Bb* females plus one-fourth from *bb* females).

The parental males also occur as ¼ *BB*, 2/4 *Bb*, ¼ *bb*. By the same line of reasoning, half of the sperm will contain *B*, half will contain *b*.

So when the ova and sperm form zygotes, we can expect the following:

		Ova	
		½ *B*	½ *b*
Sperm	½ *B*	¼ *BB*	¼ *Bb*
	½ *b*	¼ *Bb*	¼ *bb*

The offspring (in the body of the checkerboard) are immediately seen to be ¼ *BB*, 2/4 *Bb*, ¼ *bb*, the same result we obtained by more complex diagrams before.

The Hardy-Weinberg Formula

In general, we can speak of the total of all genes in ova and sperm produced by the parents in a population as constituting the **gene pool** of that population. This is a very useful concept, which we shall use frequently. In our present example, half of the genes in the gene pool are *B* and half are *b* (here we need not distinguish between *B* genes contained in ova and *B* genes contained in sperm; the same applies to *b* genes). Hence, we can generalize and summarize by saying that in a population whose gene pool is composed of ½ *B* genes and ½ *b* genes, the offspring will be expected to occur in the proportions of ¼ *BB*, 2/4 *Bb*, ¼ *bb* (and the population is in equilibrium).

Another way of expressing this is as follows:

Let p = the relative frequency of gene *B*
q = the relative frequency of gene *b*

Then $(p + q)^2$ = the expected frequencies of the different types of offspring.

In our present example:

$$
\begin{aligned}
p &= \tfrac{1}{2} \\
q &= \tfrac{1}{2} \\
(p+q)^2 &= p^2 + 2pq + q^2 \\
&= (\tfrac{1}{2})^2 + 2\tfrac{1}{2} \times \tfrac{1}{2} + (\tfrac{1}{2})^2 \\
&= \tfrac{1}{4} + \tfrac{2}{4} + \tfrac{1}{4}
\end{aligned}
$$

Note that p^2 is *BB*, $2pq$ is *Bb*, q^2 is *bb*. So we obtain at once ¼ *BB*, 2/4 *Bb*, ¼ *bb*.

Parenthetically, we may note that this formula is simply an algebraic way of expressing what is shown in the little checkerboard above. In that diagram, the sperm are of two types: $\frac{1}{2}\ B + \frac{1}{2}\ b$. In terms of our formula this is "$p + q$." In the same way, the ova are of two types: $\frac{1}{2}\ B + \frac{1}{2}\ b$. This also is "$p + q$." In filling in the checkerboard we did diagrammatically the equivalent of multiplying "$\frac{1}{2}\ B + \frac{1}{2}\ b$" by "$\frac{1}{2}\ B + \frac{1}{2}\ b$," or in other words of multiplying "$p + q$" by "$p + q$," which equals, of course, $(p + q)^2$.

The application of the squared binomial, $(p + q)^2$, to panmictic populations was stated independently and almost simultaneously by two men, Hardy and Weinberg (Hardy, 1908; Weinberg, 1908). Hence it has come to be referred to as the **Hardy-Weinberg law** or **formula.**

We have introduced the Hardy-Weinberg law by discussing a hypothetical population of guinea pigs having a gene pool in which the relative frequency of the dominant gene and the relative frequency of the recessive gene are equal—in which $p = \frac{1}{2}$ and $q = \frac{1}{2}$, in other words. Note that $p + q = 1$. This must necessarily be so; the number of dominant genes plus the number of recessive genes must equal the total number of genes of that allelic pair present in the gene pool.

However, p and q need not be equal; in fact, they probably seldom are equal in actual populations. Suppose, for example, that $p = 0.9$ and $q = 0.1$; in a panmictic population of guinea pigs having this gene pool, what proportions of the different types of offspring can be expected?

$(p + q)^2$	$=$	p^2	$+$	$2pq$	$+$	q^2
		$(0.9)^2$		$2(0.9)(0.1)$		$(0.1)^2$
		0.81		0.18		0.01
Answer:		81% *BB*		18% *Bb*		1% *bb*

Similarly, we can determine expectations with regard to any values of p and q we wish.

However, in actual investigations the situation usually is reversed. We know the phenotypes, and if we are fortunate the genotypes, of individuals in a population, and wish to determine the constitution of the gene pool—the values of p and q.

For an example, let us return to the MN blood types where, as we saw (pp. 81–84), we have the advantage of being able to distinguish heterozygotes from homozygotes by serological tests. An investigator tested 1000 inhabitants of the city of Frankfurt am Main. His findings were 52.9 percent of type MN, 27.0 percent of type M, 20.1 percent of type N (data of Laubenheimer, from Wiener, 1943–1962). Translating the data into genotypes, we have

$$27.0\% \text{ or } 0.270 = L^M L^M$$
$$52.9\% \text{ or } 0.529 = L^M L^N$$
$$20.1\% \text{ or } 0.201 = L^N L^N$$

In this group of 1000 people, what is the relative frequency of gene L^M? Of gene L^N?

$$p = \text{frequency of } L^M$$
$$q = \text{frequency of } L^N$$

Then p is composed of all the genes produced by type M people plus half of the genes produced by type MN people. Similarly, q is composed of all the genes produced by type N people plus the other half of the genes produced by type MN people. Thus,

$$p = 0.270 + \frac{0.529}{2} = 0.5345 = \text{approx.} \quad 0.53$$

$$q = 0.201 + \frac{0.529}{2} = 0.4655 = \text{approx.} \quad 0.47$$

This means that in the gene pool of these 1000 people, about 53 percent of the genes are L^M genes and 47 percent are L^N genes.

Having determined this, we can now ask a further question: Do these 1000 people in Frankfurt constitute a population in equilibrium? If they do, the Hardy-Weinberg formula, using these values of p and q, should give the same proportions of M, MN, and N individuals as those actually observed.

$$p = 0.53$$
$$q = 0.47$$

$$\begin{aligned}(p+q)^2 &= p^2 &+\ & 2pq &+\ & q^2 \\ &= (0.53)^2 &+\ & 2(0.53)(0.47) &+\ & (0.47)^2 \\ &= 0.2809 &+\ & 0.4982 &+\ & 0.2209 \\ &= 28.09\%\ L^ML^M &+\ & 49.82\%\ L^ML^N &+\ & 22.09\%\ L^NL^N\end{aligned}$$

These percentages are to be compared to the observed percentages: 27.0% L^ML^M, 52.9% L^ML^N, 20.1% L^NL^N. The agreement is fairly good (probably it would have been better if we had not rounded off the values of p and q as we did). Apparently the population is in equilibrium or nearly so. If there were any point in determining the probability of whether the difference between observed and expected values is or is not significant (i.e., is or is not caused by chance sampling error), we could apply statistical methods, such as the chi-square test. However, it would not contribute to our present line of thought.

Thus far, the examples we have given have been of populations that at least approximated genetic equilibrium at the outset. Let us see what happens when this is not the case. Here we can make good use of the population geneticist's old friend, the "uninhabited island."

Suppose that a group of 500 people decide to migrate to an uninhabited island. A serologist tests them and finds that they are assorted as follows: 50 are type M; 300 are type MN; 150 are type N. Do they constitute a population in equilibrium?

First, we calculate the values of p and q. In doing so, we shall translate numbers into percentages expressed as decimal fractions, and we shall remember that the value of p, for example, consists of all the genes contributed by M people plus half those contributed by MN people.

$$\begin{aligned} \text{Genotype } L^M L^M &= \ \ 50 \text{ (of 500)} = 0.1 \ (10\%) \\ L^M L^N &= 300 \text{ (of 500)} = 0.6 \ (60\%) \\ L^N L^N &= 150 \text{ (of 500)} = 0.3 \ (30\%) \end{aligned}$$

$$p = 0.1 + \frac{0.6}{2} = 0.4$$

$$q = 0.3 + \frac{0.6}{2} = 0.6$$

$$\begin{aligned} (p + q)^2 &= p^2 + 2pq + q^2 \\ &= (0.4)^2 + 2(0.4)(0.6) + (0.6)^2 \\ &= 0.16 + 0.48 + 0.36 \end{aligned}$$

Hence a population with these gene frequencies ($p = 0.4$; $q = 0.6$) would be expected to have the following proportions of the different types: type M, 16 percent; type MN, 48 percent; type N, 36 percent. Because the actual percentages among the migrants were 10, 60, and 30, respectively, we can see that the migrants were not a population in equilibrium at the outset.

The use of the Hardy-Weinberg formula also tells us something else. It tells us the proportions of the different types to be expected among the children of the migrants if the parents marry at random as far as the L^M and L^N genes are concerned. This is likely to be the case, for most people do not know their M-N type—and if they did, who would choose his or her marriage partner on the basis of it?

We may expect that the children born on our imaginary island will occur in the proportions of 16 percent M, 48 percent MN, 38 percent N. Do these children in their turn constitute a population in equilibrium? When they grow up and marry, what proportions of the different types will be expected among their children (the grandchildren of the original migrants)? In answering this we proceed as before:

$$p = 0.16 + \frac{0.48}{2} = 0.4$$

$$q = 0.36 + \frac{0.48}{2} = 0.6$$

$$\begin{aligned} (p + q)^2 &= p^2 + 2pq + q^2 \\ &= (0.4)^2 + 2(0.4)(0.6) + (0.6)^2 \\ &= 0.16 + 0.48 + 0.36 \end{aligned}$$

Therefore: 16% $L^M L^M$, 48% $L^M L^N$, 36% $L^N L^N$

The proportions among the grandchildren are the same as those among the children. This demonstrates that the children themselves constituted a population in equilibrium. Here we see an example of an important corollary of the Hardy-Weinberg law: *If a population is not in equilibrium, it will reach that state by one generation of random mating (panmixis).*

We note that the persistence of an equilibrium, once established, is in this case caused by the fact that the values of p and q do not change. This is to say that the relative proportions of L^M and L^N genes in the gene pool remain the same, generation after generation. We may compare the gene pool to a deck of cards. Unless someone is dishonest, the cards in the deck remain the same, game after game.

To look ahead a bit, we may ask the question: Under what conditions may the gene pool itself change? If one allele mutates to form the other occasionally, the relative proportions of the two will be altered. If the possessors of one allele have some disadvantage affecting fertility, as compared to the possessors of the other, the proportions will change from generation to generation. We are touching here on subjects that we shall discuss later in other connections. Change in the gene pool is essential if evolution is to occur (Chap. 27).

To summarize: The Hardy-Weinberg law states in algebraic form the fact that a panmictic population tends to attain genetic equilibrium, with regard to a pair of alleles, in one generation, and to maintain that equilibrium thereafter. This is a statement of an ideal situation; it assumes (a) that mating is truly random as far as the alleles in question are concerned, (b) that the alleles do not differ in their effects on viability and fertility, (c) that the mutation rate is so low that it does not affect the results significantly, and (d) that there is no significant change in the gene pool by addition or subtraction of genes as the result of migration into or out from the population (a phenomenon known as gene flow). To the extent that these conditions are not met in actual populations, these populations do not conform to the Hardy-Weinberg law.

Estimating the Frequency of Heterozygotes

Of what practical use is the Hardy-Weinberg formula? One important use lies in the estimation of the relative frequency of heterozygotes in cases in which heterozygotes cannot be detected directly. As we have noted, in many human traits heterozygotes are like one of the homozygotes in phenotype. Here we may return to the example of tyrosinase-negative albinism. People with the genotype *Cc* have as normal pigmentation, as do people with the genotype *CC*. But it is of interest to estimate how many people have the *Cc* genotype (i.e., are carriers of albinism), and here the Hardy-Weinberg formula is helpful.

A survey of Caucasians in North Carolina indicated that one

person in about 39,000 was a tyrosinase-negative albino (for Negroes the proportion was found to be 1:28,000) (Witkop, 1971, Table V).

Applying the Hardy-Weinberg formula to tyrosinase-negative albinism, we have:

$$p = \text{relative frequency of } C$$
$$q = \text{relative frequency of } c$$

Then

$$(p + q)^2 = p^2 + 2pq + q^2$$

in which

$$p^2 \text{ represents } CC$$
$$2pq \text{ represents } Cc$$
$$q^2 \text{ represents } cc$$

This means that *if* we can, for example, regard the Caucasian population of North Carolina as panmictic with regard to genes C and c, we can say that $q^2 = 1/39{,}000$. Let us examine that "if." In a truly panmictic population, being an albino or not being an albino would have no influence on one's likelihood of marriage and parenthood. That may not be true in the present instance. Even if it is not, albinos constitute such a small proportion of the population that the total gene pool will be little affected by a possible difference between them and normally pigmented people in the proportion in which genes are passed on to children. In the population as a whole, 38,999/39,000 of people are normally pigmented. Because very few of these people know before they marry whether or not they are carriers of gene c, it seems safe to say that among these normally pigmented people, mating is random as far as these alleles are concerned. (Possibly an occasional person may hesitate to marry someone who has an albino sib, but this must be rare enough so that it has little effect on the total gene pool.) We can say, then, that we are reasonably justified in applying the Hardy-Weinberg formula to recessive albinism. Hence

$$q^2 = 1/39{,}000$$
$$q = \sqrt{1/39{,}000} = \text{approximately } 1/197$$

Because

$$p + q = 1,$$
$$p = 1 - q = 1 - 1/197 = 196/197$$

Recalling that heterozygotes are represented by $2pq$;

$$2pq = 2 \times \frac{196}{197} \times \frac{1}{197} = \frac{2}{197} \text{ or approximately } \frac{1}{99}$$

In calculating this estimate we disregard the $196/197$, because it is so close to 1 that multiplying by it would make no significant difference in what is at best only an approximation.

We conclude from this that in a population in which only one person in 39,000 is a tyrosinase-negative albino, one person in 99 is a carrier. (According to Witkop, 1971—footnote to Table V—the number of tyrosinase-positive carriers would be about the same in the Caucasian population studied.) This high proportion of carriers may seem rather surprising. If it is correct, why are albinos so rare? Because the chances are small that one carrier will happen to marry another (moreover, if two carriers do marry, only one-fourth of their children, on the average, are expected to be albinos.) As we intimated previously (p. 57) and shall discuss later (Chap. 11), the chance of a carrier marrying another carrier is greatly increased if the spouse is a relative (e.g., a cousin).

Useful as the Hardy-Weinberg formula is in estimating the relative frequencies of heterozygotes in a population, direct tests for heterozygosity—such as the ones we mentioned for PKU and galactosemia—are preferable. But when such tests are unavailable, estimates can be of value, and they doubtless yield frequencies that are at least of the correct order of magnitude. As evidence on this point, let us return to the data on MN blood types in Frankfurt (p. 149). Suppose that in this case we knew the number of type N people but for some reason could not determine directly the number of type MN people. If this were the case, how closely could we estimate the number of type MN people using the formula?

We equate the percentage of type N people ($L^N L^N$) to q^2:

$$\begin{aligned} q^2 &= 0.201 \\ q &= \sqrt{0.201} = \text{approximately } 0.45 \\ p &= 1 - q = 1 - 0.45 = 0.55 \\ 2pq &= 2 \times (0.55)(0.45) = 0.495 \text{ or } 49.5\% \end{aligned}$$

Thus, use of the method would lead us to expect that 49.5 percent of the population would be MN, whereas the investigator found that 52.9 percent of them actually were. The difference of slightly less than 3 percent is not great. We may agree that the estimate is close enough so that it would be useful if we could not make actual counts of the number of carriers.

Evidence Concerning Single-Allele Inheritance

A second area in which the Hardy-Weinberg formula is useful relates to the question that has been a recurrent theme throughout the book thus far. How do we determine whether a certain characteristic in which we may be interested depends on a single gene or on a more complicated genetic arrangement? In Chapter 2, we discussed at length just how and why obtaining a 3:1 (or 1:2:1) ratio in an F_2 generation gives evidence of single gene-pair inheritance. In Chapter 7, we saw how this interpretation could be placed on pooled data of human matings if we used suitable methods of correcting for incomplete ascertainment. Now we ask a further question: How can similar interpretations be drawn from data relating to whole populations? How does single-gene inheritance manifest itself in a population? Or, more specifically, if a characteristic depends on a single recessive gene, what proportion of the population will be expected to show the corresponding phenotype (i.e., to be homozygous for the presumed recessive gene)?

Because most of the traits about which we ask this question are relatively rare, marriages between people who show the trait are so infrequent that they produce few useful data (as we noted earlier concerning the marriages of albinos to each other). Hence we can narrow the question still further: From matings in which both parents show the dominant trait, what proportion of the children will be expected to show the recessive trait? The Hardy-Weinberg formula can be applied to answer these questions (Snyder, 1934; Li, 1961; Stern, 1973).

Three types of matings are involved here: $AA \times AA$, $AA \times Aa$, and $Aa \times Aa$. We recall that in the formula,

$$\begin{aligned} AA &= p^2 \\ Aa &= 2pq \\ aa &= q^2 \end{aligned}$$

Hence we can express the frequencies of the matings as follows:

Mating	Expected Frequency
$AA \times AA$	$p^2 \times p^2 = p^4$
$AA \times Aa$	$2[p^2 \times 2pq] = 4p^3q$
$Aa \times Aa$	$2pq \times 2pq = 4p^2q^2$

In the case of the $AA \times Aa$ mating, the terms in the square brackets are multiplied by two because the mating can arise in two ways: either the father or the mother may be the AA parent.

These matings with their expected frequencies and the expected frequencies of the offspring are shown in Table 10.2. Note that in the

TABLE 10.2 Expected relative frequencies of different types of offspring from matings in which both parents show the dominant phenotype.

Mating	Frequency	Offspring		
		Total	Dominant	Recessive
$AA \times AA$	$p^2 \times p^2 = p^4$	p^4	p^4	0
$AA \times Aa$	$2\,[p^2 \times 2pq] = 4p^3q$	$4p^3q$	$4p^3q$	0
$Aa \times Aa$	$2pq \times 2pq = 4p^2q^2$	$4p^2q^2$	$3p^2q^2$	p^2q^2

first two matings all children are expected to show the dominant phenotype, while in the third mating a 3:1 ratio is expected.

What fraction of the offspring will be expected to show the phenotype determined by the recessive allele? The answer is given by the following fraction, in which the denominator represents the total number of children:

$$\frac{p^2q^2}{p^4 + 4p^3q + 4p^2q^2}$$

Dividing numerator and denominator by p^2, we have

$$\frac{q^2}{p^2 + 4pq + 4q^2} = \left(\frac{q}{p + 2q}\right)^2$$

Now we substitute $1 - q$ for p:

$$\left(\frac{q}{1 - q + 2q}\right)^2 = \left(\frac{q}{1 + q}\right)^2 \qquad (1)$$

This formula will enable us to predict the proportion of children showing the phenotype produced by the recessive allele to be expected from matings in which both parents show the phenotype produced by the dominant allele.

As an example we may use the estimate that the relative frequency of the recessive gene for tyrosinase-negative albinism is $\frac{1}{197}$ (p. 153). We substitute this for q in the formula:

$$\left(\frac{q}{1 + q}\right)^2 = \left(\frac{\frac{1}{197}}{\frac{197}{197} + \frac{1}{197}}\right)^2 = \left(\frac{\frac{1}{197}}{\frac{198}{197}}\right)^2 = \left(\frac{1}{197} \times \frac{197}{198}\right)^2 =$$

$$\left(\frac{1}{198}\right)^2 = \frac{1}{39{,}204} = 0.0000255$$

This means that if the gene for tyrosinase-negative albinism is a simple recessive, we should expect about 25 out of a million children born to normal parents to be albinos of this type. So if we made a survey of a certain population and found this to be the case, we should regard that finding as evidence of the correctness of our hypothesis as to the mode of inheritance.

We should note here that we have come nearly full circle. Recall that our value of q is based on the estimate that one person in 39,000 is an albino of the type in question. The equation gives us 1/39,204, which is almost the same. Why is there any difference at all? Aside from the arithmetic effects of rounding off fractions, the difference reflects the fact that the 1/39,000 includes offspring having one or two albino parents, the 1/39,204 does not. The closeness of the two fractions gives us numerical illustration of the fact mentioned previously that most albino children are born to parents both of whom are normally pigmented. Similar statements apply to other rare recessive traits.

At times it may be useful to ask concerning expectation in families of which one parent shows the trait produced by the dominant allele, being, for example, *AA* or *Aa,* and the other parent shows the trait produced by the recessive allele, being *aa.* This expectation is given by the formula

$$\frac{q}{1 + q} \tag{2}$$

This formula is derived in the same manner as formula (1). Both formulas are sometimes called Snyder's formulas, after Laurence H. Snyder, prominent American geneticist, who set them forth in substantially their present form (Snyder, 1934). Formula (2) is less generally useful than formula (1) because of the smaller number of matings of the type to which it applies.

Distal Hyperextensibility As an example of the use of these formulas to demonstrate the mode of inheritance of a trait not previously analyzed, we cite the study by Glass and Kistler (1953) on double-jointedness of the thumbs: **distal hyperextensibility** (Fig. 10.1). This easily observed, if rather trivial, trait may be useful in studies of genetics of human populations and in the field of physical anthropology if its precise genetic basis can be determined.

Investigation of 895 white persons and 157 Negroes indicated that the trait is quite common; it was evident in approximately one-fourth (24.7 percent) of the white population and one-third of the Negro population.

The authors investigated the possibility that distal hyperextensibility (dht) may depend on a single recessive gene for which, of course, people showing the dht phenotype are homozygous.

If a single recessive gene is involved, the 24.7 percent incidence of dht in the "white" sample can be equated to q^2 of the Hardy-Weinberg formula:

$$q^2 = 0.247 \qquad q = \sqrt{0.247} = 0.496$$

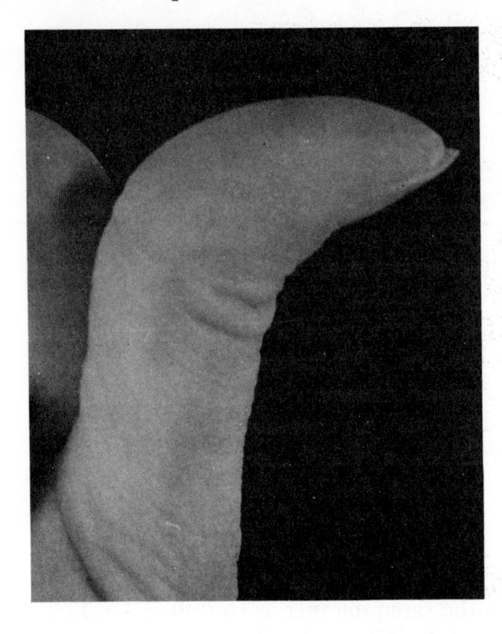

FIG. 10.1 The distally hyperextensible thumb of J. I. M. (Photo by James Benton.)

The investigators studied the offspring of various types of matings. They found 11 families in which both parents showed the trait (dht × dht). There were 24 children in these families; all but one was dht. If the trait depends on a recessive gene, *all* children in families born to two homozygous recessive parents must be homozygous recessive. How can we account for the one non-dht child in 24? The authors are probably correct in ascribing the occurrence to incomplete penetrance: The child probably was homozygous for the recessive allele but for some reason failed to develop distal hyperextensibility (p. 70).

Glass and Kistler found 48 families in which one parent showed the trait, whereas the other did not (dht × non-dht). There were 108 children; 37 were dht; 71 were non-dht. Using formula (2) and the frequency q of the presumed recessive gene, they calculated the number of dht children to be expected in such families:

$$\frac{q}{1 + q} = \frac{0.496}{1.496} = 0.3315$$

$$0.3315 \times 108 = 35.8 = \text{number of dht children expected}$$

Obviously 35.8 is very close to the 37 actually observed. Even without employing the chi-square test as the authors did, we can see the close agreement.

The investigation included 132 families in which both parents lacked the trait (non-dht × non-dht). There were 313 children: 32

were dht, 281 were non-dht. Using formula (1), the authors calculated the expectation for such families:

$$\left(\frac{q}{1 + q}\right)^2 = (0.3315)^2 = 0.1099$$

$$0.1099 \times 313 = 34.4 = \text{number of dht children expected}$$

Again this is in good agreement with the 32 dht children actually observed.

The authors also used the formulas to check the opposite hypothesis: that it is the *non-dht* gene that is recessive. Expectation calculated by the formulas differed widely from observed numbers, adding evidence that it is the *dht* gene that is recessive. Additional investigation showed that the gene is not sex-linked (Chap. 14) and that inheritance is not explainable on a basis of multiple cumulative genes (Chap. 8).

Assortative Mating

When people are influenced in their choice of mates by the genetic trait being studied, the conditions of random mating assumed by the Hardy-Weinberg formula are not met. There is evidence, for example, that tall men and women are more inclined to marry each other than they are to choose short partners. Although many exceptions are encountered, the tendency is present, and to the extent that it exists, it affords an example of assortative mating. In its most common form, assortative mating is positive—"like mating with like," although it may also be negative—unlike individuals choosing each other as mates. The mathematics of assortative mating, both positive and negative, was worked out by Dahlberg (1948), a leading human geneticist in Sweden.

What is the effect of positive assortative mating on a population? For the sake of simplicity, we shall picture a situation in which the tendency of like to marry like is complete, a condition seldom if ever encountered in actual examples. Furthermore, the effect is most clearly discernible when heterozygotes differ from both types of homozygotes, as in the MN blood types. Let us inquire into the consequences of complete assortative mating based on MN blood types. In doing so, we picture the possible, although highly improbable, situation in which type M people would marry only type M people, type MN people would marry each other exclusively, and type N people would choose only type N people as mates.

Let us start with a population having the constitution: ¼ M, 2/4 MN, ¼ N. From our previous discussion we know that if mating were random, this population would be in equilibrium, maintaining the same relative proportions generation after generation. However, in this case we postulate complete assortative mating. Under these conditions, what will be the nature of the next generation?

The type M people will mate among themselves ($L^M L^M \times L^M L^M$), producing only type M children to constitute one-fourth of this next generation.

The type N people will also produce only type N children, to constitute one-fourth of the next generation.

The type MN people will mate together ($L^M L^N \times L^M L^N$), but in this case three types of children will be produced. Of these children, one-fourth will be M, two-fourths will be MN, one-fourth will be N. The type MN people will produce one-half of the total number of children. Of this half, one-fourth will be M, thus, $\frac{1}{2} \times \frac{1}{4} = \frac{1}{8}$—therefore, $\frac{1}{8}$ of the total number of children will be of type M derived from MN parents. Similarly, $\frac{1}{2} \times \frac{2}{4} = \frac{2}{8}$ of the total number of children will be of type MN derived from MN parents, and $\frac{1}{2} \times \frac{1}{4} = \frac{1}{8}$ will be of type N derived from MN parents.

The total result will be as follows:

Type M children:
- $\frac{1}{4}$ from M parents
- $\frac{1}{8}$ from MN parents
- $\frac{3}{8}$ total

Type MN children:
- $\frac{2}{8}$ from MN parents

Type N children:
- $\frac{1}{8}$ from MN parents
- $\frac{1}{4}$ from N parents
- $\frac{3}{8}$ total

Hence, in one generation of assortative mating, the fraction of M individuals has increased from $\frac{1}{4}$ to $\frac{3}{8}$, the fraction of N individuals has increased to the same extent, but the fraction of MN individuals has *decreased* from $\frac{2}{4}$ to $\frac{2}{8}$.

What will happen if assortative mating is continued for a second generation? The production of grandchildren of the original population will occur as follows:

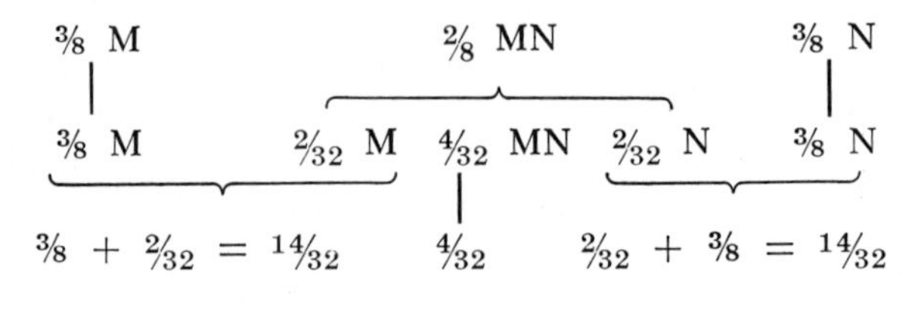

Children: $\frac{7}{16}$ M $\frac{2}{16}$ MN $\frac{7}{16}$ N

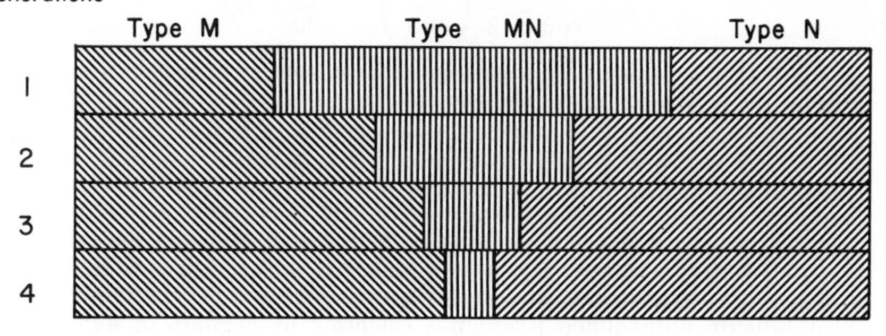

FIG. 10.2 Effect of assortative mating on the relative proportions of homozygotes and heterozygotes. Four generations of a model assuming complete, positive assortative mating based on MN blood types, and starting with a population in which ¼ are of type M, 2/4 of type MN, ¼ of type N.

Again we note that the proportions of M and N individuals have increased at the expense of the proportion of MN individuals.

The same trend will continue as long as assortative mating is practiced (Fig. 10.2). Although our model is oversimplified, it illustrates the effects of positive assortative mating: *The number of homozygotes tends to increase, the number of heterozygotes tends to decrease.* If assortative mating is only partial, the effects will be less striking than those in our model, but the tendency will be present. This tendency to change in the relative proportions of homozygotes and heterozygotes stands in marked contrast to the tendency to maintenance of the *status quo* when mating is random. Thus, the tendency of populations to reach and maintain an equilibrium is upset if assortative mating occurs.

In the next chapter we shall direct our attention to the manner in which consanguinity or inbreeding modifies expectations based on random mating.

Problems

1. A certain group of 300 people is constituted as follows: 150 *AA*, 120 *Aa*, 30 *aa*. Do these people form a panmictic population at equilibrium? If not, what number of individuals would have been expected to have had each of the three genotypes if the population *had* been in equilibrium? *Note:* When it is not known whether or not a population is in equilibrium, the values of p and q are calculated as follows. The *AA* people contribute to the gene pool only *A* genes, whereas *half* of the genes contributed by the *Aa* people are *A* genes. Hence the value of p equals the proportion of people who are *AA* plus half of the proportion of people who are *Aa;* in this case,

$$p = \frac{150}{300} + \frac{1}{2}\left(\frac{120}{300}\right) = \frac{210}{300} = 0.7$$

Similarly, the value of q equals half the proportion of people who are *Aa* plus the proportion of people who are *aa*:

$$q = \frac{1}{2}\left(\frac{120}{300}\right) + \frac{30}{300} = \frac{90}{300} = 0.3$$

Or, if you have determined the value of p, you may obtain the value of q by subtraction, because $1 - p = q$. Alternatively, if you have determined the value of q, $p = 1 - q$.

2. A hitherto uninhabited island was colonized by a group of 475 people. When these people were blood typed, 304 were found to belong to type M, 19 to type N, and 152 to type MN. Did this group of people constitute a population in equilibrium? If not, what number of individuals would have been expected to have belonged to each of the three types if the population *had* been in equilibrium?
3. During the eighteenth century a group of 100 colonists migrated to a previously uninhabited island. They had the following blood-type distribution: 34 were of type M, 58 of type MN, 8 of type N. Today their descendants number 2000. If we can assume that mutation and migration during the intervening years have not significantly altered the gene pool, how many of the 2000 will be expected to be of each blood type?
4. Is this population of 500 individuals in genetic equilibrium: 310 *AA*, 150 *Aa*, 40 *aa*? If not, what numbers would have been expected for each of the three genotypes if the population *had* been in equilibrium?
5. In terms of the Hardy-Weinberg formula, what proportion of children in a panmictic population is expected to consist of heterozygotes born to parents both of whom are heterozygous?
6. Suppose that you discover a "new" human trait that you suspect of being inherited as a Mendelian recessive. You find that one person in 10,000 shows the trait. If you are correct in your hypothesis as to the genetic basis of the trait, (a) what percentage of the children born to parents neither of whom shows the trait will be expected to show it, and (b) what percentage of the children born to parents one of whom shows the trait but the other does not will be expected to show it?
7. Suppose that a national census reveals that one person in 400 has "forked eyelashes." In a certain city of that nation there are 10,000 children. If "forked eyelashes" depends on a recessive autosomal gene, how many of the children will be expected to have forked eyelashes but to be the children of parents neither of whom shows the trait?
8. In a population panmictic for a certain pair of autosomal alleles, what proportion of the marriages is expected to occur between individuals both of whom show the dominant trait? (Answer in terms of p and q.)
9. In a certain panmictic population, 16 percent of the people are homozygous for a recessive gene. What percentage of the people in this population are heterozygous (carriers)?
10. In our example of assortative mating applied to the MN blood types (pp. 160–161), we noted that a second generation of assortative mating resulted in the proportions: $\frac{7}{16}$ M, $\frac{2}{16}$ MN, $\frac{7}{16}$ N. What would be the proportions of the three types following an additional generation of complete assortative mating?

11. What proportion of the total number of children in a panmictic population will be expected to belong to blood type N and to be the offspring of parents both of whom belong to type MN? (Answer in terms of p and q.)

12. On p. 153 we noted that among black people one person in about 28,000 is a tyrosinase-negative albino. What is the proportion of carriers (heterozygotes) in this population?

13. On pp. 155–156 we worked out the derivation of formula (1): $(q/1 + q)^2$. In the same manner, work out the derivation of formula (2), p. 157: $q/1 + q$.

References

Dahlberg, G., 1948. *Mathematical Methods for Population Genetics.* New York: Interscience Publishers.

Glass, B., and J. C. Kistler, 1953. "Distal hyperextensibility of the thumbs," *Acta Genetica et Statistica Medica,* **4**:192–206.

Hardy, G. H., 1908. "Mendelian proportions in a mixed population," *Science,* **28**:49–50. Reprinted in J. A. Peters (ed.), 1959. *Classic Papers in Genetics.* Englewood Cliffs, N.J.: Prentice-Hall, Inc. Pp. 60–62.

Li, C. C., 1961. *Human Genetics.* New York: McGraw-Hill Book Company, Inc.

Snyder, L. H., 1934. "Studies in human inheritance. X. A table to determine the proportion of recessives to be expected in various matings involving a unit character," *Genetics,* **19**:1–17.

Stern, C., 1973. *Principles of Human Genetics,* 3rd ed. San Francisco: W. H. Freeman & Company.

Weinberg, W., 1908. "Über den Nachweis der Vererbung beim Menschen," *Jahreshefte des Vereins für Vaterländische Naturkunde in Württemberg, Stuttgart,* **64**:368–382. (English translation: "On the demonstration of heredity in man," in S. H. Boyer, IV, 1963. *Papers on Human Genetics.* Englewood Cliffs, N.J.: Prentice-Hall, Inc. Pp. 4–15.)

Wiener, A. S., 1943–1962. *Blood Groups and Transfusion,* 3rd ed. Springfield, Ill.: Charles C Thomas. Reprinted–New York: Hafner Publishing Company.

Witkop, C. J., Jr., 1971. "Albinism." In Harris, H., and K. Hirschhorn (eds.), *Advances in Human Genetics, Vol.* 2. New York: Plenum Press. Pp. 61–142.

11

Consanguinity or Inbreeding

The term consanguinity refers to marriage between relatives. Without intending a pun, we point out immediately that consanguinity is a relative term. Actually all human beings are related to each other. We all had ancestors in common if we go back far enough in our study, and for people in the same racial group it is not necessary to go back very far. It has been estimated (Dahlberg, 1948), for example, that if there had been *no* consanguinity, each individual living today would have had 1 billion ancestors who lived 30 generations ago (about 750 years ago). We surmise that there were not 1 billion people in existence in the thirteenth century; certainly there were not 1 billion people for every person alive today.

How closely must two people be related for their marriage to be considered consanguineous? In modern society, the term refers principally to the marriage of first cousins. In this case, one parent of the husband is a sib of one of his wife's parents (there are rarely "double first cousins," in which case both parents of the husband are sibs, respectively, of his wife's parents). In certain times and societies, brother-sister marriage, a still closer consanguinity, has been practiced; occasionally marriage between uncle and niece occurs. Although marriage of second cousins, the children of first cousins, is also considered consanguineous, it has so little genetic effect that we shall largely ignore it. For present purposes, consanguinity will refer principally to first-cousin matings.

We shall begin by asking the question: If a man has a certain recessive gene, what are the chances that his first cousin has the same recessive gene? To give concreteness to our question, we may return to

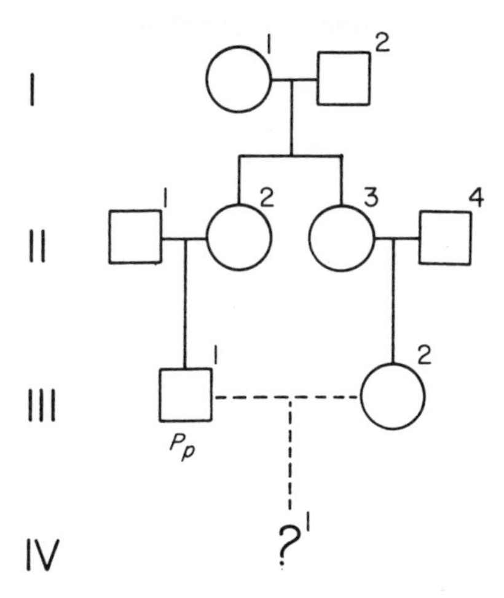

FIG. 11.1 Consanguinity. Proposed marriage of first cousins, one of whom (III-1) is known to be heterozygous for a certain pair of alleles.

our example of phenylketonuria (pp. 53–56). If a man is a carrier (heterozygous), what are the chances that his first cousin is also a carrier? The pedigree in Figure 11.1 represents such a situation. Here the siblings shared as parents by the man and his cousin are represented as sisters; the result would be the same if they were brothers or if they were a brother-sister pair (for we are not discussing a sex-linked gene; see Chap. 14).

The man in question, III-1, is represented as being heterozygous; what are the chances that his cousin, III-2, is also heterozygous?

1. What are the chances that III-1 inherited the *p* gene from his mother (the parent related to III-2) rather than from his father (the parent not related to III-2)? This chance is ½ (as far as we know he was equally likely to have inherited it from either parent). If he inherited *p* from his mother, II-2, that means that she was heterozygous, *Pp*.

2. If II-2 was heterozygous, she must have received *p* from one of her parents, I-1 × I-2. Thus their mating must have been *Pp* × *pp* (without designating which was which). If so, what are the chances that II-3, sister of II-2, is also *Pp*? Because she had an equal chance of receiving *p* or *P* from her heterozygous parent in generation I, the chance is ½.

3. If II-3 is *Pp*, what are the chances that she will pass on gene *p* to her daughter, III-2? Clearly the chance is ½.

At three points as we pass through our pedigree from III-1 to III-2, we encounter chances of ½. Because each is an independent event, the chance that all three will occur is ½ × ½ × ½ = ⅛. This means that if a man who is a carrier for a certain recessive gene marries his first cousin, the chances are one in eight that he will be marrying another carrier.

What are the chances that he will be marrying another carrier if he marries someone who is *not* his cousin? This depends on the relative frequency of the gene in the gene pool. In the case of PKU, we note that estimates of the frequency of occurrence of phenylketonuric children vary, but that one in 15,000 is a reasonable approximation (p. 55). We equate 1/15,000 to q^2 of the Hardy-Weinberg formula (see preceding chapter).

$$q^2 = 1/15{,}000$$

$$q = \sqrt{1/15{,}000} = \text{approx. } 1/122$$

$$\text{carriers} = 2pq = 2 \times \frac{121}{122} \times \frac{1}{122} = \frac{2}{122} = \text{approx. } \frac{1}{61}$$

Hence if his wife is not his cousin, there is one chance in 61 that she is a carrier. This means that in this particular case he is nearly eight times as likely to marry a carrier if he marries his cousin than if he marries "outside the family." (However, the difference in likelihood varies with difference in the gene frequency, becoming less if the gene is common, so that the random chance of marrying a carrier is greater.)

Going one step further we ask, if III-1 marries III-2, what are the chances that a child (e.g., the first one) will have PKU. From $Pp \times Pp$ matings, one-fourth of the children are expected to be pp. Hence the total chance that any one child will have PKU is $1/8 \times 1/4 = 1/32$.

For comparison we ask, if III-1 does not marry his cousin, what are the chances that he will have a child with PKU? The chance is $1/61 \times 1/4 = 1/244$ (the chance that his wife is a carrier multiplied by the chance of a pp child from $Pp \times Pp$ mating). Again we note the approximately eightfold difference.

Now we expand the discussion by asking: If III-1 is not *known* to be a carrier, what are the chances that he will have a child with PKU (a) if he marries his cousin, and (b) if he marries a woman unrelated to him? The question simply adds one more item to the computation—the chance that III-1 is a carrier. This is $1/61$. So if he marries his cousin:

chance he is a carrier = $1/61$
chance his cousin is a carrier if he is = $1/8$
chance for recessive phenotype child = $1/4$

$$\frac{1}{61} \times \frac{1}{8} \times \frac{1}{4} = \frac{1}{1952}$$

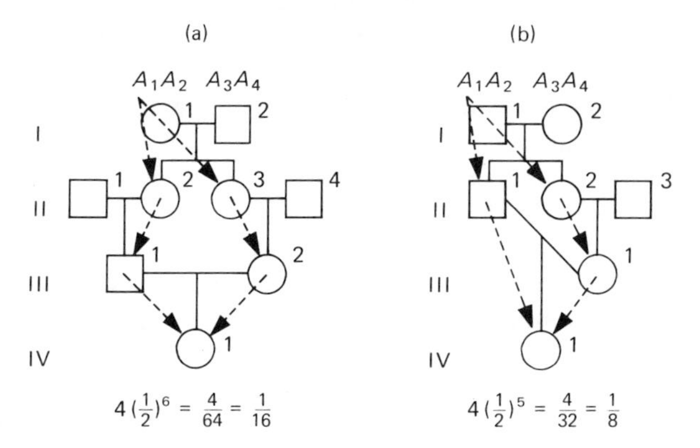

FIG. 11.2 Calculation of the coefficient of inbreeding or consanguinity. (a) For a child of first cousins. (b) For a child from an uncle-niece marriage.

If he marries an unrelated woman:

chance he is a carrier = $1/61$
chance his wife is a carrier = $1/61$
chance for recessive phenotype child = $1/4$

$$\frac{1}{61} \times \frac{1}{61} \times \frac{1}{4} = \frac{1}{14884}$$

The nearly eightfold difference in likelihood is still evident.

In our calculations so far we have made one simplifying assumption: that the *p* gene entered the pedigree only once—that if it came from I-1 or I-2 it did not *also* come from II-1 or II-4, the husbands of the sisters II-2 and II-3. With rare genes, such a simplifying assumption is usually warranted, but if the gene is common there is greater likelihood that unrelated spouses may be carriers.

Coefficient of Consanguinity or Inbreeding

We now ask a somewhat different question about a child of first cousins. What is the chance that IV-1 in Figure 11.2a will be *homozygous* for a gene received from one of the two great grandparents from whom the parents are both descended (I-1 or I-2)? The probability that an individual is homozygous and that both genes of the pair were derived from one gene possessed by one ancestor is called the **coefficient of consanguinity** or **inbreeding.** In Figure 11.2a, I-1 and I-2 were homozygous *AA;* in order to identify the genes, they are numbered. The coefficient of inbreeding of IV-1 is the probability that she is homozygous for A_1, or for any *one* of the other three. Such a person is sometimes said to be *autozygous* for the gene in question.

This coefficient can be calculated as follows. The chance that I-1 will pass along gene A_1 to II-2 is $\frac{1}{2}$. The chance that II-2 will pass it to III-1 is $\frac{1}{2}$. The chance that III-1 will pass it to IV-1 is $\frac{1}{2}$. So the total probability that IV-1 will receive A_1 by the route indicated is $\frac{1}{2} \times \frac{1}{2} \times \frac{1}{2} = \frac{1}{8}$. But IV-1 may also receive gene A_1 by another route: I-1 to II-3, to III-2, to IV-1 (second set of dashed lines). The probability of this occurring is also $\frac{1}{8}$. Hence the probability that IV-1 will receive A_1 by *both* routes (and be A_1A_1) is $\frac{1}{8} \times \frac{1}{8} = \frac{1}{64}$.

Similarly, the chance that IV-1 will be homozygous for A_2 is $\frac{1}{64}$. The chance that she will be A_3A_3 is $\frac{1}{64}$, and the chance that she will be A_4A_4 is also $\frac{1}{64}$. So the total chance that she will be homozygous for some *one* of these four genes is

$$4\left(\frac{1}{64}\right) = \frac{4}{64} = \frac{1}{16} = 0.0625$$

This is the coefficient of inbreeding of first cousins (assuming that none of the ancestors themselves were inbred). Because this computation applies to all of the genes a child of first cousins has, we expect that on the average such a child will be autozygous at one-sixteenth of his or her gene loci.

The coefficient can be calculated in various ways. One way is to draw connecting lines like the dashed arrows in Figure 11.2 and use their number as the power to which $\frac{1}{2}$ is raised, multiplying by 4 because of the four genes possessed at a locus by the two great grandparents. Thus for Fig. 11.2a,

$$4\left(\frac{1}{2}\right)^6 = \frac{4}{64} = \frac{1}{16} = 0.0625$$

Figure 11.2b shows the marriage of an uncle to his niece. Here the coefficient of inbreeding is $\frac{1}{8}$:

$$4\left(\frac{1}{2}\right)^5 = \frac{4}{32} = \frac{1}{8} = 0.1250$$

Another method of calculation is to count the number of ancestors in a circuit from the propositus (IV-1) back to the propositus (without counting the propositus). In Fig. 11.2a, such a circuit consists either of III-1, II-2, I-1, II-3, III-2, *or* of III-1, II-2, I-2, II-3, and III-2. The number in each circuit is the power to which $\frac{1}{2}$ is raised, and the product is multiplied by 2 because there are two circuits:

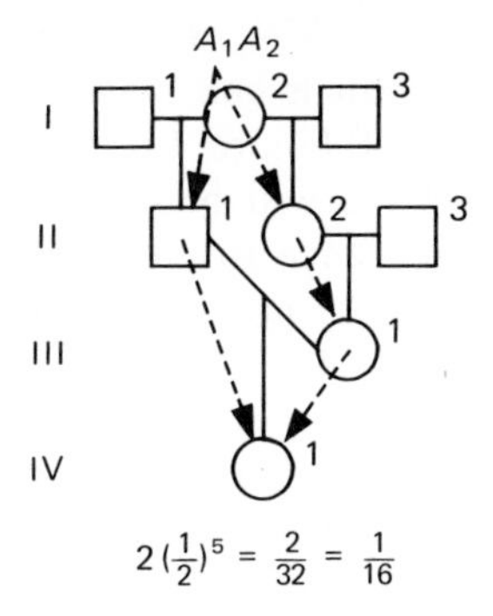

FIG. 11.3 Calculation of the coefficient of inbreeding for a child from the marriage of a man to his half-sister's daughter.

$$2\left(\frac{1}{2}\right)^5 = \frac{2}{32} = \frac{1}{16}$$

In the case of Fig. 11.2b, each of the two circuits numbers 4 (II-1, I-1, II-2, III-1 or II-1, I-2, II-2, III-1), so

$$2\left(\frac{1}{2}\right)^4 = \frac{2}{16} = \frac{1}{8}$$

These simple examples will suffice to illustrate the computation, although human relationships are frequently much more complex. For example, cousins may be the children of parents who are half-sisters rather than full sisters as in Figure 11.2. In that case they have but one parent in common, with the result that we multiply $(\frac{1}{2})^6$ by 2 instead of 4, if we use the first method, or have but one circuit instead of two if we use the second method. Also, ancestors may be somewhat inbred themselves, further complicating computations.

Reference to half-sisters brings us to the question asked in Chapter 1: "A friend of mine is desirous of marrying his half-sister's daughter; would such a relationship be too close, relative to 'blood-lines,' for normal, healthy children?" Such a marriage is diagramed in Figure 11.3, in which IV-1 is indicated as being a child from the mating. Calculating the coefficient of inbreeding as before, we find it to be $\frac{1}{16}$ or 0.0625. We note that this is the same as the coefficient of a child from a first-cousin mating. So the inquirer can be told that the risk of producing a defective child is the same as it would be if the individuals wishing to marry were first cousins. On first thought, it might seem that the risk should be greater in the case of a man marrying his half-sister's daughter. But we note that in this case the parents would have only one ancestor in common, whereas first cousins have two ancestors in common. (Fig. 11.2).

Coefficient of Relationship The **coefficient of relationship** is the probability that two individuals both have the same gene derived from the same ancestor. In terms of Figure 11.2a, we may ask: What is the probability that III-1 and III-2 will both have a certain gene possessed by one of the ancestors they have in common, such as I-1? They may both have A_1 or A_2. Counting dashed lines, we calculate that the chance that they both have A^1 is $(\frac{1}{2})^4 = \frac{1}{16}$. Similarly, the chance that they both have A_2 is $\frac{1}{16}$. So the chance that both have A_1 or both have A_2 is $2(\frac{1}{16}) = \frac{1}{8}$. We note that this is twice the coefficient of inbreeding of their child (IV-1). The coefficient of relationship of parents is always twice the coefficient of inbreeding of a child they produce.

The above calculation that the coefficient of relationship for first cousins is $\frac{1}{8}$ applies to all of their genes. This means that on the average, one-eighth of the genes of first cousins are the same because they have been inherited from the same ancestor. Every common ancestor may contribute genes to both cousins, of course.

The Effects of Consanguinity

In Terms of Single Genes We have seen that consanguinity increases the chance that homozygous offspring will be produced. But by itself it does not change the gene pool, any more than a deck of cards is changed by a change in the method of dealing. Inbreeding simply alters the genetic "hands" dealt out to the offspring.

The amount of effect consanguinity has varies with the gene frequency. This is logical, because what consanguinity does is to increase the chance that heterozygotes will marry each other. If the gene is abundant, there will be many heterozygotes, with the result that they will marry each other frequently just by chance in the absence of inbreeding.

For example, we saw in the last chapter that when the frequency, q, of a recessive gene is 0.5, then 50 percent of a panmictic population is expected to be heterozygous and 25 percent to be homozygous recessive. Since half of the people are heterozygous, the chance of their marrying each other is high, without inbreeding. Now suppose that in such a population the mating system were changed and *everyone married a first cousin.* What would be the effect on the proportion of homozygous recessives among their children ?

If all matings are first-cousin matings, the percentage of homozygous recessives among the offspring is given by the formula $q/16\ (1 + 15q)$. (See Appendix A, pp. 459–461, for the derivation and explanation of this formula.)

Applying the formula to a population in which $q = 0.5$, we have

$$\frac{q}{16}(1 + 15q) = \frac{0.5}{16}(1 + 7.5) = \frac{4.25}{16} = 0.265 = 26.5\%$$

Thus we see that a complete program of first-cousin matings would only increase the percentage of homozygous recessives from 25 percent to 26.5 percent. To put it another way: In a panmictic population of 10,000, 2500 offspring are expected to be homozygous recessive. In a population of 10,000 in which everyone marries his first cousin, 2650 offspring are expected to be homozygous recessive. (In both cases we assume that the total size of the population is not increasing from generation to generation.) Thus the increase is 150, and this increase is only 6 percent of the number of homozygous recessives (2500) we should expect if there were no inbreeding.

Parenthetically: Some cousin marriages would occur with panmixis, for random mating assumes that each woman of marriageable age is equally likely to marry any man of marriageable age whether he is her cousin or not.

Now let us look at the population in which $q = 0.1$. As we saw earlier (p. 149), if this population is panmictic, 1 percent of the offspring are expected to be homozygous recessive. What percentage would be expected to be homozygous recessive if everyone married his first cousin?

$$\frac{q}{16}(1 + 15q) = \frac{0.1}{16}(1 + 1.5) = \frac{0.25}{16} = 0.0156 = 1.56\%$$

The percentage has increased from 1 percent to 1.56 percent. Although this may seem to be a negligible increase, let us think in terms of a population of 10,000. With panmixis, 100 homozygous recessives are expected; with complete first-cousin mating, 156 homozygous recessives are expected. This is an increase of 56, which, of course, is a 56 percent increase over the 100 we would have expected to result from panmixis.

Summarizing, we note that when $q = 0.5$, complete first-cousin mating results in only a 6 percent increase in homozygous recessives. When $q = 0.1$, complete first-cousin mating results in a 56 percent increase in homozygous recessives. This illustrates the principle that the effect of inbreeding on the production of homozygous recessives varies inversely with the gene frequency.

New Mutations and Consanguinity It will be instructive to consider the extreme case, that of a new mutation that results in a gene not

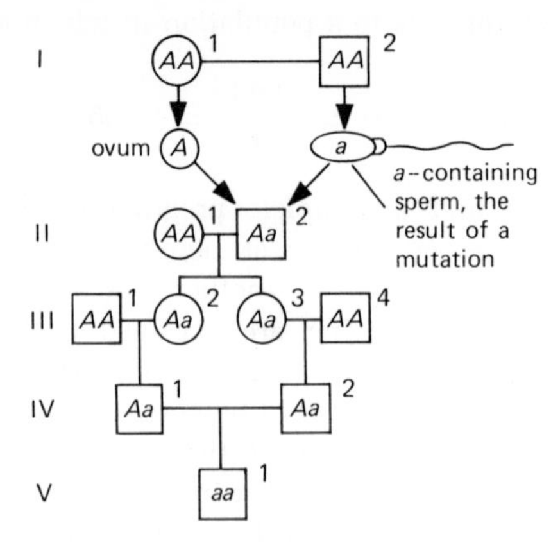

FIG. 11.4 Consanguinity as the means by which a homozygote for a "new" mutation can arise in the smallest possible number of generations.

otherwise present in a population. By way of example, imagine a population in which everyone has the genotype *AA*. Suppose that I-2 (Fig. 11.4) produced some sperm cells containing the mutant gene *a,* and that one of these sperms fertilized an ovum with resulting production of a son II-2. Such a male, in turn, *might* pass on gene *a* to his two daughters III-2 and III-3. These daughters *might* then pass the gene to their respective offspring, IV-1 and IV-2. If these latter (first cousins) marry each other, each of their children has a ¼ chance of being *aa* (V-1).

The new gene, *a,* remained hidden for four generations. If it is completely recessive, the first person who could show its phenotype would be a great-great-grandchild of the person in whom the mutation occurred, and then it could only come to light so quickly if consanguinity were involved. We recognize immediately, of course, that the new mutation may remain hidden for many generations, an *aa* individual first being produced when remote descendants of I-2 happen to marry each other. However, we have deliberately set out to demonstrate the quickest production of an *aa* individual.

We readily appreciate that the time represented by the generations separating a person in whom a mutation occurs and the first descendant to show the phenotype has a marked delaying effect on investigations such as those concerned with new mutations produced by atomic bombs. If only the great-great-grandchildren of persons who experienced the blast are likely to show the phenotypic effects—and then only if inbreeding occurs—such investigations must be long-term ones. Fortunately for investigators, although unfortunately for the victims themselves, many new mutations have some degree of dominance, meaning that they *do* produce some detectable effect in heterozygotes. Also, new sex-linked mutations do not remain hidden for as long a time as does the type indicated in our hypothetical pedigree (Chap. 14).

Increased Consanguinity among Parents of Offspring Showing Rare Recessive Traits If a gene is not completely new (as in our hypothetical model) but is nevertheless very rare, heterozygotes for the gene will seldom marry each other unless they are cousins. We stressed this point in our discussion of alkaptonuria (pp. 56–58), emphasizing that for this reason increased consanguinity between the parents of affected individuals is an important hallmark of recessive inheritance—so recognized by Garrod.

Another interesting case is afforded by a rare defect in which fibrinogen is lacking from the blood plasma. Fibrinogen is a plasma protein that is converted to fibrin in a cut or wound, forming a meshwork of fibers throughout a blood clot. Because blood cannot clot without fibrinogen, its absence is extremely serious. The defect is known as congenital **afibrinogenemia.** Fortunately, the trait is rare. Frick and McQuarrie (1954) reported that 12 cases were known in which information about the parents was available. In five of the 12 cases, the parents were first cousins. Thus, in these 12 cases, the incidence of consanguinity was over 41 percent (as compared to an incidence of about 1 percent in the population at large). Here we have a fine example of increased consanguinity as a hallmark of a rare recessive gene.

In the paper just cited, the authors presented an interesting pedigree of a case the paper described in detail (Fig. 11.5). The afibrinogenemic propositus is IV-2. His parents were first cousins. The other interesting aspect of this study is the indication that carriers can be de-

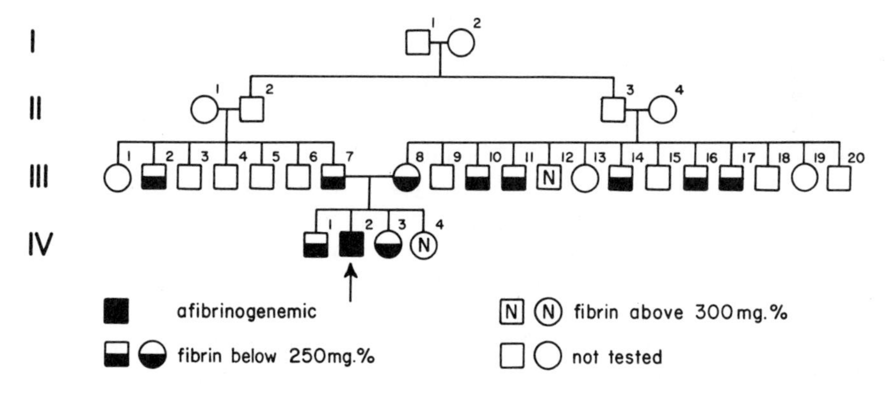

FIG. 11.5 Consanguinity in the ancestry of an afibrinogenemic child. (Redrawn from Frick, P. G., and I. McQuarrie, 1954. "Congenital afibrinogenemia," *Pediatrics,* **13**:44–58.)

tected by virtue of the fact that they have less fibrinogen, and therefore less fibrin, than do other people (although the fibrin is sufficient for the formation of adequate clots). Thus the patient's parents (III-7 and III-8) both had a low fibrin level, as did some of their sibs. From their offspring we surmise that II-2 and II-3 were heterozygous; unfortunately, they were not living at the time of the investigation and so could not be tested for fibrin level.

If we represent the rare recessive gene by *f* and its normal allele by *F*, the parents (III-7 and III-8) were $Ff \times Ff$. Then their children were, in sequence *Ff*, *ff*, *Ff*, and *FF*. The authors pointed out the interesting coincidence of this perfect tally with the 1:2:1 ratio.

Many studies have been made of the increased incidence of consanguinity among the parents of persons showing rare recessive defects. Without attempting a complete catalogue, we may mention one or two typical investigations of this kind. Neel et al. (1949) studied consanquineous matings in Japan. In the general population, first-cousin marriages occur about 6 percent of the time, but among parents of albinos the incidence of first-cousin marriage is between 37 and 59 percent. Among parents of children suffering from infantile amaurotic idiocy (Tay-Sachs disease), the incidence of first-cousin marriage is between 55 and 85 percent. Comparable figures for ichthyosis congenita are 67 to 93 percent; for congenital total color blindness, 39 to 51 percent; for xeroderma pigmentosum, 37 to 43 percent.

The authors quoted comparable figures for Caucasian populations, in which the general level of first-cousin marriage is about 1 percent. The percentages of first-cousin marriages for the parents of children affected with the five traits mentioned are as follows: albinism, 18 to 24 percent; infantile amaurotic idiocy, 27 to 53 percent; ichthyosis congenita, 30 to 40 percent; congenital total color blindness, 11 to 21 percent; xeroderma pigmentosum, 20 to 26 percent.

It is interesting that while increased consanguinity is characteristic of Caucasian parents of children with Tay-Sachs disease, consanguinity is not strikingly elevated among Jewish parents of children with the disease. Tay-Sachs disease is much more common among Jews, especially those of Eastern European ancestry (Ashkenazi Jews) than it is in the non-Jewish population. In the United States the frequency of heterozygotes is estimated to be 2.6 percent among Ashkenazi Jewish individuals, but only about 0.29 percent among non-Jewish individuals (O'Brien, 1972). As a consequence, unrelated Jewish heterozygotes are much more likely to marry each other by chance than are unrelated non-Jewish heterozygotes. This illustrates the general principle that the effect of inbreeding varies inversely with the gene frequency (p. 171). Tay-Sachs disease will claim our attention again in our discussion of genetic counseling (Chap. 28).

Total Risk A brief search of the medical literature will convince one that the list of abnormalities dependent upon recessive genes that come to attention most frequently as a result of the marriage of relatives is seemingly endless. But each of these defects, taken singly, is very rare. What is the overall picture? In general, how much more "dangerous" is it to marry a relative than to marry an unrelated person?

Answering this question is difficult because of the variables involved and the varying degrees of inbreeding found in actual populations. But most investigations reveal a significant increase in the risk of producing children abnormal or below normal in some respects. For example, Slatis, Reis, and Hoene (1958) studied 109 consanguineous marriages in the Chicago region. They concluded that the rate of abnormalities among children of consanguineous marriages was somewhat less than twice as great as the rate among children of unrelated parents, "but if only major abnormalities are considered, the added risk of consanguinity is exceedingly great, i.e., 8 of 192 consanguineous children have had serious abnormalities, whereas 0 of 163 living control children have suffered serious abnormalities."

The most extensive study of consanguinity ever made was connected with, although separate from, investigation of genetic effects of atomic radiation on the people of Japan (Schull and Neel, 1965, 1972—with papers listed there). These studies were so extensive that we cannot even summarize them here. We content ourselves with an example: In Hiroshima, in 1384 families of which the parents were not related, 3.55 percent of the children died within the first year, whereas in 532 families of which the parents were first cousins, 6.12 percent of the infants died within the first year. The difference is 72 percent of the rate for unrelated parents.

In general summary of investigations on Japanese populations, Schull and Neel (1972) stated: "Fetal loss and death prior to the age of reproduction are both estimated at approximately 4% more in the children of a first-cousin marriage as compared with children resulting from the marriage of unrelated parents."

The investigations included other traits besides differential mortality. Analysis of inbreeding effect was applied to data on many physical defects and diseases, growth rates, bodily measurements, bone development, dental characteristics, serological traits, neuromuscular and mental traits, school performance, and the like. In some cases, there seemed to be no inbreeding effect (e.g., stature, weight, bone dimensions and ossification, and so on). In other cases, an effect was observable. Most interesting, perhaps, were analyses of major physical defects. "Taken at face value, the data indicate an increase in children with major physical defect from approximately 8.5% in the controls to 11.7% in the children of first-cousin marriages, a 37.6% relative increase." Interestingly, also, the analysis of school data revealed "a consistent and significant depression in performance in all subjects with inbreeding" (Schull and Neel, 1965).

Many studies have been made on the effects of inbreeding. They differ in size, method of ascertainment, thoroughness of analysis, and in many other respects. But most of them show that consanguinity *is* accompanied by increased risk of producing defective children.

Genetic Isolates Sometimes relatively small populations form a more or less closed community, for many generations marrying within their own group. Such a population is called a **genetic isolate.** Sometimes there may be actual, geographical isolation, as when a population is isolated in a remote mountain valley, or on an island, with poor means of transportation to the outside world. Sometimes the isolating factor is social or religious. Whatever the cause of isolation, genetic isolates exhibit increased consanguinity as compared to the general population.

For example, Jackson et al. (1968) studied a community of Old Order Amish people in a county in Indiana. They found that of 276 matings, in 14 the relationship of the parents was that of first cousins or closer, in 109 the relationship was between that of first and second cousins, and in 35 between that of second and third cousins, and in 12 between that of third and fourth cousins. In the remainder of matings, the parents were unrelated, but the coefficient of inbreeding for the entire population was 0.0195, which falls between that for offspring of first cousins (0.0625; see above) and that for offspring of second cousins (0.0156).

As we would expect from this amount of inbreeding, genetic isolates are likely to show an increased incidence of otherwise rare, inherited defects. Thus, these authors commented concerning the community they studied: "The increasing percentage of consanguineous marriages most likely accounts for the increasing number of cases of limb-girdle muscular dystrophy, in recent generations."

Among other examples, the studies of McKusick et al. (1964a, b) on another inbred Amish population (Lancaster County, Pennsylvania) are pertinent. They found 30 sibships with a total of 52 cases of a usually very rare form of dwarfism (Ellis-van Creveld syndrome, characterized by polydactyly and various physical defects). The syndrome is inherited on the basis of a recessive gene which was probably inherited from one early ancestor of the group. The number of cases almost equals the number previously reported in all of the medical literature.

Problems

1. In the gene pool of a certain population, the frequency, q, of the recessive gene b is 0.001 (0.1 percent). If a man having the genotype Bb marries his first cousin, what is the chance that their first child will have the genotype bb? If the man marries a woman unrelated to him, what is the chance that their first child will have the genotype bb? Hence, how many times more likely is production of a bb child if the man marries a first cousin that it is if he marries an unrelated woman?
2. In the population mentioned in Problem 1, what percentage of the offspring will be bb under a system of random mating? What percentage of the offspring would be bb if all marriages were first-cousin matings? By how many times is the occurrence of bb children increased through a uniform practice of first-cousin mating?
3. Assuming that the frequency, q, of the gene for tyrosinase-negative albinism is $1/197$, if a normally pigmented man picked at random in the population marries his niece, his sister's daughter, what is the chance that their first child will be an albino? (Assume that the sister's husband is not heterozygous.) What would be the chance for an albino child if this man married a first cousin? From the standpoint of producing recessive-phenotype children, is it more "dangerous" to marry one's niece or one's first cousin? How many times more?
4. Members of a pair of identical (monozygotic) twins have the same genotype. Identical-twin brothers married identical-twin sisters. The brothers were albinos. One couple produced a normally pigmented daughter and the other couple a normally pigmented son. If these first cousins marry each other, what is the chance that their first child will be an albino?

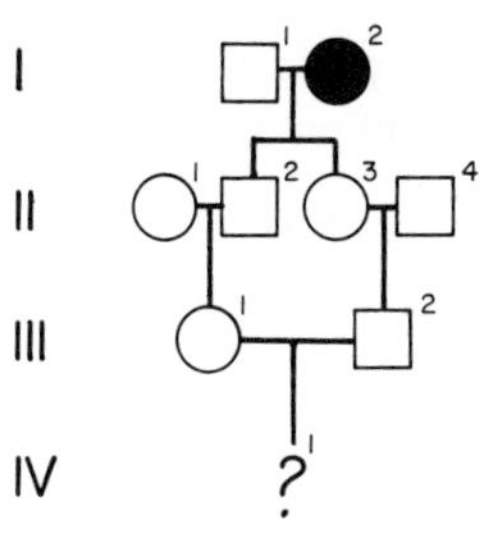

5. In the accompanying pedigree chart, I-2 is an albino; all other persons in the first three generations have normal pigmentation. What is the chance that IV-1, yet unborn, will be an albino? (Assume that II-1, and II-4 are not heterozygous.)
6. In the accompanying pedigree chart, if IV-1, known to be heterozygous Aa, marries her second cousin, IV-2, what is the chance that their first child will have the genotype aa? (Assume that gene a enters the kindred only once.) What is the coefficient of inbreeding (consanguinity) of V-1? What is the coefficient of relationship of IV-1 and IV-2?

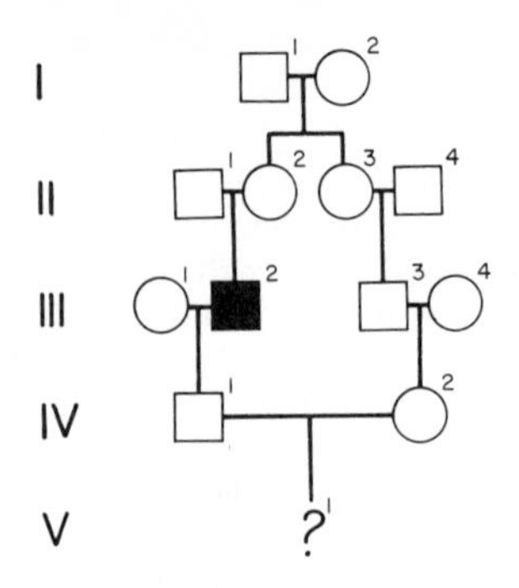

7. In the accompanying pedigree, III-2 is a tyrosinase-negative albino; all other persons in the first four generations have normal pigmentation. What is the chance that V-1, yet unborn, will be albino? (Assume that I-1, II-4, III-1, and III-4 are not heterozygous.)

8. John Doe's mother has a brother who is a tyrosinase-negative albino, the only person in the kindred who is. If John marries his cousin, the daughter of his mother's sister, what is the chance that their first child will be an albino? (Assume that the gene for albinism does not enter the kindred from unrelated spouses in later generations.)
9. A man marries his first-cousin's daughter, and they produce a son.
   (a) Draw a chart of this pedigree.
   (b) Calculate the coefficient of inbreeding of this son.

References

Dahlberg, G., 1948. *Mathematical Methods for Population Genetics.* New York: Interscience Publishers.

Frick, P. G., and I. McQuarrie, 1954. "Congenital afibrinogenemia," *Pediatrics,* **13**:44–58.

Jackson, C. E., W. E. Symon, E. L. Pruden, I. M. Kaehr, and J. D. Mann, 1968. "Consanguinity and blood group distribution in an Amish isolate," *American Journal of Human Genetics,* **20**:522–527.

McKusick, V. A., J. A. Egeland, R. Eldridge, and D. E. Krusen, 1964a. "Dwarfism in the Amish. I. The Ellis-van Creveld syndrome," *Bulletin of the Johns Hopkins Hospital,* **115**:306–336.

McKusick, V. A., J. A. Hostetler, J. A. Egeland, and R. Eldridge, 1964b. "The distribution of certain genes in the Old Order Amish," *Cold Spring Harbor Symposia on Quantitative Biology,* **29**:99–114.

Neel, J. V., M. Kodani, R. Brewer, and R. C. Anderson, 1949. "The incidence of consanguineous matings in Japan," *American Journal of Human Genetics,* **1**:156–178.

O'Brien, J. S., 1972. "Ganglioside storage diseases." In Harris, H., and K. Hirschhorn (eds.), *Advances in Human Genetics,* Vol. 3. New York: Plenum Press. Pp. 39–98.

Schull, W. J., and J. V. Neel, 1965. *The Effects of Inbreeding on Japanese Children,* New York: Harper & Row.

Schull, W. J., and J. V. Neel, 1972. "The effects of parental consanguinity and inbreeding in Hirado, Japan. V. Summary and interpretation," *American Journal of Human Genetics,* **24**:425–453.

Slatis, H. M., R. H. Reis, and R. E. Hoene, 1958. "Consanguineous marriages in the Chicago region," *American Journal of Human Genetics,* **10**:446–464.

12

Multiple Alleles: ABO Blood Groups

In Chapter 5 we introduced the idea of multiple alleles in our discussion of the abnormal hemoglobins (pp. 47–52). There we saw that a series of genes may be the alleles of each other: each may occupy alternatively a certain specific locus in a certain chromosome. Because chromosomes occur in pairs, one person has only two members of a series of multiple alleles, no matter how long that series may be. The genes controlling the ABO blood groups form one of the best known and studied examples of multiple alleles.

Genetics of the ABO Blood Groups

The "classic" blood groups were discovered by Karl Landsteiner at the beginning of this century. In his initial investigation, he collected blood samples from 22 individuals and separated the cells from the fluid portion or serum (Landsteiner, 1901). Then by cross-matching (mixing the serum of one person with the blood cells of another person), he found that the serum of some people contains antibodies that will clump or agglutinate the red blood cells of other people. By such cross-matching, he classified his 22 people into three groups:

Group A—the serum of these people reacted with cells of people in group B, but not with cells of people in group A.

Group B—the serum reacted with cells from group A, but not with cells from group B.

Group C (or, as we call it today, group O)—the serum of these people agglutinated cells of people in both group A and group B, but group C cells were not agglutinated by group A or B serums.

Landsteiner's initial sample of 22 people did not include a person whose serum failed to agglutinate all three of the other kinds of cells (groups A, B, and C) and whose cells were agglutinated by serums from all three of these groups. Such people were soon discovered, however; today we classify them as group AB.

We note immediately some differences between these groups and the MN types we discussed earlier (pp. 81–84). (1) People do not have naturally occurring antibodies against antigen M or N when they lack the antigen in their cells (e.g., a type M person does not have anti-N antibodies in his serum). On the other hand, people *do* have naturally occurring antibodies against antigen A or B when they lack the antigen in their cells. (2) People have in their red blood cells antigen M, antigen N, or both antigens; no one lacks both. On the other hand, some people *do* lack both A and B (people of group O). Table 12.1 summarizes the

TABLE 12.1 ABO blood groups.

Blood group	Antigens in red blood cells	Antibodies in serum
O	neither	anti-A and anti-B
A	A	anti-B
B	B	anti-A
AB	A and B	neither

relationships of cellular antigens A and B to antibodies in the serum.

Antigens A and B on red blood cells are complex macromolecules of the type called glycolipids, composed of five different sugar molecules joined to fatty acids (Watkins, 1966). The specificity of the antigen is apparently determined by the carbohydrate (sugar) portion. As with the inborn errors discussed in Chapter 5, the number of enzyme-controlled steps between the gene and the completed glycolipid is probably relatively small. Furthermore, the chemical difference between antigen A and antigen B is so small that we can readily imagine that a gene mutation might change production of one into production of the other, or that production of either might have arisen by mutation of a precursor gene.

As we noted earlier, test fluids for detecting M and N are prepared by inoculating rabbits, which then form antibodies. Test fluids for A and B may be prepared in the same manner, but human serums are normally employed. From Table 12.1 we can readily see that serum from a person who belongs to group A forms a test fluid that will react with cells containing B, and serum from a person belonging to group B forms a test fluid for cells containing A. Using these two serums, we can determine the group to which any person belongs, by testing his red blood cells as shown in Table 12.2.

TABLE 12.2 ABO blood groups—Reactions of cells with antiserums.

If red blood cells contain	Reaction with antiserums		Blood group of donor of the cells
	Anti-A	Anti-B	
neither A nor B	−	−	O
A only	+	−	A
B only	−	+	B
both A and B	+	+	AB

+ = clumping of cells.
− = no clumping of cells.

The von Dungern-Hirszfeld Hypothesis That multiple alleles formed the genetic basis of the ABO groups was not immediately perceived. Some ten years after Landsteiner's discovery, von Dungern and Hirszfeld (1910) hypothesized that the groups are determined by two pairs of genes that assort independently after the manner we discussed on pp. 32–34. We may represent the hypothesized genes as follows:

A = presence of antigen A
a = absence of antigen A
B = presence of antigen B
b = absence of antigen B

Individual genotypes would then be as follows:

group A = *AAbb* or *Aabb*
group B = *aaBB* or *aaBb*
group AB = *AABB* or *AaBB* or *AABb* or *AaBb*
group O = *aabb*

We note parenthetically that it is the antigens A and B for whose inheritance we have to account. The antibodies in the serum (anti-B and anti-A) accompany the antigens, but do not seem to be inherited separately. Discussion of the origin of the antibodies is not pertinent here (see Wiener, 1943–1962).

This hypothesis accounted for many observed facts, but not for all of them. For example, it would explain why two group O parents can have only group O children (*aabb* × *aabb*). But according to the hypothesis, two group AB parents could have a group O child (*AaBb* × *AaBb* mating, if both parents produce *ab* gametes and those gametes form an *aabb* zygote; Fig. 8.1, p. 99). Actual observation shows, however, that a group AB parent, no matter to whom mated, does not in fact have group O children (or vice versa, for that matter).

The theory ran into further trouble when the distribution of blood groups in whole populations was studied. Data of this kind have been accumulated in tremendous number. In 1925, Bernstein pointed out that if the theory were valid, the following equation should be correct:

$$(\overline{A} + \overline{AB}) \cdot (\overline{B} + \overline{AB}) = \overline{AB}$$

This may be translated as follows: In a population, the number of people who belong to group A added to the number who belong to group AB, if multiplied by the number who belong to group B added to the number who belong to group AB, should equal the number who belong to group AB. We may not understand at first glance why this equation should apply if the two-gene-pair theory is correct. The equation is derived by simple algebra, employing reasoning of the type encountered in applications of the Hardy-Weinberg formula. The derivation of the equation is given in Appendix B (pp. 461–462). The results of large numbers of investigations have shown that this equation does not represent the actual situation encountered in populations.

We have presented the von Dungern-Hirszfeld hypothesis to illustrate the way in which science grows by discarding old hypotheses, which once seemed logical and satisfactory, when new evidence proves them no longer adequate.

The Multiple-Allele Hypothesis The multiple-allele hypothesis was proposed by Bernstein (1925). According to it, the basic groups we have described depend upon a series of three alleles. These are designated in various ways by different writers; we shall follow the precedent of those who employ the initial *I* (for **isoagglutinogen**—a normally occurring antigen) with appropriate superscripts. Thus:

I^A = gene for antigen A
I^B = gene for antigen B
I^O = gene determining neither antigen B nor antigen B

In Figure 12.1, we represent by two straight lines the pair of chromosomes concerned. The short cross line represents the locus in question. This locus in any one chromosome may be occupied by any one of the three genes. The figure shows the genotypes possible in persons belonging to each group.

Does the multiple-allele hypothesis explain observed facts better than did the theory of two independently assorting gene pairs? We noted

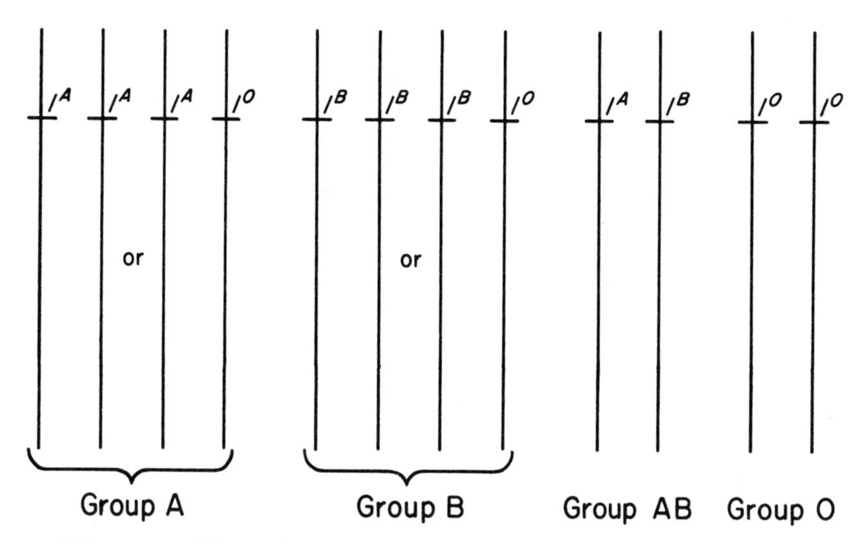

FIG. 12.1 All possible pairings of the blood group alleles I^A, I^B, and I^O. Each pair of vertical lines represents a pair of chromosomes. The short cross line represents the locus of the multiple alleles concerned.

above that people belonging to blood group AB do not, in fact, have children belonging to group O. The present theory, unlike the former one, explains this:

$$I^A I^B \times I^A I^B$$

Germ cells: $I^A \to I^A$, $I^B \to I^B$ (crossed)

	$I^A I^A$	$I^A I^B$	$I^A I^B$	$I^B I^B$
Group	A	AB		B

In fact, even if one parent belongs to group O, it is still impossible for a group AB parent to have a group O child:

$$I^A I^B \times I^O I^O$$

Germ cells: $I^A \to I^O$, $I^B \to I^O$ (crossed)

	$I^A I^O$	$I^A I^O$	$I^B I^O$	$I^B I^O$
Group	A		B	

This last diagram also demonstrates the converse: a group O

parent (even when married to a group AB person) cannot have a group AB child.

Here the multiple-allele theory agrees with observation. What does the study of populations reveal as to the adequacy of the multiple-allele theory?

We answer this question by calculating the relative proportions of the different groups to be expected in a panmictic population *if* the theory is correct. To do this, we employ the Hardy-Weinberg formula extended to include three alleles.

Let p = relative frequency of gene I^A
q = relative frequency of gene I^B
r = relative frequency of gene I^O
$p + q + r = 1$

Then:

$$(p + q + r)^2 = p^2 + 2pq + 2pr + q^2 + 2qr + r^2$$

This means that in a panmictic population:

the relative frequency of I^AI^A persons is p^2
the relative frequency of I^AI^O persons is $2pr$
the relative frequency of I^BI^B persons is q^2
the relative frequency of I^BI^O persons is $2qr$
the relative frequency of I^AI^B persons is $2pq$
the relative frequency of I^OI^O persons is r^2

Hence the frequency of:

group A persons $(I^AI^A + I^AI^O)$ is $p^2 + 2pr$
group B persons $(I^BI^B + I^BI^O)$ is $q^2 + 2qr$
group AB persons (I^AI^B) is $2pq$
group O persons (I^OI^O) is r^2

Group A + group O $= p^2 + 2pr + r^2 = (p + r)^2$
Group B + group O $= q^2 + 2qr + r^2 = (q + r)^2$

Then:

$$p + r = \sqrt{\bar{A} + \bar{O}}$$

(Where $\bar{A}$ is the proportion of group A people, $\bar{O}$ the proportion in group O)

and

$$q + r = \sqrt{\bar{B} + \bar{O}}$$

Because

$$p + q + r = 1,$$

$$p = 1 - (q + r) = 1 - \sqrt{\bar{B} + \bar{O}} \quad (1)$$

$$q = 1 - (p + r) = 1 - \sqrt{\bar{A} + \bar{O}} \quad (2)$$

$$r = \sqrt{\bar{O}} \quad (3)$$

Hence:

$$p + q + r = (1 - \sqrt{\bar{B} + \bar{O}}) + (1 - \sqrt{\bar{A} + \bar{O}}) + \sqrt{\bar{O}} = 1$$

This is usually simplified and written as

$$\sqrt{\bar{O} + \bar{B}} + \sqrt{\bar{O} + \bar{A}} - \sqrt{\bar{O}} = 1$$

(The square root of the frequency of group O persons added to group B persons, plus the square root of the frequency of the group O persons plus group A persons, minus the square root of the frequency of group O persons, equals 1.)

This is the expectation; do observed facts fit? Tests on large numbers of populations, including thousands of individuals, demonstrate that the expectation expressed by the formula fits very precisely indeed (see Wiener, 1943–1962, for detailed results). Agreement has been so close and so uniformly attained that the multiple-allele theory is now established beyond reasonable doubt.

Some of the tests of the theory have involved applications of the Hardy-Weinberg formula that are of theoretical interest to us. To use an example given by Wiener (1943–1962), we may ask: If the theory is correct, how frequently should the combination of group O mother and group O child be encountered in a population?

Group O children can arise from group O mothers in the following ways:

			Expected Frequency
Mother I^OI^O × Father I^OI^O	$r^2 \times r^2$	=	r^4 (all children O)
Mother I^OI^O × Father I^AI^O	$r^2 \times 2pr$	=	$2pr^3$ (½ of children O)
Mother I^OI^O × Father I^BI^O	$r^2 \times 2qr$	=	$2qr^3$ (½ of children O)

Hence the combination of group O mother and group O child has the expectation

$$r^4 + \frac{2pr^3}{2} + \frac{2qr^3}{2} = r^4 + pr^3 + qr^3 = r^3(r + p + q) = r^3$$
$$(\text{since } p + q + r = 1)$$

So an investigator collects data on a population and computes the value of r (which equals $\sqrt{\bar{O}}$ as we saw above). Then if the theory is correct, the frequency of the combination group O mother and group O child should approximate r^3. Repeated testing in this manner has substantiated the correctness of the theory.

The example chosen above is simply a sample. We can ask questions about expectation concerning other parent-child combinations, and then test expectation against actual findings. For example, in what proportion of cases would we expect a group O mother and group A child? Expectation may be calculated in the manner of the preceding example.

Medico-Legal or Forensic Applications

With the multiple-allele theory firmly established, we can now determine what types of children are possible, and what types impossible, in any mating in which we may be interested. For example, we determined that when one parent is AB and the other is O, the children may be A or B, but cannot be AB or O. Many books (e.g., Snyder and David, 1957; Wiener, 1943–1962) contain tables from which one can read quickly the types of "possible" and "impossible" offspring from every type of mating, but we can easily calculate expectations for any mating in which we may be interested.

We can readily see that the ability to determine such expectations enables us to decide some disputes concerning family relationships. In most actual instances, these are cases of disputed paternity.

In one instance, for example, a woman hailed a man into court, claiming that he was the father of her child. Serological tests were performed. The mother belonged to group A, the child to group B; the accused man belonged to group O. Hence his innocence was established. Even without writing out the genotypes, we understand that the actual father must have contributed antigen B (which the mother did not have). The accused man did not have this antigen either. The mother's genotype was I^AI^O (note that if she had been I^AI^A she could not have had a group B child). The child was I^BI^O (he must have inherited I^O from his mother). The accused man was I^OI^O; hence he could not have an I^BI^O child by an I^AI^O woman. The true father must have been I^BI^B or I^BI^O or I^AI^B.

Suppose that the accused man *had* belonged to group B. What would that have indicated? Actually nothing, taken by itself, because any group B man could have been the father. The evidence in these cases is of an excluding variety: In a proportion of cases men wrongly accused of paternity can prove their innocence by means of blood tests. Of course, if they are *not* wrongly accused, the tests will be of no avail to them, and in some cases, as suggested, even if they are innocent the blood tests will not help them prove they are. The chance of proving innocence is increased by making use, in addition to the A and B antigens, of the M and N antigens discussed earlier, of the various Rh antigens to be discussed presently, and of a variety of other antigens known to serologists. When one set of antigens does not prove innocence, another set may do so.

If, for example, the mother is of group A, type M, and her child is of group O, type M, we know that an accused man of group O, type N, could not have been the father. The A and O groups do not show his innocence, but the MN types do. The child had the genotype $L^M L^M$; he received one L^M from his mother, and he received the other L^M from his father. The accused man, being $L^N L^N$, could not have contributed that second L^M.

As we bring more and more antigens to bear on a given case, we increase the chances of proving innocence. We also increase the chances of proving guilt. It is customary to say that the blood tests cannot prove guilt—that any man of the designated type could have been the father. This is true, but if we increase the number of antigens and still find that the accused man agrees in every detail with the genetic constitution that the true father must have had, we progressively decrease the likelihood that some *other* man of this exact genetic constitution may have been present under the proper circumstances to father the child in question.

In this respect, almost the ultimate has been reached in cattle, in which some 40 pairs of antigens are known. The chance that two unrelated bulls will agree perfectly with regard to all these pairs is of the order of $(1/2)^{40}$—a vanishingly small chance. For all practical purposes, an individual bull can be "finger printed" by means of his blood, and the fact that he sired a certain calf established beyond all reasonable doubt. This is obviously important with pedigreed cattle.

The number of identified human antigens has increased so rapidly in recent years that the positive individual identification possible today in cattle may soon be attainable in man.

Not all medico-legal cases involve disputed paternity. Sometimes babies are wrongly identified in hospitals, or at least parents think that they have been. From the standpoint of genetics, a particularly instructive case occurred in a large midwestern city some years ago. Two ladies, whom we shall call Mrs. Smith and Mrs. Jones, went to the same hospital and gave birth to baby girls on the same day. A few days later, they left the hospital on the same day, taking their babies with them.

When she reached home, Mrs. Smith discovered that her baby was labeled "Jones," and Mrs. Jones discovered that her baby was labeled "Smith." What to do? Were the babies correctly labeled but somehow accidently exchanged in the flurry of leaving the hospital? Or were the labels incorrect, so that each mother had, indeed, taken home her own baby? Solving the problem was not helped by the fact that both mothers claimed the same baby.

Blood tests were performed and cleared up the whole matter. By fortunate chance, Mr. Smith was AB and Mrs. Smith was O. The baby they took home, labeled "Jones," was group O. As we know, such a parent-child combination is impossible. The baby labeled "Smith" (whom the Joneses had taken home) belonged to group A, so the Smiths could have been her parents.

Mr. and Mrs. Jones both belonged to group O, as did the baby marked "Jones" (whom the Smiths took home). Thus, the Joneses could not have been the parents of the baby labeled "Smith" (group A).

Evidently, then, the labels were correct. The families exchanged the babies and everyone was satisfied. The happy outcome was due in large measure to the fact that one parent happened to belong to group AB. Under other circumstances, such clear-cut conclusions might not have been possible.

Applications to Anthropology

We noted in our discussion of distal hyperextensibility of the thumb (p. 157) that geneticists and physical anthropologists are constantly seeking human traits that have a clearly established, and preferably simple, genetic basis. Such traits are of great value in comparing populations and in tracing the relationships of different groups to one another. Because of the relatively direct relationship between genes and the antigens they produce or control, red blood cell antigens form the best traits of this kind known at present.

The anthropological literature is filled with serological data. Blood tests have been performed on an amazing number of varied populations from all parts of the earth. Readers interested in the blood group distribution of almost any people will find pertinent data in the huge tables compiled by Wiener (1943–1962) and Mourant (1954).

Extensive discussion of the anthropological implications of the blood groups is found in Boyd's *Genetics and the Races of Man* (1950). At present, we can find space for only a few examples to illustrate the manner in which serological data can be useful in the study of human diversity and the formation of races.

At the outset we may note that, on the whole, although populations differ in the *proportions* in which the different blood groups occur, rarely, if ever, is one antigen completely lacking in one population. A comparison of the blood group distribution of Western Europeans with that of Eastern Asiatics demonstrates this. A typical study of Englishmen in London shows the following distribution (Boyd, 1950):

$$\text{group O} = 47.9\%;\ A = 42.4\%;\ B = 8.3\%;\ AB = 1.4\%$$

We may compare these figures with results of a study of a Chinese population:

$$O = 34.2\%;\ A = 30.8\%;\ B = 27.7\%;\ AB = 7.3\%$$

We note that a smaller proportion of Chinese have antigen A as compared to the number of Englishmen who have it (30.8% + 7.3% = 38.1% of Chinese have it, alone or combined with B; whereas 42.4% + 1.4% = 43.8% of Englishmen have it, alone or combined with B). The largest contrast is in the possession of antigen B. Among Englishmen, 8.3% + 1.4% = 9.7% have B, alone or combined with A; among Chinese, 27.7% + 7.3% = 35% have B, alone or combined with A.

A genetically preferable way of saying this is to use formulas (1), (2), and (3) (p. 184) to calculate the frequencies of genes I^A, I^B, and I^O in the two gene pools.

In this manner, the data concerning the English population sample reveal a gene pool in which the approximate gene frequencies are (Boyd, 1950)

$$p = 0.250 \qquad q = 0.050 \qquad r = 0.692$$

whereas the data concerning the Chinese population sample reveal a gene pool in which the approximate frequencies are

$$p = 0.220 \qquad q = 0.201 \qquad r = 0.587$$

The *q*'s tell the story: among the English, 5 percent of the genes are I^B, whereas among the Chinese, 20 percent of the genes are I^B. Geneticists look forward to the day when *all* differences between populations, and racial groups, can be expressed in terms of gene frequencies.

England and China are at the two extremes of the Eurasian continent. Long ago it was pointed out (by H. and L. Hirszfeld; see Wiener, 1943–1962) that there are progressive differences in populations lying between these two extremes. As one travels eastward on the continent of Eurasia, one successively encounters populations having less and less antigen A and more and more antigen B. Although this geographic trend is interesting, we are still far from explaining how it originated. We might imagine that antigen A originated in Europe and antigen B in the Far East, and that as people migrated back and forth, the I^A gene gradually spread eastward and I^B spread westward. Such a hypothesis has the appeal of simplicity, but it leaves many questions unanswered. Much evidence suggests that both antigens A and B are very ancient; in fact, they are probably older than man himself, because they are found in subhuman primates such as the great apes (see Moody, 1970) and even in some lower mammals. They may have arisen by mutation of genes more than once in evolutionary history, a fact that complicates our attempts to trace that history by serological means.

At times, however, blood grouping may enable an investigator to determine the affinities and probable origin of some groups of people. The following well-known case is of value because we have independent knowledge of the history of the group in question. Several hundred years ago a group of gypsies migrated from their native India to Hungary. They lived among the Hungarian people but did not intermarry with them (thus they constituted a genetic isolate—p. 176). Serological tests demonstrate that they still retain the blood group distribution of their ancestors (Table 12.3), which differs markedly from that of their Hungarian neigh-

TABLE 12.3 ABO blood group frequencies of Hungarians, Hungarian gypsies, and people in India.

	Percent in group			
	O	A	B	AB
Gypsies (in Hungary)	34.2	21.1	38.4	8.5
People in India	31.3	19.0	41.2	8.5
Hungarians	31.0	38.0	18.8	12.2

From a larger table in Wiener, A. S., *Blood Groups and Transfusion*, 3rd ed., 1943. Courtesy of the author and of Charles C Thomas, Publisher, Springfield, Ill.

bors. If we did not know the history of these people, but concluded from the blood group distribution that they were of Indian, or at least Asiatic, origin, we should have drawn a correct conclusion. In cases in which we do not know the history, then, conclusions based on the blood groups may frequently be of value.

Students of human genetics and evolution are very interested in the exchange of genes between populations, and in the related problem of how mutations are distributed from the center in which they first occur. Gene pools are mingled whenever diverse peoples come into contact, either through migration or even through such casual contact as when soldiers of one country are stationed for a time in another country. The mixed phenotypes of the children produced give evidence of the genetic mixing, although many traits of coloring, facial features, and other characteristics by which we usually distinguish one people from another have not yet been thoroughly analyzed genetically. In this respect, blood groups offer an advantage, as we have already noted. An interesting case in point is afforded by the Australian aborigines and their neighbors to the north (Fig. 12.2). These neighbors possess the B gene (I^B), but the gene seems to be lacking among the natives of Australia except for a few individuals in the extreme northern portion (Cape York Peninsula and the region adjacent to the Gulf of Carpentaria). Birdsell (1950) concluded that the infiltration of I^B into these regions is to be explained by contacts between the aborigines and (a) Papuan populations from the islands in the strait between Australia and New Guinea, and (b) Malays of Indonesia, who in

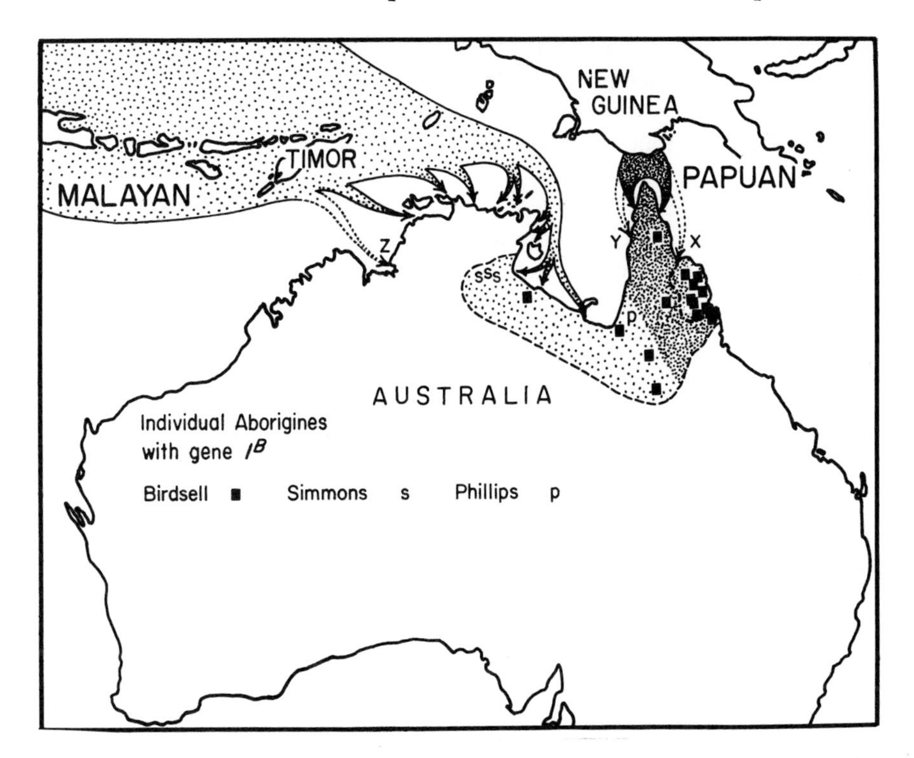

FIG. 12.2 The northern coast of Australia with neighboring islands. Shading shows regions in which some of the aborigines possess antigen B. The shaded arrows show probable routes by which visitors in times past brought the I^B gene to Australia. (After Birdsell, J. B., 1950. "Some implications of the genetical concept of race in terms of spatial analysis," *Cold Spring Harbor Symposia on Quantitative Biology,* **15**:259–311.)

early historic times are known to have sent fishing fleets into Australian waters.

Incidentally, the student of human blood groups has many interesting riddles to solve. How shall we explain the fact that Australian aborigines originally lacked I^B, if our evidence is correct? And how shall we explain the fact that American Indians are almost lacking in I^B (some groups tested lack it entirely; others have gene pools in which q is about 1 percent)? The ancestors of the American Indians presumably migrated from Asia across a land bridge spanning what is now Bering Strait, yet modern Asiatics have a high percentage of I^B! Did the ancestors of the Indians migrate before I^B became abundant in Asiatic gene pools? Did the ancestors belong to a different racial stock from that represented by the modern inhabitants of Asia? Perhaps chance, in the form of "genetic drift" (pp. 417–423), determined that the ancestral migrants included very few individuals having I^B. Perhaps differential susceptibility to disease played a role (see pp. 209–212).

These are typical of the fascinating problems presented to the student of blood group distributions.

Some Allied Considerations We have spoken of the genes concerned in the ABO blood groups as constituting a series of three multiple alleles: I^A, I^B, I^O. Actually there are four or five, and possibly more. For many years it has been known that antigen A is not a single entity but that there are two serologically distinct forms of it, called A_1 and A_2. Correspondingly, there are two "A genes": I^{A1} and I^{A2}. Additional variants of the A antigen have also been discovered, but they are rarer.

Recognition of subgroup A_1 and subgroup A_2 increases the list of multiple alleles to four. Persons in subgroup A_1 may have any one of three genotypes: $I^{A1}I^{A1}$, $I^{A1}I^{A2}$ (I^{A1} is dominant to I^{A2}), or $I^{A1}I^O$. People in subgroup A_2 are $I^{A2}I^{A2}$ or $I^{A2}I^O$. The AB group is similarly divided: $I^{A1}I^B$ or $I^{A2}I^B$.

These subgroups are useful in medico-legal tests, increasing the chance of proving innocence in paternity cases, for example. They are also of interest anthropologically. In most populations studied, group A_1 is much more numerous than group A_2. In a typical study of Englishmen, the A_1 people outnumbered the A_2 people about 4:1. Some peoples seem to lack A_2 completely, for example, Australian aborigines, Chinese, Japanese, American Indians, and inhabitants of Pacific islands (Boyd, 1950). In the peoples of Africa, like those of Europe, subgroup A_2 is well represented. With so much diversity, the subgroups may contribute significantly to studies of human relationships.

The red blood cells of people belonging to group O, lacking both antigens A and B, can usually be demonstrated to have an antigen known as the **H substance.** This can be shown by suitable reagents. The evidence indicates that presence of this substance depends upon a gene *H*, separate from the series of ABO multiple alleles. With rare exceptions, people belonging to the other blood groups also have this gene, although their cells do not usually react with anti-H antiserum. (Blood of subgroup A_2 regularly reacts.) This failure to react has been interpreted in various ways. According to one idea, the H substance is converted into the A or B antigen when I^{A1} or I^B is also present (Watkins, 1966). Or perhaps genes I^{A1}, I^B, and *H* compete for a common precursor substance needed to produce the antigens, and I^{A1} and I^B are stronger competitors than is *H* (Wiener, 1970). People lacking gene *H* (genotype *hh*) are extremely rare, and as we would expect from the discussion in the preceding chapter, there is a high degree of consanguinity in the parentage of known cases (Race and Sanger, 1968).

Secretors

Antigens A and B are found not only in red blood cells but also in the cells of other tissues in the body. In some people the antigens are found in such secretions of the body as the saliva. It seems that all possessors of A or B have an alcohol-soluble form of these antigens, but some people have a water-soluble form as well. Such people are called **secretors;** persons who have the antigens only in their cells are called **nonsecretors.**

Genetic analysis leads to the conclusion that a single gene difference is involved, secretion of the antigens being dependent on a dominant gene, *Se*. Homozygotes for the recessive allele (*se se*) are nonsecretors. Although this conclusion is of interest in itself, the most meaningful question for our present consideration is: What evidence supports this conclusion? How would we go about determining whether or not it is correct?

One way of answering the question begins with determining the relative frequency, q, of the presumed recessive gene. Investigations of 1118 persons in Liverpool showed that 254 or 22.72 percent of them were nonsecretors (Race and Sanger, 1968).

From this, we obtain:

$$q^2 = 0.2272 \qquad q = \sqrt{0.2272} = 0.4767$$

Therefore

$$p = 1 - 0.4767 = 0.5233$$

Table 12.4 presents data from an investigation of 185 families. Note that when both parents are nonsecretors, all the children are, too. This agrees with the hypothesis that nonsecretors are homozygous for a recessive gene.

TABLE 12.4 Inheritance of the secretor trait.

Parents	No. of families	Children		
		Secretor	Nonsecretor	Total
Secretor × Secretor	105	241	33	274
Secretor × Nonsecretor	62	103	67	170
Nonsecretor × Nonsecretor	18	0	42	42

From Wiener, A. S., *Blood Groups and Transfusion,* 3d ed., 1943. Courtesy of the author and of Charles C Thomas, Publisher, Springfield, Ill.

The first line of the table presents the most numerous cases—105 families in which both parents are secretors. If our hypothesis that secretion depends on a dominant gene is correct, we have lumped together here three parental combinations: (a) both parents homozygous; (b) one parent homozygous, one heterozygous; (c) both parents heterozygous. From such grouped parentage, what proportion of recessive (nonsecretor) children would be expected?

It was to answer such questions that we developed formula (1) in Chapter 10 (p. 156). The proportion expected is

$$\left(\frac{q}{1+q}\right)^2$$

Let us assume that the value of q given by the Liverpool data is typical of the gene pools to which these families belong. Then,

$$\left(\frac{q}{1+q}\right)^2 = \left(\frac{0.4767}{1.4767}\right)^2 = (0.3228)^2 = 0.1042 = 10.42\%$$

Thus, 10.42 percent of the children would be expected to be nonsecretors. The total number of children in these families was 274 (Table 12.4); 10.42 percent of 274 is 28.6. Reference to the table shows that 33 nonsecretor children were actually found. Because 33 and 28.6 are not far apart, we are probably justified in concluding that the evidence supports the hypothesis of single gene-pair inheritance with the secretor gene dominant.

Perhaps, on the other hand, someone might object that the agreement is not close enough to support the hypothesis. Is there any way in which we can bolster our argument? We may ask: How frequently would one expect to find the observed deviation (between 33 and 28.6) arising just by chance? If we investigated a larger number of families, might not the discrepancy between observation and expectation disappear? These 105 families are just a sample of all families in which both parents are secretors. Perhaps the deviation observed arises from the fact that the 105 families are not completely typical of all families of this kind. In other words, perhaps an error of sampling is involved. How can we determine whether or not the observed deviation is probably a chance occurrence caused by a sampling error in our data?

Long ago, Karl Pearson invented a test to help in such a situation. It is called the **chi-square test.** Although it is by no means the only, or necessarily the best, means of testing statistical significance, it is one of the simplest and one of the most widely used tests. In Appendix C (pp. 463–465) the test is explained and applied to the present data. We find that with 274 children, obtaining 33 of one type when we expected 28.6 would occur by chance about 40 percent of the time. In other words, if we had a large number of investigations, each involving 274 children, 40 percent of the investigations would be expected to show as much deviation as this, just by random chance. Hence there is a good likelihood that chance alone is involved in the deviation between 33 and 28.6 and that this deviation is not significant (i.e., is not significant evidence against our hypothesis). In this manner, the chi-square test enables us to feel more confident of the validity of conclusions based on data that are limited in quantity, as they always are in human genetics.

We must not gain the impression that we have considered all the evidence supporting the conclusion that secretion of A and B in fluids such as saliva depends on a dominant gene. This example has been included to illustrate the type of thinking involved when geneticists approach a problem of this kind.

The alleles *Se* and *se* determine a pair of contrasting phenotypes that can be employed, along with the blood groups themselves, in investigations of families, populations, and racial or ethnic groups.

Problems

1. A widow belonging to blood group A had two children, of groups O and B, respectively. What must have been the genotype and phenotype of the father? Also give the genotypes of the widow and her children.
2. A sibship was composed of four children belonging, respectively, to groups A, B, O, and AB. Give the genotypes and phenotypes of the parents of the sibship.
3. To what blood groups may children from each of the following matings belong: A × O; B × O; A × A; AB × O; A × AB; B × B?
4. Give all the different matings that may give rise to a child belonging to group AB.
5. In the gene pool of a certain population the relative frequencies of the blood group genes are as follows: $p = 0.30$; $q = 0.06$; $r = 0.64$. In this population, what percentage of the people belongs to each of the four groups?
6. If the multiple-allele theory of ABO blood group inheritance is correct, how frequently should the combination of group O mother and group A child be encountered in a population? (Answer in terms of p and r, and then in terms of the gene frequencies given in Problem 5.)
7. A baby belongs to blood type M, group O, and is a secretor. Which of the following pairs of parents could *not* have been the parents of this baby? Explain in each case.
   (1) M, A, nonsecretor × M, B, secretor
   (2) M, AB, nonsecretor × M, O, secretor
   (3) M, O, secretor × N, O, secretor
   (4) M, A, nonsecretor × M, O, nonsecretor
   (5) M, A, secretor × M, A, secretor
8. The baby in a paternity dispute was found to be A_2, MN. The mother was A_1, M. The accused man was A_1, N. What, if any, conclusions can be drawn?
9. In a panmictic population, what proportion of the mother-child combinations will be expected to consist of group B mothers and group A child?

(Answer in terms of p, q, and r, and then in terms of the gene frequencies given in Problem 5.)

10. An investigation of Italian people in Sardinia revealed that among them the following approximate gene frequencies are found:

p (frequency of I^A) = 18%
q (frequency of I^B) = 11%
r (frequency of I^O) = 71%

In this population, what percentage of the marriages would be expected to occur between group A men and group O women? (Answer first in terms of the Hardy-Weinberg formula applied to these groups, and then utilize the above percentages to obtain the final answer.)

11. A woman accused of kidnapping claimed that she was the mother of the child in question. Blood tests showed that she belonged to group AB, the child to group O. What conclusions were warranted?

In working the following problems, consult Appendix C (pp. 463–465)

12. From Table 12.4 we learn that an investigation of 62 families in which one parent was a secretor the other a nonsecretor showed that 103 of the children were secretors and 67 were nonsecretors. If the nonsecretor trait depends on a recessive gene, how many of this total of 170 would we expect to be secretors, how many nonsecretors? (In computing this, use the values of p and q given on p. 193.) Apply the chi-square test to your findings. What is the probability that the observed deviation from expected numbers can be attributed to chance? Is the deviation significant evidence against the hypothesis that nonsecretors are homozygous for a recessive gene? What would you suggest as a next step in investigating the hypothesis?

13. In discussing inheritance of distal hyperextensibility of the thumb, we noted that of 108 children born to dht × non-dht parents, 35.8 would be expected to be dht if the latter condition depends on a recessive gene (p. 158). The actual data obtained by Glass and Kistler gave 37 dht children, 71 non-dht children. Use the chi-square test to determine the probability that the deviation from expectation is attributable to chance. Is the deviation significant evidence against the hypothesis?

14. If you worked Problem 1 in Chapter 10 correctly (p. 161), you found that the population at equilibrium would have consisted of 147 *AA*, 126 *Aa*, 27 *aa*. The respective numbers actually observed were 150 *AA*, 120 *Aa*, 30 *aa*. Use the chi-square test to determine whether or not the deviation is significant evidence against the hypothesis that the population is in equilibrium. What do you conclude and why?

15. Apply the chi-square test to the results obtained in Problem 4 of Chapter 10 (p. 162). What is the probability that the observed deviation from expected numbers is attributable to chance? Is the deviation significant evidence against the hypothesis that the population is in equilibrium?

References

Bernstein, F., 1925. "Zusammenfassende Betrachtungen über die erblichen Blutstrukturen des Menschen," *Zeitschrift für induktive Abstammungs- und Vererbungslehre,* **37**:237–270.

Birdsell, J. B., 1950. "Some implications of the general concept of race in terms of spatial analysis," *Cold Spring Harbor Symposia on Quantitative Biology,* **15**:259–311.

Boyd, W. C., 1950. *Genetics and the Races of Man.* Boston: Little, Brown & Company.

Erskine, A. G., 1973. *Principles and Practice of Blood Grouping.* St. Louis: C. V. Mosby Co. (A valuable source of information on blood groups.)

Landsteiner, K., 1901. "Ueber Agglutinationserscheinungen normalen menschlichen Blutes," *Wiener klinische Wochenschrift,* **14**:1132–1134.

Moody, P. A., 1970. *Introduction to Evolution,* 3rd ed. New York: Harper & Row.

Mourant, A. E., 1954. *The Distribution of Human Blood Groups.* Springfield, Ill.: Charles C Thomas.

Race, R. R., and R. Sanger, 1968. *Blood Groups in Man,* 5th ed. Philadelphia: F. A. Davis Company.

Snyder, L. H., and P. R. David, 1957. *The Principles of Heredity,* 5th ed. Boston: D. C. Heath & Company.

von Dungern, E., and L. Hirszfeld, 1910. "Ueber Vererbung gruppenspezifischer Strukturen des Blutes. II," *Zeitschrift für Immunitätsforschung,* **6**: 284–292.

Watkins, W. M., 1966. "Blood-group substances," *Science,* **152**:172–181.

Wiener, A. S., 1943–1962. *Blood Groups and Transfusion,* 3rd ed. Springfield, Ill.: Charles C Thomas. Reprinted—New York: Hafner Publishing Company.

Wiener, A. S., 1961, 1965, 1970. *Advances in Blood Grouping,* Vols. I, II, and III. New York: Grune & Stratton, Inc.

13

Multiple Alleles: Rh and Other Antigens

Rh, the "Rhesus Factor"

Discovery of another red blood cell antigen was announced by Landsteiner and Wiener in 1940. These authors had inoculated rabbits and guinea pigs with suspensions of red blood cells from rhesus monkeys. The antibodies formed by the rabbits and guinea pigs agglutinated red blood cells of rhesus monkeys. These investigators then used the antisera to test human red blood cells. They found that among 448 white inhabitants of New York City, 379 (84.6 percent) had cells that reacted with the antibodies, whereas 69 (15.4 percent) had cells that did not. People whose cells were agglutinated by the antibodies were called **Rh-positive;** those whose cells were not agglutinated were called **Rh-negative** ("Rh" refers to Rhesus).

Landsteiner and Wiener then investigated the inheritance of the new antigen, using methods of analysis much like those we have discussed in connection with the secretor gene investigations (see their paper of 1941). They demonstrated that the presence of the antigen Rh is determined by a dominant gene *R*. Thus Rh-positive people have the genotype *RR* or *Rr* whereas Rh-negative people have the genotype *rr*.

Subsequent investigation by many people revealed that the genetic situation is much more complicated than the simple mode of inheritance just stated. Not all Rh-positive people are exactly alike in the antigens or "blood factors" they possess, and neither are all Rh-negative people. On the basis of extensive investigations, Wiener formulated the theory that a series of multiple alleles is involved, as in the ABO groups.

The whole subject has become complex. In a book such as this one, which emphasizes principles, there is no space for many of the details, and therefore a simplified explanation must suffice. In our discussion of the MN types (pp. 81–82) and the ABO groups (pp. 180–181), we showed in each case how one's blood cells could be typed by the use of two antisera, forming test fluids. To demonstrate the essentials of the Rh system, we need three antisera. In Table 13.1, these three test fluids are called anti-**Rh**$_0$ or anti-D, anti-**rh′** or anti-C, anti-**rh″** or anti-E. The reason for the double designation in each case will be evident as we proceed. The original antisera employed by Landsteiner and Wiener were of a specificity similar to anti-**Rh**$_0$ or anti-D. Hence people whose cells are clumped by this antiserum are "Rh-positive," as that term is usually employed, and people whose cells do not react with this antiserum are "Rh-negative."

Table 13.1 shows all the possible combinations of reactions between cells and those three antisera. The first horizontal line, for example, represents a case in which the cells being tested do not react with any of the three antisera. Possessors of such cells have a phenotype designated as rh. The second line of the table shows reaction with anti-**rh′** or anti-C but not with the other two antisera. Persons who have such cells are said to be of the rh′ phenotype.

TABLE 13.1 The eight Rh phenotypes demonstrable with three antisera.

Reactions with antisera			Phenotypes	
anti-**Rh**$_0$ or anti-D	anti-**rh′** or anti-C	anti-**rh″** or anti-E		
−	−	−	rh	Rh-negative
−	+	−	rh′	
−	−	+	rh″	
−	+	+	rh_y	
+	−	−	Rh_0	Rh-positive
+	+	−	Rh_1	
+	−	+	Rh_2	
+	+	+	Rh_z	

+ = cells react with antiserum.
− = cells do not react with antiserum.

What are the genotypes that produce the phenotypes given in the table? The eight phenotypes can be explained on the basis of various combinations of eight multiple alleles, designated as follows:

r or (dce)
r' or (dCe)
r'' or (dcE)
r^y or (dCE)
R^0 or (Dce)
R^1 or (DCe)
R^2 or (DcE)
R^Z or (DCE)

The designations on the left are those of Alexander S. Wiener who, as noted above, was one of the discoverers of Rh. The symbols on the right are those of the Fisher-Race system. Each set of parentheses may be considered a genetic unit comprised of three subunits represented by the three enclosed letters. We shall discuss the postulated nature of these units and subunits presently.

Let us see how the system works. We note from Table 13.1 that people with the rh phenotype have red blood cells that do not react with any of the three antiserums (first horizontal row). Such people have the genotype *rr* or (*dce*) (*dce*).

People of the rh′ phenotype (second row) may be homozygous *r′r′* or heterozygous *r′r;* (*dCe*) (*dCe*) or (*dCe*) (*dce*). Note that in the Fisher-Race designation a capital (uppercase) letter appears wherever a plus sign occurs in the table.

People who belong to rh″ may be *r″r″* or *r″r;* (*dcE*) (*dcE*) or (*dcE*) (*dce*).

People of the rh_y phenotype may have any one of five genotypes; three of them are r^yr^y [(*dCE*)(*dCE*)], r^yr [(*dCE*)(*dce*)], *r′r″* [(*dCe*) (*dcE*)]. Readers can easily calculate the other possibilities, remembering that the phenotype will result from any combination of genes that produces cells that will react with anti-**rh′** and anti-**rh″** but will fail to react with anti-$\mathbf{Rh_0}$.

All of the types mentioned so far are Rh-negative. The last four rows of the table represent various Rh-positive phenotypes. In the first place, note that the cells of Rh-positive people always react with anti-$\mathbf{Rh_0}$ or anti-D, whereas the cells of Rh-negative people do not react with this antiserum. Among the Rh-positive people, those who belong to type Rh_0 have the genotype R^0R^0 orR^0r; (*Dce*) (*Dce*) or (*Dce*) (*dce*).

People with the Rh_1 phenotype may have the genotype R^1R^1 or R^1r; (*DCe*) (*DCe*) or (*DCe*) (*dce*). They may also have other genotypes such as R^0r'; (*Dce*) (*dCe*).

The Rh_2 phenotype may arise from the R^2R^2 or R^2r [(*DcE*) (*DcE*) or (*DcE*) (*dce*)] genotypes, but other pairs of genes may produce it [for example, R^0r'' (*Dce*) (*dcE*)].

Similarly, a person belonging to type Rh_Z may have the genotype R^ZR^Z or R^Zr [(DCE) (DCE) or (DCE) (*dce*)]. He may also have the genotype R^zR^0 [(*DCE*) (*Dce*)], R^1R^2 [(*DCe*) (*DcE*)], and so on. Dr. Wiener lists 14 genotypes that Rh_Z people may have.

Table 13.1 gives the results of tests with only three antisera. Use of additional antisera reveals further complexity, but the principle of multiple-allele inheritance as it applies to Rh is demonstrated adequately by the series of eight multiple alleles we have discussed.

We turn to the problem presented by the two systems of designating genotypes. According to the Wiener system, each gene is thought to produce one antigen, but the different antigens vary in the number of "blood factors" (serological specificities) they possess. For example, gene r^y produces an antigen with two blood factors, one that causes the antigen to react with anti-**rh′**, and another that causes the antigen to react with anti-**rh″**. On the other hand, some of the other alleles (e.g., R^0) are thought to produce an antigen with only one of these three specificities, and R^Z to produce an antigen with all three.

What do the symbols in the Fisher-Race system represent? As originally conceived, the three letters were postulated to represent three genes tightly linked together by being in the same chromosome, the reason for enclosing them within parentheses (Chap. 15). Each gene was considered responsible for production of a single antigen with a single serological specificity. In those days genes were thought of as units that could not be subdivided. But as years passed, intensive investigation of the genetics of bacteria and viruses gave us the picture of a gene we presented in Chapter 4: a length of DNA composed of many—even hundreds—of subunits (nucleotides). On p. 45 we defined a gene in terms of function, concluding that it is "that portion of DNA that codes for one polypeptide chain." And we noted that the term *cistron* is frequently used for a gene defined in this manner.

Accordingly, the parentheses used in the Fisher-Race system may be taken to represent a length of DNA composed of a large number of "mutational sites" (Race and Sanger, 1968). These are sites where genetic changes and even crossing over (pp. 29–31) may occur. As we noted in our discussion of the abnormal hemoglobins (pp. 47–50), change in a single nucleotide in the DNA molecule may cause important change in the hemoglobin polypeptide chain for which that section of DNA codes. *D, C,* and *E* in the Fisher-Race system are now considered to represent three such sites in the portion of DNA coding for the Rh antigen.

Do these sites constitute portions of *one* gene? If so, each pair of parentheses is the equivalent of one letter in the Wiener system of nomenclature. At the present time it is impossible to answer this question decisively. The method used in experimental genetics to distinguish whether two sites are in the same cistron or in different ones is called the *cis-trans* test. Discussion of it would be out of place here, but will be found in more advanced textbooks of general genetics. Suffice it to say that what has been taken to be the equivalent of such a test indicates that *C* and *E* are probably in one cistron (gene) (Race and Sanger, 1968). But no comparable evidence is available with regard to *D,* so the question remains open.

It is to be hoped that in the not-distant future, it will be possible to extract the Rh antigen from red blood cells and analyze its chemical composition, together with the chemical differences between one form of the antigen and another (between Rh_0 and Rh_Z substance, for example).

Then it will be important to isolate and analyze the enzyme that makes possible production of the antigen in its various forms. Recalling our definition of "gene," we ask: Can changes in one polypeptide chain in such an enzyme be expected to account for all the differences in serological specificities actually found? An affirmative answer seems possible, even probable, when we recall the many abnormal hemoglobins that are produced by simple alterations of the beta chain of that protein molecule. Indeed, an affirmative answer was suggested in a theory concerning the Rh locus advanced by Edwards (1968).

Incompatibility

Rh Incompatibility Rh-negative people lack certain antigens found in the red blood cells of Rh-positive people, but the serum of Rh-negative people normally does not contain antibodies against these antigens (isoantibodies). In this, Rh is like MN and unlike ABO.

However, Rh-negative people can be stimulated to form Rh antibodies—by blood transfusion, for example. If an Rh-negative person receives blood from an Rh-positive person, the introduced red blood cells may stimulate the Rh-negative recipient to form antibodies. These may do no damage at the time, but if, later on, the Rh-negative person receives a second transfusion of Rh-positive blood, the antibodies formed previously may clump and destroy the newly introduced Rh-positive cells, leading to a variety of undesirable and perhaps serious effects.

Soon after the discovery of Rh, it was realized that Rh-negative women can be stimulated in another way to form antibodies. If such a woman marries an Rh-positive man, she may become pregnant with an Rh-positive fetus. (The marriage is either mother *rr* × father *RR* or mother *rr* × father *Rr*; if the father is homozygous, the baby is sure to be Rh-positive (*Rr*); if the father is heterozygous, the chance that the baby is Rh-positive is ½. Here we use *R* to represent any one of the "capital *R* genes" in Wiener's series of multiple alleles, *r* to represent any one of the "small *r* genes.") In some cases, but by no means in all, an Rh-positive fetus may stimulate its Rh-negative mother to form antibodies.

In these instances, the intact red blood cells of the fetus, with their Rh antigens, must find their way into the mother's bloodstream and then stimulate her antibody-forming mechanisms. The circulatory systems of fetus and mother are separate. Maternal and fetal bloods come into contact in the placenta, but even there a thin membrane keeps them separate. It seems necessary to postulate that at times lesions (holes) must develop in this membrane, permitting fetal red blood cells to "leak" into the mother's blood stream. Richardson-Jones (Levine, 1958) demonstrated the presence of Rh-positive cells in the circulation of Rh-negative mothers during the last three months of pregnancy. He estimated a ratio of 1:2000–1:5000 fetal Rh-positive cells to maternal Rh-negative ones.

If the pregnancy is a first one, the antibodies may form slowly and not reach sufficient concentration to do any harm. Indeed the greatest likelihood of sensitization of the mother seems to be at the time of birth, when a considerable quantity of fetal blood cells may leak into the mother's circulation and stimulate the formation of antibodies. Then if this first pregnancy is followed by a second one, involving a second Rh-positive fetus, unfortunate results may ensue. Blood cells escaping from this second fetus may increase the mother's production of antibodies and certain of these, called blocking or incomplete antibodies, may pass through the placenta into the fetus (Fig. 13.1). There they coat fetal red blood cells; these are subsequently destroyed by the fetus, with resulting anemia and the production of a harmful compound called bilirubin. This is a red bile pigment formed from the hemoglobin of destroyed red blood cells. An elevated level of bilirubin may result in a condition known as kernicterus, in which nerve centers are destroyed. The general term for the hemolytic anemia is *erythroblastosis fetalis.* (The name refers to erythroblasts, the nucleated cells from which red blood cells develop.)

Because of these difficulties, marriages between Rh-negative women and Rh-positive men are called incompatible matings, a term that is perhaps ill-chosen. Such matings are not at all rare. In a population in which 85 percent of the men are Rh-positive, an Rh-negative woman is much more likely to marry an Rh-positive man than she is an Rh-negative man. Such marriages represent about 13 percent of all marriages in Caucasian populations, and in most cases no difficulties are encountered. Why not?

1. The small size of human families is one reason. As we have noted, although a first Rh-positive child may stimulate his mother to form antibodies, he may suffer no ill effects himself because antibodies are not present in sufficient quantity. Many families consist of no more than one child.

2. Furthermore, if the father is heterozygous (*Rr*), the first child may be Rh-negative and therefore fail to stimulate antibody formation. A second child, even if he is Rh-positive, will encounter no antibodies in

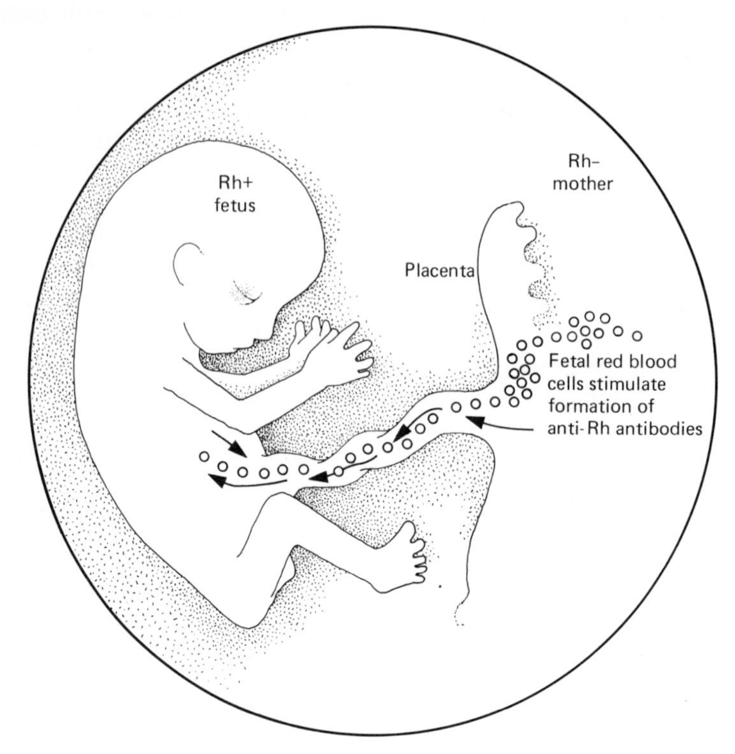

FIG. 13.1 Stimulation of a sensitized Rh-negative mother by Rh-positive fetal red blood cells escaping through a "leak" in the placenta. The mother is stimulated to form antibodies, certain of which pass through the placenta into the fetus. See text.

his mother's blood. In fact, the second child may also be Rh-negative, and so may a third or fourth child. (The chance of three successive Rh-negative children from an $Rr \times rr$ mating is $\frac{1}{2} \times \frac{1}{2} \times \frac{1}{2} = \frac{1}{8}$—not a small chance at all.) Even if the first child is Rh-positive and stimulates antibody production, if the second child is Rh-negative the antibodies will not act on it.

3. Again, women may well differ in their sensitivity to introduced Rh antigens. Pregnancy with an Rh-positive fetus causes not more than 4 percent of Rh-negative women to produce Rh antibodies (Wiener, *personal communication*). Hence one woman may form antibodies very readily, whereas another woman may form few, if any, antibodies under the same circumstances. This is comparable to the differences between people in their sensitivity to plant pollens—some people have symptoms of asthma or hay fever, others do not. In this case, there is evidence that genetic factors are concerned. Future research may well demonstrate that Rh-negative women have genetically determined differences in their sensitivity to Rh antigens introduced into their systems.

4. Medical science is constantly improving methods of preventing antibody formation or alleviating the effects of it. We stated above that an Rh-negative mother is most likely to be stimulated to antibody formation at the time of birth of an Rh-positive baby (through leakage of the baby's blood into the mother's circulation). If, *before* she has started to develop antibodies, such a mother is injected with gamma globulin containing anti-$\mathbf{Rh}_0$ (anti-D) antibodies (trade name: RhoGAM), active antibody production by her will be prevented (Freda et al., 1967).

Furthermore, if an infant is born who is in danger of developing erythroblastosis, his blood with its dangerous level of bilirubin may be removed and "safe" blood substituted by transfusion. Or the bilirubin may be decomposed in the infant's body in a variety of ways, such as by placing the baby in an incubator bathed with fluorescent light. New medical procedures are constantly being developed.

5. The parents may be "ABO incompatible," thus reducing the likelihood that their Rh incompatibility will be detrimental to production of normal offspring.

ABO Incompatibility As soon as the role of Rh incompatibility in causing erythroblastosis was known, investigators began to take a closer look at the ABO blood groups, hitherto regarded as a biological curiosity of no clinical importance, aside from problems concerned with blood transfusion. If a fetus containing Rh antigens can be damaged by anti-Rh antibodies in its mother's blood, might not a fetus containing antigen A be damaged by anti-A antibodies in its mother's blood? This situation may arise when, for example, the mother belongs to group O, the father to group A. We recall that in this case the anti-A antibodies are already present in the mother's blood. Perhaps the normal concentration is usually not great enough to cause difficulty, but in some cases the mother may be stimulated to form more, until a concentration is reached that will harm the fetus. Such antibodies might be particularly destructive to the group A fetus because the A antigen is present in tissue cells as well as in the red blood cells (whereas the Rh antigens are confined to red blood cells) (Levine, 1943). Hence anti-A antibodies in sufficient concentration might be widely destructive throughout the fetal body.

For these reasons, investigators look to ABO incompatibility to explain some of the previously mysterious early miscarriages and spontaneous abortions. Not infrequently, embryos are lost very early, often before the mother realizes that she is pregnant.

What evidence do we have that ABO incompatibility may play a role in such cases?

Long ago Hirszfeld, one of the pioneers in blood group research, noticed that in A $\times$ O marriages there are fewer group A children when the mother is O and the father A than there are when the mother is A and the father O (Levine, 1943). He attributed the difference to miscarriages and stillbirths when mother and fetus differed in blood group. This early clue was not pursued further until the discovery of the dramatic role of Rh in producing hemolytic disease in the newborn.

In investigations of this kind, we ask: What proportion of the children from A × O matings are expected to belong to group A? The answer depends on the gene frequencies in the population to which the families belong. We may apply the Hardy-Weinberg formula, as it relates to blood groups, to the problem (pp. 184–185).

Let us ask first concerning families in which the father is O, the mother A. Here the mother may be homozygous or heterozygous. We may write out the expected frequencies as we did previously:

Expected Frequency

Mother I^AI^A × Father I^OI^O, $p^2 \times r^2 = p^2r^2$ (all children A)
Mother I^AI^O × Father I^OI^O, $2pr \times r^2 = 2pr^3$ (½ of children A)

Thus the expected frequencies of all children in these families are $p^2r^2 + 2pr^3$. Of these children, $p^2r^2 + pr^3$ belong to group A. Hence the proportion of group A children is

$$\frac{p^2r^2 + pr^3}{p^2r^2 + 2pr^3} = \frac{pr^2(p + r)}{pr^2(p + 2r)} = \frac{p + r}{p + 2r}$$

Therefore, if we know the values of p and r (frequencies of genes I^A and I^O, respectively), for a given population, we can calculate the proportion of the children of A mother and O father marriages that will be expected to belong to group A.

Now if incompatibility of blood group between mother and fetus did no harm, the same proportion should be found in families in which the mother belongs to group O, the father to group A.

Many investigations have indicated that this is contrary to fact, that when the mother is O, and the father A, there are fewer A children than expected. As an example, we may cite the investigation of Matsunaga (1955) on 2709 Japanese families, including 6360 children. He found that in families in which the mother was A and the father O, group A children occurred in the proportion expected with the gene frequencies characterizing Japanese people. On the other hand, when the mother was O and the father A, there were fewer A children than expected, the deficiency amounting to about 14 percent of all children. Similarly, a deficiency of about 10 percent was found in group B children produced from O mother × B father matings (as compared to B mother × O father matings).

Matsunaga found an added bit of evidence through study of miscarriages. The rate of miscarriage is significantly higher when the mother is O and the father A than it is when the mother is A and the father is O.

This study of Japanese families is particularly interesting because the findings are not complicated by Rh incompatibility. More than 99 percent of Japanese people are Rh-positive, so Rh incompatibility is extremely rare among them.

ABO incompatibility is said to occur whenever a mother lacks an A or B antigen possessed by the father. In this sense, all the following matings are incompatible (the mother is given first in each one): O × A; O × B; O × AB; B × A; A × B; B × AB; A × AB. In our population, such matings constitute nearly 35 percent of all matings. Evidence from many investigations indicates that such incompatibility contributes to early spontaneous abortion and miscarriage, but it is by no means the only cause.

At times ABO incompatibility may cause erythroblastosis in newborn infants as well. Wiener has presented evidence that a third type of antibody, anti-**C**, is present in the serum of group O mothers, in addition to anti-**A** and anti-**B** (Wiener, 1953). Anti-**C** is believed to be specific for a blood factor (serological specificity) **C** shared by antigens A and B (**C** here is not to be confused with the **C** of the Fisher-Race Rh nomenclature). It is postulated that anti-**C** passes through the placenta more readily than do anti-A and anti-B and "appears to be the antibody most often responsible for A-B-O hemolytic disease of the newborn" (Wiener and Ward, 1966). (Not all workers agree that anti-**C** is an entity separate from anti-A and anti-B. For discussion of the pros and cons of the question, see Prokop and Uhlenbruck, 1969, and Socha and Wiener, 1973.)

That the serological incompatibility between wife and husband may operate in other ways than through incompatibility between *fetus* and mother is suggested by the investigation of Cohen (1970a). She found that ABO incompatibility between wife and husband increases the chance of early loss of fetuses regardless of whether or not the fetus itself has an antigen lacking in the mother. "Thus, rather than selection directed specifically against ABO incompatible fetuses, embryos, zygotes, or gametes and in favor of compatibles, there appears to be selection directed against the overall 'biological fitness' of the matings — possibly against the gametes of the fathers whose ABO type is incompatible with their spouses, or possibly against the products of conception of these incompatible fathers and their spouses, but, in any case, independent of the genotype of the concepti."

Rh Incompatibility and ABO Incompatibility Combined Our principal reason for discussing ABO incompatibility lies in the interesting relationship between it and Rh incompatibility. Conclusive evidence indicates that *ABO incompatibility may prevent the development of the harmful effects of Rh incompatibility.* This, then, is another reason why Rh-negative mothers married to Rh-positive fathers do not produce more erythroblastotic infants than they do.

How does this protective action work? We may explain the leading theory by taking the case of a group O, Rh-negative woman married to a group A, Rh-positive man. A zygote produced may inherit genes I^A and R from the father. This means that the embryonic red blood cells will contain antigens A and Rh. If these cells leak into the mother's circulation, they may be destroyed by the mother's anti-A antibodies, or, perhaps more probably, they may stimulate the mother to form more anti-A antibodies than she normally has, and these antibodies may then destroy the cells from the embryo (containing A and Rh). In any event, the reaction involving antigen A is pictured as destroying the cells before the Rh antigen can stimulate the mother to form anti-Rh antibodies. In severe cases, of course, the embryo is injured to such an extent that an early miscarriage occurs, but even when this does not happen, the fetal cells containing A and Rh are eliminated without stimulating the mother to form anti-Rh antibodies.

This is the theory proposed initially by Race; what sort of evidence indicates its correctness? If it is correct, an increased proportion of mothers who *do* form anti-Rh antibodies should be ABO-compatible with their husbands, as compared to the general population of wives. (Because they are ABO-compatible, the Rh-positive blood cells from the fetus are not destroyed, and so can stimulate the mother to form anti-Rh antibodies.) Investigators usually turn the statement around and say that among mothers who have formed anti-Rh antibodies there should be a *decreased* proportion of ABO-*incompatible* matings (if the theory is correct).

Many investigations demonstrate that such a decrease actually occurs. Typical of such studies is that of Reepmaker and his colleagues (1962) in Holland. These authors summarized the results of their predecessors, and added data from 1742 additional women who had developed anti-Rh antibodies; 18.5 percent of them were incompatibly mated with regard to ABO groups. Recall that, in general, about 35 percent of matings are ABO-incompatible.

Particularly striking was the complete absence of matings in which the mother was of group O, the father of group AB. Of 1742 matings, 25 would be expected, by chance, to be of this type. The fact that none were seems to indicate that because all the children of an O $\times$ AB mating inherit either A or B from their father, all of them fail to stimulate formation of Rh antibodies by their mothers. (Hence matings of this type were absent from the data based on 1742 mothers who *did* develop antibodies.)

Thus we see that an Rh-negative woman is more likely to develop anti-Rh antibodies as a result of fetal stimulation if her marriage is ABO-compatible than she is if the marriage is ABO-incompatible. However, a mother may develop anti-Rh antibodies even when the marriage is ABO-incompatible. Reepmaker et al. (1962) have suggested that this occurs only if the father is heterozygous; in this case, group O, Rh-positive fetuses can be produced and may stimulate the mother to form anti-Rh antibodies (whereas group A or B, Rh-positive fetuses, when produced, do not cause anti-Rh antibody formation).

It is probable that the mechanism we have described does not account for all the factors involved in protection afforded by double incompatibility. This is suggested by the results of Cohen (1970b), who presented statistical evidence that in some manner, Rh incompatibility protects against the harmful effects of ABO incompatibility. She found that when wife and husband were incompatible for both Rh and ABO, there were fewer fetal deaths in proportion to live births than there were when the parents were incompatible for Rh only or for ABO only. What mechanism can we visualize by which Rh incompatibility might protect against ABO incompatibility? Because such fetuses are doubly heterozygous (e.g., I^AI^ORr), is heterosis involved (pp. 338–339)?

In our discussion of Rh incompatibility, we have spoken in every case of Rh-negative mothers. The term refers to mothers who lack the blood factor $\mathbf{Rh_0}$ (or D). Sometimes Rh-positive mothers (possessing this factor) may nevertheless develop antibodies and have erythroblastotic children. When this occurs, the antigen that stimulates the antibody formation is likely to be **rh′** (or C). The fetus has it, and the mother, lacking it, forms antibodies against it. The antigen at fault, however, may be **Hr**, one of a group of antigens whose occurrence bears a sort of reciprocal relationship to the occurrence of the Rh antigens, and whose mention serves to remind us of the many facets of Rh for which there is no space in an introductory discussion (see Wiener, 1943–1962; Wiener, 1954; Race and Sanger, 1968; Wiener and Wexler, 1958).

Blood Groups and Disease

We have seen that erythroblastosis fetalis is connected with Rh incompatibility, and that the underlying basis of the connection between them has been carefully worked out. Are there other diseases in which susceptibility is connected with blood group constitution?

Hundreds of investigations have been directed toward the answering of this question. Most of these are concerned with the ABO blood groups and their connection, if any, to susceptibility to diseases. For the most part these investigations are of a statistical nature, with computation of the percentage of patients having a certain disease who belong to each of the different blood groups compared to the percentage of persons in a normal control population who belong to each of the different groups. Taking a hypothetical disease—the "x syndrome"—we might obtain such results as these:

	Group A	**Group O**
Patients with "x syndrome"	50%	40%
Normal controls	43%	47%

Taken at face value, such results might be regarded as an indication that people in group A are more susceptible to the "x syndrome" than are people in group O (because the percentage of group A people who have the syndrome is greater than the percentage of group A people among the normal controls). The data are subjected to statistical analysis, frequently with computation of an index number expressing the relative difference between the blood group distribution of patients and controls.

Vogel (1970) presented a table summarizing results of many such investigations. As an example we take the most extensively investigated condition: cancer of the stomach. Vogel included 101 investigations totalling 55,434 patients and 1,852,288 controls. On the average, the percentage of group A individuals among the patients exceeded the percentage of group A individuals among the controls. (We should note that averages based on many investigations include some individual investigations in which no correlation between blood group and susceptibility to disease was found.) The index number obtained was 1.22 (an index number of 1 would indicate that patients and controls did not differ in blood group distribution). This may be taken to indicate that group A people are about 22 percent more likely to develop stomach cancer than are group O people. Similar investigations of duodenal ulcer may be interpreted to indicate that group O people are some 35 percent more likely to have the condition than are group A people. (For more examples, see McConnell, 1966, and Vogel, 1970.)

We were careful to state that the observed correlations "may be" interpreted to indicate greater susceptibility of one group or another to some pathological condition. *Need* they be so interpreted? Correlations do not of themselves prove that a cause-and-effect relationship exists. They only *suggest* such a relationship. But to prove the relationship we need investigations of a different type. Can we determine differences between group O people and group A people in biochemistry and physiology—differences that could account for differences in susceptibility? We must have answers to questions of this type before we can be sure how to interpret such correlations as those mentioned. Thoughtful reading of Wiener (1970) is recommended in this connection.

Polymorphism The fact that mankind is divided into blood groups is our most thoroughly analyzed example of human polymorphism, a term that means "many forms" and refers to the fact that even members of the same race are divided into different genetically defined groups. But it is one thing to discover polymorphism and even work out its genetic basis, quite another to explain *why* it is present, and how it arose. In Chapter 27 we shall discuss natural selection as a force in evolution and shall see that it occurs whenever individuals of one type survive better and leave more offspring on the average than do individuals of some other type. Natural selection has been invoked to explain some polymorphisms; did it operate in the dividing of mankind into blood groups? Some people conclude that it did and search for evidence that under some circumstances belonging to one blood group rather than another might make a difference in the likelihood that one would survive and leave children. Such evidence might be concerned with differences between blood groups in susceptibility to diseases that were likely to be fatal in childhood in the days before modern medicine.

Smallpox is one such disease. Fatality among primitive peoples was high and it still is high among people largely deprived of medical services. Vogel and Chakravartti (1966) investigated smallpox patients during epidemics in rural India. In all they studied 437 "suffering patients," to use their term, very few of whom had been vaccinated. As controls they employed siblings of the patients. They found that "within the patient group, severe forms and fatal outcome were significantly more frequent in groups A and AB as compared to groups B and O."

These investigators also studied survivors of smallpox epidemics. They found that among them there was a much higher percentage of groups B and O than there was in patients at the time of an epidemic. This was taken to indicate that more of the A and AB patients had died. Also, severe smallpox scars were more frequent in survivors belonging to groups A and AB than they were in survivors belonging to groups B and O.

In the light of their investigations and those of others, Vogel and Chakravartti concluded that "the influence of the blood group seems to be confined to severe manifestations in unvaccinated persons who are living under natural, primitive conditions without medical care."

People who belong to groups A and AB have antigen A and so lack antibodies against this antigen, while people who belong to groups B and O have anti-A antibodies. Under some circumstances, might these antibodies be of help in enabling the body to fight the disease? The hypothesis has been advanced that the smallpox virus may possess an A-like antigen. If so, the anti-A antibodies possessed by people in groups B and O might aid in the body's fight against the disease—aid sufficiently to make a significant difference in severity and even mortality under the primitive conditions just mentioned. Possession of such an A-like antigen by the virus has yet to be demonstrated conclusively, however; we look forward with anticipation to results of further research.

Other Red Blood Cell Antigens

In addition to the antigens we have discussed so far, the red blood cell is a veritable treasure trove of other antigens, some of them common, some rare. When inheritance is analyzed, we usually find that the genetics is of simple Mendelian variety, the presence of the antigen being dominant to its absence. In some cases there are interrelationships between the antigens or their genes. Thus, people who possess an antigen known as Le^a ("Lewis A") seem always to be nonsecretors. Space will not permit even a sample of the vast literature on these matters (see Race and Sanger, 1968; Wiener and Wexler, 1958).

Problems

1. A husband had the phenotype rh_y, his wife the phenotype rh. They had two children, one of rh′ phenotype, the other of rh″ phenotype. Give the genotypes of the parents according to both systems of notation.
2. A husband was Rh-negative, his wife Rh-positive. Their first child was Rh_0, their second child Rh_2. Would it be possible for these parents to produce an Rh-negative child? Explain.
3. Four sisters had the phenotypes Rh_z, Rh_0, rh_y, and rh, respectively. Give the genotypes and phenotypes of their parents.
4. On p. 201 we noted that persons having the phenotype Rh_Z may have any one of 14 genotypes. We listed four of them. What are the other ten?
5. In the gene pool of a certain population the relative frequencies of the ABO blood group genes are as follows: $p = 0.30$; $q = 0.06$; $r = 0.64$. (a) In families of which the mother belongs to group A and the father to group O, what percentage of the children will be expected to belong to group A? (b) In families of which the mother belongs to group B and the father to group O, what percentage of the children will be expected to belong to group B?
6. In terms of gene frequencies, p, q, and r, in families in which the mother is group AB and the father group A, what proportion of the children will be expected to belong to group B?
7. Which of the following matings is *least* likely to give rise to a child who will suffer erythroblastosis as a result of Rh incompatibility?

	Mother	Father
(a)	Rh-negative, group AB	Rh-positive, group O
(b)	Rh-negative, group A	Rh-positive, group O
(c)	Rh-negative, group B	Rh-positive, group A
(d)	Rh-negative, group A	Rh-positive, group A

8. We noted that among group O women in Holland who had developed anti-Rh antibodies, none were the wives of group AB husbands. In a population having the gene frequencies given in Problem 5, what percentage of marriages would be expected to occur between group O women and group AB men?
9. The first child of a pair of Rh-positive parents is Rh-negative. What is the probability that a second child will be Rh-positive? That the next three children in succession will be Rh-positive?
10. A survey of blood groups in Micronesia indicated that the following gene frequencies are found in that population:

$$p \text{ (frequency of } I^A) = 0.19$$
$$q \text{ (frequency of } I^B) = 0.11$$
$$r \text{ (frequency of } I^O) = 0.70$$

If we make a survey of Rh-negative Micronesian women belonging to group O, what *percentage* of them would we expect to find incompatibly mated with regard to antigen A? (Note: Because the study is confined to Rh-negative group O women, you need not compute their frequency of occurrence in the population. Concentrate on their husbands.)

11. In a group of mothers who *have* developed anti-Rh antibodies, would we expect that a larger *or* a smaller proportion of them would be ABO incompatible with their husbands, as compared to women in general?
12. John Doe's red blood cells: (a) react with anti-Rh_0 (anti-D) antiserum; (b) do not react with anti-rh′ (anti-C) antiserum; (c) react with anti-rh″ (anti-E) antiserum. Using both the Wiener multiple-allele and the Fisher-Race notations, give five genotypes that John Doe may have.
13. Which of the following pairs of parents could *not* be the parents of a child with the rh_y phenotype?
    (a) $Rh_0 \times$ rh′
    (b) rh′ × rh′
    (c) rh″ × Rh_2
    (d) $Rh_1 \times$ Rh″
    (e) $Rh_Z \times$ rh
14. The following chart shows the results of serological tests on the principals in a paternity suit. Could the man have been the father of the child? Give genotypic evidence in support of your answer.

	anti-Rh_0 anti-D	anti-rh′ anti-C	anti-rh″ anti-E
Mother	−	+	+
Child	−	+	−
Accused man	−	−	−

References

Cohen, B. H., 1970a. "ABO and Rh incompatibility. I. Fetal and neonatal mortality with ABO and Rh incompatibility: some new interpretations," *American Journal of Human Genetics,* **22**:412–440.

Cohen, B. H., 1970b. "ABO and Rh incompatibility. II. Is there a dual interaction in combined ABO and Rh incompatibility?" *American Journal of Human Genetics,* **22**:441–452.

Edwards, J. H., 1968. "The Rhesus locus," *Vox Sanguinis,* **15**:392–395.

Freda, V. J., J. G. Gorman, W. Pollack, J. G. Robertson, E. R. Jennings, and J. F. Sullivan, 1967. "Prevention of Rh isoimmunization. Progress report of the clinical trials in mothers," *Journal of the American Medical Association,* **199**:390–394.

Landsteiner, K., and A. S. Wiener, 1941. "Studies on an agglutinogen (Rh) in human blood reacting with anti-rhesus sera and with human isoantibodies," *Journal of Experimental Medicine,* **74**:309–320.

Levine, P., 1943. "Serological factors as possible causes in spontaneous abortions," *Journal of Heredity,* **34**:71–80.

Levine, P., 1958. "The influence of the ABO system on Rh hemolytic disease," *Human Biology,* **30**:14–28.

Matsunaga, E., 1955. "Intra-uterine selection by the ABO incompatibility of mother and foetus," *American Journal of Human Genetics,* **7**:66–75.

McConnell, R. B., 1966. *The Genetics of Gastro-intestinal Disorders.* New York: Oxford University Press.

Prokop, O., and G. Uhlenbruck, 1969. *Human Blood and Serum Groups.* New York: John Wiley & Sons, Inc.

Race, R. R., and R. Sanger, 1968. *Blood Groups in Man,* 5th ed., Philadelphia: F. A. Davis Company.

Reepmaker, J., L. E. Nijenhuis, and J. J. Van Loghem, 1962. "The inhibiting effect of ABO incompatibility on Rh immunization in pregnancy: A statistical analysis of 1,742 families," *American Journal of Human Genetics,* **14**:185–198.

Socha, W. W., and A. S. Wiener, 1973. "Problem of blood factor **C** of A-B-O system," *New York State Journal of Medicine,* **73**:2144–2156.

Vogel, F., 1970. "ABO blood groups and disease," *American Journal of Human Genetics,* **22**:464–475.

Vogel, F., and M. R. Chakravartti, 1966. "ABO blood groups and smallpox in a rural population of West Bengal and Bihar (India)," *Humangenetik,* **3**:166–180.

Wiener, A. S., 1943–1962. *Blood Groups and Transfusion,* 3rd ed. Springfield, Ill.: Charles C Thomas. Reprinted—New York: Hafner Publishing Company.

Wiener, A. S., 1953. "The blood factor **C** of the A-B-O system, with special reference to the rare blood group C," *Annals of Eugenics (Human Genetics),* **18**:1–8.

Wiener, A. S., 1954. *Rh-Hr Blood Types.* New York: Grune & Stratton, Inc.

Wiener, A. S., 1970. "Blood groups and disease," *American Journal of Human Genetics,* **22**:476–483.

Wiener, A. S., and F. A. Ward, 1966. "The serological specificity (blood factor) **C** of the A-B-O blood groups," *American Journal of Clinical Pathology,* **46**:27–35.

Wiener, A. S., and I. B. Wexler, 1958. *Heredity of the Blood Groups.* New York: Grune & Stratton, Inc.

14

Sex Linkage

Sex Chromosomes

In Chapter 3 we discussed the behavior of chromosomes during the process of meiosis. Males and females are alike with regard to the chromosomes discussed in that chapter (Fig. 3.1, p. 25). Such chromosomes are called **autosomes** to distinguish them from a pair of chromosomes with regard to which the sexes differ. Man, as mentioned previously, has 23 pairs of chromosomes; 22 of these are pairs of autosomes; one consists of a pair of **sex chromosomes.**

The sex chromosomes are of two kinds, called **X chromosomes** and **Y chromosomes.** Females have two X chromosomes and no Y chromosome. Males have one X chromosome and one Y chromosome. In man, the X chromosome is much larger than the Y chromosome, but despite the size difference, the fact that they synapse together in primary spermatocytes indicates that they constitute a homologous pair of chromosomes.

Figure 14.1 shows meiosis in a male, starting at the primary spermatocyte stage. The diagram is similar to Figure 3.4 (p. 29) except that in addition to the two pairs of autosomes in each spermatogonium, an X chromosome and a Y chromosome are also included. We note that these chromosomes, like the others, duplicate themselves, and pair in synapsis in the primary spermatocyte. When this cell divides to form two secondary spermatocytes, the X chromosome goes to one secondary spermatocyte, the Y chromosome to the other. When the spermatids are formed by division of the secondary spermatocytes, the chromatids separate. The outcome of the meiosis is that two of the four spermatids, and hence sperm,

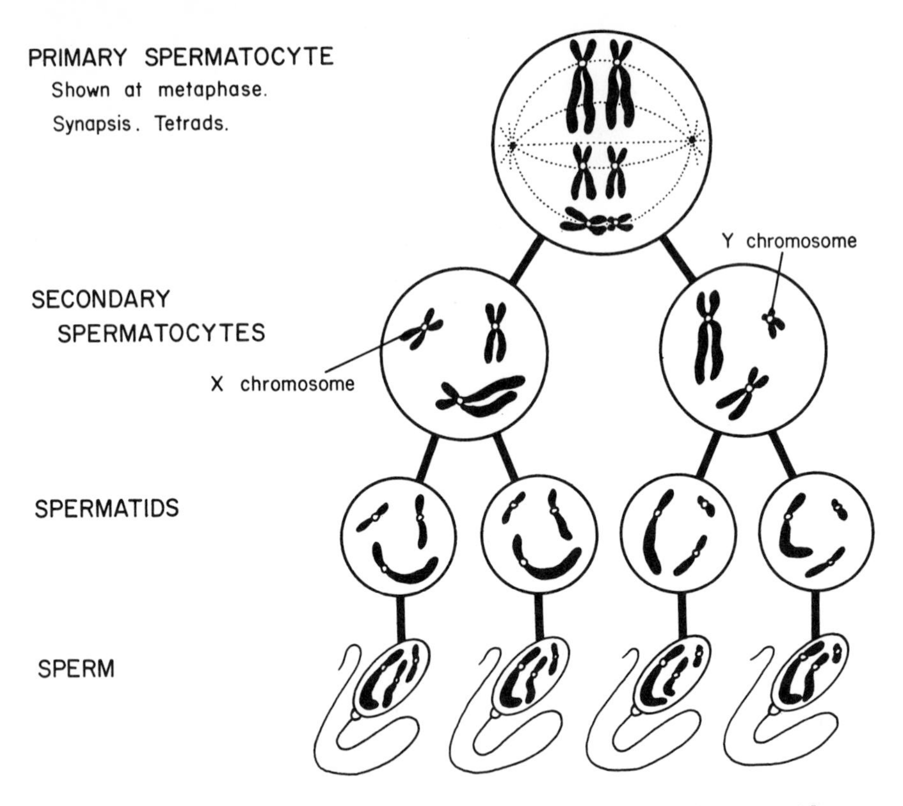

FIG. 14.1 Meiosis in the male (spermatogenesis) as in Fig. 3.4 with the addition of X and Y chromosomes. In the primary spermatocyte the X and Y chromosomes exhibit terminal (end-to-end) synapsis.

developing from a primary spermatocyte contain X chromosomes and two contain Y chromosomes. Thus males produce X-containing and Y-containing sperms in equal numbers.

Because females have two X chromosomes, meiosis does not differ from meiosis of the autosomes (Fig. 3.7, p. 33). Each ovum contains one X chromosome.

In fertilization there are two possibilities: (1) An ovum, containing X, may be fertilized by a sperm containing X, in which case the zygote normally develops into a female. (2) An ovum, containing X, may be fertilized by a sperm containing Y, in which case a male is normally produced. Thus the sex of an individual is normally determined at the time of fertilization, although the sexual development of the individual may be modified subsequently by other influences. We shall discuss sex determination in Chapter 17.

Like the autosomes, the sex chromosomes contain genes, although there seem to be few in the tiny Y chromosome. Genes in the Y chromosome are said to be **Y-linked,** those in the X chromosome, **X-linked.** Usually when the term "sex-linked" is used, it refers to X-linked genes unless the Y chromosome is specified.

X-Linked Inheritance

Of the X-linked human genes, the best known are those for red-green color blindness, the most common form of partial color blindness. (Complete or total color blindness is inherited as a recessive, autosomal trait and so does not enter into the present discussion.)

There are two main types of red-green color blindness, each with several variants. The main types are (a) **deuteranopia,** in which there is confusion in distinguishing red, yellow, and green, and (b) **protanopia,** in which difficulty in distinguishing red predominates. Because deuteranopia ("deutan color blindness") is the more common of the two, we shall be referring to it when we use the term "color blindness" in the present discussion. Studies in Europe indicate that about 6 percent of men have deutan color blindness, and about 2 percent have the protan form (McKusick, 1971).

If a woman inherits a gene for (deutan) color blindness from one parent and a gene for normal vision from the other, she will have normal vision. This fact demonstrates that the gene for color blindness is recessive. We may represent it by *d* (for deutan), its normal allele by *D*.

What evidence do family studies afford to indicate that those genes are indeed in the X chromosomes?

1. *If a woman is color-blind, all her sons are color-blind, but none of her daughters are* (unless her husband is color-blind). If the gene is in the X chromosome, this would work out as follows (using initials X and Y to represent the chromosomes):

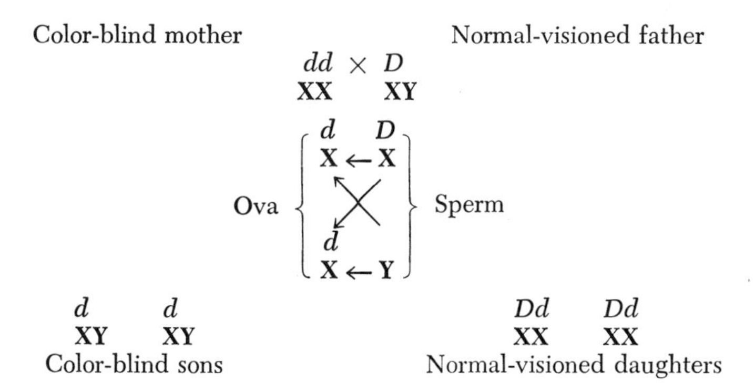

Note that the mother must be homozygous if she shows the trait, because *d* is recessive.

Note also that the father, having the gene D in his X chromosome, has no allele for it in the Y chromosome. He is said to be **hemizygous**—he has only one allele of a pair (hence he cannot be either homozygous or heterozygous for X-linked genes).

2. *A woman cannot be color-blind unless her father is* (and unless her mother also has the gene).

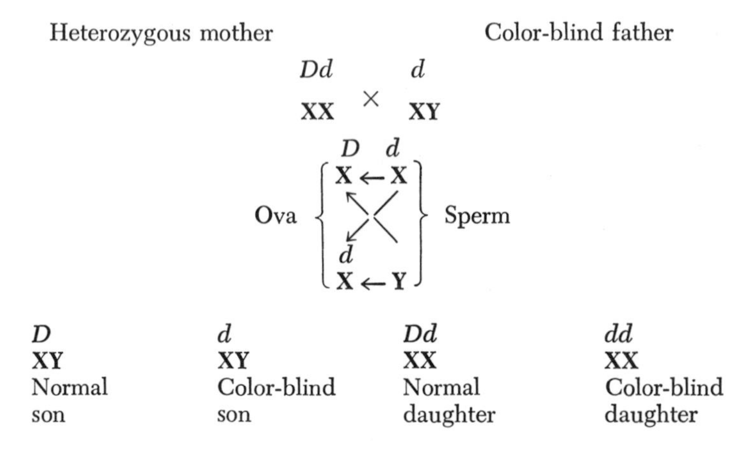

Note that every daughter receives one X chromosome from her father. Unless this X chromosome contains the gene d (in which case the father himself is color-blind), she will not be color-blind.

Color-blind daughters can arise only when a color-blind man happens to marry a woman who is heterozygous (or herself color-blind). Color-blind women are quite rare, therefore, in comparison to the number of color-blind men. During my years as a teacher of human genetics, only one girl in my classes has confessed to being color-blind. She had a sister who was color-blind and a brother who was *not* color-blind—an unusual combination. She gave me the facts with regard to her family; Figure 14.2 presents these data, although the order of birth of the children in some of the sibships may not be shown correctly.

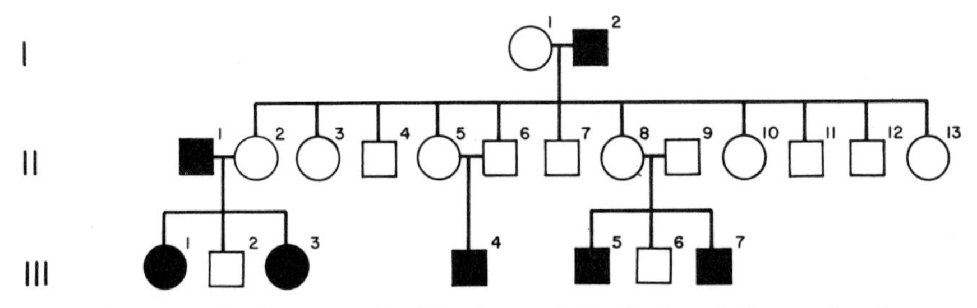

FIG. 14.2 Red-green color blindness. A kindred including a family composed of two color-blind daughters and one noncolor-blind son. Solid squares and circles represent color-blind individuals.

The girl herself (the propositus) is III-1, with her brother and sister as mentioned. Her father, II-1, was color-blind, as he must necessarily have been to have color-blind daughters. As we have indicated, the mother, II-2, must have been heterozygous to have had color-blind daughters, but in this case her heterozygosity is also indicated by the fact that her father, I-2, was color-blind. I-2 had six daughters; all of them must have been heterozygous; two of them proved that they were by producing color-blind sons.

We have stated that a family consisting of two color-blind daughters and a noncolor-blind son is an unusual combination. How often would such a sibship of three be expected when a color-blind man married a heterozygous woman (II-1 × II-2)? The genotypes are given in the preceding diagram, which indicates that 1/4 of the children are expected to be noncolor-blind sons, 1/4 color-blind sons, 1/4 noncolor-blind daughters, and 1/4 color-blind daughters.

Following the practice set forth in Chapter 7, we may let

c = chance for a color-blind daughter = 1/4
d = chance for a noncolor-blind son = 1/4

Then sibships of three are represented by the expression

$$(c + d)^3 = c^3 + 3c^2d + 3cd^2 + d^3$$

In this equation the "$3c^2d$" item represents families consisting of two color-blind daughters and one noncolor-blind son. The chance of such a family occurring is: $3 \times (1/4)^2 \times 1/4 = 3/64$ or about 1 in 21. This not such a very small chance after all.

3. *Half of the sons of heterozygous mothers are expected to be color-blind,* regardless of the genotype of the fathers. This is illustrated in the diagram just given. The father contributes a Y chromosome to all his sons. Of course, as with all Mendelian expectations, the 1:1 ratio of color-blind to normal sons is most closely approximated when data regarding sons of many heterozygous (carrier) mothers are pooled.

4. *Fathers do not pass on the gene for color blindness to their sons, but only to their daughters.* This follows from the fact that fathers transmit a Y chromosome, not an X, to all their sons (see diagram on p. 218). X-linked inheritance is sometimes called "crisscross inheritance" because males transmit the gene to female offspring, and it is, in turn, the male offspring of the latter who exhibit the trait.

Types of Inheritance That May Be Confused with X-Linkage

As indicated above, men show the phenotype of a recessive X-linked gene when they inherit the gene from only one parent (the mother), whereas women show the phenotype only if they inherit the gene from both parents. This difference explains in large measure why X-linked traits are shown much more commonly by men than by women. Indeed, unlike Figure 14.2, many pedigrees of X-linked traits show only males exhibiting the phenotype. (In such kindreds no affected male happened to marry a heterozygous female, or if such a marriage did occur it did not produce homozygous female offspring.) However, we must beware of the temptation to jump to a conclusion here. If we assemble a pedigree and find only males exhibiting a trait, that finding does not in itself constitute proof that the trait depends on a recessive X-linked gene. What other genetic situations might result in only males showing a certain phenotype?

1. The trait might depend on *autosomal* genes but be **sex-limited.** This means that females may have the genes but for some reason do not show the phenotype. Secondary sex characteristics are an example; in this case the male and female sex hormones are important in determining which characteristics shall appear in the individual. To take an example from cattle, bulls may have the genetic constitution for high milk production, but obviously they do not produce large quantities of milk; milk production is sex-limited to females. Among human beings, production of a heavy beard is normally limited to males, although there is no reason to doubt that females may transmit the genes concerned.

Note that with regard to sex-limited traits, males and females may be genotypically alike. Both may be *aa,* for example (recall that the genes are in the autosomes), but for some reason, hormonal or otherwise, the genotype *aa* produces a phenotypic effect in only one sex.

2. The trait might depend on autosomal genes that are **sex-influenced** or **sex-controlled.** This means that in one sex the gene produces its phenotypic effect only in homozygotes. Horns in sheep are an example of this.

If the gene for the development of horns is designated by the initial *H* and the gene for hornlessness by *h,* three genotypes are possible: *HH, Hh, hh.* Individuals of *both sexes* having the genotype *hh* are hornless. On the other hand, males having the genotype *Hh* are horned, whereas females having the genotype *Hh* are hornless. (It is as if in females the *h* becomes dominant to the *H*. We sometimes say that sex-influenced genes, e.g., *H,* are dominant in males and recessive in females.)

Although we need more research on the subject, there is some evidence that "pattern baldness" in human beings depends on a sex-influenced gene, *B*. Women having the genotype *BB* are bald, but women having the genotype *Bb* are not bald, whereas men of both genotypes *BB* and *Bb* are bald.

Note that both sex-limited and sex-influenced genes, being autosomal, may be passed on from fathers to sons, as well as to daughters. In this respect they differ from X-linked genes.

The phenotypic expression of both sex-limited and sex-influenced genes is restricted in some way by the sex of the individual. Sex hormones are probably involved in the examples we have given. On the other hand, sex in itself really has no effect on the expression of X-linked genes. A man is not color-blind because he is a male, but merely because he has the gene *d* in his X chromosome, with no allele to "cover it up" in the Y chromosome. A woman's sex hormones have no influence on whether she is color-blind or not; if she has the genotype *dd,* she is color-blind; if her genotype is *DD* or *Dd,* she is not. These genes in the X chromosome, then, really have nothing to do with sex. Judging from studies of lower animals where our knowledge is more complete than it is in man, the varied and miscellaneous assemblage of genes in the X chromosome seems to be there by historical accident. If we had knowledge of the evolution of chromosomes, we should be in a position to understand why some genes happen to be in X chromosomes, others in autosomes.

In summary, the genetic phenomena of sex limitation and sex influence result in pedigrees in which phenotypic expression of autosomal genes occurs predominantly in one sex. Predominant expression of the trait in the male sex is most easily confused with X-linked inheritance, of course. Perhaps the most easily applied means of distinguishing X linkage from sex limitation and sex influence is the fact that fathers do not transmit X-linked genes to sons, whereas inheritance from father to son does occur in the other cases. Nevertheless, in the absence of ability to arrange matings, distinguishing between the three types of inheritance is sometimes difficult when we encounter a "new" trait whose genetics is as yet unknown.

3. Finally, we should mention that pedigrees in which persons showing the phenotype in question are all males may occur just by *chance* (no X linkage, sex limitation, or sex influence being involved). This is particularly likely to happen if the pedigree includes small numbers of individuals showing the phenotype. For example, suppose that in a certain pedigree five people show the phenotype in which we are interested. If the phenotype depends on a simple autosomal gene, uninfluenced by sex in any way, what are the chances that all five of the people will be males? In our general population the ratio of male births to female births is about 105:100. This is so near 1:1 that for present purposes we may say that the chance for a male birth is ½, the chance for a female birth is ½. The chance that all five of the people showing the phenotype will be males is $(\frac{1}{2})^5$ or $\frac{1}{32}$. This is not really a small chance. Of course, if there are ten people showing the phenotype and all are males, the likelihood of this happening just by chance is only $(\frac{1}{2})^{10}$ or $\frac{1}{1024}$. Nonetheless, it *could* happen. Human geneticists are well aware of the danger of placing undue emphasis on isolated pedigrees that happen to "look good"—pedigrees that come to an investigator's attention because of their "curiosity value." Here

again we can see the importance of including large numbers of individuals whenever possible, and of analyzing as many different kindreds as possible.

X Linkage and the Hardy-Weinberg Formula In employing the Hardy-Weinberg formula we use the initial p to represent the relative frequency of a dominant gene, the initial q to represent the frequency of its recessive allele. We then square the algebraic sum of these $(p + q)^2$. Why do we square this binomial? Because two parents are involved in the production of zygotes, and hence of the offspring that develop from them.

This same principle applies to X-linked inheritance and the production of *female* offspring. Each female has two X chromosomes, one received from the mother, one from the father. Returning to our example of color blindness, we may let p represent the frequency of gene D, and q the frequency of d.

Then, *among females,* $p^2 + 2pq + q^2$ represents the expected frequency of (a) homozygous noncolor-blind individuals (p^2), (b) heterozygous noncolor-blind individuals ($2pq$), and (c) color-blind individuals (q^2). If, for example, $p = 0.95$ and $q = 0.05$, then $q^2 = 0.0025$. This means that we shall expect to find that 0.25 percent of all women in this population are color-blind. Incidentally, a much larger proportion will be carriers (heterozygous): $2pq = 2 \times (0.95) \times (0.05) = 9.5$ percent.

What proportion of *males* will be color-blind? A male inherits the gene for color blindness from only one parent, his mother. His father contributes neither D nor d, as we have seen. Thus, every male who develops from an ovum containing D is noncolor-blind, and every male who develops from an ovum containing d is color-blind. What are the relative frequencies of these two kinds of ova? The frequency of D-containing ova is p, the frequency of d-containing ova is q. Hence it follows that the frequency of noncolor-blind men is p, the frequency of color-blind men is q (not q^2 as in women, who receive genes concerned with color vision from *both* parents).

Here we find mathematical expression of the fact mentioned previously that color-blind men are much more numerous than color-blind women. If q is 0.05, we expect 5 percent of the men to be color-blind, but only 0.25 percent of the women to be $[q^2 = (0.05)^2 = 0.0025]$. In other words, the percentage of color-blind women is expected to be the square of the percentage of color-blind men. Actual data indicate that this relationship is approximated. Sources of error include the fact, mentioned previously, that there is more than one type of X-linked partial color blindness, and the fact that some heterozygous women have poor enough color discrimination so that they may be classified as color-blind [in other words, dominance of *D* over *d* is not always complete; recall our discussion of partial penetrance (pp. 70–72)].

The frequency, *q*, of a recessive, X-linked gene, can be approximated by merely counting the proportion of men in a population who show the phenotype. Thus, if we find that 5 percent of men are colorblind, we can conclude that the frequency of gene *d* is approximately 0.05 in that population. This is because men have only one of each of their X-linked genes. The validity of computing *q* by this means rests on a number of assumptions, however. The most important is probably the assumption that males who show the trait in question have viability equal to that of males who do not show it. This may well be the case with color blindness, but males who have the X-linked trait of hemophilia (see below) are not as viable as are males who have the normal allele. Accordingly, a mere count of the number of living hemophiliacs would be likely to give too low a value for the frequency of the gene in the gene pool. And for various other X-linked genes, equality of viability would be an unproved assumption. Clearly, then, a method of calculating gene frequencies that includes an estimate of the gene frequencies in women, as well as in men, is to be preferred. A formula for estimating the frequencies of X-linked genes, taking the genes of both sexes into account, was worked out by Haldane (1963). See Appendix D (pp. 465–466) for a statement of the formula and an example of its application.

The Xg Blood Groups

In our discussion of color blindness, we listed several criteria for distinguishing X-linked inheritance from autosomal inheritance in pedigree study. Application of the Hardy-Weinberg formula adds another means of obtaining evidence as to whether or not a trait is X-linked. We may ask such questions as this: If the trait is X-linked, what proportion of the sons and of the daughters, born in families of which the father shows the trait but the mother does not, will be expected to show the trait? Or again, in families in which neither parent shows the trait, what proportion of the sons will be expected to show it? What proportion of the daughters will be expected to do so? The answers to the questions will depend on the frequency of the gene in the population, as well as on the dominance or recessiveness of that gene. Then, having determined the expectation, we can see how closely actual findings agree with expectation, and thereby support our hypothesis concerning the mode of inheritance.

A particularly instructive example of this means of attacking the problem is furnished by the **Xg blood groups,** first called to the attention of the scientific world in 1962 (Mann et al., 1962). A certain "Mr. And.," a patient who had received many blood transfusions, was found to possess in his serum an antibody against a previously unknown antigen. The red blood cells of some people reacted with this antibody [they were called Xg(a+)], whereas the cells of other people did not [they were called Xg(a—)]. The antigen possessed by cells that did react was named **Xg**^{a}, and the corresponding antibody in Mr. And.'s serum was called anti-**Xg**^{a}. The gene for the presence of the antigen was designated as *Xg*^{a}, and the gene for its absence was called Xg.

Accumulated data on white populations in Britain, North America, and northern Europe indicate that 35.4 percent of males are Xg(a—), while only 10.7 percent of females are (Race and Sanger, 1968). Or, to reverse the statement, only 64.6 percent of males are Xg(a+), while 89.3 percent of females are. This discrepancy between the sexes suggests that the genes are X-linked, with Xg^{a} being the dominant member of the pair. This hypothesis has received extensive investigation.

Qualitative Investigations Qualitative investigations consist of studies of family groups: 50 families with a total of 104 children in the initial investigation. These are represented in Figure 14.3, classified by the type of mating. We now know of many more families (Noades et al., 1966, listed 1339).

Figure 14.3 illustrates the following criteria to be expected if gene Xg^{a} is dominant and X-linked:

1. *All daughters of an* Xg(*a*+) *man will be* Xg(*a*+), because his single X chromosome carries the gene Xg^{a} and all his daughters receive it.

2. Because women *may* be heterozygous, in pooled data *some of the sons of* Xg(*a*+) *mothers will be* Xg(*a*+), *some will be* Xg(*a*—).

3. *All sons of an* Xg(*a*—) *mother will be* Xg(*a*—), because her genotype is XgXg.

4. *Fathers do not pass on* Xg^{a} *to sons* (but see below).

As corollary criteria we note that (a) mothers, as well as daughters, of Xg(a+) men will be Xg(a+), and (b) fathers, as well as sons, of Xg(a—) women will be Xg(a—) (Mann et al., 1962).

All these criteria are illustrated by Figure 14.3. In the intervening years many more families have been studied, with almost perfect agreement with expectation. Interestingly, there are a few exceptions not attributable to errors of testing or uncertainty as to paternity. The most interesting are three families in which a total of six Xg(a+) sons were born to Xg(a—) mothers married to Xg(a+) men, thus forming exceptions to our criterion 4 above.

The suggestion has been made that crossing over may occasionally occur between the X chromosome and the Y chromosome. As we shall note in the next chapter, the short arms of the two chromosomes come together in synapsis. If such crossing over occurred, the Xg^{a} gene *might* be trans-

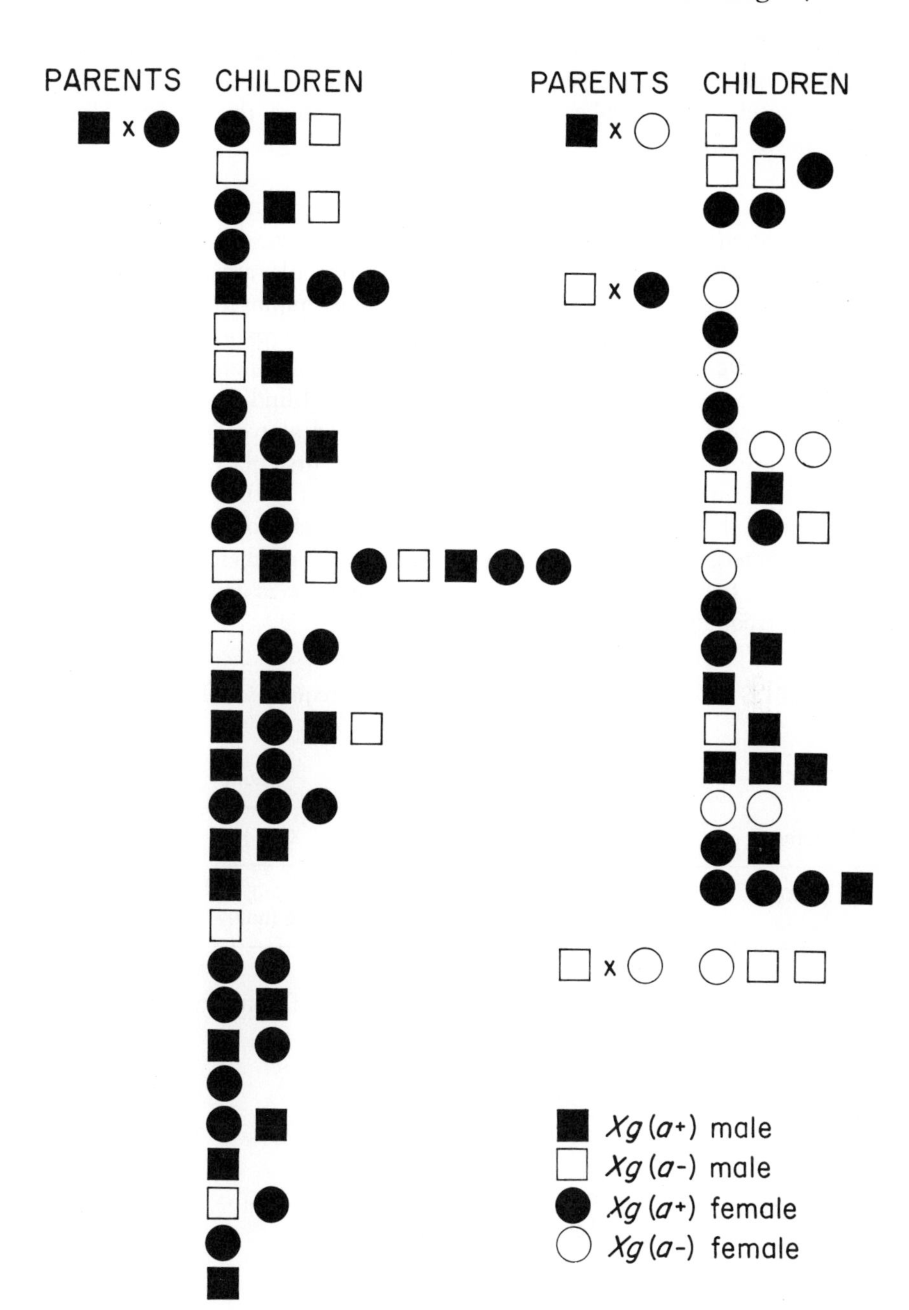

FIG. 14.3 The Xg blood groups. Fifty sibships classified by nature of the parents. (Redrawn from Mann, J. D., et al., 1962. "A sex-linked blood group," *Lancet,* **1**:8–10.)

ferred to the Y chromosome. Then a man possessing such an Xg^a-containing Y chromosome could pass it on to sons, who would then be Xg(a+) despite receiving an Xg-bearing X chromosome from the mother (Race and Sanger, 1968.)

Quantitative Investigations In quantitative investigations the principles expressed by the Hardy-Weinberg formula are applied to the analysis of data. We can illustrate the methods employed by asking the question: *If* Xg^a *is a dominant X-linked gene, what proportion of the sons of Xg(a+) women will be expected to be Xg(a+)?* Then, having determined the expectation, we can see how well actual findings agree with it.

In computing the expectation we proceed as we did in the computation given in Table 10.2 (p. 155). In the first place, we note that because the fathers contribute only Y chromosomes to their sons, we need not include them. We then compute the frequencies of the homozygous and the heterozygous mothers and the proportion of their sons expected to be of the two types (Table 14.1). In the table, p = frequency of Xg^a, and q = frequency of Xg.

TABLE 14.1 Expectations concerning the frequencies of Xg(a+) women and their sons.

		Sons		
Mothers	Frequency	Total	Xg(a+)	Xg(a−)
Xg^aXg^a	p^2	p^2	p^2 (all)	0
Xg^aXg	$2pq$	$2pq$	pq (half)	pq

From the table we find that the relative frequency of Xg(a+) sons is expected to be

$$\frac{p^2 + pq}{p^2 + 2pq} = \frac{p(p + q)}{p^2 + 2pq} = \frac{p^*}{p^2 + 2pq}$$

*Because $p + q = 1$.

Because the denominator is the same as the expected frequencies of Xg(a+) women, the formula becomes in effect:

$$\frac{p}{\text{proportion of women who are Xg (a+)}}$$

(Noades et al., 1966; Race and Sanger, 1968).

To illustrate how this calculation can be applied, we shall use the data of Chown, Lewis, and Kaita (1964) on 294 white families, mainly Canadian.

Applying Haldane's formula (Appendix D), the authors computed that the gene frequencies are

$$\begin{aligned} Xg^a &= 0.6475 = p \\ Xg &= 0.3525 = q \end{aligned}$$

Then p^2 is 0.4193, and $2pq$ is 0.4565, the sum, 0.8758 representing expectation as to the fraction of women who are Xg(+).

$$\frac{p}{p^2 + 2pq} = \frac{0.6475}{0.8758} = 0.7393$$

Thus 73.93 percent of the sons of Xg(a+) women in this population will be expected to be Xg(a+). How does expectation compare with results found?

The data included 255 Xg(a+) mothers who produced a total of 333 sons. How many of these would be expected to be Xg(a+)? $0.7393 \times 333 = 246.2$. The number of Xg(a+) sons actually found was 249. Hence observation agreed well with expectation based on the hypothesis concerning mode of inheritance.

This example indicates the thinking involved in such analyses. In practice, expectations with regard to daughters as well as sons are included. This means including the fathers and computing expectations for each of the following categories: (1) mother Xg(a+) and father Xg(a+); (2) mother Xg(a+) and father Xg(a—); (3) mother Xg(a—) and father Xg(a+); (4) mother Xg(a—) and father Xg(a—). This was done by Chown et al. (1964). Noades et al. (1966), analyzed the results of the various surveys of Xg made prior to that date, and their analysis was summarized by Race and Sanger (1968).

The Xg blood groups provide us with a particularly instructive example of the means employed in testing the hypothesis that a given gene is X-linked.

Hemophilia

Another well-known X-linked trait in man is hemophilia ("bleeder's disease"). The clotting of human blood is the result of a complex chain reaction involving several organic compounds. One of these is fibrinogen; we have already discussed afibrinogenemia, the hereditary lack of this substance (pp. 173–174). On the other hand, the blood of hemophiliacs fails to clot because some other substance necessary for clotting is functionally deficient (Ratnoff and Bennett, 1973). There are no conclusive statistics as to what proportion of male births is that of hemophiliacs. The frequency probably lies somewhere between 1/10,000 and 1/100,000 (Li, 1964).

Hemophilia A In this most common form of hemophilia (between 80 and 90 percent of hemophiliacs), the deficiency is in a serum substance known as antihemophilic globulin or **factor VIII.** The presence of the substance in functional form depends on a dominant gene, which we shall represent by A^H. Presence of a nonfunctional variant depends on the recessive allele a^H.

At one time it was thought that a woman could not be hemophilic, that homozygosity for the recessive gene might be lethal. More recently, however, investigators have found women who are hemophilic. One reason that they are so rare lies in the fact that they can be produced only when a husband is hemophilic and his wife is heterozygous ($a^hY \times A^Ha^H$). Hemophilia is such a disabling condition that until recently most hemophilic boys died young, or, at any rate, did not marry and become fathers. Not only do hemophiliacs bleed seriously from the slightest wound, but the simplest surgical operation, even the extraction of a tooth, precipitates a major crisis. Moreover, they suffer from internal bleeding and hemorrhages in the tissues even when no wounds have occurred. Recently medical science has come to their rescue. Transfusions of blood containing antihemophilic globulin will alleviate symptoms temporarily, as will extracts of the globulin prepared from blood plasma. Additional means of treatment will doubtless be found; thus in the future we may expect more hemophilic boys to live to maturity, marry, and produce children. If the wives happen to be carriers, we may expect an increased number of hemophilic daughters among these children. Hemophiliacs are rare, however, and the chance that one will happen to marry a carrier is small—unless he marries a relative.

Another difficulty, which we might not anticipate, lies in establishing the diagnosis beyond reasonable doubt. Various conditions other than hemophilia prevent normal clotting of the blood (e.g., afibrinogenemia). Consequently an investigation must demonstrate that a woman thought to be a hemophiliac has exactly the same deficiency as that possessed by male hemophiliacs. Fortunately a whole battery of tests is now available for determining whether or not a person, male or female, actually has hemophilia rather than some other condition in which blood clotting is abnormal. Application of these tests has revealed at least three women who seem to have true hemophilia, and more will doubtless be found in the future.

As we might expect from earlier discussions, consanguinity increases the likelihood that a hemophilic man will marry a heterozygous woman. Such a marriage is shown in Figure 14.4; III-5 was a hemophilic man who married his first cousin (III-2), daughter of his mother's sister. Evidently both mothers (II-4 and II-6) were heterozygous and transmitted the gene a^H to III-2 and III-5, respectively.

Of the daughters in generation IV, Merskey (1951) tested IV-4 and IV-11 intensively to demonstrate that they actually had true hemophilia. The daughters IV-1 and IV-9 had symptoms similar to those of their more intensively analyzed sisters. IV-7 had many of the symptoms; she is especially interesting because she had four sons, all hemophiliacs. All the sons of a homozygous hemophilic woman would be expected to be hemophilic.

As would be expected, none of the sixth generation have inherited hemophilia from their hemophilic fathers (although VI-4 and VI-5 must be carriers).

Note that I-2 is indicated as having been hemophilic. There seems to be some doubt about this. He lived long ago, and the investigators were dependent on family hearsay. Even if he were, we may be sure that II-1, II-2, and II-3 were hemophilic, not because of that fact but because their mother, I-1, was a carrier.

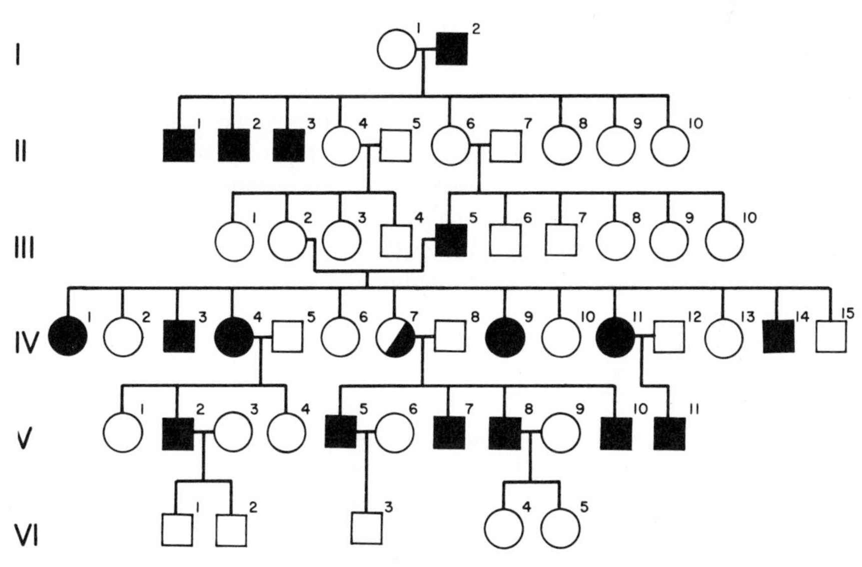

FIG. 14.4 Inheritance of hemophilia in a kindred including affected females. Black squares and circles represent affected individuals. IV-7 may have been hemophilic; she had some of the symptoms. (Abbreviated and redrawn from Merskey, C., 1951. "The occurrence of haemophilia in the human female," *Quarterly Journal of Medicine,* **20**:299–312.)

Hemophilia has received much attention because of its occurrence in European royalty. Queen Victoria must have been heterozygous for the gene because among her large family were a son with hemophilia and two daughters who had hemophilic sons. From the two carrier daughters the gene found its way into the royal families of Russia and Spain, respectively. Fortunately, it is absent from the present royal family in Great Britain. Prince Philip is not hemophilic, and Queen Elizabeth's inheritance from Queen Victoria comes through a line of nonhemophilic males: Edward VII, George V, and George VI.

How did Queen Victoria acquire the gene? Because there is no clear record of "bleeders" in her ancestry, it is possible, and perhaps probable, that the gene originated as a new mutation in the sperm or ovum that combined to form the zygote from which she developed. Further discussion of this historically interesting example is found in Iltis (1948), and McKusick (1965).

Hemophilia B Hemophilia B, also X-linked, is rarer than hemophilia A. Its basis is deficiency in a blood constituent known as **plasma thromboplastin component** (PTC) or **factor IX.** (It also goes by the names of "PTC deficiency," "factor IX deficiency," or "Christmas disease," this latter referring to the surname of one of the families in which it was first recognized.) We shall use B^H as the symbol for the dominant gene on which production of functional plasma thromboplastin component depends, b^H as the symbol for the recessive allele responsible for failure to produce the functional component. In both types of hemophilia, serological tests demonstrate that the respective blood constituents are present, even though they are in nonfunctional form (Ratnoff and Bennett, 1973).

In Chapter 15 we shall utilize these and other X-linked genes as we discuss the mapping of the X chromosome.

Y-Linked Inheritance

Any genes that may be found in the Y chromosome will be passed directly from a father to *all* his sons but to *none* of his daughters. This means that all the sons of a father who shows the trait will also show the trait. (Males are hemizygous for the genes because of the fact that the X chromosome does not contain alleles for them.)

Women, never normally possessing a Y chromosome, will not show the trait or possess the gene for it. Hence they cannot pass the gene on to any of their sons (or daughters).

Direct inheritance through males (grandfathers to fathers to sons, and so on), but never through females characterizes Y-linked inheritance. The criteria, therefore, are clear-cut. Can we find human examples of this type of inheritance? A time-honored example has been that of the so-called porcupine men, whose skin developed rough scales and bristle-like protrusions suggesting a porcupine's quills. Earlier accounts indicated that the trait was passed on from fathers to all their sons, and never to daughters, for seven generations in one English family. Penrose and Stern (1957–1958) reexamined all available evidence and concluded that much of the information concerning details was uncertain. Probably not all the sons of an affected man showed the trait, and it seems likely that some of the daughters did show it. Hence the authors concluded that the trait behaved as if it were caused by a rare autosomal dominant gene, chance determining that most of the affected individuals were males (recall our discussion on p. 221).

Stern (1957) reviewed 17 human traits that have been proposed at one time or another as possible examples of Y-linked inheritance. In some cases the evidence indicated clearly that Y linkage was not the basis of inheritance; in other cases the data conformed to expectation for Y-linked inheritance but could equally well be examples of inheritance through a dominant autosomal gene whose expression was sex-limited to males. Recalling our discussion of sex-limited traits (pp. 220–221), we can understand the difficulty of distinguishing between Y-linked inheritance and sex limitation of the expression of an autosomal gene. Perhaps the best means of distinguishing between the two is on the basis of transmission through *females.* A Y-linked gene is not transmitted by females to their sons. On the other hand, an autosomal gene, even one whose phenotype is expressed only in males, is freely transmitted by females to their sons. This is a good test but, unfortunately, many of the proposed pedigrees are too fragmentary to permit definite decisions by its use.

The most probable example of Y-linked inheritance available at present is that of hypertrichosis or hairiness of the pinna of the ear. The trait refers to a growth of prominent hairs on the surface of the pinna and along the rim of the ear (Fig. 14.5). Originally described in an Italian kindred, the trait has been investigated in kindreds in India (Gates, 1960; Sarkar et al., 1961; Dronamraju, 1960; Gates et al., 1962; Stern, Centerwall, and Sarkar, 1964).

Many family studies of the trait have been made. On the whole, the two criteria are met: (1) All sons of an affected male show the trait. (2) Females do not show the trait or transmit the gene. The trait is variable in the amount of hair developed, some males developing only three or four hairs on the pinna of the ear. The fact that occasionally the son of an affected male does not develop the trait may be an example of lack of penetrance (pp. 70–72). In some families the hair does not develop until the men are between 20 and 30 years of age, and hence younger men cannot be classified accurately.

It seems wisest to suspend judgment at the present time. The

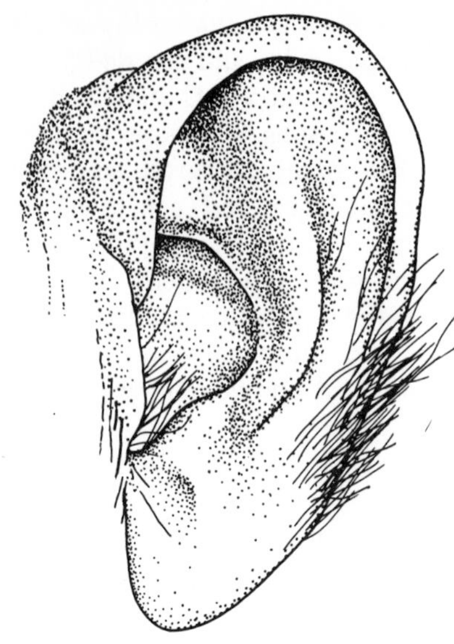

FIG. 14.5 Hypertrichosis of the pinna of the ear. (Drawn from a photograph in Dronamraju, K. R., 1960. "Hypertrichosis of the pinna of the human ear, Y-linked pedigrees," *Journal of Genetics,* **57**: 230–243.)

trait *may* be Y-linked, yet the possibility of autosomal inheritance has not been excluded "beyond the shadow of a doubt." The case is instructive as an example of the great difficulty frequently encountered in determining how human traits are inherited.

Examples of Y-linked inheritance have been found in fishes and in insects but, so far, not in mammals, even in such genetically well-known ones as mice and guinea pigs. Hence, if hypertrichosis of the pinna of the ear does depend on a Y-linked gene, it will form the first well-established example among mammals.

The human Y chromosome is very small (Fig. 17.5, p. 275). Accordingly, we might not expect it to contain many genes. As we shall see in our discussion of sex determination (Chap. 17), the mammalian Y chromosome has important male-determining properties, presumably dependent on the genes it contains.

Partial or Incompletely Sex-Linked Inheritance?

We noted previously that in primary spermatocytes the X and Y chromosomes synapse together but do so in a fashion unlike that of the pairs of autosomes. Homologous autosomes pair together *side by side,* corresponding gene loci in the two being brought close together. Contrariwise, the X and Y chromosomes pair together *end to end* (terminal synapsis, Fig. 14.1), the ends of the short arms of the two being in contact. Does the fact that these ends pair together at all indicate that they contain corresponding gene loci between which crossing over can occur as it does between members of a pair of autosomes? We have no clear evidence on which to base an answer to this question. Students of human chromosomes differ in their conclusion concerning it. Suspended judgment seems wise at present.

If there were gene loci possessed by both the X and Y, with occasional crossing over, this would make possible transferring a gene from one chromosome to the other. We have noted that such a possibility would provide a convenient explanation for those rare Xg(a+) sons who have Xg(a−) mothers and Xg(a+) fathers (p. 224). A gene so transferred would not exhibit the behavior we expect of X linkage or that we expect of Y linkage. The term "partial" or "incomplete sex linkage" has been applied to such inheritance. It would differ from autosomal inheritance in that when, for example, equal numbers of sons and daughters might be expected to show a dominant autosomal trait, some families would have more daughters than sons showing it (if the gene were in the father's X chromosome), and some families would have more sons than daughters showing it (if the gene were in the father's Y chromosome).

The whole subject involves so much uncertainty that we scarcely seem justified in devoting more space to it. More extensive discussion is found in Neel and Schull (1954).

Problems

1. A color-blind man marries a woman with normal vision whose father was color-blind. (a) What is the chance that their first child will be a color-blind son? A color-blind daughter? (b) If they have four children, what is the chance that two will be color-blind sons, two will be noncolor-blind daughters?
2. A sibship consists of a color-blind son, a noncolor-blind daughter, and a color-blind daughter. Give the phenotypes and genotypes of the parents.
3. If one male in 120 shows a defect dependent on a recessive, X-linked gene, what proportion of the females will be expected to exhibit the defect? What proportion of the females will be expected to be normal but heterozygous?
4. If 36 percent of men show a *dominant* X-linked trait, what percentage of the women will be expected to show the trait?
5. In the accompanying pedigree chart, I-1 and III-1 have hemophilia. All other individuals are nonhemophilic. What are the chances that IV-1, yet unborn, will be hemophilic?

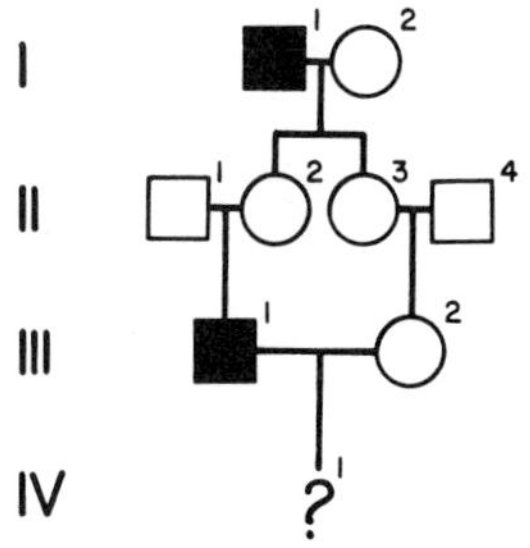

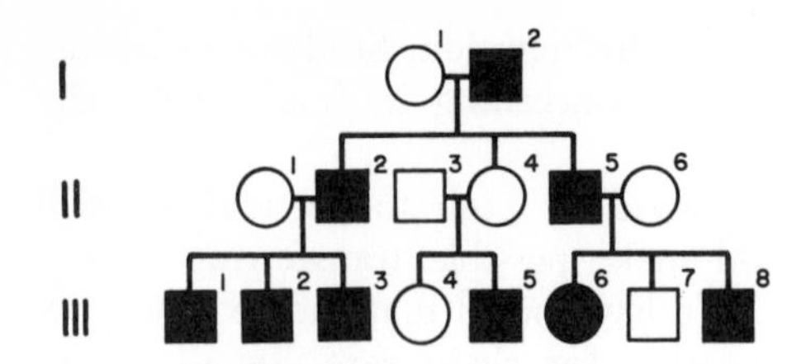

6. In the accompanying pedigree chart, solid squares and circles represent persons who show a certain trait.
   (a) Could the trait depend on a recessive autosomal gene?
   (b) Could the trait depend on a recessive X-linked gene? Give the genotypes all individuals would have if this were the mode of inheritance.
   (c) Could the trait depend on a Y-linked gene? Give as many reasons as possible for your answer.
   (d) Could the trait depend on an autosomal gene having expression limited to males?
   (e) Could the trait depend on a sex-influenced autosomal gene (dominant in males, recessive in females)? Give the genotypes all individuals would have if this were the mode of inheritance.
7. In a certain family neither husband nor wife was bald, but their three sons developed pattern baldness. Give the genotypes of the parents.
8. A sibship consisted of two daughters, one Xg(a+), the other Xg(a−). Give the genotypes and phenotypes of the parents.
9. If an Xg(a+) man marries an Xg(a+) woman whose father was Xg(a−), what types of sons and daughters will be expected and in what proportions?
10. Ability to taste phenylthiocarbamide (PTC) depends on a dominant autosomal gene, *T*, inability to taste the chemical on its recessive allele, *t*. An Xg(a+) man who is a taster but whose mother was a nontaster marries an Xg(a−) woman who is a nontaster. Give expectations with regard to their children.
11. An investigation revealed that in a certain population the frequency, *p*, of the Xg^a gene is 0.75. In this population, what proportion of the sons of Xg(a+) women will be expected to be Xg(a+)? (See p. 226.)
12. In a certain population 20 percent of the men show a phenotype dependent on a recessive X-linked gene; that is, they are "affected." In pooled data from the marriages of unaffected women to unaffected men, what percentage of the children will be expected to be (a) unaffected sons, (b) affected sons) (c) unaffected daughters, (d) affected daughters?
13. About 6 percent of the men in our population have deuteranopia, caused by a recessive, X-linked gene.
    (a) What percentage of women are heterozygous for this gene?
    (b) III-1 in the accompanying pedigree is deuteranopic. What is the probability that I-2 is heterozygous for the gene?

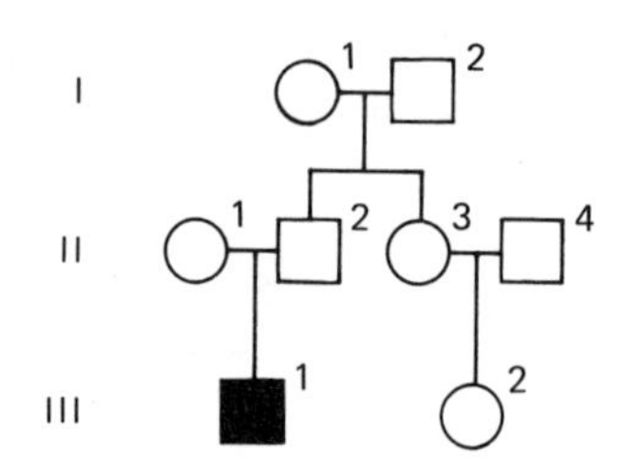

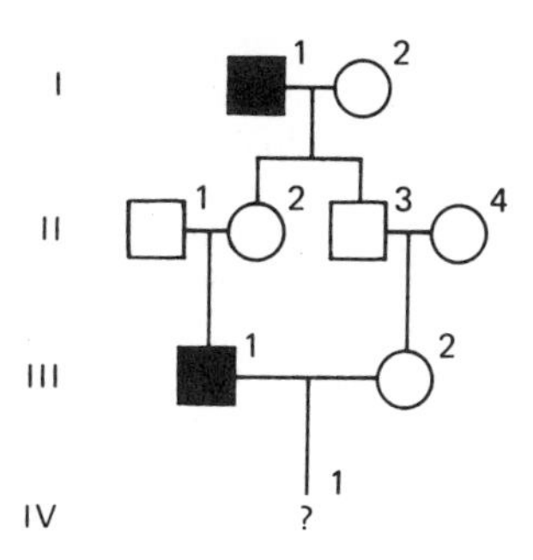

14. In the accompanying pedigree chart, I-1 and III-1 have deuteranopia. All other individuals have normal color vision. What are the chances that IV-1, yet unborn, will be a deuteranopic girl? (Employ the estimate of the frequency of deuteranopic males used in Problem 13.)

15. Chown et al. (1964) found 87 families in which the mother was Xg(a+) and the father was Xg(a−). They produced 120 sons and 118 daughters. Utilizing information on pp. 226–227, compute (a) the number of sons expected to be Xg(a+) and the number expected to be Xg(a−), and (b) the number of daughters expected to be Xg(a+) and the number expected to be Xg(a−). The observed numbers were 95 Xg(a+) sons, 25 Xg(a−) sons; 90 Xg(a+) daughters, 28 Xg(a−) daughters. How well did observation agree with expectation in this case?

References

Chown, B., M. Lewis, and H. Kaita, 1964. "The Xg blood group system: Data on 294 white families, mainly Canadian," *Canadian Journal of Genetics and Cytology,* **6**:431–434.

Dronamraju, K. R., 1960. "Hypertrichosis of the pinna of the human ear, Y-linked pedigrees," *Journal of Genetics,* **57**:230–243.

Gates, R. R., 1960. "Y-chromosome inheritance of hairy ears," *Science,* **132**: 145.

Gates, R. R., M. R. Chakravartti, and D. R. Mukherjee, 1962. "Final pedigrees of Y chromosome inheritance," *American Journal of Human Genetics,* **14**: 363–375.

Haldane, J. B. S., 1963. "Tests for sex-linked inheritance on population samples," *Annals of Human Genetics,* **27**:107–111.

Iltis, H., 1948. "Hemophilia, 'the Royal Disease,' " *Journal of Heredity,* **39**: 113–116.

Li, C. C., 1964. "The hemophilia gene in the population," In Brinkhaus, K. M. (ed.), *The Hemophilias.* Chapel Hill, N.C.: University of North Carolina Press. Pp. 393–404.

Mann, J. D., A. Cahan, A. G. Gelb, N. Fisher, J. Hamper, P. Tippett, R. Sanger, and R. R. Race, 1962. "A sex-linked blood group," *Lancet,* **1**:8–10.

McKusick, V. A., 1965. "The royal hemophilia," *Scientific American,* **213**: 88–95.

McKusick, V. A., 1971. *Mendelian Inheritance in Man,* 3rd ed. Baltimore: The Johns Hopkins Press. (Note added in proof: A 4th edition was published in 1975.)

Merskey, C., 1951. "The occurrence of haemophilia in the human female," *Quarterly Journal of Medicine,* **20**:299–312.

Neel, J. V., and W. J. Schull, 1954. *Human Heredity.* Chicago: University of Chicago Press.

Noades, J., J. Gavin, P. Tippett, R. Sanger, and R. R. Race, 1966. "The X-linked blood group system Xg. Tests on British, northern American, and northern European unrelated people and families," *Journal of Medical Genetics,* **3**:162–168.

Penrose, L. S., and C. Stern, 1957–1958. "Reconsideration of the Lambert pedigree (Ichthyosis hystrix gravior)," *Annals of Human Genetics,* **22**: 258–283.

Race, R. R., and R. Sanger, 1968. *Blood Groups in Man,* 5th ed. Philadelphia: F. A. Davis Company.

Ratnoff, O. D., and B. Bennett, 1973. "The genetics of hereditary disorders of blood coagulation," *Science,* **179**:1291–1298.

Sarkar, S. S., A. R. Banerjee, P. Bhattacharjee, and C. Stern, 1961. "A contribution to the genetics of hypertrichosis of the ear rims," *American Journal of Human Genetics,* **13**:214–223.

Stern, C., 1957. "The problem of complete Y-linkage in man," *American Journal of Human Genetics,* **9**:147–169.

Stern, C., W. R. Centerwall, and S. S. Sarkar, 1964. "New data on the problem of Y-linkage of hairy pinnae," *American Journal of Human Genetics,* **16**: 455–471.

15

Mapping the X Chromosome

When we say that a gene is sex-linked, we mean that the gene is in one of the sex chromosomes. When we say that two genes are linked, we mean that they are in the same chromosome. This may or may not be a sex chromosome.

Obviously, if two genes are X-linked, they are also linked to each other. For example, we saw in the preceding chapter that deutan color blindness and hemophilia A are both X-linked. This means that each is represented in the X chromosome by a locus that may be occupied by the abnormal gene or by its normal allele. We should like to know where these loci are, and how far apart they are.

In order to obtain an estimate of the relative positions of these two loci, we must find families in which both deuteranopia and hemophilia A are present. Such families are rare, but they do exist. Males in such families may be of four genotypes. Recall from the preceding chapter that d represents the deutan gene, a^H the gene for hemophilia A. Because two genes are found in the same chromosome, we shall enclose the symbols for them in parentheses. Y represents the chromosome.

(da^H)Y = color-blind, hemophilic male
(dA^H)Y = color-blind, nonhemophilic male
(Da^H)Y = noncolor-blind, hemophilic male
(DA^H)Y = male with normal color vision and blood clotting

Females have two X chromosomes, and either of these may be any one of the four types of X chromosomes listed for males. Hence women may have 4 $\times$ 4 or 16 different genotypes with regard to these two loci. Of most interest for present purposes are women who are heterozygous for both pairs of genes, having either the genotype (DA^H) (da^H) or the genotype (Da^H) (dA^H). In the first instance the dominant gene of one pair is linked to the dominant gene of the other pair (and the recessive genes of both pairs are linked together). This is called **coupling linkage** (or linkage in *cis* phase). In the second case the dominant gene of one pair is linked to the recessive gene of the other pair, and vice versa: **repulsion linkage** (*trans* phase).

What shall we expect if a doubly heterozygous woman having the genes in coupling linkage marries a man normal in both traits?

$$(DA^H)\,(da^H) \quad \times \quad (DA^H)\mathrm{Y}$$

$$\text{Ova}\begin{Bmatrix}(DA^H) \leftarrow (DA^H)\\ (da^H) \leftarrow \quad \mathrm{Y}\end{Bmatrix}\text{Sperm}$$

Offspring:	$(DA^H)(DA^H)$	$(DA^H)(da^H)$	$(DA^H)\mathrm{Y}$	$(da^H)\mathrm{Y}$
Phenotypes:	Normal daughters		Normal son	Color-blind, hemophilic son

Notice the sons especially; their phenotypes (for X-linked traits) are determined by the genes received from the mother. Half the sons will be expected to have normal color vision and normal blood clotting, half to be color-blind and hemophilic. This is what actually occurs for the most part. Occasionally, however, a son will be found to have normal color vision but be hemophilic, and occasionally a son will have the converse combination: color blindness but no hemophilia. How can such occasional occurrences be explained? Evidently the son with normal color vision and hemophilia has an X chromosome of the constitution (Da^H), and the other son has a chromosome of the constitution (dA^H). Both came from the mother. How can a mother with (DA^H) (da^H) chromosomes produce (Da^H) and (dA^H) chromosomes? **Recombination** must have occurred.

The mechanism of this recombination is the *crossing over* mentioned in Chapter 3 in our discussion of meiosis. At A in Figure 15.1 we represent a tetrad composed of two duplicated X chromosomes in synapsis in a primary oöcyte (see Fig. 3.4, p. 29). This is represented as occurring in a woman with the (DA^H) (da^H) constitution. Note that a crossover between two of the chromatids is shown.

At D in Figure 15.1 we see the four kinds of X chromosomes formed when the pairs of chromatids separate. Every ovum receives one of these X chromosomes. Ova that receive chromosome No. 1 or chromosome No. 4 have the same constitution they would have had if no crossing over had occurred, but ova that receive chromosome No. 2 or No. 3 have the recombinations. Ovum No. 2, if fertilized with a Y-containing sperm

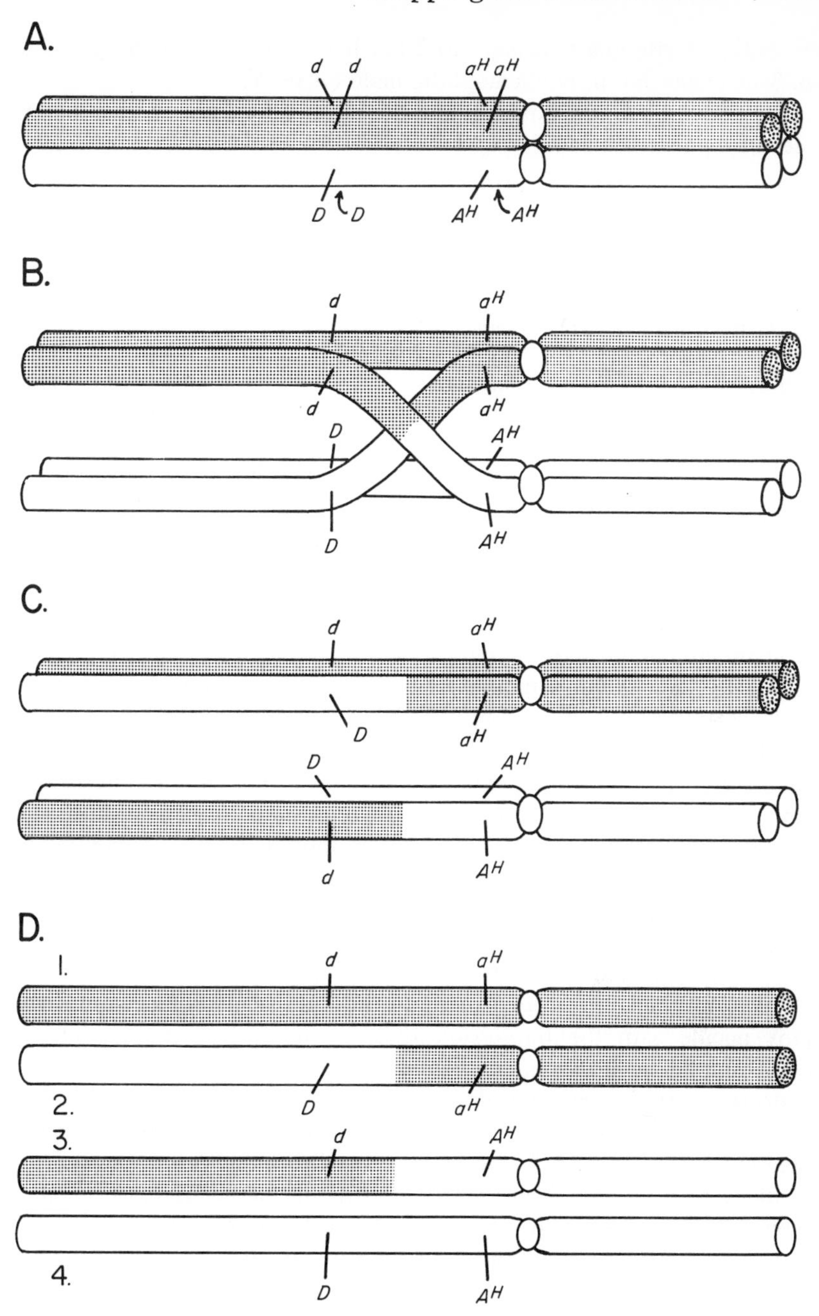

FIG. 15.1 Chromosomal basis of recombination of X-linked genes, using the genes for deuteranopia and hemophilia A as examples. The genes are represented as being located in the long arm of the X chromosome. Location in this arm has not been established with certainty.

cell, will give rise to a noncolor-blind but hemophilic male (Da^H)Y, ovum No. 3 to a color-blind, nonhemophilic male (dA^H)Y.

What shall we expect if the doubly heterozygous mother has the genes in repulsion linkage?

$$(Da^H)\ (dA^H)\quad \times \quad (DA^H)Y$$

Ova $\begin{cases} (Da^H) \\ (dA^H) \end{cases}$ $\begin{matrix} (DA^H) \\ Y \end{matrix}$ Sperm

Offspring:	$(Da^H)(DA^H)$	$(dA^H)(DA^H)$	(Da^H)Y	(dA^H)Y
Phenotypes:	Normal daughters		Noncolor-blind, hemophilic son	Color-blind, nonhemophilic son

Concentrating on the sons again, we note that most of them would be expected to have normal color vision and be hemophilic, *or* to be color-blind and not be hemophilic. These are the phenotypes arising when recombination does not occur. Occasional crossing over, with resulting recombination, would give rise to sons who are noncolor-blind and nonhemophilic, (DA^H)Y, or color-blind and hemophilic, (da^H)Y.

In sum, when a mother is (Da^H) (dA^H), most of her sons are (Da^H)Y or (dA^H)Y. Or to put it another way, the rare combinations (resulting from crossing over) in the offspring of a mother with coupling linkage are frequent combinations (arising in the absence of crossing over) in the offspring of a mother with repulsion linkage. Because there is no reason why coupling-phase mothers should be more common than repulsion-phase mothers in a population, there is no reason why coupling-phase sons, (DA^H)Y and (da^H)Y, should be more abundant than repulsion-phase sons, (Da^H)Y and (dA^H)Y, in that population.

We mention this matter to dispel a widespread misunderstanding of linkage. When we say that deuteranopia and hemophilia A are linked, many people gain the impression that if this is true most color-blind people should be hemophilic, or vice versa. Now we see why this idea is incorrect. In linkage, the dominant allele of one pair may be found combined with the dominant allele of the other pair or with the recessive allele of that other pair. In individual families more definite associations are likely to be found. For example, when the mother is (DA^H) (da^H), a color-blind son is much more likely to be hemophilic than he is to be nonhemophilic (he can be the latter only if crossing over has occurred). On the other hand, when the mother is (Da^H) (dA^H), a color-blind son is much more likely to be nonhemophilic than he is to be hemophilic.

Let us be a little more specific: *How much* more likely is occurrence of a son not involving recombination than is occurrence of a son whose genotype arose as a result of recombination? Or, conversely, how frequently does crossing over occur between the loci of these two genes? This question is of importance because frequency of recombination is the basis of mapping the chromosome. Thus if we found that recombination seldom occurs, we would conclude that the deuteranopia and hemophilia A loci are close together. But if we found that recombination is frequent, we would conclude that the loci are farther apart.

Maximum-Likelihood Methods Computing frequency of recombination is complicated. The complexity arises largely from the small number of individuals included in these investigations. Computations of percentages of recombination are relatively simple and direct when large numbers are available, as in genetical investigations of fruit flies (*Drosophila*). By contrast, a student of human genetics might find three or four families in which the mother is doubly heterozygous (termed "informative families"), and in these families there might be ten sons of whom one showed recombination of the traits being studied. What is the frequency of recombination in this case? Our first impulse is to answer: $1/_{10}$ or 10 percent (0.10).

But can we be at all sure that another investigator finding ten sons in informative families will also find that one, but only one, of them is a recombinant? Or that if we eventually learn of 100 sons in informative families, 10 (rather than 7 or 14, perhaps) will be recombinants? Obviously, we cannot be sure. And so we attempt to estimate the *likelihood* that 10 percent is the "true" percentage of recombination. To do this we apply the mathematics of probability, computing what percentage has the maximum likelihood of being the correct one, and also computing a range of probabilities.

Most widely employed is the *lod* score method of Morton (1955, subsequently modified somewhat by that author and others). Essentially this method asks with regard to a certain informative family: Is this sequence of children most likely to have been produced (a) if the gene loci under consideration are not linked, (b) if the loci are linked but with 1 percent of recombination, (c) if there is linkage with 5 percent recombination, (d) if there is 10 percent recombination, and so on? We might find, for example, that linkage with a recombination frequency of 0.10 (10 percent) has the maximum likelihood of being the correct explanation, and in addition that there is 95 percent probability that the "true" frequency lies somewhere between 0.01 and 0.20. These are called the confidence limits; when the range between them is small, we feel more confident that our estimate is correct than we do when the range is wide.

There are other corrections to be applied in estimating the true frequencies of recombination, but our simplified discussion may suffice to indicate why complicated computations are needed. The mathematics of them would be out of place in a brief elementary discussion (interested readers will find more detail in Levitan and Montagu, 1971).

In the case of deuteranopia and hemophilia A, study of a large kindred was originally interpreted to indicate a recombination frequency of 6 percent (Whittaker, Copeland, and Graham, 1962), but reanalysis indicated a maximum likelihood that the true percentage is nearer 12 (McKusick, 1964), so that frequency is generally employed in mapping the X chromosome. This percentage agrees well with the 10 percent computed originally for recombination between color blindness and hemophilia (Haldane and Smith, 1947). In these earlier studies the type of color blindness was not always specified, and the studies were made before the distinction between hemophilia A and hemophilia B had been recognized. Because hemophilia A is about four times as common as hemophilia B, it is probable that the earlier studies dealt principally with hemophilia A.

Protanopia and Hemophilia B Whittaker et al. (1962) studied a kindred having the two X-linked traits, protanopia and hemophilia B. Because recombination occurred in about 50 percent of the sons in the one informative family, they concluded that these gene loci are far apart in the chromosome.

Deuteranopia and G6PD Another locus near the deuteranopia locus is the "G6PD locus." The symbol **G6PD** stands for *glucose-6-phosphate dehydrogenase deficiency*. Glucose-6-phosphate dehydrogenase is an enzyme that most people possess. Persons who are deficient in the enzyme develop hemolysis (destruction of red blood cells, causing anemia) when they take antimalarial drugs such as primaquine, or when they eat the fava bean. Recombination between the loci for deuteranopia and G6PD has been estimated to occur less than 4 percent of the time (Siniscalco, Filippi, and Latte, 1964).

Protanopia and Deuteranopia It was originally thought that protanopia and deuteranopia, the two types of color blindness, might depend on a pair of allelic genes. Later evidence showed that recombination occurs at times, indicating that the genes are not alleles but are closely linked. Analysis of results of several investigations indicates that the recombination percentage is in the neighborhood of 9.5 (Arias and Rodríguez, 1972).

Mapping the Chromosome Can we use the data we have collected to start a map of the X chromosome? In doing so we translate percentages into "map units":

deuteranopia–hemophilia A	=	12 map units
deuteranopia–G6PD	=	4 or less
deuteranopia–protanopia	=	about 9.5

These data may indicate the following sequence along the chromosome:

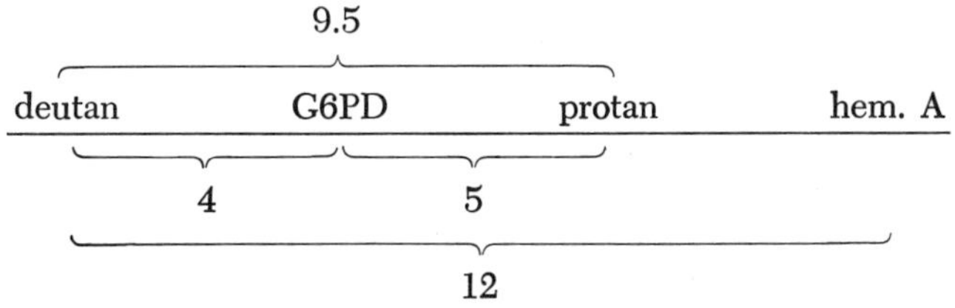

9.5

deutan G6PD protan hem. A

4 5

12

Here we place the gene for G6PD between the genes for the two types of color blindness. This is in accord with the suggestion of Siniscalco et al. (1964). Doing so seems reasonable at present, but there is always the chance that future investigations may prove it incorrect.

This map does not agree well with the finding of Boyer and Graham (1965) that the loci of the G6PD and hemophilia A genes are close together, perhaps as close as 4 map units. Our map would lead us to expect about 8 units. But in this investigation the confidence limits (see above) are sufficiently wide that 8 units is a not improbable estimate.

Double Crossing Over We have spoken of families in which the mother is doubly heterozygous as "informative families." Still more informative are families in which the mother is heterozygous for *three* X-linked traits. One such family has been studied (Graham et al., 1962). There were five sons in the family and the triple heterozygosity was evidenced by the fact that (a) some sons had hemophilia B, some did not; (b) some sons had protan color blindness, some did not; and (c) some sons were Xg(a+), whereas some were Xg(a−). The various ways in which the traits were combined in the five sons are represented in Figure 15.2.

The most interesting aspect of this family lies in the fact that it presents the first clear case of *double* recombination (crossing over) known in man. In double crossing over, chromosomes in synapsis break and recombine at *two* points (instead of only one as in Fig. 15.1). This

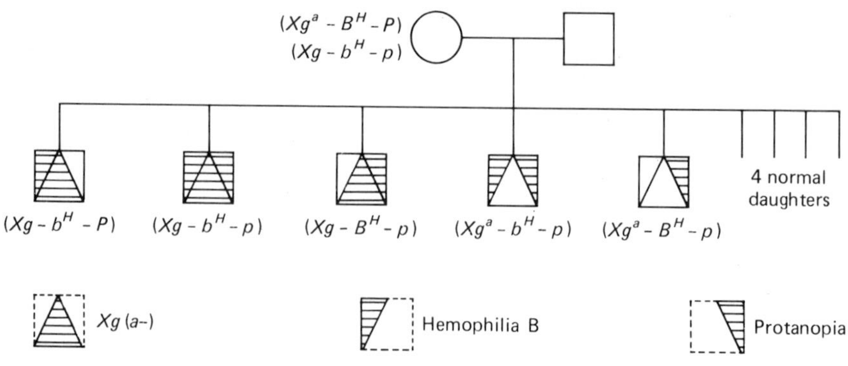

FIG. 15.2 Double crossing over. Five sons of a mother heterozygous for three pairs of genes. A suggested sequence of the genes is indicated, but we do not know whether or not the b^H locus is between the other two loci. In the mother, the gene B^H may have been in the chromosome with the other two dominant genes, or it may have been in the chromosome with the genes for Xg(a−) and protanopia. In either case, double crossing over occurred. (After Graham, J. B., H. L. Tarleton, R. R. Race, and R. Sanger, 1962. "A human double cross-over," *Nature*, **195**:834.)

phenomenon is well known in plants and lower animals, and there was no reason to doubt its occurrence in man. Still, the actual demonstration of an example is of interest. Triple heterozygosity, as in the present family, is essential for such a demonstration.

The mother was normal in vision and blood clotting, and belonged to the Xg(a+) blood group. Study of her ancestry seemed to indicate that the genes for color blindness and hemophilia were both inherited from her mother. This would suggest coupling linkage. It is not known whether the Xg gene came from her father or her mother, but the principles illustrated will be the same in either case, so let us assume that this recessive allele also came from her mother. The genotype of the mother can then be represented as follows: (Xg^a–B^H–P) (Xg–b^H–p). Here b^H is the gene for hemophilia B; p is the gene for protanopia (B^H and P, respectively, are their normal alleles); Xg^a and Xg are the Xg blood group genes (Chap. 14). Our placing of gene b^H in the middle is arbitrary, but is justified in view of the present uncertainty as to its linkage relationships.

Assuming that the type of linkage and the arrangement of genes are as indicated, how was the X chromosome of each of the sons produced? The chromosome of the first son, (Xg–b^H–P), must have arisen by crossing over between the hemophilia B locus and the protanopia locus. The chromosome of the second son, (Xg–b^H–p), was inherited intact, without crossing over.

The chromosome of the third son, (Xg–B^H–p), could have arisen only if crossing over occurred at two points: (a) between the Xg locus and the hemophilia B locus, and (b) between the hemophilia B locus and the protanopia locus. This, then, would be the double crossing over; it is diagramed in Figure 15.3.

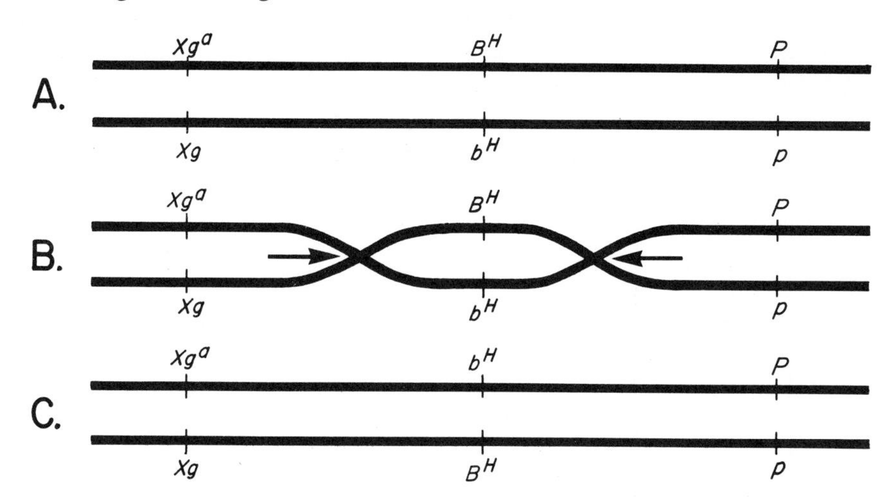

FIG. 15.3 Portions of two chromatids in synapsis, with double crossing over and consequent double recombination of genes (as in Fig. 15.2). The chromatids that do not cross over are omitted (Fig. 15.1). A. Chromatids as originally constituted. B. Chiasma formation. The chromatids break and recombine at the points marked by the arrows. C. Chromatids resulting from the double crossing over.

The chromosome of the fourth son, (Xg^a–b^H–p), arose by crossing over between the Xg locus and the hemophilia B locus. The chromosome of the fifth son, (Xg^a–B^H–p), resulted from crossing over between the hemophilia B locus and the protanopia locus. Notice that in a sense it is the reciprocal of the chromosome possessed by the first son.

We should emphasize that our designation of the third son as the double-crossover individual rests on two assumptions: (a) that the linkage in the mother is coupling, for all three pairs; and (b) that the hemophilia B locus is in the middle. If the protanopia locus is in the middle, both the first and fifth sons are double crossovers and the third and fourth sons are single crossovers. Because double crossing over is a rare event, the law of parsimony might bid us to regard the postulated occurrence of two double crossovers in five individuals as evidence against the hypothesis that the protanopia locus is in the middle. However, we need more substantial evidence than this for the mapping of the three loci.

We should note that the frequency with which crossing over occurred in this family indicates that the three loci are far apart in the X chromosome.

X Chromosome Map

In Figure 15.4 we present a tentative and provisional diagram of the relative positions of some of the gene loci in the X chromosome. In general design the diagram is like those in Berg and Bearn (1968) and Race and Sanger (1968).

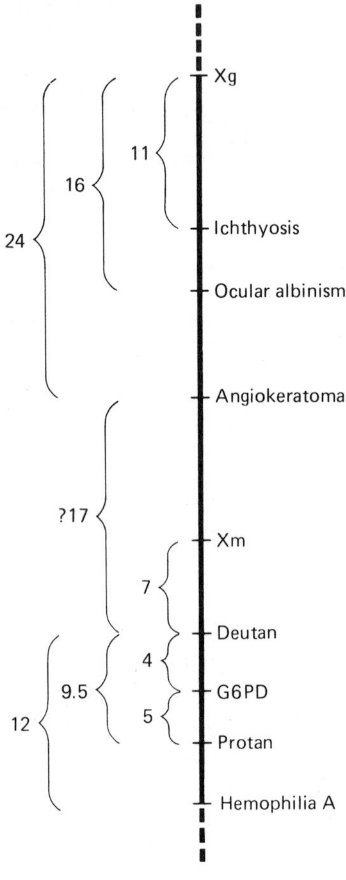

FIG. 15.4 Tentative map of a portion of the X chromosome. Only the gene loci for which we have the best estimates of relative position are included.* See text.

We should note at the outset that the line represents only a portion of the chromosome—the ends of the chromosome are not shown. We note also that the loci tend to form two groups: those grouped around the color blindness–G6PD–deuteranopia genes we have been discussing and those that are within measurable distance of the Xg locus. As we

*Sources of the numbers, other than sources given in the text:
Xg–ichthyosis: Adam et al., 1969; Went et al., 1969.
Xg–ocular albinism: Pearce et al., 1968; Hoefnagel et al., 1969.
Xg–angiokeratoma–deutan: Johnston et al., 1969.
Xm–deutan: Berg and Bearn, 1968.

shall see in Chapter 19, evidence from somatic cell hybridization experiments indicates that the locus of the G6PD gene is in the long arm of the X chromosome. Are both groups of loci in the long arm? At present, w cannot be sure, but with the new methods of mapping chromosomes now available (Chap. 19), evidence should soon be forthcoming.

The map includes some unfamiliar terms. Ichthyosis is an abnormality characterized by thickened, scaly skin. In ocular albinism the reduction of pigment is confined to the eyes, in contrast to the types of albinism we discussed earlier. Angiokeratoma (Fabry's disease) involves serious abnormalities of skin, blood vessels, and kidneys. The symbol Xm designates a blood protein detectable by means of an antiserum formed in rabbits. The gene (Xm^a) for presence of the protein is dominant to the gene for its absence (Berg and Bearn, 1968).

Many investigations have indicated that the distance between the Xg locus and the color blindness–G6PD–deutan group is so great as to be "unmappable" (e.g., Siniscalco et al., 1966). This means that recombination occurs 50 percent of the time, or close to it. This is the amount of recombination to be expected with independent assortment (pp. 32–34) when the genes are not linked at all (Fig. 3.8, p. 34; note that half the gametes containing *A* also contain *B,* while half of them contain *b*). Chromosome maps are built by adding together shorter distances representing smaller amounts of recombination.

We note that our map includes one investigation that indicates a connection between the two groups of loci. Johnston et al. (1969) found about 17 percent recombination between angiokeratoma and deuteranopia. But the confidence limits were wide, indicating that we cannot yet be certain of this distance.

This map will undoubtedly be altered as new investigations bring new knowledge. Some of the estimated distances will be found to be incorrect. In some instances even the sequences of genes along the chromosome may be incorrect. We know of many more X-linked genes than are given places on this map (e.g., hemophilia B). Further investigations will enable us to place their loci in correct position relative to the loci already mapped. In other words, this map is only a beginning. (See Renwick, 1961, 1971, for a fuller discussion of chromosome mapping.)

Genetic Markers

Some reader may ask: What of it? Why are investigations of this kind important? Increased understanding of our own genetic mechanisms is the greatest justification of such studies. In addition, there are possible practical applications in the fields of preventive medicine and genetic counseling. Geneticists are eager to discover as many genes as possible that may serve as **marker genes.** Ideally, a good marker gene should produce a trait that is clearly defined, not harmful, and present at birth. Such a gene would afford evidence that a certain chromosome is present, with its genetic contents (insofar as crossing over does not alter these genetic contents). To be of maximum utility, the marker gene should be located near a gene having harmful effects, so near that crossing over would seldom occur. Then the marker would serve as warning that the harmful gene is present, even though the latter might not produce its harmful effects until long after birth.

As our knowledge of the genetic contents of the X chromosome expands, Xg blood group genes may well prove to be such useful markers. Suppose that eventually a harmful gene is discovered, having its locus close to the Xg locus. Suppose also that this gene produces a severe crippling effect, either physical or mental, in people of middle age, but has no effect in earlier years. We may assume that the gene is recessive and designate it by *c* (for "crippling").

Let us suppose that a normal woman who is Xg(a+) is heterozygous for the hypothetical crippling gene, with linkage as follows: ($Xg^a c$) (XgC). We see readily that any son of hers who is Xg(a+) is very likely to have the gene *c* also, and thus suffer its ill effects as he grows older, unless therapy of some kind is possible.

If therapy is possible, knowledge that a son is Xg(a+) will enable the parents to have medical treatment begun before any harmful symptoms appear.

But suppose that no therapeutic measures are known. Then the parents of an Xg(a+) son will be faced with the tragic likelihood that he will become disabled as he grows older. For the good of his own morale, the boy himself probably should not be told what awaits him. There is always the chance that before he reaches middle age medical science may discover means of preventing or alleviating the symptoms.

Fortunately the picture has a brighter side also. Any son in this family who is Xg(a−) may be almost certain that he will not become crippled later in life, and that any descendants he may have will be free of the harmful gene.

We have mentioned genetic counseling as well as preventive medicine. We shall discuss this subject further in Chapter 28, but here we simply raise the question of whether a woman having the ($Xg^a c$) (XgC) genotype should have children. In the first place, how might she know, prior to producing sons, that she has the genotype? This might be possible if her parents could be studied—her father especially. Because one of her X chromosomes came from him, his phenotype would reveal whether the two genes in that chromosome were in coupling or repulsion linkage. (This is sometimes called the "grandfather method." Investigators of X linkage include the father of the doubly heterozygous woman whenever possible.)

In the present instance, we shall suppose that the woman's father was Xg(a+) and developed crippling as he advanced in years. (The other X chromosome came from the mother who, we shall suppose, was Xg(a—) and had no history of the crippling in her ancestry.) If there are no known therapeutic measures, the genetic counselor can simply advise the inquirer of the statistical probability that half of any sons she produces will be crippled (in middle age). Because the chance that her first child, for example, will be a boy is also ½, the total chance that this first child will develop the crippling is ¼. Half of her daughters will, like herself, be carriers. The value judgment as to whether or not to have children must be made by the woman herself. Should she take the chance? How serious is the crippling?

If therapeutic measures are possible, the problem is less acute. All her sons will be normal if she sees to it that the Xg(a+) sons are given the preventive treatment.

Problems

1. In a certain sibship half the children *of both sexes* had deuteranopia and hemophilia A; half were not color-blind or hemophilic. Give the genotypes of the children and their parents, assuming that crossing over is not involved.
2. A woman of normal phenotype whose father had normal color vision and had hemophilia A married a man who had deuteranopia and hemophilia A. They had two sons and two daughters. One son had normal color vision and hemophilia A; the other son had deuteranopia and was not hemophilic. The daughters married and produced several sons apiece (grandsons of the original couple). What proportion of these grandsons would be expected to have *both* deuteranopia and hemophilia A? Give the genotypes of all individuals. (Assume that crossing over had not occurred in the grandmother.)
3. Let us suppose that the imaginary trait "forked eyelashes" depends on a recessive X-linked gene, *f*.

   A normal woman whose father had deuteranopia and forked eyelashes married a normal man. They had seven sons. Three sons had deuteranopia and forked eyelashes; three were not color-blind and had unforked eyelashes; one was not color-blind and had forked eyelashes.

   Another woman had unforked eyelashes and was Xg(a+). Her father was Xg(a+) and had forked eyelashes; her mother's brother had unforked eyelashes and was Xg(a−). This woman's husband had unforked eyelashes and was Xg(a+). They had eight sons. Two sons had forked eyelashes and were Xg(a−); two had forked eyelashes and were Xg(a+); three had unforked eyelashes and were Xg(a−); one had unforked eyelashes and was Xg(a+).

   Give the genotypes of both pairs of parents and of their sons. Insofar as conclusions can be drawn from so few data, approximately where is the "*f* locus" on the X-chromosome map (p. 246)?
4. Suppose that in the case of the family diagramed in Fig. 15.2 the *p* locus were in the middle, and that the mother had the linkage relationships: (B^H – p – Xg^a) (b^H – P — Xg). Then which son or sons would have rep-

resented the occurrence of double crossing over? Which ones single crossing over? Which ones no crossing over?

5. If the diagram on p. 243 is correct, about how much crossing over would we expect to find if we did an investigation on families in which the mothers were heterozygous for both protanopia and hemophilia A?
6. Judging from Figure 15.4, about how much crossing over would we expect to find if we did an investigation on families in which the mothers were heterozygous for both ichthyosis and angiokeratoma?
7. The accompanying pedigree chart represents a family in which the X-linked traits of deuteranopia and G6PD deficiency are present. Give the genotype of the mother, indicating whether coupling or repulsion linkage is most probably present in her X chromosomes. How can we account for II-3, who is both deuteranopic and G6PD deficient? Give his genotype. (In giving genotypes indicate linkages, and use the following symbols: d = deuteranopic; D = not deuteranopic; $g6pd$ = G6PD deficient; $G6PD$ = not G6PD deficient.)

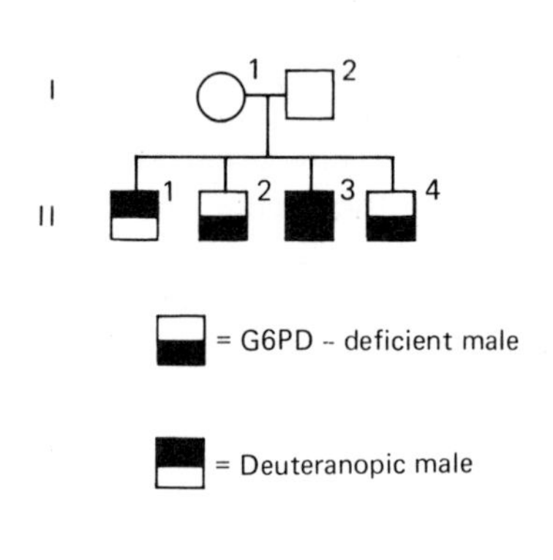

References

Adam, A., L. Ziprkowski, A. Feinstein, R. Sanger, P. Tippett, J. Gavin, and R. R. Race, 1969. "Linkage relations of X-borne ichthyosis to the Xg blood groups and to other markers of the X in Israelis." *Annals of Human Genetics,* **32**:323–332.

Arias, S., and A. Rodríguez, 1972. "New families, one with two recombinants for estimation of recombination between the deutan and protan loci," *Humangenetik,* **14**:264–268.

Berg, K., and A. G. Bearn, 1968. "Human serum protein polymorphisms," *Annual Review of Genetics,* **2**:341–362.

Boyer, S. J., and J. B. Graham, 1965. "Linkage between X chromosome loci for glucose-6-phosphate dehydrogenase electrophoretic variation and hemophilia A," *American Journal of Human Genetics,* **17**:320–324.

Graham, J. B., H. L. Tarleton, R. R. Race, and R. Sanger, 1962. "A human double cross-over," *Nature,* **195**:834.

Haldane, J. B. S., and C. A. B. Smith, 1947. "A new estimate of the linkage between the genes for colour-blindness and haemophilia in man," *Annals of Eugenics (Human Genetics),* **14**: 10–31.

Hoefnagel, D., F. H. Allen, Jr., and M. Walker, 1969. "Ocular Albinism and Xg." *Lancet,* **1**:1314.

Johnston, A. W., P. Frost, G. L. Spaeth, and J. H. Renwick, 1969. "Linkage relationships of the angiokeratoma (Fabry) locus," *Annals of Human Genetics,* **32**:369–374.

Levitan, M., and A. Montagu, 1971. *Textbook of Human Genetics*. New York: Oxford University Press.

McKusick, V. A., 1964. *On the X chromosome of Man*. Washington, D.C.: American Institute of Biological Sciences.

Morton, N. E., 1955. "Sequential tests for the detection of linkage," *American Journal of Human Genetics,* **7**:277–318.

Pearce, W. G., R. Sanger, and R. R. Race, 1968. "Ocular Albinism and Xg." *Lancet,* **1**:1282–1283.

Race, R. R., and R. Sanger, 1968. *Blood Groups in Man,* 5th ed. Philadelphia: F. A. Davis Company.

Renwick, J. H., 1961. "Elucidation of gene order." in Penrose, L. S. (ed.), *Recent Advances in Human Genetics*. Boston: Little, Brown & Company. Pp. 120–138.

Renwick, J. H., 1971. "The mapping of human chromosomes," *Annual Review of Genetics,* **5**:81–120.

Siniscalco, M., G. Filippi, and B. Latte, 1964. "Recombination between protan and deutan genes; Data on their relative positions in respect to the G6PD locus," *Nature,* **204**:1062–1064.

Siniscalco, M., G. Filippi, B. Latte, S. Piomelli, M. Rattazzi, J. Gavin, R. Sanger, and R. R. Race, 1966. "Failure to detect linkage between Xg and other X-borne loci in Sardinians," *Annals of Human Genetics,* **29**:231–252.

Went, L. N., W. P. de Groot, R. Sanger, P. Tippett, and J. Gavin, 1969. "X-linked ichthyosis: linkage relationship with the Xg blood groups and other studies in a large Dutch kindred." *Annals of Human Genetics,* **32**:333–345.

Whittaker, D. L., D. L. Copeland, and J. B. Graham, 1962. "Linkage of color blindness to hemophilias A and B," *American Journal of Human Genetics,* **14**:149–158.

16

Autosomal Linkage and Chromosome Mapping

Investigations of linkage in the X chromosome have a great advantage over investigations of autosomal linkage. Because males are hemizygous, the genetic contributions of their mothers are clearly revealed, without regard to dominance. The Y chromosome from the father does not obscure the genetic contribution from the mother. Hence linkage and recombination in the mother are clearly revealed by the phenotypes of her sons. It was no accident that in the fruit fly, *Drosophila,* mapping of the X chromosome proceeded more rapidly than did mapping of the autosomes. Nevertheless, with suitably planned experiments, mapping of autosomes is also perfectly feasible.

As background for our discussion of autosomal linkage in man, we shall describe the method of studying autosomal linkage in the fruit fly.

Method of Studying Autosomal Linkage in *Drosophila*

Normal or "wild-type" fruit flies have red eyes and long wings. One mutant gene is known that changes the color of the eyes to purple. Because this gene is recessive, we may represent it by the initial p and its normal allele by the initial P. Another mutant gene results in the wings being short, functionless stubs. This recessive gene, called vestigial, we may represent by v, its normal allele by V.

We are interested in determining whether or not these genes are linked to each other (they are not sex-linked). One way of starting is to produce homozygous stocks, one stock homozygous for the wild-type traits—red eyes and long wings, and the other stock homozygous for purple eyes and vestigial wings. Then we interbreed the two stocks to produce double heterozygotes:

	PPVV	×	*ppvv*
Germ cells	*PV*		*pv*
Zygotes		*PpVv*	

Because of dominance, these hybrids have red eyes and long wings.

As the simplest means of investigating linkage in the hybrid females, we mate them, *not* to hybrid males but to purple-eyed, vestigial-winged males: *PpVv* females × *ppvv* males. In this way we gain the advantage that nature provides in the case of the X chromosome: Genetic phenomena of linkage and recombination in the female parent are not obscured by the genetic constitution of the male (for he contributes only recessive genes). This is a back cross or test cross (p. 16). What types of offspring will occur and in what proportions will they occur?

First let us answer the question *on the assumption that the genes are not linked.*

PpVv × *ppvv*

Ova { *PV* → *pv*, *Pv* → *pv*, *pV* → *pv*, *pv* → *pv* } Sperm

Genotypes:	*PpVv*	*Ppvv*	*ppVv*	*ppvv*
Phenotypes:	red, long	red, vestigial	purple, long	purple, vestigial
Proportions:	25%	25%	25%	25%

Thus we see that, in the absence of linkage, independent assortment of genes (pp. 32–34) will occur in the formation of ova so that all four possible phenotypic combinations will be present in equal proportions.

On the other hand, if the genes *are* linked, what shall we expect? We shall expect to find the same four types of offspring, but *in very unequal proportions.* Specifically, we shall expect that the offspring having phenotypes like their grandparents (the mother's parents) will far outnumber those having phenotypes that represent recombination of the grandparental characteristics.

We recall that the original stocks were homozygous *PPVV* and *ppvv*. Hence, *if* the genes were linked, this must have been coupling linkage: (*PV*) (*PV*) and (*pv*) (*pv*), respectively. Their hybrid daughters, *PpVv,* would have the same coupling linkage: (*PV*) (*pv*). The back cross would then be as follows:

$$PpVv \times ppvv$$
$$(PV)\ (pv) \qquad (pv)\ (pv)$$

Ova				Sperm
	Noncrossover gametes	(PV) —	(pv)	
		(pv) —	(pv)	
	Crossover gametes	(Pv) —	(pv)	
		(pV) —	(pv)	

Genotypes:	$(PV)\ (pv)$	$(pv)\ (pv)$	$(Pv)\ (pv)$	$(pV)\ (pv)$
Phenotypes:	red, long	purple, vestigial	red, vestigial	purple, long
	No recombination (noncrossovers)		Recombination (crossovers)	

Data on actual experiments of this type were summarized by Bridges and Morgan, two of the pioneers in *Drosophila* genetics (Bridges and Morgan, 1919). Combining results of several investigations by three geneticists, they found that of a total of 13,601 back-cross offspring, 88.2 percent showed no recombination, whereas 11.8 percent did show recombination, divided almost equally between the two crossover types.

As we have seen, if there were no linkage the four types would have been expected in equal numbers; the results obtained were therefore interpreted to mean that the genes are linked, by being in the same autosome, and that crossing over between the loci of the genes occurs 11.8 percent of the time. This percentage was interpreted to mean that the loci are 11.8 "map units" from each other in the chromosome.

In the above experiment we deliberately started with genes in coupling linkage. We could just as easily have started with genes in repulsion linkage: (P*v*) (*p*V). If we had done that, the results would have been the same, except that what in our experiment were the non-crossover types would have been the crossover types, and vice versa.

Let us carry our experimentation one step further. Another recessive mutant gene in *Drosophila* causes the body to be black instead of the normal gray. We may represent the genes by *b* and *B*, respectively. Are these genes linked to the other two pairs we have been discussing? For variety we shall plan our experiment so that the genes, if they are linked, will be in repulsion phase. In other words, we begin by producing homozygous gray-bodied, purple-eyed flies and homozygous black-bodied, red-eyed flies:

	$BBpp$	$\times$	$bbPP$
	$(Bp)\ (Bp)$		$(bP)\ (bP)$
Germ cells	(Bp)		(bP)
Zygotes		$(Bp)\ (bP)$	

The doubly heterozygous females are then mated to black-bodied, purple-eyed males:

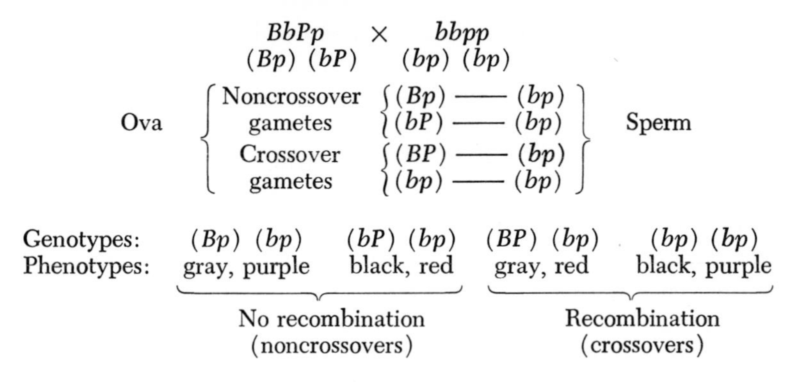

Again, if the genes for body color and eye color are *not* linked, the four types of back-cross offspring will be expected to occur in equal numbers. If they *are* linked, the two "grandparental types" (gray-bodied, purple-eyed, and black-bodied, red-eyed) should predominate. When the experiments of this type were performed, 48,931 offspring were raised (Bridges and Morgan, 1919). Of these offspring, only 6.2 percent were of the two recombination types. This is interpreted to indicate that the genes are indeed linked—with 6.2 percent crossing over. The further interpretation is that the loci of the genes for body color and eye color are 6.2 map units apart.

Now we can begin to map this autosome. In doing so we designate the loci by the mutant genes (always remembering that on occasion each locus may be occupied by the corresponding "wild-type" allele). The first experiment showed us that *p* and *v* are 11.8 units apart. Now we find that *p* and *b* are 6.2 units apart.

Representing the chromosome by a line, we may map these relationships in two ways:

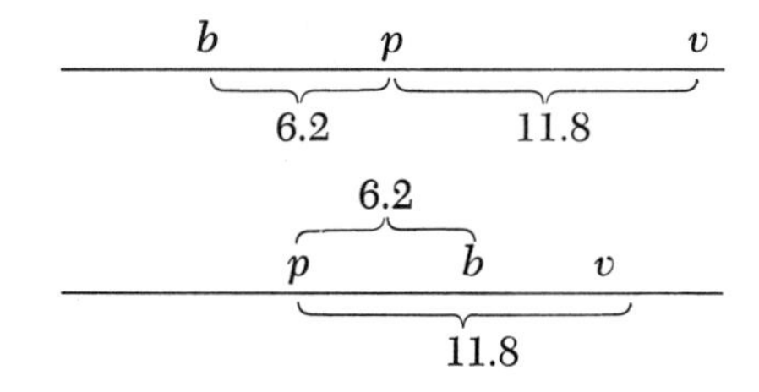

As far as present data indicate, either arrangement is equally likely. How can we decide which is correct? By determining the amount of crossing over between *b* and *v*. If the first arrangement is correct, the amount of crossing over between them should be about the sum of 6.2 plus 11.8, or about 18 percent. If the second arrangement is correct, the amount of crossing over between *b* and *v* should be 11.8 minus 6.2, or about 5.6 percent.

When the experiments were performed, over 20,000 back-cross offspring were produced; of these, 17.8 percent were crossover types (Bridges and Morgan, 1919). Therefore, we see that the first arrangement shown is the correct mapping of the three loci *b, p,* and *v*.

This gives a taste of the means by which autosomes were mapped in the early days of chromosome mapping. Many experiments, carefully planned, added one gene after another to the maps until the detailed maps of the *Drosophila* chromosomes found in all genetics textbooks were built up. More complex experiments than the ones we have described add data more rapidly. For instance, a "three-point test," using females simultaneously heterozygous for black body, purple eye, and vestigial wing, would enable us in one experiment to do what we required three experiments to accomplish: show the sequence and distances between the loci *b, p,* and *v*. Such experiments are commonplace in *Drosophila* genetics, but they are far beyond anything attainable as yet in human genetics.

Autosomal Linkage in Man

We have seen that study of autosomal linkage in experimental organisms involves carefully planned breeding experiments and the raising of progeny in large numbers. In human genetics neither of these procedures is available. How, then, is it possible to learn anything about human autosomal linkage? Here, as elsewhere in human genetics, instead of experimentation we must substitute mathematical analysis of such data as are available. In fact, the study of autosomal linkage in man is so completely the province of the biometricians (mathematical biologists) that I hesitate to discuss it in a book of this kind. Yet mathematics, in this context, is a tool, and one can understand something of the work for which a tool is intended without understanding much about the tool, or without being able to wield it.

In our discussion of linkage in the X chromosome, we noted that because of the two types of linkage (coupling and repulsion), linked genes are not more likely to be found together in a population than they are to be found separated. If *A* and *B* are linked genes, their respective alleles being *a* and *b*, individuals with the chromosome (*AB*) are not more likely to be found than are individuals with the chromosome (*Ab*) or (*aB*). This statement is not completely accurate, but it is correct in emphasizing that there is nothing about linkage itself that would cause one type to be more common than another. The actual proportions of the different types depend on the *gene frequencies,* not on linkage itself. We could show mathematically that a population at equilibrium will have the frequencies of different genotypes (*AABB, AaBB, AaBb,* and so on) determined by the gene frequencies, whether or not there is linkage. When the *A*'s and *a*'s are in one pair of chromosomes, the *B*'s and *b*'s in another pair, they assort independently according to the second Mendelian law. When the *A*'s (or *a*'s) and *B*'s (or *b*'s) are in one pair of chromosomes, they do not assort independently in this way, but eventually crossing over will result in their being as thoroughly mixed and recombined as if they *had* undergone independent assortment.

Thus, we cannot detect genetic linkage by noting what characteristics seem to occur together *in a population.* However, as we suggested with linkage in the X chromosome, we *can* gain some clue as to linkage by noting what characteristics seem to occur together *in individual families.* This fact forms the basis of the different methods of attacking the problem. Several methods have been devised, and each has its own advantages. The method most easily discussed without recourse to the mathematics is the sib-pair method of Penrose.

The Sib-Pair Method To illustrate the principles of the sib-pair method, we shall employ a model utilizing again the imaginary trait of forked eyelashes. This time, we shall imagine that the trait depends on a dominant gene, *F*, and "unforked eyelashes" on its recessive allele, *f*. We shall state at the outset that the locus of this allelic pair is in the same chromosome as the locus (or loci) of the Rh gene (or genes). Now the problem is how would someone who did not know that linkage is present determine it from the data of the model?

Our model starts with ten pairs of parents. To keep things as simple as possible, we shall imagine that one parent of each pair is Rh-positive and has forked eyelashes, but is heterozygous for both traits, and that the other parent is Rh-negative and has unforked eyelashes. Thus all ten matings are of the type: *FfRr* × *ffrr*. (We recognize that this is exactly the "back-cross" mating that experimental geneticists use in investigating linkage.)

Now, *if there were no linkage,* what offspring would we expect from these matings?

FfRr × *ffrr*

Germ cells { *FR* —— *fr*, *Fr* —— *fr*, *fR* —— *fr*, *fr* —— *fr* } Germ cells

Offspring: ¼*FfRr* = forked eyelashes, Rh-positive
¼*Ffrr* = forked eyelashes, Rh-negative
¼*ffRr* = unforked eyelashes, Rh-positive
¼*ffrr* = unforked eyelashes, Rh-negative

We would expect the four possible combinations in equal numbers.

That, in fact, is what we find in our "population" of ten families. Each pair of parents has four children, and when we pool the latter we discover:

10 children: forked eyelashes, Rh-positive
10 children: forked eyelashes, Rh-negative
10 children: unforked eyelashes, Rh-positive
10 children: unforked eyelashes, Rh-negative

There is nothing about these data to suggest linkage! However, we have just stated that we should not expect linkage to be revealed by such pooled data. What we need to do is to analyze the data, family by family. Table 16.1 presents the data in this way. The families (sibships) are designated by letters of the alphabet, and the four children in each family are designated by Roman numerals across the top of the table.

TABLE 16.1 The sib-pair method of testing for autosomal linkage, applied to an imaginary model.

Sibship	I Lashes	I Rh	II Lashes	II Rh	III Lashes	III Rh	IV Lashes	IV Rh
A	Forked	+	Unforked	−	Forked	+	Forked	+
B	Forked	−	Forked	−	Unforked	+	Unforked	+
C	Forked	+	Forked	+	Forked	+	Unforked	−
D	Unforked	−	Unforked	−	Unforked	−	Forked	+
E	Forked	−	Forked	−	Forked	−	Forked	+
F	Unforked	+	Forked	−	Unforked	+	Unforked	+
G	Forked	+	Unforked	−	Unforked	−	Forked	−
H	Unforked	−	Unforked	−	Forked	+	Unforked	−
I	Unforked	+	Forked	+	Unforked	+	Unforked	+
J	Unforked	+	Forked	−	Forked	−	Forked	−

The name **sib-pair method** suggests that we analyze the data by pairing the children within each family. The children in a sibship of four may be paired arbitrarily in six different ways: I with II, I with III, I with IV, II with III, II with IV, and III with IV. Let us apply this to sibship A:

I with II: sibs are unlike with regard to eyelashes and unlike with regard to Rh.
I with III: sibs are alike with regard to eyelashes and alike with regard to Rh.
I with IV: sibs are alike in both traits.
II with III: sibs are unike in both traits.
II with IV: sibs are unlike in both traits.
III with IV: sibs are alike in both traits.

We can conveniently tabulate our data in a "fourfold table":

		Eyelashes	
		Alike	Unlike
Rh	Alike	xxx 30 pairs	2 pairs
	Unlike	4 pairs	xxx 24 pairs

The "x's" in the table represent tabulations of the sib-pair comparisons we have just made for sibship A. In three pairs the sibs were alike with regard to both traits, in three pairings they were unlike with regard to both traits. As far as they go, these data suggest linkage, with the four children having the genotypes: I, (*FR*) (*fr*); II, (*fr*) (*fr*); III, (*FR*) (*fr*); IV, (*FR*) (*fr*). [Recall the fact that the parents are *FfRr* × *ffrr*. The data suggest that their genes are linked: (*FR*) (*fr*) × (*fr*) (*fr*).]

If we now analyze the other nine sibships in the same way, we obtain the total tabulation indicated above in the fourfold table: in 30 pairings both sibs alike with regard to both traits; in 24 pairings both sibs unlike with regard to both traits; in 6 pairings the sibs alike with regard to one trait but unlike with regard to the other one. The fact that within each family there is such great regularity in sib-pairs' agreeing or disagreeing in *both* traits strongly suggests that linkage is present.

The six pairings showing exceptions to this regularity involve child IV of family E and child IV of family G. The first three children of family E evidently have the genotype (*Fr*) (*fr*). In contrast to family A, repulsion linkage is apparently present here, the parents having the genotypes (*Fr*) (*fR*) × (*fr*) (*fr*). If this is correct, child IV, with the genotype (*FR*) (*fr*), must be the product of recombination through crossing over. Similarly, in family G (where coupling linkage is present), child IV would be a crossover individual.

The data of the model suggest that crossing over is infrequent and hence that the loci for eyelash traits and for Rh are close together in an autosome.

This model has taken us about as far as we can go into the sib-pair method without mathematics (for a good introduction to the mathematics of the method, see Li, 1961, pp. 131–141). Perhaps our greatest oversimplification was in starting with only *FfRr* × *ffrr* families. In practice, an investigator would often be unable to tell whether an Rh-positive parent with forked eyelashes was *FfRr,* or *FFRr,* or *FfRR.* A moment's reflection will indicate that an *FfRR* × *ffrr* mating will give the same offspring whether linkage is or is not present. In either case all offspring will be Rh-positive and half of them will have forked eyelashes, half will have unforked ones. In other words, crossing over does not lead to recombination in such an individual.

Thus many matings that an investigator may include in his analysis do not yield data of significance. However, because other matings do yield significant data, the method works despite lack of the precise regularity shown by our model.

The sib method has the advantage that it can be applied even when knowledge concerning some sibs in a sibship is unavailable. We note that it utilizes data concerning only one generation. This is an advantage if information cannot be obtained about the parents. But obviously it would be of value to include the parents whenever possible. This is done in the *lod* score method discussed briefly in the preceding chapter. It is the method most widely employed in the study of autosomal linkage.

Human Linkage Groups

A series of genes linked together constitutes a linkage group. In organisms like *Drosophila,* which have been subjected to thorough genetic analysis, there are found to be as many linkage groups as there are pairs of chromosomes. This is exactly what we would expect if each linkage group is indeed the assemblage of genes contained in one chromosome.

For man, with his 22 pairs of autosomes, we expect eventually to demonstrate that there are 22 linkage groups (plus the linkage group of X-linked genes, and probably a short group of Y-linked genes). Because of the evident difficulties, we are far from our goal in this respect.

Most of the linkage groups so far known consist of two loci found to be linked together, with a certain amount of crossing over. We know of only a few of these small "groups," but the number is slowly increasing as more and more traits, especially biochemical ones, are being discovered and their genetics are being analyzed. Instead of attempting to list all linked pairs known at the moment, we shall direct attention to the three most interesting groups, including the only groups consisting of three loci that are known at the time this is written (more complete lists will be found in Race and Sanger, 1968, and Levitan and Montagu, 1971).

Rh–Elliptocytosis The locus of the Rh genes is closely linked to the locus of a gene that has a mutant allele causing the red blood cells to be elliptical in shape (instead of circular). Two genes at different loci have been found to produce elliptical red blood cells; one of the two is linked to the Rh locus with a recombination frequency of about 3 percent.

ABO–Nail-Patella Syndrome–Adenylate Kinase We are already familiar with the ABO blood group genes (Chap. 12). The nail-patella syndrome (NP) is identified by abnormalities of the fingernails, and by skeletal abnormalities, especially of the ilium (one of the bones of the pelvic girdle) and of the patella (kneecap). Adenylate kinase is an enzyme concerned with the release of energy within cells. It exists in two slightly differing forms (isozymes) identifiable by electrophoresis: AK-1 and AK-2. The genes, AK^1 and AK^2 are alleles, and dominance is absent, heterozygotes having both forms of the enzyme (thus inheritance resembles that of the MN blood types, Chap. 7).

The NP and AK loci are very closely linked, while the ABO locus is slightly farther from these two, about 12 percent recombination being estimated for the ABO-NP linkage (Schleutermann et al., 1969). The ABO–AK linkage is closely similar, 10 percent recombination being the estimate of Wendt et al., 1971. So at the moment it is best to defer judgment as to whether the gene sequence is ABO:NP:AK or ABO:AK:NP.

Myotonic Dystrophy–Secretor–Lutheran Of these three, the secretor gene is already familiar to us (pp. 193–195). Lutheran (a surname, with no reference to the church) is a red blood cell antigen, comparable in many respects to the other cell antigens we have discussed. Myotonic dystrophy (Dm) is a disabling condition characterized by muscle wasting, cataracts, and other defects. The gene concerned seems to be dominant to its normal allele.

Harper et al. (1972) found the most probable frequency of recombination between Dm and Secretor to be 11 percent, that between Dm and Lutheran to be 24 percent. If the Secretor locus is in the middle, subtraction would indicate a probable Lutheran–Secretor recombination percentage of 13. This agrees well with the estimate of Cook (1965). Hence we may tentatively map the loci as follows:

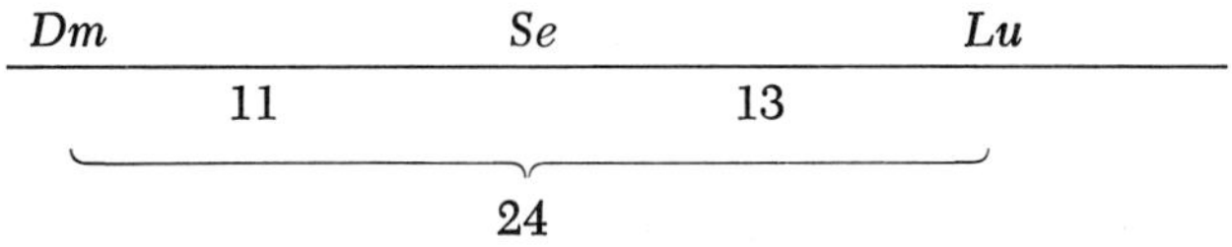

Dm *Se* *Lu*

11 13

24

For a somewhat detailed general discussion of chromosome mapping see Renwick, 1971.

Usefulness of Linkage Studies

In the face of so much difficulty, why do human geneticists persist in the study of autosomal linkage? The best reason is the desire for knowledge. The dream of an eventual 22 maps of human autosomes would be attractive even if such maps had no practical use.

In the preceding chapter we discussed the value, in preventive medicine and in genetic counseling, of marker genes in the X chromosome. Marker genes in autosomes may be similarly useful. The situation is a bit more complicated, in that the genetic contribution of fathers to their sons, as well as to their daughters, must be taken into account, as it need not be in X-linked traits. Yet the same principles of potential usefulness apply.

Dominant Genes First we shall consider the problem of a dominant gene that produces a harmful effect later in life, but has no effect during childhood. The gene for Huntington's chorea (Huntington's disease) (pp. 66–69) is a good example. We shall imagine a family in which one parent develops the tragic symptoms some time after the children have been born. Naturally the parents wonder if any of the children will develop the symptoms as they grow older. If there were a marker gene closely linked to *H* (the gene for Huntington's chorea), it would help in answering this question. Such a marker might be a blood group gene (let us call it *Bd*) that produces an antigen easily identified by serological testing.

The choreic parent might have the genes in coupling linkage, (*Bd H*) (*bd h*), and the other parent have the genotype (*bd h*) (*bd h*). In that case any child who had the antigen would be very likely to have the harmful gene (while recombination is always possible, it would be highly unlikely if linkage is close).

On the other hand, the choreic parent might have the genes in repulsion linkage: (*Bd h*) (*bd H*). Then the children *lacking* the antigen would be very likely to have gene *H*. Evidently, study of the choreic parent's ancestry would be needed to determine the type of linkage present.

In either case, presence or absence of the antigen produced by the marker gene would yield information of value. But, as with the hypothetical gene for crippling discussed in the preceding chapter (p. 248), possession of the information would raise some difficult questions. Should children indentified as highly likely to have the harmful gene be told, as they approach maturity? What would be the effect on one's morale of knowing in advance that severe and eventually fatal mental and physical deterioration were in store for one? Perhaps it would be better not to know. Yet if the child, reaching maturity, had the knowledge, he or she could decide whether or not to produce children, each one of whom would have a 50 percent chance of becoming choreic. Should the information be passed on, or should it not?

There is, of course, a brighter side of the picture: the test would also identify those children who almost surely did *not* have gene *H,* and hence would neither become choreic themselves nor transmit the gene to their own offspring in turn.

The value of a marker gene in this connection will be lessened if means are eventually developed for detection of the presence of gene *H* itself in childhood. We shall discuss this later in connection with means of detecting genetic carriers (pp. 446–447).

There is another possible way of utilizing marker genes closely linked to harmful dominant genes. Sometimes a diagnosis can be made long before birth, in the early fetus. By a process known as amniocentesis (pp. 451–453), some of the amniotic fluid surrounding the fetus can be withdrawn and studied. This fluid contains substances produced by the fetus and also sloughed-off fetal cells that may yield information of value.

We noted above that the locus of the secretor gene seems to be quite closely linked to the locus of the gene for myotonic dystrophy. There is evidence that the secretor status of the fetus can be determined in early pregnancy. Under favorable circumstances, then, it might be possible to inform a woman whether or not the fetus was likely to develop into a severely crippled child (Harper et al., 1972). This information would call for a difficult decision: Should the fetus be aborted, or should it not? As we shall see later (pp. 453–454), there is no possibility of reaching unanimous agreement on the question of utilizing abortion.

Recessive Genes Certain genes are harmless in heterozygotes but harmful, even lethal, in homozygotes. In order to avoid producing abnormal children, it would be desirable that two such heterozygotes not marry each other (or if they did marry, not produce children). But if the gene is recessive and heterozygotes are normal, it is frequently difficult to tell who the heterozygotes are.

Suppose that we could demonstrate that in a certain kindred a dominant gene for a red blood cell antigen (we shall use *Bd* again) is closely linked to such a recessive gene, *a* (for "abnormality"). Then heterozygotes in that kindred would have the genotype (*Bd a*) (*bd A*), and two people found to possess the antigen could be warned of the danger of marrying each other, or if they did marry, of having children. The two people concerned might be members of the same kindred, or members of different kindreds shown to have the (*Bd a*) (*bd A*) genotype. In Chapter 11 we found that if a heterozygote marries his first cousin, the chance is ⅛ that he will be marrying another heterozygote. With the imagined knowledge of linkage, we could change this ⅛ to virtual certainty (if linkage is close).

In some other kindreds, of course, the linkage would be (*Bd A*) (*bd a*). In such kindreds the *absence* of the antigen would identify carriers of the harmful recessive gene.

Of even greater value than reliance on a marker gene would be ability to identify *Aa* heterozygotes by some direct test, as we mentioned earlier and will discuss more fully in Chapter 28.

Trait Association Versus Linkage Earlier in the chapter we called attention to the fact that linked genes are not necessarily found together in a population more frequently than they are found separately. This is because of the two types of linkage: coupling and repulsion. In this connection it might be of value to turn the matter around and ask: What *are* some of the genetic reasons that traits are found associated together?

1. In the first place, a gene may have more than one effect. More complete knowledge may demonstrate that this is a true of all genes, but at present we call genes that demonstrably have more than one effect **pleiotropic genes.** For example, the gene for white eye color in *Drosophila* not only removes the pigment from the eye (leaving it white), but also changes the color of the testicular membrane, changes the shape of the spermatheca, and affects the length of life. Hence these traits are found to be associated because they all result from the action of one gene.

2. Sex limitation (pp. 219–221), frequently dependent on sex hormones, which in turn reflect genetic constitution, affords another familiar example. Beards are associated with bass voices more frequently than they are with soprano voices.

3. Another cause of observed association of characters results from the nature of the gene pool. Over much of the earth, brown skin, brown hair, and brown eyes occur together. This does not mean that they are all the effect of one gene, or that the genes are linked. It simply means that the gene pools of these people have the genes in high proportion, with correspondingly low frequencies of the genes producing nonbrown eyes or red hair, for example. This is sometimes called **stratification.** Sometimes the gene pool of an isolated group or of a caste or clan may have an unusual constitution, resulting in an unusual association of characteristics. Stratification is not to be confused with genetic linkage.

Problems

1. In addition to the recessive genes for black body, purple eyes, and vestigial wings in *Drosophila,* another recessive gene, *c,* causes the wings to be curved instead of straight.

   Black-bodied, curved-winged flies were mated to homozygous gray-bodied, straight-winged flies, and the F_1 females were back-crossed to black-bodied, curved-winged males. Progeny produced totaled 62,679; of these 14,237 (22.7 percent) were recombination types, the result of crossing over. What were the genotypes and phenotypes of the two recombination types?

On the chromosome map (p. 255) how far from the locus of gene *b* is the locus of gene *c*? Is gene *c* to the "left" of gene *b* or to the "right" of it? In attempting to answer this last question, the experimenters produced females heterozygous for straight wing and long wing (*CV*) (*cv*), and mated them to curved-winged, vestigial-winged males. Of 1720 progeny, 141 (8.2 percent) were recombination types. With this added bit of information, place gene *c* on the chromosome map. (Note: The total map distance obtained by adding short intermediate distances is greater than the *apparent* map distance indicated by an experiment in which only the amount of crossing over between the genes at the extremes of a series of genes is measured. When genes are far apart, some of the crossing over between them is double crossing over, Fig. 15.3, p. 245. As the figure indicates, double crossing over leaves the genes at the ends unchanged in their linkage relationships and so reduces the *apparent* amount of crossing over that has occurred.)

2. About what percentage of crossing over would you expect to observe if you mated purple-eyed, curved-winged flies to homozygous red-eyed, straight-winged ones, and then back-crossed the F_1 females to purple-eyed, curved-winged males? Diagram the experiment, giving genotypes and phenotypes of all individuals, and indicating linkage correctly.
3. Another imaginary human trait is "eyebrowlessness." The accompanying table presents data concerning five families of three children each. In these

Family	Child I		Child II		Child III	
	Eyebrows	Eyelashes	Eyebrows	Eyelashes	Eyebrows	Eyelashes
A	present	forked	absent	unforked	present	forked
B	absent	forked	present	unforked	present	unforked
C	absent	unforked	absent	forked	absent	unforked
D	present	forked	present	forked	absent	unforked
E	present	unforked	absent	forked	present	unforked

families the following traits are being studied: eyebrows present versus eyebrows absent; eyelashes forked versus eyelashes unforked. Apply the sib-pair method to the data in the table to determine whether they form evidence for *or* against the hypothesis that the two loci are linked.

4. The nail-patella syndrome arises through the action of a dominant gene, which we may designate as *N*.

(a) A certain woman has the syndrome and belongs to blood group B. Her father lacked the syndrome and belonged to group O. If she marries a man of group O who lacks the syndrome, which two of the following types of offspring will be expected to be the more numerous, which two the less numerous? (1) Group B, with syndrome. (2) Group B, without syndrome. (3) Group O, with syndrome. (4) Group O, without syndrome. Diagram the mating, giving the genotypes of all individuals and indicating linkage correctly.

(b) A certain man has the nail-patella syndrome and belongs to group A. His mother had the syndrome and belonged to group O; his father lacked the syndrome and belonged to group A. If this man marries a woman of group O who lacks the syndrome, what types of offspring will be expected and with what relative frequencies? Diagram the mating, giving genotypes and phenotypes of all individuals, and indicating linkage correctly.

5. Suppose we are fortunate enough to find a family in which the mother is heterozygous for blood group A, for the nail-patella syndrome, and for adenylate kinase (p. 261). We shall further suppose that in the mother one chromosome of the pair concerned has genes I^A, N, and AK^1, while the homologous chromosome has genes I^O, n, and AK^2. The father is homozygous for group O, for lack of the syndrome, and for AK-1. Judging from information given in the text, is a child in this family who has the nail-patella syndrome more likely to belong to blood group O *or* to be heterozygous AK^1/AK^2?
6. In our general population, why do not most people with the nail-patella syndrome belong to the same blood group?

References

Bridges, C. B., and T. H. Morgan, 1919. "Contributions to the genetics of *Drosophila melanogaster*. Part II. The second-chromosome group of mutant characters," *Carnegie Institution of Washington,* Publication No. **278**: 125–304.

Cook, P. J. L., 1965. "The Lutheran-Secretor recombination fraction in man: A possible sex difference," *Annals of Human Genetics,* **28**:393–401.

Harper, P. S., M. L. Rivas, W. B. Bias, J. R. Hutchinson, P. R. Dyken, and V. A. McKusick, 1972. "Genetic linkage confirmed between the locus for myotonic dystrophy and the ABH-secretion and Lutheran blood group loci," *American Journal of Human Genetics,* **24**:310–316.

Levitan, M., and A. Montagu, 1971. *Textbook of Human Genetics.* New York: Oxford University Press.

Li, C. C., 1961. *Human Genetics.* New York: McGraw-Hill Book Company, Inc.

Race, R. R., and R. Sanger, 1968. *Blood Groups in Man,* 5th ed. Philadelphia: F. A. Davis Company.

Renwick, J. H., 1971. "The mapping of human chromosomes," *Annual Review of Genetics,* **5**:81–120.

Schleutermann, D. A., W. B. Bias, J. L. Murdoch, and V. A. McKusick, 1969. "Linkage of the loci for the nail-patella syndrome and adenylate kinase," *American Journal of Human Genetics,* **21**:606–630.

Wendt, G. G., H. Ritter, I. Zilch, G. Tariverdian, I. Kindermann, and G. Kirchberg, 1971. "Genetics and linkage analysis on adenylate kinase," *Humangenetik,* **13**:347–349.

17

Sex Determination, and Changes in Number of Sex Chromosomes

In our discussion of sex chromosomes (Chap. 14), we noted that females normally have two X chromosomes, and that males have one X chromosome and one Y chromosome. Does the distribution of the chromosomes determine the sex of the individual? What is the role of the sex chromosomes in determining sex?

At the outset we should note that if the sex chromosomes determine sex, the determination will occur at the time of fertilization and will depend on whether an X-bearing sperm or a Y-bearing sperm fertilizes the ovum. Direct evidence on this point is not available, but indirect evidence indicates that sex is determined very early in the embryonic life of the individual. In some animals the fertilized ovum gives rise to an embryonic mass, which then divides to form several or many individual offspring. In this case, all the offspring come from the same ovum, and all are of the same sex. This indicates that sex was determined in the ovum stage, or at least before the subdivision of the embryonic mass. Armadillos afford a mammalian example. Normally four armadillos are born at a time and all four are of the same sex. Investigations of the embryology have demonstrated that the four come from one ovum (that is, they are monozygotic). A similar situation is found in human "identical" twins, who are always of the same sex, having come from a single ovum. Here, however, we must beware of circular reasoning, for twins would not be classified as identical (monozygotic) in the first place unless they *were* of the same sex. In every case, the conclusion that a pair of twins arose from a single ovum is only an inference, not a matter of direct observation.

Problems concerned with classifying twins will be discussed in Chapter 23.

As frequently happens, abnormalities give us valuable clues about normal functions. What happens when individuals have an abnormal number of X and Y chromosomes? If such abnormalities interfere with normal sex determination, then we may reach some conclusions concerning the roles of these chromosomes. First, however, we may well inquire into the means by which abnormal numbers of X and Y chromosomes arise.

Nondisjunction

In our discussion of meiosis (Fig. 3.4, p. 29; Fig. 14.1, p. 216), we stressed the normal separations of chromosomes as the process proceeds. In the primary spermatocyte, for example, each tetrad consists of a pair of chromosomes that have duplicated themselves to form two chromatids held together by a centromere. When the primary spermatocyte divides to form the two secondary spermatocytes, each of the latter receives one pair of chromatids from each tetrad. This normal separation of chromosomes is called **disjunction.** The failure of the chromosomes to separate is called **nondisjunction.**

The upper part of Figure 17.1 shows nondisjunction of the X chromosomes in the first meiotic division of a female. For simplicity, only 2, instead of 22, pairs of autosomes are shown in the primary oöcyte in addition to the X chromosomes. The autosomes are shown undergoing normal disjunction, but the X chromosomes fail to separate, both entering the first polar body. Thus the ovum is left with autosomes only.

The reverse may also occur—the two X chromosomes remaining in the secondary oöcyte, the first polar body receiving only autosomes (upper part of Fig. 17.2). The resulting ovum then has two X chromosomes.

Nondisjunction also occurs in males. Figure 17.3A shows an example in which nondisjunction of the sex chromosomes in the first meiotic division leads eventually to production of two kinds of sperm cells: those containing *both* an X and a Y chromosome, and those containing no sex chromosomes.

Figure 17.3B shows an example in which normal disjunction occurs in the first meiotic division, but nondisjunction occurs in the second division. This leads to production of sperm containing (a) two X chromosomes, (b) two Y chromosomes, and (c) no sex chromosomes. Rarely nondisjunction may occur in both divisions of a primary spermatocyte.

XO

The symbol XO designates an individual who has the normal complement of autosomes, but only one sex chromosome, an X. The "O,"

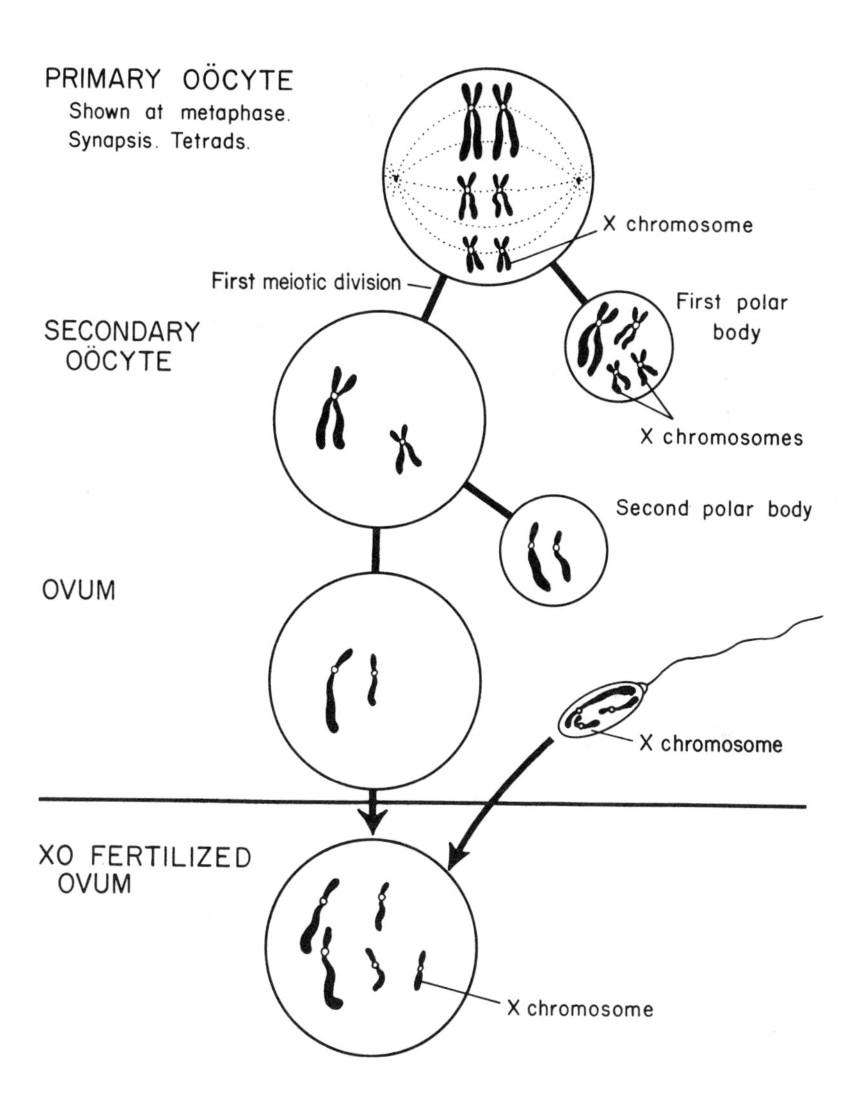

FIG. 17.1 Above the line: Nondisjunction of X chromosomes in the first meiotic division, producing an ovum containing the normal haploid number of autosomes but no X chromosome. Below the line: Fertilization of such an ovum by a sperm cell containing an X chromosome, to produce an XO zygote.

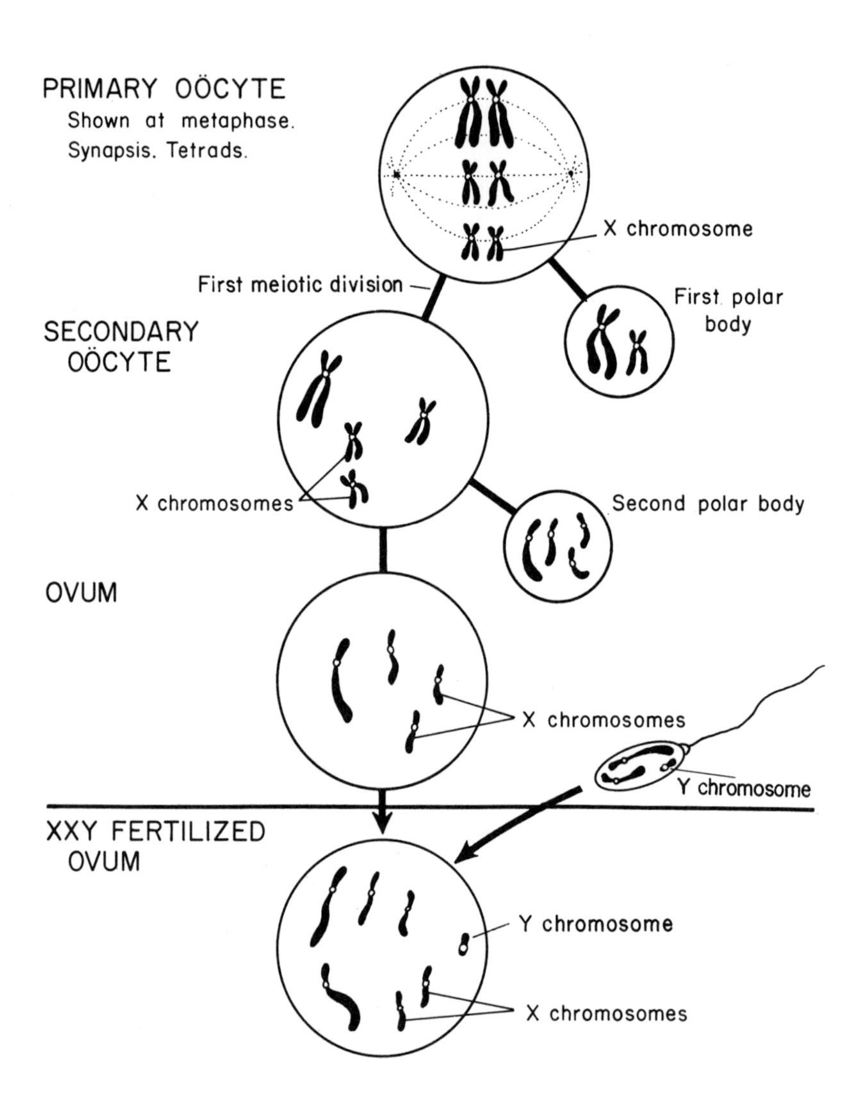

FIG. 17.2 Above the line: Nondisjunction of X chromosomes in the first meiotic division, producing an ovum containing the normal haploid number of autosomes but two X chromosomes. Below the line: Fertilization of such an ovum by a sperm containing a Y chromosome, to produce an XXY zygote.

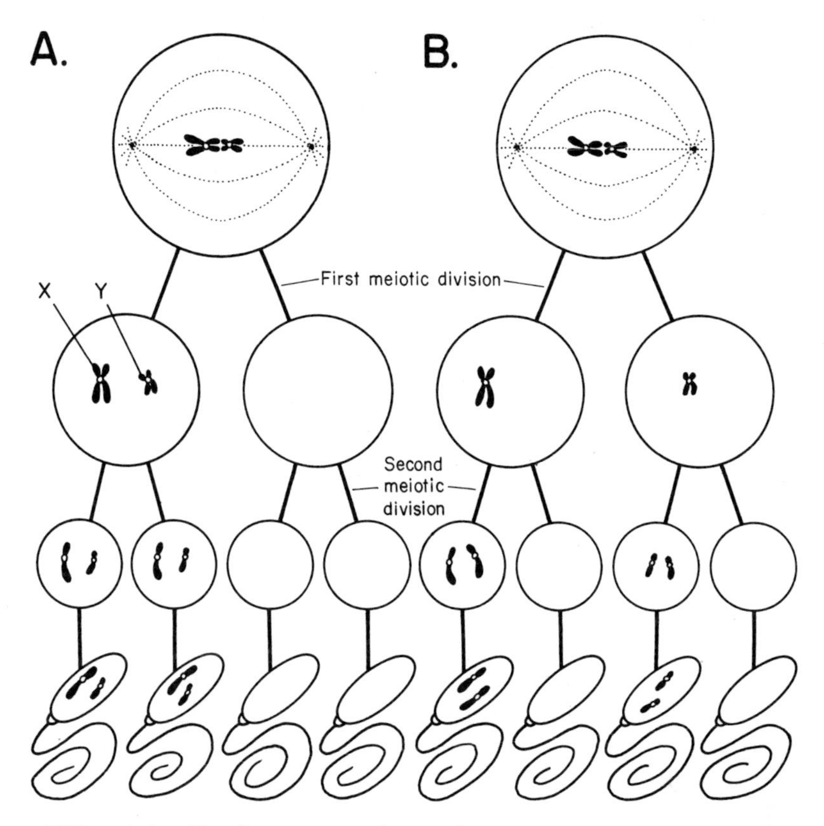

FIG. 17.3 Nondisjunction of sex chromosomes in meiosis in the male. Autosomes are omitted although they are present in all cells. Both A and B start with a primary spermatocyte having the sex chromosomes in terminal synapsis (see Fig. 14.1, p. 216, in which the stages are named). A. Nondisjunction in the first meiotic division. Resulting sperm contain either (1) both and X and a Y chromosome or (2) no sex chromosomes. B. Nondisjunction in the second meiotic division. Resulting sperm contain either (1) two X chromosomes, or (2) two Y chromosomes, or (3) no sex chromosomes. In this portion of the diagram, both secondary spermatocytes are shown undergoing nondisjunction simultaneously; this need not occur.

therefore, really represents zero. The individual has a total of 45 chromosomes in each cell instead of the normal 46.

Origin How can such a constitution arise? There are two possibilities: Either the fertilized ovum never contained more than one X, or at first it was XX (or XY) and subsequently lost one X (or the Y). Both manners of origin may well occur.

The first possibility involves nondisjunction in the production of germ cells. The lower part of Figure 17.1 shows an ovum containing no X chromosome being fertilized by a sperm cell containing an X chromosome. In this case the individual would inherit the X-linked traits of the father.

At other times a normal ovum containing an X may be fertilized by a sperm cell containing no sex chromosome (produced as shown in Fig. 17.3). Such an XO individual would inherit X-linked traits from the mother.

The second possibility for manner of origin of the XO constitution involves an error in the mitosis undergone by the fertilized ovum (zygote) or some cell derived from it. If a zygote contained two X chromosomes (or an X and a Y), but one of those failed to line up correctly on the spindle, and so was lost, the two cells produced by the division would both be XO (Fig. 17.4A). On the other hand, if the mitotic error occurred a little later in development, the embryo would be a mixture of two kinds of cells: those that were XX (or XY) and those that were XO

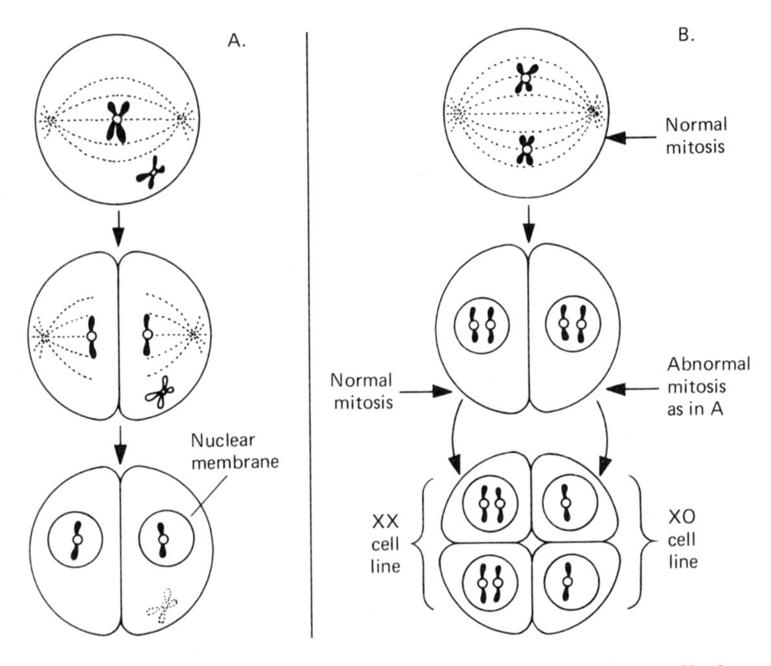

FIG. 17.4 A. Abnormal mitosis, producing XO daughter cells from an XX fertilized ovum. One X is not included in mitosis, remains in the cytoplasm when the nuclear membrane forms, and is destroyed by enzymes. B. Production of an XX/XO mosaic. At the two-cell stage, one cell undergoes abnormal mitosis as in A and so gives rise to an XO cell line. The abnormal mitosis may occur at a later stage, in which case the cells in the two lines will be unequal in number.

(Fig. 17.4B). Such an individual is called a **mosaic** or **mixoploid.** There is an increase in proportion of XO individuals who are mosaics as compared to other people.

Figure 17.4A shows the abnormal mitosis as occurring in an XX zygote. In this case the chromosome lost might be either the one that came from the father or the one that came from the mother, although some investigators conclude that it is more likely to be the one from the father. But the cell undergoing the abnormal mitosis might have the constitution XY, the Y chromosome being lost. In this case the remaining X chromosome certainly came from the mother.

Studies of X-linked traits in XO individuals and their parents indicate that the single X chromosome may in fact come from either parent, although there is some evidence from such studies that it more frequently comes from the mother than from the father (review in Hamerton, 1971, Vol. 2).

There is an increased tendency, also, to monozygotic twinning in sibships in which XO individuals occur (Nance and Uchida, 1964). Twinning of this type may be regarded as an accident that occurs to the zygote, or at least to a very early derivative of it.

Finally, there is no tendency of older mothers to have more XO offspring than do younger mothers. As we shall see in the next chapter, likelihood of producing a child with Down's syndrome increases with the age of the mother, and is interpreted to indicate increased tendency to nondisjunction in older mothers. For these reasons many investigators conclude that the XO constitution more frequently arises by an accident that occurs to a fertilized ovum than it does by nondisjunction in the production of sperm or ovum before they unite in fertilization.

Characteristics of XO Individuals Is one X chromosome by itself sufficient to cause an individual to be female? The answer is: *partially.* Such individuals are feminine in general body structure, but their one big lack is functional ovaries. Their ovaries are either missing entirely or are reduced to what are called "streak ovaries," bits of tissue almost completely lacking oöcytes, ova, and their associated cells. Hence, this condition is called **ovarian** (gonadal) **dysgenesis.** Another name is **Turner's syndrome.**

Interestingly, XO fetuses have normal ovaries until the third month of gestation; after that the number of öocytes in them decreases. A few öocytes may still be present at birth, but none are found when the age of puberty is reached. Evidently, one X is sufficient to determine that the gonads shall be ovaries, but a second X is necessary if normal function is to develop, including stimulation of the genital organs and secondary sex characteristics to reach maturity.

Girls with the XO constitution do not undergo the bodily changes normal girls experience at puberty. Their genital organs remain juvenile, and their breasts do not develop. They do not menstruate.

Girls with this syndrome are short in stature, seldom exceeding 4 feet, 9 inches. And they may have a variety of other bodily peculiarities. They are not usually retarded mentally.

Not all girls with the symptoms of Turner's syndrome have the XO constitution. Some of them have two X chromosomes, but in many cases it can be shown that the second X is defective in some way. (See Hamerton, 1971, Vol. 2, for a more complete discussion of all the points we have mentioned.)

Turning from man to mouse, we might note that XO mice are *fertile* females, although frequently their fertility is below normal.

Abortuses A considerable proportion of fetuses that are lost by spontaneous abortion are abnormal in some way. Many of them have abnormal chromosomal constitutions, and of these, 20 percent or more are XO fetuses. Spontaneous abortion is not always the unmitigated calamity mothers frequently consider it. Many of the lost fetuses could not have developed into normal children.

XXY

The lower part of Figure 17.2 shows an ovum containing two X chromosomes (as the result of nondisjunction) being fertilized by a normal sperm containing a Y chromosome. Studies of sex-linked traits indicate that this is a common manner of origin of the XXY constitution (Race and Sanger, 1969). At other times this constitution may arise when a sperm containing both an X and a Y (produced as shown in Fig. 17.3) fertilizes a normal ovum.

What is the sex of an XXY individual? Are the two X's sufficient to cause the individual to be female? What is the influence of the Y chromosome?

That the Y has potent male-determining properties is indicated by the fact that XXY individuals are males. Yet the two X's are not without effect because these males have very small testes that do not produce sperm, and frequently they have feminine development of the breasts. The condition is known as **Klinefelter's syndrome.** Frequently these boys have some degree of mental retardation.

Figure 17.5 shows the chromosomes of a person with Klinefelter's syndrome. Such a picture is called a **karyotype.** It was made by making an enlarged photographic print and cutting out the picture of each chromosome. These pictures were then matched into pairs and arranged in rows as shown. Karyotypes will be discussed more fully in the next chap-

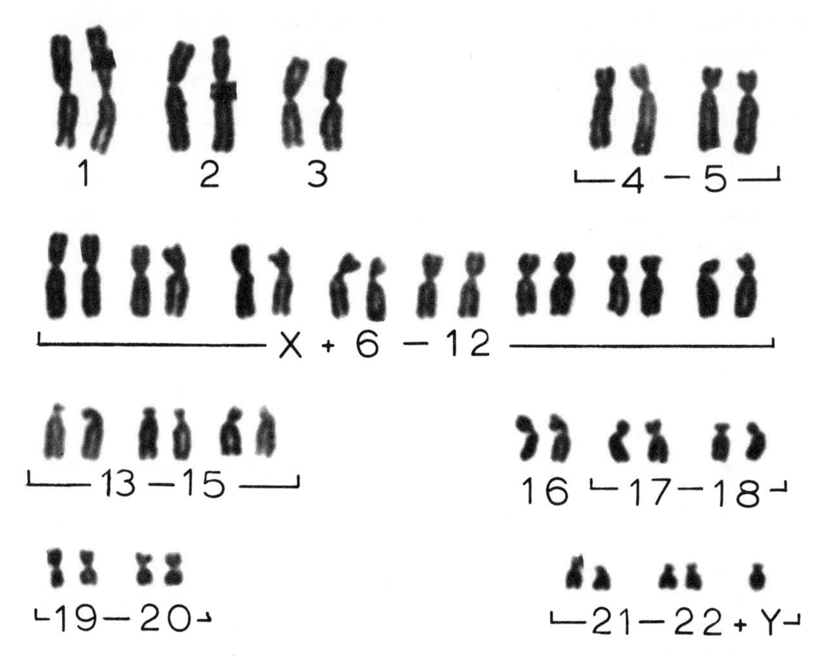

FIG. 17.5 Chromosomes (karyotype) of a person with Klinefelter's syndrome: 47 chromosomes, including two X chromosomes and one Y chromosome. (Courtesy of Victor A. McKusick, M.D.)

ter. Here we note that the group consisting of "medium-sized" chromosomes is composed of the X's and the autosomes numbered 6 through 12, inclusive. Thus this individual has two X's and one Y (lower right-hand corner).

It is interesting to note that the Y chromosomes of different animals differ in importance as determiners of maleness. In the fruit fly, *Drosophila,* for example, an XXY individual is a *female.* Evidently the Y chromosome does not contain important male-determining genes for fruit flies. Further evidence is afforded by XO flies, which are *males* despite the absence of a Y chromosome (contrast the situation in human beings and in mice). XO males are sterile, however; apparently the Y chromosome contains genes for male fertility. Experimentation in *Drosophila* indicates that the X chromosome contains female-determining factors or genes and that the *autosomes* contain the male-determining genes. The female-determining genes of two X chromosomes outweigh the effect of the male-determining genes in a normal set of autosomes, but the female-determining "strength" of one X chromosome is unable to do this.

Some boys with symptoms of Klinefelter's syndrome have two X's, but no observable Y. Why, then, are they not females? There are various possible answers. Perhaps these individuals are really mosaics of XX-containing and XXY-containing cells, the Y in these latter cells being sufficient to cause the degree of maleness observed although they have not been found in the tissues used for chromosomal study, or have been eliminated during fetal development. Perhaps the region of the Y chromosome important for male determination has been translocated (Chap. 18) onto an X. Some evidence supports such a hypothesis. There are other possible explanations, but discussion of them and of various other chromosomal constitutions that give rise to symptoms like those of Klinefelter's syndrome would lead us too far afield (see Hamerton, 1971 Vol. 2; Levitan and Montagu, 1971).

XYY

Figure 17.3B shows production of sperm cells that contain two Y chromosomes, the result of nondisjunction in the second meiotic division. If such a YY sperm fertilizes a normal ovum, a zygote with the XYY constitution is produced.

The principal effect of the extra Y chromosome seems to be increase in stature, XYY boys being tall for their age, and becoming men over 6 feet tall. Mental retardation is common, although intelligence may be normal. Often the reproductive organs are somewhat abnormal.

Perhaps because many of the studies of XYY men have been made on criminals being treated in prisons and other correctional institutions, the belief is common that these men tend to be overly aggressive and have criminal tendencies. There is no doubt that this characterizes some of them, but others have been found who are not aggressive and have no criminal tendencies. It is best to withhold judgment until we have more data on the chromosomal constitution of tall men who are not institutionalized (Breslau and Bell, 1970; Pitcher, 1971; Hook, 1973). Data available to date seem to indicate little tendency of XYY men to pass this constitution on to their sons (Melnyk et al., 1969).

Greater Numbers of X's and Y's

In Figure 17.2 we see an XX ovum being fertilized by a Y-containing sperm. If fertilization is by an X-containing sperm, the resulting zygote will be XXX. We might anticipate that such an individual would be a "super female," and indeed that term has been used. Such women may be normal, but many of them are abnormal, physically or mentally or both. They are also likely to be infertile, although some of them are fertile and produce normal children.

Hamerton (1971, Vol. 2) reported that five women were known having four X chromosomes—the XXXX constitution. All were physically normal but had severe mental retardation. Apparently, this is an example of "too much of a good thing." Having X's is good, but having too many of them upsets the delicate balances that must be maintained if an embryo is to develop into a normal individual.

Males with three X chromosomes—the XXXY constitution—have symptoms much like those in Klinefelter's syndrome and are mentally retarded. Males with four X chromosomes—XXXXY—are mentally retarded and suffer severe physical abnormalities. An increase in the number of Y chromosomes is also deleterious and likely to result in mental retardation as well as physical abnormality. About 40 known cases of the XXYY constitution were reported by Hamerton (1971, Vol. 2), a reference containing a wealth of information on anomalies of the sex chromosomes.

Intersexes and Sex Determination

People whose genital organs are neither those of a normal male nor those of a normal female are termed intersexes or hermaphrodites. The observable ambiguities of structure usually reflect inner abnormalities of the reproductive system. A person having an ovary on one side of the body and a testis on the other, or having an ovotestis—a gonad containing a mixture of testicular and ovarian tissue—is termed in medical parlance a **true hermaphrodite.** (To a zoologist like the present author, the term is not entirely acceptable because he thinks that the only *true* hermaphrodites are creatures like earthworms having both male and female systems that are *fully functional.* Ancient Greco-Roman mythology to the contrary notwithstanding, human "true" hermaphrodites cannot function both as males and as females. Frequently they cannot function effectively as either one.) A person having ovaries combined with ambiguous genital organs that are most like those of a male, or having testes combined with genital organs resembling those of a female, is termed a **pseudohermaphrodite.**

In the medical literature we can find descriptions of almost every imaginable combination of male and female organs, external and internal (see Hamerton, 1971, Vol. 2; Overzier, 1963; Jones and Scott, 1971). In any given individual, some or all of the organs are reduced in size and are nonfunctional. Sometimes these ambiguous developments reflect presence of chromosomal abnormalities, sometimes not. Our present interest in them arises from the fact that the abnormal developments aid in understanding normal development.

The puzzle has many missing pieces, but the broad outlines of a picture are traceable. In the early fetus, the gonads are undifferentiated, capable of developing into either an ovary or a testis. The undifferentiated gonad consists of an outer **cortex** and an inner **medulla.** If no Y chromosome is present, the X chromosome causes the cortex to develop into an ovary, the medulla becoming reduced. As we noted in connection with Turner's syndrome, a single X is capable of doing this, but a second X is needed if the ovary is to continue to develop normally. On the other hand, if a Y chromosome is present, the medulla is stimulated to become a testis, the cortex becoming reduced. As boys with Klinefelter's syndrome demonstrate, however, presence of more than one X along with the Y in some manner inhibits the testis so that it does not produce sperm cells.

Once established as an ovary or as a testis, the gonad normally controls subsequent sexual development by controlling the production of sex hormones. Everyone has both male sex hormones (**androgens**) and female sex hormones (**estrogens**). In normal females the proportion of estrogen greatly exceeds the proportion of androgen; in normal males the reverse is true. Normal females and males form the extremes of a graded series of proportions in which the balance between estrogen and androgen may not be such as to make possible normal development of either sex. And the problem is complicated by the fact that a sex hormone must not only be present in proper quantity, it must be present at the time in fetal development when the organs in question are developing, and the cells in these organs ("target cells") must be capable of responding to the hormone.

This latter point is illustrated by people with the XY constitution who are nevertheless completely feminine in body form, often voluptuously so. They are always raised as girls and may marry as such. The vagina is normal but ends blindly, not connecting to a uterus (cf. normal organs, Fig. 17.6). There are no ovaries. Because testes are present, the condition is called **testicular feminization.** The testes are frequently abnormal and they are retained within the body instead of descending into a scrotum as in normal males. This is as complete an example of sex reversal as can be found in humans. But male sex hormones are present at levels comparable to those of normal men. Why, then, is the body so strongly feminine? This is interpreted as an instance in which the target cells in the body that should respond to androgens are for some reason insensitive to them, and so the estrogens produce their effects. This insensitivity of the target cells seems to be an inherited trait, the mode of inheritance suggesting that the gene is X-linked or is an autosomal dominant with expression sex-limited to males (p. 220).

We have suggested some of the variables that are involved in sex determination. But there are many others, known and unknown. Autosomal genes are doubtless involved. For example, experiments on goats have demonstrated the existence of an autosomal gene that will stimulate the undifferentiated gonad to become a testis in the absence of a Y chromosome. If this occurs in man, it may explain why some XX individuals have testes (Hamerton, 1971, Vol. 2). We have already mentioned mosaicism; this is common among intersexes, and it doubtless occurs in individuals in whom it has not been detected. Commonly the chromosomes

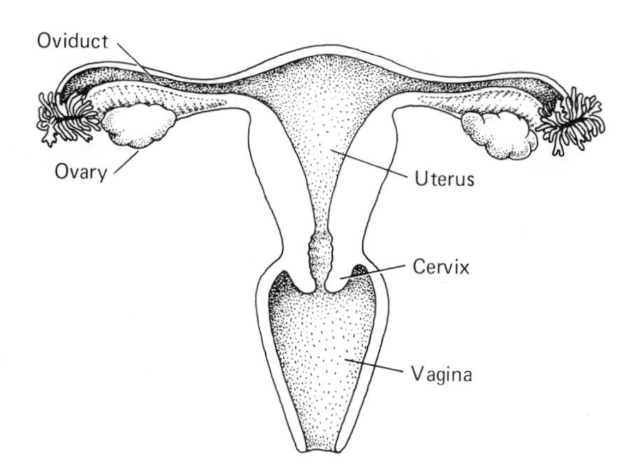

FIG. 17.6 Normal female reproductive organs. (From Keeton, W. T., 1967. *Biological Science,* New York: W. W. Norton & Company.)

in only one tissue of an individual, frequently in the white blood cells, have been studied. Other tissues may differ, and the chromosomal constitutions of these other tissues may have influenced sexual development.

The subject of sex determination and its anomalies is a large one. We have mentioned here only a few aspects of it. For fuller discussion see especially Hamerton (1971, Vol. 2) and McKusick (1964).

X-chromatin or Barr Bodies

An interesting and useful discovery is recognition of the fact that nuclei of cells that are not undergoing mitosis usually reveal the sex of the individuals from which the cells were derived. Suppose, for example, that we scrape a few cells from the lining of the cheek of a woman (using a wooden tongue depressor, perhaps). We spread the cells on a microscope slide and stain them, preparing what is called a *buccal smear.* We shall find a deeply stained particle or mass attached to the inside of the nuclear membrane of many of the cells (Fig. 17.7A). This particle is called **X-chromatin,** or a **Barr body.** (This latter based on the name of its discoverer; Barr, 1960). A similar buccal smear from the mouth of a man is devoid of such Barr bodies, or nearly so.

Here is a means of distinguishing male tissue from female tissue even when chromosomes as such are not visible. The mucous membrane of the mouth was taken merely as an example; in general, tissues of all kinds from a female possess Barr bodies in cell nuclei, whereas nuclei of male tissues lack such bodies. An individual whose cells have a Barr body is said to be "X-chromatin positive" (or sometimes simply "chromatin

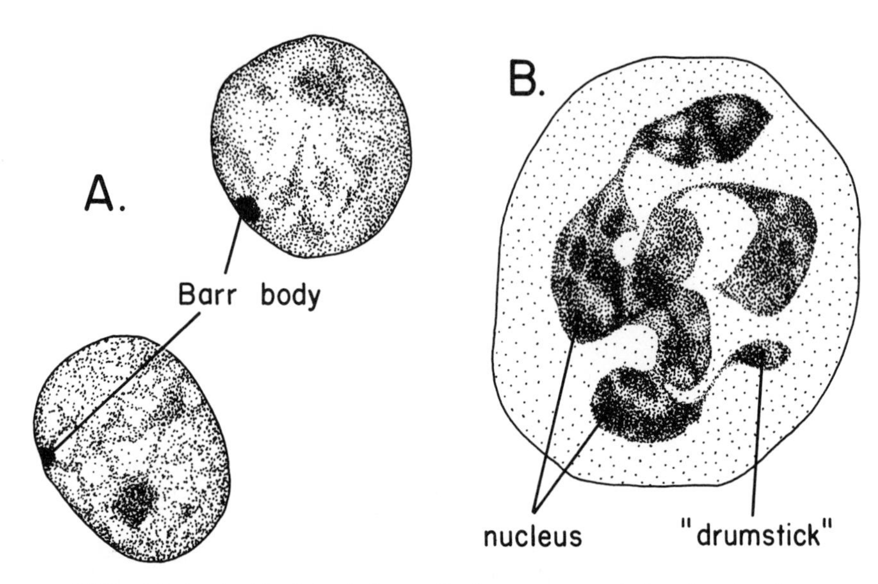

FIG. 17.7 A. Nuclei of skin cells from a normal female, showing X-chromatin bodies (Barr bodies). B. Leucocyte (white blood cell) from a normal female. Note the irregular nucleus with the tiny appended "drumstick." (Drawn from photomicrographs in Barr, M. L., 1960. "Sexual dimorphism in interphase nuclei," *American Journal of Human Genetics*, **12**: 118–127.)

positive"), while an individual whose cells lack a Barr body is called "X-chromatin negative" ("chromatin negative").

Of what is the Barr body composed? Evidence indicates that it consists of *one* of the two X chromosomes that normal female cells contain. According to this idea, when a cell is not undergoing mitosis, the substance of one X, like that of the autosomes, is in an attenuated form and hence not visible. The other X remains in a compact form, so that it stains deeply and is identifiable as a Barr body. If more than two X's are present, all but one become Barr bodies.

As a result, there is good correlation between the number of Barr bodies and the number of X chromosomes. This is of great help to investigators; when they determine the number of Barr bodies, by relatively simple methods, they know how many X's to expect when they study the chromosomes—by investigations of a more complicated type. As a rule, the maximum number of Barr bodies is one less than the number of X chromosomes (some cells may have fewer Barr bodies). Thus, *no* Barr bodies are found in cells of individuals having the XY or XYY constitution (phenotypic males) or XO (phenotypic juvenile females, as noted earlier). One Barr body characterizes XX (normal female) cells, XXY cells (Klinefelter's syndrome), and XXYY cells. A maximum of two Barr bodies is found in XXX and XXXY cells. And the rule holds for greater numbers of X chromosomes.

As we should expect, sex chromosome mosaics have cells differing in number of Barr bodies. For example, the XO/XX mosaic mentioned earlier would have some cells with no Barr body, some cells with one Barr body. A XXX/XXY mosaic would have some cells with two Barr bodies, some cells with one such body, and so on.

Figure 17.7B shows a drawing of a white blood cell from a female. The irregular stained portion is the nucleus, with its appendage, called a "drumstick," which is found in the white blood cells of women but not in those of men. This forms another trait by which cells of females differ from those of males.

The Lyon Hypothesis

Because females have two X's, while males have only one, why do not X-linked genes produce twice as great an effect in females as they do in males? We noted, for example, that people normally have an X-linked gene (*G6PD*) that determines production of the enzyme glucose-6-phosphate dehydrogenase. Why do not homozygous normal females produce twice as much of this enzyme as do normal males? The fact that they do not, suggests some type of **dosage compensation.**

In line with a hypothesis advanced by Lyon (1962), we may theorize that in mammals dosage compensation arises from the fact that one of a female's two X chromosomes remains inactive. If this is correct, both male and female cells have one active X chromosome. It was a logical step from this hypothesis to the suggestion that the genetically inactive X forms the compact Barr body observed in female cells. Subsequently, much evidence has been accumulated in support of this combined theory, although not all experimental findings are explainable by it.

One of the best ways of demonstrating the hypothesis is by the study of heterozygous females. If an X-linked trait is one that can be observed at the cellular level, heterozygous females may be expected to have two kinds of cells: (a) those in which the X containing one allele is the active one, and (b) those in which the X containing the other allele is active. Lyon's own research was with mice. For example, a female mouse may be heterozygous for certain X-linked color genes and have a mottled appearance, apparently because one gene is active in cells composing certain areas of the skin, while its allele is active in cells of other areas. The "decision" as to which X shall be inactive in a given cell (and all cells derived from it by subsequent cell divisions) must be made early in embryonic development. And the "decision" is assumed to be at random, at least when both X's are normal in structure.

More direct evidence can be obtained when the X-linked trait is production of an enzyme whose presence in cells can be tested. Such an enzyme is hypoxanthine-guanine phosphoribosyl transferase (HGPRT). This is an enzyme normally present and active in our cells. The gene is an X-linked dominant. A recessive allele of the gene is sometimes found. Males having this allele in their single X show extremely small activity of the enzyme. As a result they suffer from the **Lesch-Nyhan syndrome,** which includes elevated levels of uric acid, mental retardation, spastic cerebral palsy, compulsive self-mutilation (e.g., chewing of the fingers), and other abnormalities.

The mother of such a son is necessarily heterozygous. Does she have two kinds of cells, those in which the enzyme is active and those in which it is inactive (because the X with the normal allele is the one "turned off")? The evidence is that she does.

This can be demonstrated by taking a sample of the heterozygous woman's cells and growing them in tissue culture. Commonly a bit of skin tissue is taken by biopsy, and the fibroblasts in the connective tissue are cultured. They will grow readily under suitable conditions, and can then be studied to determine their HGPRT activity.

Cells must have this enzyme if they are to make use of hypoxanthine in the process of producing nucleic acids. So hypoxanthine is made radioactive by incorporating radioactive hydrogen (^3H) into its structure. When this radioactive hypoxanthine is added to a culture of fibroblasts, those cells that have HGPRT will take in and utilize the hypoxanthine, while those cells that lack the enzyme or have it in inactive form will not. Which cells these are can be demonstrated by making an **autoradiograph.** The cell culture is placed in contact with a photographic film. Those cells that have incorporated the radioactive material will "take their own picture"—the region of the film with which they are in contact will be exposed (as it would be by a beam of light) and so will turn black when the film is developed. The film in contact with the nonradioactive cells will not be exposed in this manner. In this way the fact that a mother of a Lesch-Nyhan boy has two kinds of cells relative to HGPRT activity has been demonstrated (Rosenbloom et al., 1967).

Essentially the same technique has been applied to **clones** of fibroblasts. These are clusters or colonies of cells, the cells in each cluster having arisen from one original cell by repeated cell divisions. Hence cells in a clone are duplicated copies of this cell. Clones have the advantage of being larger and more easily studied than are the original single cells. Figure 17.8 shows an autoradiograph of several clones of cells from the skin of a heterozygous woman. The fact that some clones have incorporated radioactive hypoxanthine while others have not is clearly evident. Migeon (1971) analyzed 4215 such clones, finding that about half of them had HGPRT activity, half did not.

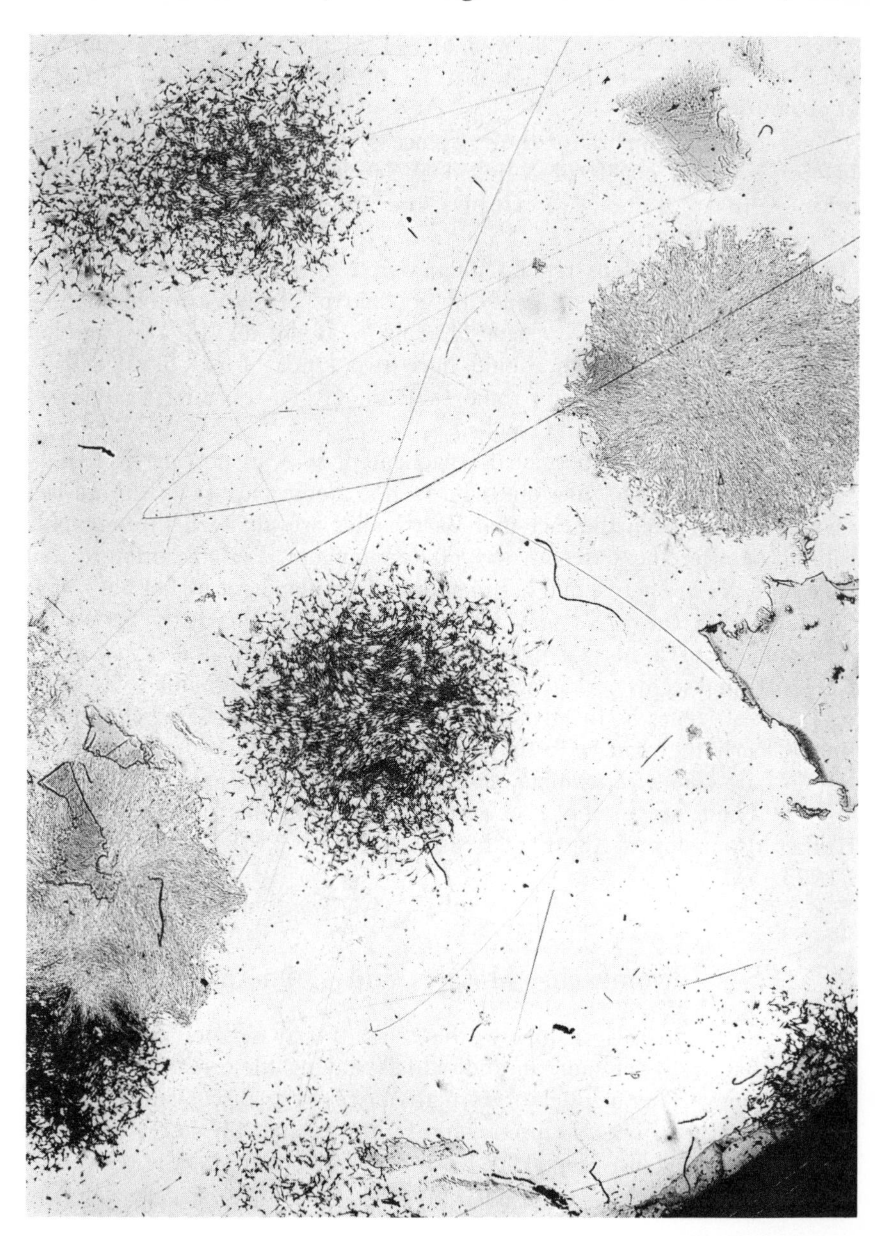

FIG. 17.8 Photomicrograph of clones of cells. The original cells came from the skin of the mother of a boy with the Lesch-Nyhan syndrome. Clones that incorporated radioactive hypoxanthine are black, those that did not are light-colored. (Courtesy of Barbara R. Migeon, M. D. From Migeon, B. R., 1971. "Studies of skin fibroblasts from 10 families with HGPRT deficiency, with reference to X-chromosomal inactivation," *American Journal of Human Genetics,* **23**: 199–210.)

Clones have also been analyzed to determine the *amount* of HGPRT activity present (not just its presence or absence) (Migeon et al., 1968).

Similar support for the existence of two types of cells in women heterozygous for a pair of X-linked alleles has been obtained for some other X-linked traits (e.g., G6PD, and the Hunter-Hurler syndrome; references in Migeon, 1971).

Many questions remain unanswered, however. When a chromosome is inactivated, are *all* genes in it inactive? Answers are conflicting. For example, does the Xg locus "Lyonize"? If so, an Xg(a+) mother who has an Xg(a—) son should have two kinds of red blood cells—some that are Xg(a+) and some that are (Xg(a—). Investigations of this have yielded conflicting results.

If one X chromosome is inactivated, why is not an XXY individual a normal male, like one who is XY? Some clue as to the answer may be gained from the fact that Barr bodies are not seen in embryonic cells until after the sixteenth day of development. Perhaps prior to that time both X's are active and act to suppress development of normal testes. Similarly, two active X's may be necessary during this early period for development of a normal ovary, thus explaining why the ovary of an XO individual seems normal at first but fails to mature normally.

Can genes be inactivated even in the absence of Barr body formation? Clearly they can be in the much-studied fruit fly, *Drosophila,* which has no Barr bodies yet exhibits dosage compensation.

These are examples of some of the questions being asked. For further discussion of the hypothesis, see Lyon (1968), and Hamerton (1971, Vol. 1).

Sex Chromosome Mosaics and Chimeras

So far in this chapter we have mentioned mosaicism repeatedly, noting that it is common in individuals having abnormal numbers of chromosomes. The method of formation we have discussed is that of abnormal mitosis, one X chromosome failing to line up correctly on the spindle and being lost (Fig. 17.4). An XX/XO mosaic was the result.

Nondisjunction in Mitosis Figure 17.9 shows nondisjunction occurring when a fertilized ovum divides, with production of an XXX/XO mosaic. These mosaics have been observed.

One such individual was studied by Jacobs et al. (1960). She was of low normal intelligence with many of the symptoms of Turner's syndrome (of feminine phenotype but lacking vagina, uterus, and breast development; no menstruation). Study of buccal smears revealed that

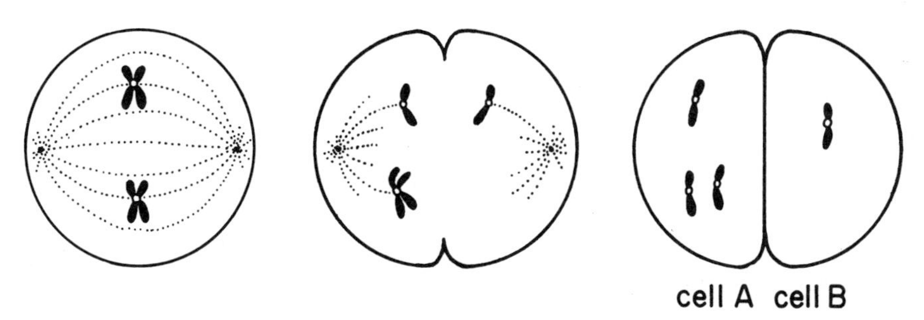

FIG. 17.9 Origin of an X-chromosome mosaic by nondisjunction in the first mitotic division of a fertilized ovum containing two X chromosomes. Autosomes have been omitted from the diagram.

37 percent of the mucosal cells had two Barr bodies. Chromosome counts were made on four samples of living tissue: (1) bone marrow from the sternum; (2) white blood cells; (3) skin from the left shoulder; (4) skin from the right leg. Most of the cells had either 45 or 47 chromosomes. Disregarding a few cells with different chromosome numbers, the chromosome counts on the four samples were as follows:

Tissue	Chromosome Number	
	45	47
1. Sternal marrow	67	0
2. Blood	71	11
3. Skin, left shoulder	18	28
4. Skin, right leg	15	20
Total	171	59

Thus we see that, on the whole, nearly three times as many cells had 45 chromosomes as had 47, but that the different samples varied greatly in this respect. For example, no sternal marrow cells had 47 chromosomes, whereas in the skin samples the number of cells containing 47 chromosomes outnumbered those containing 45.

Cells containing 45 chromosomes were found to lack one of the medium-sized chromosomes. This was considered to be an X chromosome and so the constitution of these cells was regarded as XO (44 autosomes plus one X chromosome). Cells containing 47 chromosomes had an additional medium-sized chromosome and were considered to be of XXX constitution (44 autosomes plus three X chromosomes). The finding of some cells with two Barr bodies added weight to this conclusion.

The authors suggested that her mosaicism may have originated by nondisjunction in the first division of the fertilized ovum, as in Figure 17.9. It is, of course, entirely possible that nondisjunction could occur in a mitosis in some later stage of embryonic development. However, nondisjunction at a later stage would affect fewer of the cells of the body, perhaps only the cells of a single organ or part of an organ, as compared to the widespread effect in the present example. Evidently in her case the proportion and distribution of XO cells were sufficient to produce the symptoms of Turner's syndrome as in an XO individual who is not a mosaic.

Much more complex mosaics are known (e.g., XX/XXY, XO/XXY, XO/XX/XXX), and new ones are continually being discovered. But our example demonstrates sufficiently the general principle of mosaicism resulting from nondisjunction in mitosis.

Double Inheritance from One or Both Parents Individuals with such double inheritance consist of a mixture of two or more kinds of cells and are sometimes called mosaics. But the term mosaic is best confined to individuals having two or more cell lines arising from the same zygote, as in our previous examples. Now we are picturing a situation in which there are two zygotes, or their genetic equivalents, and the term **chimera** is more appropriate.

Some of the most striking cases of genetic chimerism are individuals having some cells with two X chromosomes, some cells with an X and a Y: the XX/XY constitution. How can such an individual be produced? The most direct way would be for two sperm cells, one with an X, one with a Y, to fertilize separately two female nuclei. There are various ways in which this could happen. Two ova might be separately fertilized and then fuse together to give rise to one embryo. An ovum and a polar body might be separately fertilized and then fuse together. There are other possibilities. As likely an explanation as any would be an ovum containing two nuclei (one a retained polar body?) each of

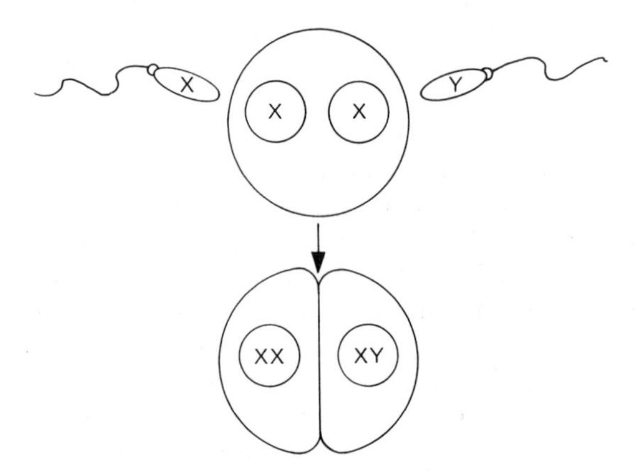

FIG. 17.10 Production of an XX/XY chimera through fertilization of a binucleate ovum by two sperm cells, followed by division of the cytoplasm.

which was separately fertilized (Fig. 17.10). Such binucleate ova have been observed.

Aside from the XX/XY constitution itself, is there genetic evidence that these individuals really have double inheritance from one or both parents? To answer this question we shall use one example from a number of cases that have been studied (Park, Jones, and Bias, 1970, listed 11).

This was a child with apparently normal mentality who was raised as a boy and psychologically oriented as a boy (Fig. 17.11) (Corey et al., 1967). On the right side he had an ovary with tiny oviduct and uterus; they were removed. On the left side he had an apparently normal testis that had descended to the scrotum. (Thus he was a "true hermaphrodite," p. 277). Four skin biopsies showed widely varying mixtures of XX and XY cells.

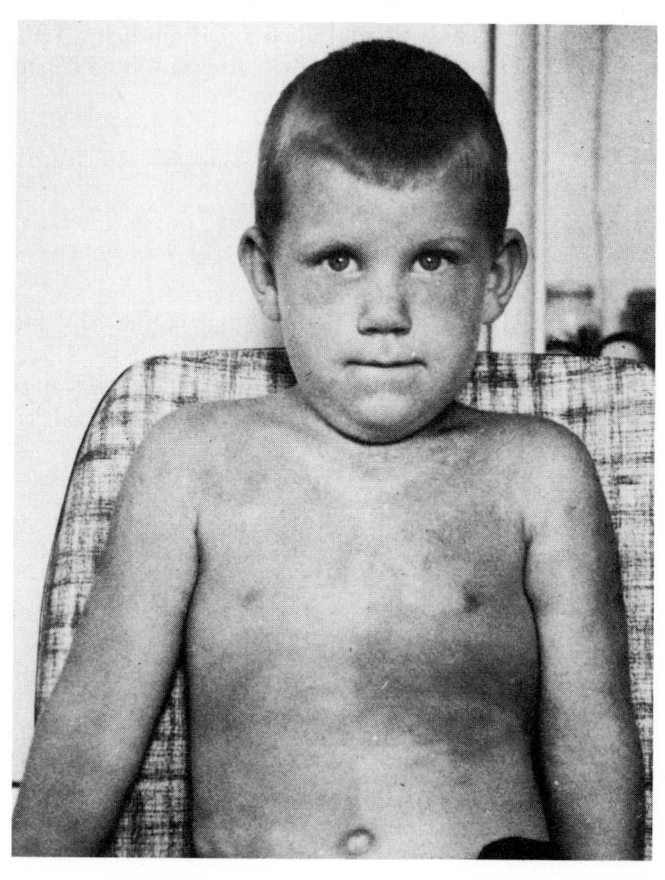

FIG. 17.11 Child who exhibits XX/XY chimerism. Note the differences in amount of pigmentation of different areas of the skin. (Courtesy of James R. Miller, Ph.D. From Corey, M. J., et al., 1967. "A case of XX/XY mosaicism," *American Journal of Human Genetics,* **19**:378–387.)

The authors studied 12 blood traits in the child and his parents. We select two of them for attention. The mother belonged to blood group A_1, the father to group O. In the child, 60 percent of the red blood cells were A_1, 40 percent were O. The genetic situation must have been:

	Mother A_1	×	Father O
Genotypes	$I^{A_1}I^O$		I^OI^O
Child's cells	O		A_1
	I^OI^O		$I^{A_1}I^O$

This shows that the mother contributed her I^O gene to production of one kind of cells, her I^{A_1} gene to the other kind.

Double contribution from the father was shown by the Lutheran blood groups (already mentioned in connection with the mapping of autosomes, Chap. 16). These depend upon a pair of alleles, Lu^a and Lu^b, that determine cell antigens Lu^a and Lu^b, respectively. Phenotypes and genotypes are as follows:

Phenotypes	**Genotypes**
Lu(a+b−)	$Lu^a\ Lu^a$
Lu(a+b+)	$Lu^a\ Lu^b$
Lu(a−b+)	$Lu^b\ Lu^b$

Thus inheritance is essentially like that of the MN blood types (Chap. 7).

In the present instance the mother was Lu(a−b+), the father Lu(a+b+). In the child, 60 percent of the red blood cells were Lu(a+b+), 40 percent were Lu(a−b+).

	Mother Lu(a−b+)	×	Father Lu(a+b+)
Genotypes	$Lu^b\ Lu^b$		$Lu^a\ Lu^b$
Child's cells	Lu(a+b+)		Lu(a−b+)
	$Lu^a\ Lu^b$		$Lu^b\ Lu^b$

This child was somewhat unusual in that his chimerism showed in his skin, in the abdominal region particularly (Fig. 17.11). Some areas were more heavily pigmented than were others. The lighter areas had a much higher proportion of XX cells than did the darker areas, although the significance of this fact is not evident.

Problems

1. How many Barr bodies would we expect to find in the cells of people with

the following sex-chromosome constitutions: XXXY, XX/XXY mo- XXYY; saic; XO/XX/XXX mosaic; XO/ XYY mosaic.

2. If an XXY individual *were* to produce sperm cells, what kinds would he produce with regard to sex-chromosome content? If he married a normal woman, and no nondisjunction were involved in the production of her ova, what types of offspring would be expected?
3. If in Fig. 17.2, nondisjunction of the X chromosomes had occurred in *both* the first and second meiotic divisions, what type of ovum would have been produced? What would have resulted from its fertilization by a Y-bearing sperm cell? What would be the total number of chromosomes (autosomes plus sex chromosomes) in this fertilized ovum? How could an unfertilized ovum containing three X chromosomes be produced by the female represented in the diagram?
4. If in Fig. 17.3, A, nondisjunction of the X and Y chromosomes had occurred in *both* first and second meiotic divisions, what types of sperm cells would have been produced? If these sperm cells fertilized normal ova, what types of offspring would have arisen?
5. If in Fig. 17.3, A, nondisjunction of the X chromosomes had occurred as shown but nondisjunction of the Y chromosomes had occurred in *both* divisions, what types of sperm cells would have been produced, and what would have resulted from their fertilization of normal ova?
6. How can an XYY individual be produced by parents having the normal constitution of sex chromosomes, through occurrence of a single nondisjunction?
7. A woman who had blood type MN (p. 82) was married to a man of type N. From blood typing a child of this couple, an investigator concluded that the child had received double inheritance from the mother. What did the investigator observe that led him to conclude this?
8. What four types of offspring would an XXX woman mated to a normal man be most likely to produce? Give genotypes of the offspring and also phenotypic designations.
9. How many Barr bodies would one expect to find in the cells of an individual with the karyotype XXXXY? The karyotype XXYYY?
10. Suppose you discover that "enzyme J" exists in two forms, separable by electrophoresis: isozyme J^1 and isozyme J^2, and that these depend upon a single pair of genes. You take a biopsy from a woman who is heterozygous for this pair, grow the cells in tissue culture, and establish many clones from the latter. If the genes concerned are X-linked, what would you expect to observe when you analyzed the clones to determine each one's isozyme content?
11. One way in which the XXYY genotype can be produced is (a) first divisional nondisjunction in meiosis in the father, followed by (b) nondisjunction of the Y chromosomes only in the second meiotic division in the father, the resulting XYY sperm then fertilizing a normal ovum. Give another set of nondisjunctional occurrences that could result in production of an XXYY child by normal parents.
12. A child with XX/XY mosaicism is found to have two kinds of red blood cells relative to the Rh antigens: 40 percent of the cells are Rh-negative (as the term is usually used), while 60 percent are Rh-positive. The father is Rh-negative. What is the most probable explanation for the situation in the child?

References

Barr, M. L., 1960. "Sexual dimorphism in interphase nuclei," *American Journal of Human Genetics,* **12**:118–127.

Breslau, N. A., and B. Bell, 1970. "XYY, health and law," *Southern Medical Journal,* **63**:831–836.

Corey, M. J., J. R. Miller, J. R. MacLean, and B. Chown, 1967. "A case of XX/XY mosaicism," *American Journal of Human Genetics,* **19**:378–387.

Hamerton, J. L., 1971. *Human Cytogenetics; Vol. 1, General Cytogenetics; Vol. 2, Clinical Cytogenetics.* New York: Academic Press.

Hook, E. B., 1973. "Behavioral implications of the human XYY genotype," *Science,* **179**:139–150.

Jacobs, P. A., D. G. Harnden, W. M. Court Brown, J. Goldstein, H. G. Close, T. N. MacGregor, N. Maclean, and J. A. Strong, 1960. "Abnormalities involving the X chromosome in women," *Lancet,* **I**:1213–1216.

Jones, H. W., Jr., and W. W. Scott, 1971. *Hermaphrodites, Genital Anomalies and Related Endocrine Disorders,* 2nd ed. Baltimore: Williams & Wilkins Company.

Keeton, W. T., 1967. *Biological Science.* New York: W. W. Norton & Company.

Levitan, M., and A. Montagu, 1971. *Textbook of Human Genetics.* New York: Oxford University Press.

Lyon, M. F., 1962. "Sex chromatin and gene action in the mammalian X chromosome," *American Journal of Human Genetics,* **14**:135–148.

Lyon, M. F., 1968. "Chromosomal and subchromosomal inactivation," *Annual Review of Genetics,* **2**:31–52.

McKusick, V. A., 1964. *On the X Chromosome of Man.* Washington, D.C.: American Institute of Biological Sciences.

Melnyk, J., H. Thompson, A. J. Rucci, F. Vanasek, and S. Hayes, 1969. "Failure of transmission of the extra chromosome in subjects with 47,XYY karyotype," *Lancet,* **II**:797–798.

Migeon, B. R., 1971. "Studies of skin fibroblasts from 10 families with HGPRT deficiency, with reference to X-chromosomal inactivation," *American Journal of Human Genetics,* **23**:199–210.

Migeon, B. R., V. M. Der Kaloustian, W. L. Nyhan, W. J. Young, and B. Childs, 1968. "X-linked hypoxanthine guanine phosphoribosyl transferase deficiency: Heterozygote has two clonal populations," *Science,* **160**:425–427.

Nance, W. E., and I. Uchida, 1964. "Turner's syndrome, twinning, and an unusual variant of glucose-6-phosphate dehydrogenase," *American Journal of Human Genetics,* **16**:380–390.

Overzier, C. (ed.), 1963. *Intersexuality.* New York: Academic Press.

Park, I., H. W. Jones, Jr., and W. B. Bias, 1970. "True hermaphroditism with 46,XX/46,XY chromosome complement. Report of a case," *Obstetrics and Gynecology,* **36**:377–387.

Pitcher, D. R., 1971. "The XYY syndrome," *British Journal of Hospital Medicine,* **5**:379–393.

Race, R. R., and R. Sanger, 1969. "Xg and sex-chromosome abnormalities," *British Medical Bulletin,* **25**:99–103.

Rosenbloom, F. M., W. N. Kelley, J. F. Henderson, and J. E. Seegmiller, 1967. "Lyon hypothesis and X-linked disease," *Lancet,* **II**:305–306.

18

Normal Complement of Chromosomes

Modern investigations of human chromosomes began in 1956 when the normal number of them (46) was first clearly demonstrated (Tjio and Levan, 1956).

Method The cells to be studied are grown in tissue culture. The cells most widely used are the white blood cells known as lymphocytes, with their large nuclei. But other tissues, sampled by removal of a bit by biopsy, are also employed, such as marrow from the sternum and fibroblasts (connective tissue cells) from the deep layer of the skin. Under suitable conditions the cultured cells will grow and divide actively, undergoing normal mitosis.

The drug colchicine is added to such a culture. It inhibits spindle formation in the cells, with the result that mitosis progresses to the metaphase stage and stops there. In this way many cells can be produced with their chromosomes all at metaphase. Treatment with hypotonic (watery) solution swells the cells and causes dispersion of the chromosomes. The cells are placed on microscope slides and flattened by drying or by applying pressure to the cover slip of the slide. The cells are fixed and stained (see Hamerton, 1971, Vol. 1, for further discussion of methods).

When a cell with the chromosomes nicely dispersed (not overlapping) is found, an enlarged photograph is made (Fig. 18.1). Then the picture of each chromosome is cut out, and the pictures are arranged

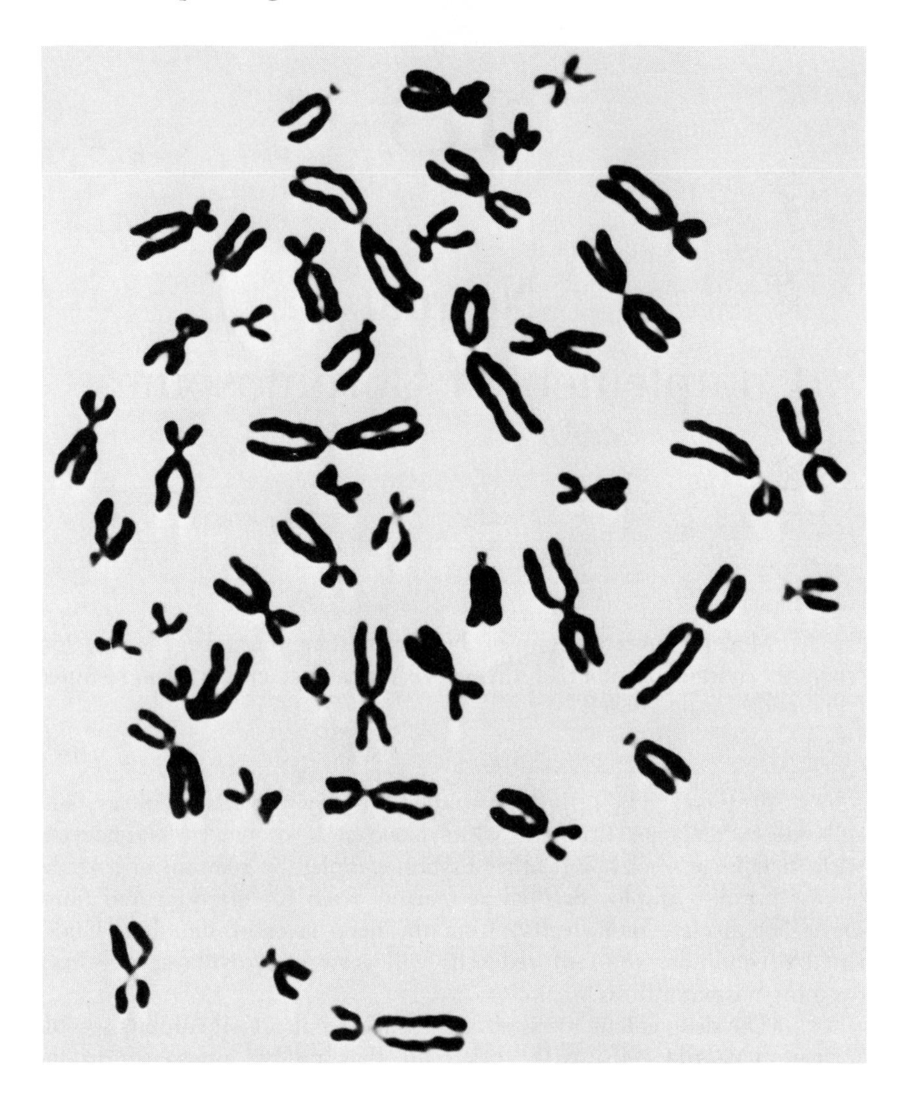

FIG. 18.1 Human chromosomes at the metaphase stage of mitosis. (Courtesy of Ronald C. Picoff, M.D., and Elizabeth M. Hier.)

in homologous pairs and in sequence from the largest to the smallest ones. Such a diagramatic arrangement of the chromosomes is called a **karyotype** (e.g., Fig. 17.5, p. 275).

Classification Figure 18.2 is an idealized diagram (**idiogram**) of the normal complement of chromosomes in a male, although only one

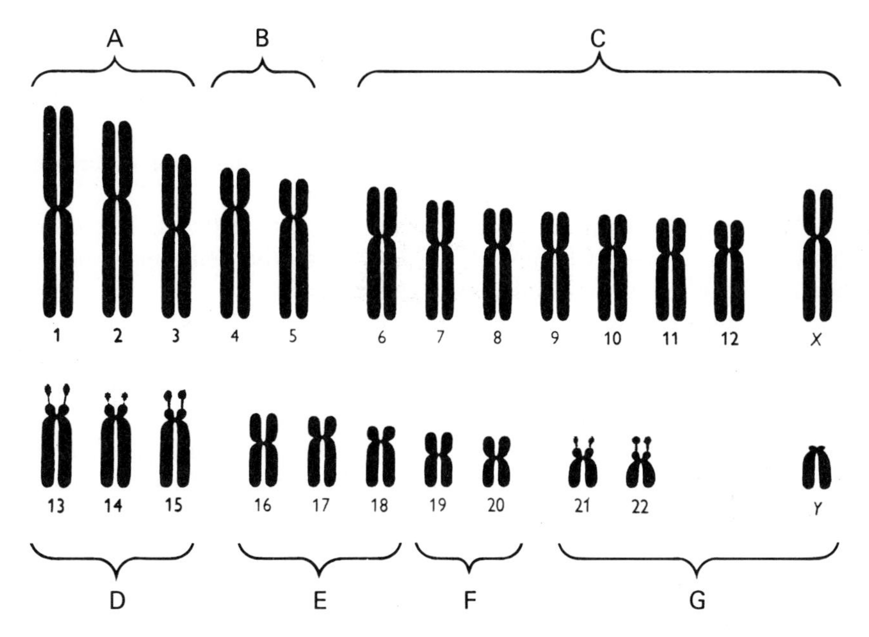

FIG. 18.2 Idealized diagram of the human chromosomes. [Modified from "A proposed standard system of nomenclature of human mitotic chromosomes (Denver, Colorado)," *Annals of Human Genetics,* **24**: 319, 1960].

member of each pair of chromosomes is shown. We note immediately that each chromosome is represented by a *double* structure. This is because each chromosome had duplicated itself and was at the metaphase stage of mitosis (Fig. 3.1, p. 25). If mitosis had continued, the centromere joining the two chromatids would have divided, releasing the chromatids from each other (Fig. 3.1D).

As the result of a series of international conferences (the first held in Denver in 1960), workers in the field have adopted a uniform system of naming the chromosomes (see *Chicago Conference,* 1966; *Paris Conference,* 1971). As indicated in Figure 18.2, the chromosomes are numbered and divided into seven groups.

Group A consists of the three largest chromosomes (Nos. 1, 2, 3), the largest one being some 10 microns in length. The centromeres are approximately median in position (metacentric).

Group B is composed of chromosomes No. 4 and No. 5. These are slightly smaller than group A chromosomes, and the centromere is not median, with the result that each chromosome has a short arm and a long arm (submedian or submetacentric).

Group C includes chromosomes numbered 6 through 12, and the X chromosome. The centromere is submedian in position.

Group D consists of chromosomes Nos. 13, 14, and 15. These three chromosomes are nearly as large as the smaller ones in group C, but they are distinguished by having the centromere so near one end that the short arm is very short indeed. Such a chromosome is called acrocentric. As indicated in the Figure, all have satellites, tiny masses of chromatin attached to the short arm by a slender connection.

Group E comprises chromosomes Nos. 16, 17, 18. These are shorter chromosomes. Chromosome No. 16 has the centromere nearly median; the other two have submedian centromeres.

Group F consists of two small chromosomes (Nos. 19 and 20) having nearly median centromeres.

Group G includes Nos. 21 and 22, and the Y chromosome. These are the smallest chromosomes. They are acrocentric, and Nos. 21 and 22 have satellites. The Y chromosome is typically larger than the others, but the size varies. Interestingly, this size difference is often hereditary, passed on from father to son.

Nomenclature The Chicago conference (1966) adopted a shorthand method of designating the chromosomal constitutions of individuals. According to it, a normal female is given the designation 46,XX. The *total* number of chromosomes is given first. Then the sex chromosomes are listed, separated from the total number by a comma. A female with Turner's syndrome (p. 268) is 45,X. A normal male is 46,XY. An individual with Klinefelter's syndrome (p. 274) is 47,XXY. An XYY individual (p. 276) is 47,XYY.

For mosaics, all types of cells are listed, the types being separated by diagonal lines. Thus an XO/XX mosaic (p. 284) is 45,X/46,XX, and an XX/XY mosaic is 46,XX/46,XY.

In the next chapter we shall discuss abnormalities in number and structure of autosomes. In the most common form of Down's syndrome (p. 318), for example, there is an extra group G chromosome usually designated as No. 21. Thus a female with Down's syndrome is 47,XX,+G, or 47,XX,+21. On the other hand, the designation 45,XX,—C indicates a female lacking one group C chromosome. Note that the + or — precedes the initial or number.

Sometimes the number of autosomes is normal, but an arm of a chromosome appears to be abnormally long or short. The short arm is designated by a "p," the long arm by a "q." Thus the notation 46,XYCp+ indicates a male having a group C chromosome with a lengthened short arm, and 46,XY,15q— indicates a male having a No. 15 chromosome with a shortened long arm. Here the + or — follows the p or q. (For further details, see *Chicago Conference,* 1966; *Paris Conference,* 1971).

Identification of Chromosomes

While identification of chromosomes by the group to which they belong can be done readily by experienced workers, identification of chromosomes *within* a group is more difficult. Small size differences and small differences in the position of the centromere are utilized when possible, as are a few structural features. For example, chromosome No. 1 frequently has a constriction in the long arm not far from the centromere, while the other two chromosomes in this group do not. But on the whole, chromosomes within a group look strikingly similar as seen in the usual karyotype preparation. Additional means of identification are needed and, fortunately, have been found.

Autoradiography One of these methods involves the labeling of chromosomes with radioactive material in much the manner mentioned in the preceding chapter (p. 282). In preparation for mitosis, DNA is synthesized by the chromosomes. A most characteristic constituent of DNA is thymine (p. 39). The thymine is rendered radioactive by incorporating into it radioactive hydrogen (^3H), the resulting compound being called tritiated thymidine.

Chromosomes synthesizing DNA utilize this tritiated thymidine. Then the chromosomes are placed in contact with photographic film, and wherever the radioactive material is located the film is exposed and a black spot appears when it is developed.

The use of this method to identify chromosomes depends upon the fact that chromosomes differ in the time at which synthesis of DNA occurs. Some chromosomes, or parts of chromosomes, carry on this synthesis—and hence the incorporation of tritiated thymidine—later than others; they are said to be "late labeling." By careful timing of the addition of tritiated thymidine to a cell culture, late labeling can be used as a means of identification.

Thus, one of the C group chromosomes is late-labeling. This is considered an X chromosome, and to be the one that forms a Barr body. If more than two X chromosomes are present, all but one of them are late-labeling (and form Barr bodies, as we mentioned in the preceding chapter).

As Figure 18.2 indicates, the three group D chromosomes are almost identical in appearance. But they have different labeling patterns (Fig. 18.3). One of them is late-labeling over considerable portions of the long arms; the number 13 is assigned to it. Another, called No. 14, is late-labeling in regions around the centromere. The one called No. 15 completes DNA synthesis earlier than the other two and hence has only a slight amount of late labeling. In this way each of the group D chromo-

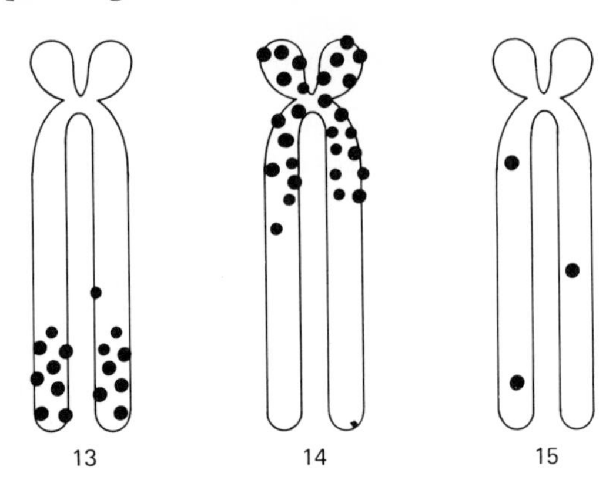

FIG. 18.3 Use of autoradiography to distinguish among the three group D chromosomes. Black spots show regions of late labeling with tritiated thymidine. The satellites have been omitted.

somes can be identified, and when an abnormal chromosome of this group is encountered, investigators can determine just which of the three is the abnormal one (Giannelli and Howlett, 1966; Bloom and Gerald, 1968).

Banding Patterns As we noted in Chapter 3, the chromosomes of the fruit fly, *Drosophila,* can be analyzed in detail because of the fortunate circumstance of the distinctive patterns of bands exhibited by the salivary gland chromosomes of the larva (Fig. 3.2, p. 27). Students of human chromosomes have long hoped for such a valuable means of identifying chromosomes, and starting with 1970 the hope has been realized (Caspersson et al., 1970a).

The method first employed consists of staining the chromosomes with **quinacrine mustard** (or quinacrine dihydrochloride: Atebrin), a chemical that fluoresces when exposed to ultraviolet light. The stain binds to the DNA of some regions of the chromosomes more freely than to others, and thus a chromosome presents brightly fluorescing bands alternating with dark bands (Fig. 19.2, p. 306). The pattern so produced can be photographed, or the intensity of fluorescence in each region can be measured, giving a "profile" of fluorescence extending from one end of a chromosome to the other. These patterns and profiles are distinctive for each chromosome and so provide a more positive means of identification than had been available previously.

Figure 18.4 gives a schematic drawing of the most characteristic patterns and demonstrates the distinctiveness of them (black areas in the drawing represent fluorescent regions). We note, for example, how clearly the previously confusing C group chromosomes (6–12, and X) can be

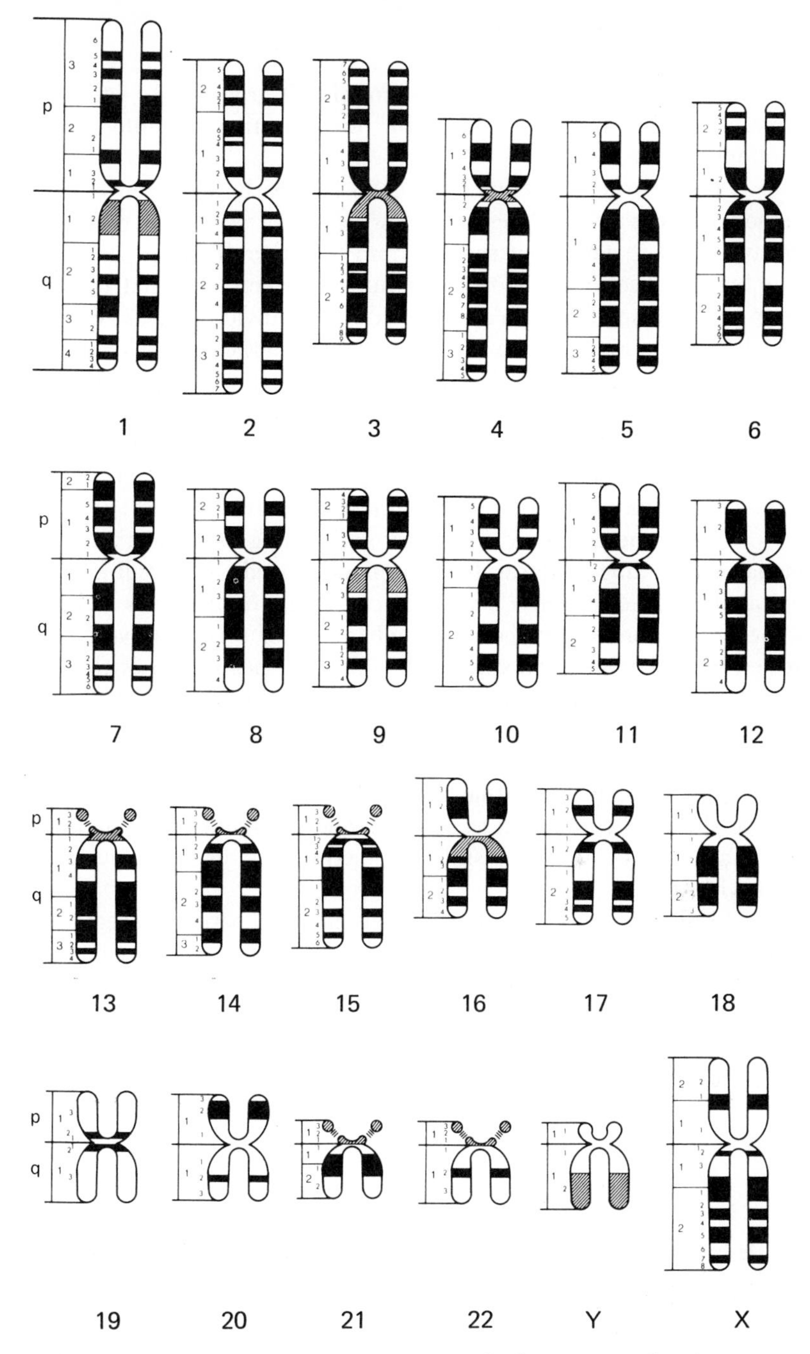

FIG. 18.4 Diagramatic representation of chromosome banding as observed with the quinacrine mustard and Giemsa methods. Black areas represent the bands that stain. Gray areas represent variable bands. (From *Paris Conference (1971): Standardization in Human Cytogenetics*. Birth Defects: Original Article Series, VIII: 7, 1972. The National Foundation, New York.)

distinguished from each other. And for the D group (13, 14, 15), the fluorescence patterns supplement the identification by autoradiography mentioned above.

One of the first chromosomes to be clearly identified by the new method was the Y, the long arm of which fluoresces with particular intensity so that it stands out sharply from the other G group chromosomes. This fact has been useful in determining with certainty that the extra G group chromosome found in a suspected XYY individual (p. 276) is really a Y. We note also that one of the other chromosomes in this group has a long arm that fluoresces brightly, while the third chromosome in the group has a long arm that fluoresces little. The one with the fluorescent arm is the one that is present in triplicate in Down's syndrome (p. 318), and is assigned the number 21 (Caspersson et al., 1970b; O'Riordan et al., 1971; Mikkelsen, 1971), the other being called No. 22.

Since the discovery of fluorescent banding, other methods of demonstrating banding patterns have been developed. The well-known biological stain called **Giemsa stain** will cause banding patterns to be visible if the chromosomes have been treated in one of several ways (e.g., pretreated with sodium hydroxide, with controlled heating, or with digestion of proteins present; Drets and Shaw, 1971; Dutrillaux et al., 1972). With minor variations, the sequence of bands revealed is the same as that demonstrated by the fluorescence techniques (Dutrillaux et al., 1972). Giemsa staining has the advantage that it is simple and inexpensive, and requires no special optical equipment as do techniques using ultraviolet light (Drets and Shaw, 1971). Also the slides are permanent, while fluorescent stain fades with time.

It has been found that similar banding patterns become visible if the chromosomes are treated with urea (Shiraishi and Yosida, 1972). Other techniques will doubtless be developed. From now on, workers with human chromosomes will be much more certain of positive identification of chromosomes than they could be prior to 1970 (see Miller, Miller, and Warburton, 1973).

Problems

1. Describe the chromosomal constitution of each of the following individuals:
(a) 47,XX,+22
(b) 45,XX,−18
(c) (46,XY,Xp−
(d) 46,XY/47,XXY,Bq−
(e) 47,XY,+14p+

2. Using the standardized nomenclature, give the correct designation for each of the following:
(a) Female whose cells show two Barr bodies.

(b) Male having an extra chromosome in the E group.
(c) Male having many of the symptoms of Klinefelter's syndrome but whose cells show two Barr bodies.
(d) Female with the symptoms of Turner's syndrome, but whose cells have two X chromosomes, one having a shortened short arm.
(e) Male with Klinefelter's syndrome and an additional G group chromosome.
(f) Mosaic having the following three cell lines: (1) two X chromosomes, and autosomes normal; (2) one X and one Y chromosome, and a No. 5 chromosome with a shortened long arm; (3) two X and one Y chromosomes, and an additional No. 21 chromosome.

References

Bloom, G. E., and P. S. Gerald, 1968. "Localization of genes on chromosome 13: Analysis of two kindreds," *American Journal of Human Genetics,* **20**:495–511.

Caspersson, T., L. Zech, C. Johansson, and E. J. Modest, 1970a. "Identification of human chromosomes by DNA-binding fluorescent agents," *Chromosoma,* **30**:215–227.

Caspersson, T., M. Hultén, J. Linsten, and L. Zech, 1970b. "Distinction between extra G-like chromosomes by quinacrine mustard fluorescence analysis," *Experimental Cell Research,* **63**:240–243.

Chicago Conference: Standardization in Human Cytogenetics. 1966. Birth Defects: Original Article Series, **II**:2. New York: The National Foundation.

Drets, M. E., and M. W. Shaw, 1971. "Specific banding patterns of human chromosomes," *Proceedings of the National Academy of Sciences, USA,* **68**:2073–2077.

Dutrillaux, B., C. Finaz, J. de Grouchy, and J. Lejeune, 1972. "Comparison of banding patterns of human chromosomes obtained by heating, fluorescence, and proteolytic digestion," *Cytogenetics,* **11**:113–116.

Giannelli, F., and R. M. Howlett, 1966. "The identification of the chromosomes of the D group (13–15) Denver: An autoradiographic and measurement study," *Cytogenetics,* **5**:186–205.

Hamerton, J. L., 1971. *Human Cytogenetics; Vol. 1, General Cytogenetics; Vol. 2, Clinical Cytogenetics.* New York: Academic Press.

Mikkelsen, M., 1971. "Identification of G group anomalies in Down's syndrome by quinacrine dihydrochloride fluorescence staining," *Humangenetik,* **12**:67–73.

Miller, O. J., D. A. Miller, and D. Warburton, 1973. "Application of new staining techniques to the study of human chromosomes," in *Progress in Medical Genetics,* Vol. IX. Pp. 1–47.

O'Riordan, M. L., J. A. Robinson, K. E. Buckton, and H. J. Evans, 1971. "Distinguishing between the chromosomes involved in Down's syndrome (Trisomy 21) and chronic myeloid leukaemia (Ph^1) by fluorescence," *Nature,* **230**:167–168.

Paris Conference (1971): Standardization in Human Cytogenetics. Birth Defects: Original Article Series, **VIII**:7, 1972. New York: The National Foundation.

"A proposed standard system of nomenclature of human mitotic chromosomes (Denver, Colorado)," *Annals of Human Genetics,* **24**:319, 1960.
Shiraishi, Y., and T. H. Yosida, 1972. "Banding pattern analysis of human chromosomes by use of a urea treatment technique," *Chromosoma,* **37**:75–83.
Tjio, J. H., and A. Levan, 1956. "The chromosome number of man," *Hereditas,* **42**:1–6.

19

Hybridization of Somatic Cells

Hybridization Usually when we think of hybrids, we think of sexual reproduction between parents differing in some way, even in extreme cases belonging to different species. The cells involved are an ovum and a sperm cell, the products of **meiosis,** produced, respectively, by individuals that differ genetically. The hybrids with which we are concerned in this chapter are not produced by the fusion of a sperm with an ovum, but by fusion of two somatic cells—cells such as fibroblasts (from the connective tissue of the deep layer of the skin) or lymphocytes (a type of white blood cell). These cells are grown in tissue culture in laboratory glassware. Reproduction is by **mitosis.** Under suitable conditions such cells will fuse together. While this fact had been known for some time, it was not until 1960 that its possibilities for genetic research were realized. At that time mouse cells that differed genetically and in chromosomal constitution (karyotype) were fused. Five years later, cells from different species were fused together (for accounts of thc early history, see Ephrussi, 1972, and Harris, 1970). As a result of this fusion, the genetic elements of the "parent" species are combined or mixed. Interestingly, somatic cells from widely differing species can be hybridized—species so widely separated that *sexual* reproduction between them is completely impossible, such as human and mouse cells, even human cells and *Drosophila* or mosquito cells.

We need hardly add that, while a zygote formed by fusion of sperm and ovum develops into a complete individual, a cell formed by the fusion of two somatic cells does not. Nevertheless, it presents us with an important tool for research. As we have repeatedly emphasized, a major handicap of the student of human genetics is his inability to do genetic experiments with people. But he can do as he likes with human cells in tissue culture, and thereby gain some of the advantages enjoyed by researchers working with bacteria, bacterial viruses, fungi, and the like.

Hybrid Cells Hybrid cells formed by fusion of cells from different species are especially useful because the chromosomes from the two "parents" are visibly different.

In order to do genetic experiments with cells we must, of course, find phenotypic differences that are detectable at the cellular level. Useful phenotypic differences are mainly biochemical ones relating to enzymes, or sometimes to surface antigens such as those that determine the blood groups. For example, if a certain enzyme is present in one "parent" cell, but not in the other, we have a means of studying the inheritance of the enzyme. Or both "parent" cells may have the enzyme, but it may differ slightly in chemical structure in the "parent" cells. In that case we say that the "parents" have different **allozymes** of the enzyme. (The term "isozyme" is used for slightly differing forms of an enzyme found *within* a single species.) Allozymes, like isozymes, are usually detected because their molecules migrate at different rates in an electrical field, so that they are identifiable by electrophoresis.

Enzymatic differences between cells may be manifested in various ways. The cells from the "parent" species may differ in resistance to drugs or antibiotics. (In experiments, such differences may be utilized to eliminate an unwanted type of cell, selecting for its alternative.) Or the cells may differ in the temperatures at which they grow most readily. Again, the cells may differ in their ability to utilize materials provided in the culture medium: to use a food substance (energy source), or to synthesize some vital cell constituent, such as DNA, from materials provided in the culture medium. For example, the experimenter may provide a medium that necessitates that cells have two enzymes in order to synthesize DNA. Suppose that cells of one "parent" type have the first enzyme but not the second, while cells of the other "parent" type have the second enzyme but not the first. If the experimenter places a mixture of these cells in this medium, only hybrid cells (having both enzymes) will survive. Techniques of this kind provide a practical means of isolating and identifying hybrid cells.

The amount of cell fusion in cultures is increased by adding a virus of the para-influenza group of viruses: the Sendai virus. But further discussion of the techniques employed in experimentation with somatic cells would be out of place here (see references cited in this chapter).

Somatic cell genetics—some aspects of which make use of cells that are not hybrid—is a rapidly developing branch of human genetics, contributing importantly at the biochemical level. We may mention especially its usefulness in understanding the biochemical bases of diseases, and in making possible diagnosis of disease utilizing cells, even those derived from an unborn fetus (Chap. 28). Another field is that of cell differentiation: the problem of how cells, all derived from a single fertilized ovum, become different from each other during embryonic development so that some become brain cells, others liver cells, others skin cells, and so on. And here we have the related problem of how and why some cells become malignant: cancerous. Somatic cell genetics contributes to these fields and others, but our primary interest at present is in its contribution to our ability to determine which genes are in which chromosomes.

In Which Chromosome Is This Gene?

To answer this question, use is made of hybrid cells formed by the fusion of human and rodent cells, the rodent in most cases being the common laboratory mouse (*Mus musculus*). Such hybrid cells have the advantage that human and mouse chromosomes are visibly different from each other, and that the individual chromosomes of both can be identified using the fluorescent or Giemsa staining discussed in the preceding chapter.

Advantage is also taken of another fact. Human somatic cells have 46 chromosomes, mouse cells have 40. Thus a cell formed by fusion of the two has 86. But this condition does not persist. For reasons not definitely known, most or all of the human chromosomes are lost during the first few divisions of such cells. Complete sets of the mouse chromosomes remain in cells that continue to survive. In later divisions of these cells there is little further change in chromosome number. So from single cells of this kind, the experimenter can establish clones (p. 282). All clones will have a complete complement of mouse chromosomes, but some will have in addition a very small number of human chromosomes, in some cases only one. If clones only exhibit activity of a certain enzyme when a certain chromosome is present, we have evidence that the gene for the enzyme is in that chromosome. As a result of this process, a type of genetic segregation is accomplished in the absence of the process (meiosis) that gives rise to genetic segregation in sexual reproduction.

A Typical Experiment Suppose that a certain enzyme, which we shall call "enzyme A," is needed in the synthesis of a nucleotide forming a constitutent of DNA. Imagine further that we have a culture of mouse cells unable to synthesize this enzyme (these cells must be supplied with the nucleotide already formed) and a culture of normal human cells able to synthesize the enzyme. We place the human and mouse cells together on culture medium lacking the nucleotide, adding Sendai virus to promote cell fusion. Some hybrid cells are formed (Fig. 19.1). While the mouse cells cannot grow and divide on the medium, the hybrid cells can do so, having received the gene for "enzyme A" from the human "parent" cell. (The human cells also continue to grow, but frequently do so less vigorously than do the hybrid cells. Various techniques are available for separating the two so that the desired type of cell can be isolated. "Growth" here, as with bacteria, means formation of colonies of cells, not increase in size of individual cells.)

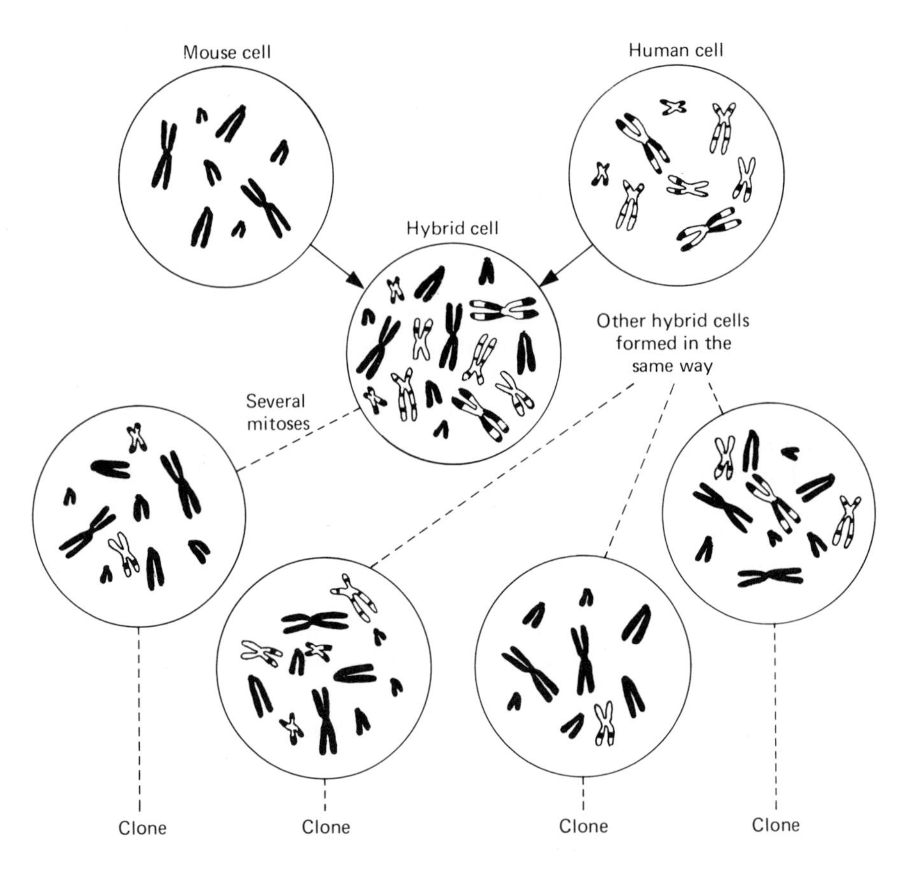

FIG. 19.1 Determining the chromosomal location of a certain gene (see text). For practical reasons only eight chromosomes (four pairs) are shown in each "parent" cell, although mouse cells have 40 (20 pairs) and human cells have 46 (23 pairs). Also, to distinguish clearly which are mouse chromosomes, these are shown in solid black, although they also show banding patterns when suitably stained.

As we indicated above, the hybrid cells at first lose human chromosomes, but later the number reaches relative stability. If one of the chromosomes lost has the gene for "enzyme A," the hybrid cell dies. So the surviving hybrid cells are those possessing the chromosome containing this gene. Cells of this type can be placed in separate culture dishes, where they will produce pure cultures (clones) that can be tested for enzyme activity, as well as for chromosomal constitution.

Figure 19.1 shows four different surviving cells derived from cells formed by fusion. All have the full complement of mouse chromosomes, but the number of human chromosomes varies in the different ones. If we look closely, we see that these cells all have a certain human chromosome: the one shown as having only one dark band, and that one located midway in one arm. The four have no other human chromosome in common. Evidently, the gene for "enzyme A" is located in this chromosome.

In many cases a "negative check" on this finding can be made. The possibility of doing so varies with the nature of the enzyme, but sometimes an experimenter can add substances to the culture medium that will *kill* cells possessing the enzyme in question, as mentioned earlier. Under these conditions, only cells *lacking* the enzyme will survive. It will then be found that *none* of the surviving cells possess the chromosome we have identified as the location of the gene for that enzyme.

We have presented this experiment as "typical," but various experimental techniques are employed. For example, an experimenter may start with mouse cells unable to synthesize one enzyme, and human cells unable to synthesize another enzyme. Then if the culture medium is such that both enzymes must be present, both mouse and human cells will fail to grow, the hybrid cells being the only cells to survive and grow. Other experimental procedures are also utilized (see Ruddle, 1972, 1973; Ephrussi and Weiss, 1969).

Gene for TK in Chromosome 17 Chromosome No. 17 was the first autosome for which possession of a specific gene was clearly demonstrated. The gene is that for the production of **thymidine kinase (TK)**. This enzyme is involved in the synthesis of nucleotides from which DNA is built. These nucleotides can be formed in two ways: (1) *de novo* from sugars and amino acids, and (2) by "salvage" of materials (nucleosides) previously formed (see Ephrussi, 1972; Ephrussi and Weiss, 1969). Thymidine kinase is one of the enzymes needed if this "salvage pathway" is to operate.

Before the advent of fluorescent staining, investigators used cell hybridization methods to determine that the TK gene is in an E-group chromosome (Nos. 17 and 18). Subsequently, use of quinacrine staining demonstrated that the chromosome concerned is No. 17 (Fig. 19.2; Miller et al., 1971). Other studies using electrophoresis demonstrated that it was the human type of TK that was present, not the type characteristic of mice.

Location Within the Chromosome Use of fluorescent and Giemsa staining enables investigators to identify such chromosomal aberrations as translocations and deletions (pp. 313–316). In some cases this makes

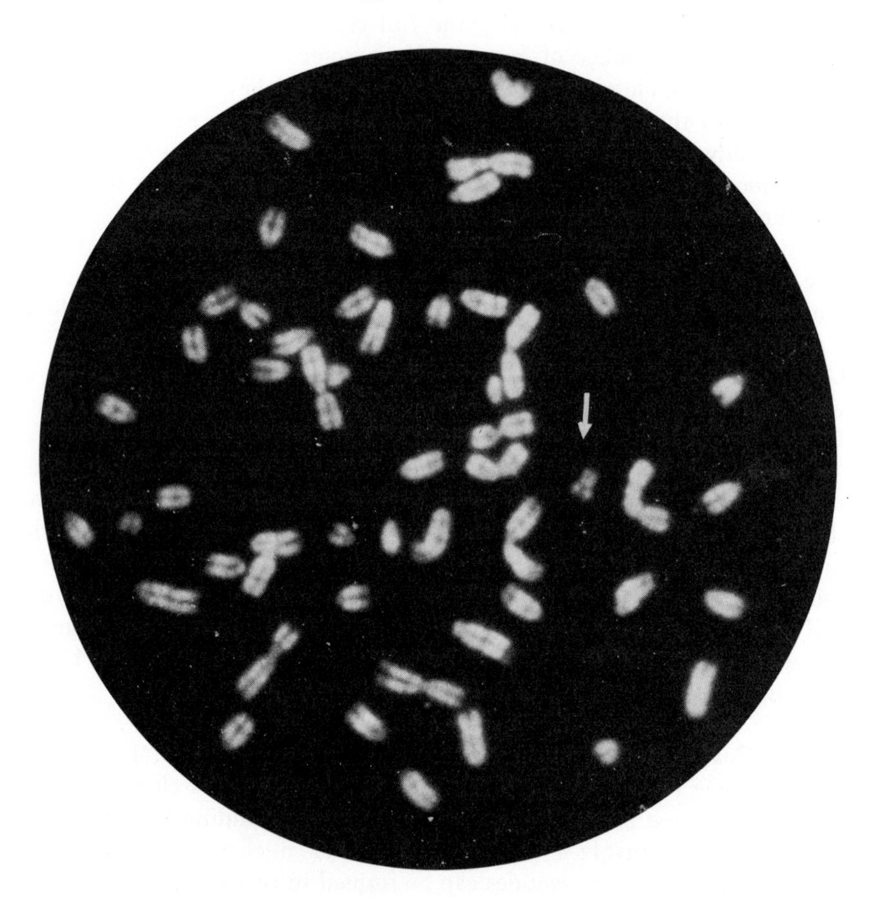

FIG. 19.2 Chromosomes in a hybrid cell of a clone arising from fusion of a human cell with a mouse cell. Fluorescent staining. The only human chromosome remaining in the cells of this clone is No. 17 (indicated by the arrow). Because the clone could synthesize thymidine kinase (TK), the gene for this enzyme is evidently in chromosome No. 17. See text. (Courtesy Orlando J. Miller, M.D. From Miller et al., 1971. "Human thymidine kinase gene locus: Assignment to chromosome 17 in a hybrid of man and mouse cells," *Science,* **173**:244–245. Copyright © 1971 by the American Association for the Advancement of Science.)

possible determination of where in the chromosome a gene is located. Thus Boone, Chen, and Ruddle (1972) found cells in which the long arm of chromosome No. 17 had been translocated to a mouse chromosome. Because they found that the translocation carried the TK gene with it, the investigators had evidence that the TK gene is in the long arm of chromosome No. 17. Refinement and extension of this type of analysis will give us more accurate chromosome mapping than has been possible with methods previously available.

Other Genes in Other Chromosomes The determination of which gene is in which chromosome is proceeding rapidly. Any listing we might make would be out of date by the time this book is published. So we content ourselves with mentioning that Ruddle (1973) published a table listing 29 enzymes, with gene locations in 15 of the 22 pairs of autosomes, and the X chromosome. Three chromosomes (Nos. 1, 11, 12) were indicated as containing three known genes, the X chromosome as containing four of them. It is undoubtedly only a question of time until every chromosome will be known to contain the gene for some specific enzyme. And the number of genes known per chromosome will increase concurrently.

Using cells having a translocation of most of the long arm of an X chromosome to a No. 14 chromosome (Fig. 19.3), Ricciuti and Ruddle (1973a) obtained evidence that the genes for HGPRT (p. 282), G6PD (p. 242), and PGK (phosphoglycerate kinase) are located in the long arm of the X chromosome. And Ricciuti and Ruddle (1973b) cited additional evidence leading to the tentative conclusion that of the three loci the one for PGK is nearest the centromere. The suggested sequence is: centromere. . . PGK. . . HGPRT. . . G6PD (see also Ruddle and Kucherlapati, 1974). This illustrates the manner in which cell hybridization techniques will provide much more accurate chromosome maps than those we have at present.

Genes Located by Inference Suppose that cell hybridization demonstrates that the gene for a certain enzyme is in a certain chromosome. Suppose also that family (pedigree) studies have indicated that the gene for that enzyme is linked to some other gene whose phenotype is not expressed in the cells being grown in tissue culture. Then we have evidence that this other gene is also in that certain chromosome. For example, this type of evidence indicates that the loci of two blood group genes (Rh and Duffy) are in chromosome No. 1 (Ruddle et al., 1972; Ruddle, 1973). That the Rh locus is in chromosome No. 1 has been confirmed by a combined serological and fluorescent-staining investigation of a man

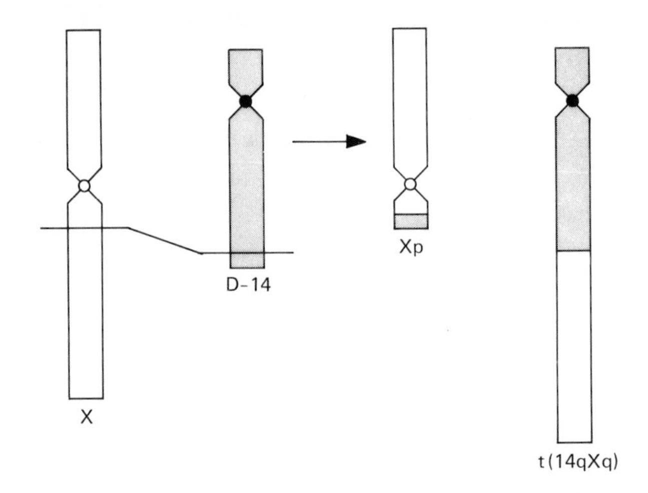

FIG. 19.3 Reciprocal translocation of most of the long arm of the X chromosome to the long arm of autosome No. 14 of the D group, and of a short piece of the long arm of No. 14 to the X chromosome. "t(14qXq)" designates the chromosome incorporating most of the long arms. Experiments demonstrated that it contained the X-linked genes for PGK, G6PD, and HGPRT (see text). Because it also contained the autosomal gene for NP (nucleoside phosphorylase), the experiment demonstrated that this gene is in chromosome No. 14. (Redrawn from Ricciuti, F., and F. H. Ruddle, 1973a. "Assignment of nucleoside phosphorylase to D-14 and localization of X-linked loci in man by somatic cell genetics," *Nature New Biology*, **241**: 180–182.)

having a chromosomal aberration that involved chromosome No. 1. This investigation also indicated the probability that the locus is in the short arm (Marsh et al., 1974).

This illustrates how the methods of classical genetics and those of somatic cell genetics combine to give us constantly increasing knowledge of where genes are located.

Problems

1. A cultured strain of mouse cells was deficient for "enzyme x." Somatic cell hybrids were produced by fusion of these mouse cells with human cells that were normal for "enzyme x." In one of these hybrid cells all human chromosomes were eliminated except for No. 5. A clone produced from this hybrid cell was found to be deficient for "enzyme x." What conclusion should be drawn?
2. Mouse somatic cells that cannot synthesize an essential enzyme were fused with normal human cells capable of forming the enzyme. From the hybrid cells thus formed, 13 clones were isolated as follows. Cells in each clone

had a *full complement of mouse chromosomes plus the human chromosomes* listed:

Clone (a): 1, X, 19, 21
(b): 5, 18
(c): X
(d): 1, 14, X
(e): 10, Y, 20
(f): 5, 14, 22, Y
(g): X, Y, 22
(h): 22, X
(i): 7, 8, 14, 22
(j): 3
(k): 3, 19
(l): 1, 9, 14, X
(m): 17

When the cells from the clones were placed in medium lacking the essential enzyme, only cells from clones (d), (f), (i), and (l) grew. This affords evidence that the gene for production of the enzyme is in which human chromosome?

References

Boone, C., T. R. Chen, and F. H. Ruddle, 1972. "Assignment of three human genes to chromosomes (LDH-A to 11, TK to 17, and IHD to 20) and evidence for translocation between human and mouse chromosomes in somatic cell hybrids," *Proceedings of the National Academy of Sciences (USA),* **69**:510–514.

Ephrussi, B., 1972. *Hybridization of Somatic Cells.* Princeton, N.J.: Princeton University Press.

Ephrussi, B., and M. C. Weiss, 1969. "Hybrid somatic cells," *Scientific American,* **220**:No. 4, 26–35.

Harris, H., 1970. *Cell Fusion.* Cambridge, Mass.: Harvard University Press.

Marsh, W. L., R. S. K. Chaganti, F. H. Gardner, K. Mayer, P. C. Nowell, and J. German, 1974. "Mapping human autosomes: Evidence supporting assignment of Rhesus to the short arm of chromosome No. 1," *Science,* **183**:966–968.

Miller, O. J., P. W. Allderdice, D. A. Miller, W. R. Breg, and B. R. Migeon, 1971. "Human thymidine kinase gene locus: Assignment to chromosome 17 in a hybrid of man and mouse cells," *Science,* **173**:244–245.

Ricciuti, F., and F. H. Ruddle, 1973a. "Assignment of nucleoside phosphorylase to D-14 and localization of X-linked loci in man by somatic cell genetics," *Nature New Biology,* **241**:180–182.

Ricciuti, F. C., and F. H. Ruddle, 1973b. "Assignment of three gene loci (PGK, HGPRT, G6PD) to the long arm of the human X-chromosome by somatic cell genetics" *(in press).*

Ruddle, F. H., 1972. "Linkage analysis using somatic cell hybrids." In Harris, H., and K. Hirschhorn (eds.), *Advances in Human Genetics, Vol.* 3. New York: Plenum Press. Pp. 173–235.

Ruddle, F. H., 1973. "Linkage analysis in man by somatic cell genetics," *Nature,* **242**:165–169.

Ruddle, F. H., and R. S. Kucherlapati, 1974. "Hybrid cells and human genes," *Scientific American,* **231**: 36–44.

Ruddle, F., F. Ricciuti, F. A. McMorris, J. Tischfield, R. Creagan, G. Darlington, and T. R. Chen, 1972. "Somatic cell genetic assignment of peptidase C and the Rh linkage group to chromsome A-1 in man," *Science,* **176**: 1429–1431.

20

Chromosomal Aberrations

In Chapter 17 we discussed some abnormalities connected with abnormal numbers of X and Y chromosomes. In this chapter we shall discuss the general types of chromosomal abnormalities, emphasizing especially those of the autosomes.

In the broad sense, the word "mutation" may be used to include any changes in the chromosomes and their constituent genes. In Chapter 4 we discussed mutations that involve changes in the structure of the DNA molecule. These are frequently called **gene mutations** or **point mutations,** to distinguish them from mutations that involve visible changes in the chromosomes, called **chromosomal mutations** or, better, **chromosomal aberrations.** At the present time, we cannot see the changes involved in gene mutations; they are too small to be seen even with the electron microscope. However, if more powerful electron microscopes are developed—as seems to be in the offing—these minute changes in molecular structure may become visible. Perhaps the best way to distinguish between gene mutations and chromosomal aberrations is to say that gene mutations consist of changes of structure *within* molecules of DNA, whereas chromosomal aberrations consist of changes of structure involving aggregates of great numbers of such molecules. The distinction may prove to be more or less artificial, yet it is useful.

Chromosomal aberrations are of two types: (1) changes in the number of chromosomes and (2) changes in the structure of individual chromosomes.

Changes in number may be of two orders: (1) changes in terms of complete haploid sets and (2) changes in numbers involving less than a haploid set. A haploid set is the complement of chromosomes found in a germ cell following meiosis. A human spermatogonium, for example, contains 46 chromosomes comprised of 23 pairs; 46 is the **diploid number.** A human sperm cell contains 23 unpaired chromosomes, one from each of the 23 pairs. Thus 23 is the **haploid number** and 23 unpaired chromosomes constitutes a haploid set.

Abnormal Numbers of Chromosomes

Aneuploidy Aneuploidy is change in number by less than a haploid set. In theory, the change in number may be either the loss or the gain of one or more chromosomes. But because normal functioning is dependent on a full and balanced complement of chromosomes, loss of a chromosome is likely to be lethal. An individual having one chromosome less than normal is called a **monosomic.** In Chapter 17 we described one example: XO individuals, having one X chromosome but having neither a second one nor a Y chromosome. We noted that many XO fetuses undergo spontaneous abortion, and that those surviving are abnormal in characteristic ways.

Loss of any one of the autosomes is usually lethal, although a few extremely abnormal children lacking one of a tiny "group G" chromosomes (p. 294) have been found.

Addition of a single chromosome to the normal diploid set is called **trisomy.** The initial n is frequently used to designate the number of chromosomes in a haploid set. Thus normal individuals with two of each kind of chromosome are $2n$. Trisomic individuals are $2n + 1$. In plants, especially, tetrasomic individuals ($2n + 2$) and even aneuploids of higher order are known, but we shall not be concerned with them.

The XXX woman we described in Chapter 17 was a trisomic (44 autosomes plus three X chromosomes). Presently we shall describe a case of trisomy involving an autosome: Down's syndrome or trisomy 21. Here we may note that the mechanism by which aneuploidy occurs is usually nondisjunction. In Chapter 17 we diagramed nondisjunction of the X chromosomes. Autosomes undergo nondisjunction at times in the same manner. In this way germ cells having an extra chromosome arise, and trisomic individuals are produced when such a germ cell unites with one having the normal haploid number.

Polyploidy Increase in the number of chromosomes by complete haploid sets is called **polyploidy.** Many human liver cells are **tetraploid,** $4n$, meaning that they have four complete haploid sets of chromosomes instead of the usual two sets. Such a tetraploid liver cell would arise when a cell prepares for mitosis by duplicating its chromosomes but then fails to divide: The 46 chromosomes originally present have duplicated themselves, making a total of 92, all of which remain in the one cell instead of being distributed to two daughter cells. Recall that in man $n = 23$; hence $4n = 92$.

In plants, especially, entire individuals may be polyploid. They arise from germ cells that are *diploid* ($2n$) instead of the normal haploid (n) in constitution. In Figure 20.1 we illustrate how such a diploid ovum may be produced by what we may call roughly "complete nondisjunction" —complete failure of chromosomes in the primary oöcyte to separate. Failure of the primary oöcyte to form a first polar body would produce the same result. Similarly, if meiotic division fails to occur, diploid pollen grains may be produced as well.

Although polyploidy is important in plants, it is rare in animals, apparently because it upsets the mechanism of sex determination. We noted in Chapter 17 that an increase in the number of X chromosomes is usually detrimental. Polyploidy is not found in normal human individuals except as it occurs in the cells of some tissues of the body, such as liver, bronchial epithelium, and amnion. It is one of the most frequent types of chromosomal aberrations found in spontaneous abortions (Hamerton, 1971). Böök and Santesson (1960) studied a very abnormal individual who had 69 chromosomes in most of his cells: three of each kind of autosome, two X chromosomes, and a Y chromosome. Hence most of his cells were triploid, $3n$.

Structural Aberrations

Turning to chromosomal aberrations that consist of structural changes in individual chromosomes, we note that these may be classified into four categories: (1) deletions, (2) duplications, (3) inversions, and (4) translocations. To these should perhaps be added a fifth: formation of isochromosomes.

From experiments, geneticists have amassed much knowledge concerning structural changes in single chromosomes of plants and animals, especially of the fruit fly, *Drosophila,* with its "giant" banded salivary gland chromosomes (p. 27).

1. **Deletion** or deficiency. A chromosome may fragment and then recombine, but with a portion missing. The chromosome is thereby shortened. The distinctive banding of the *Drosophila* salivary gland chromosomes makes possible the identification of the missing portion. Now that banding patterns of human chromosomes have been demonstrated (Chapter 18), similar precision may be expected in identification of human deletions.

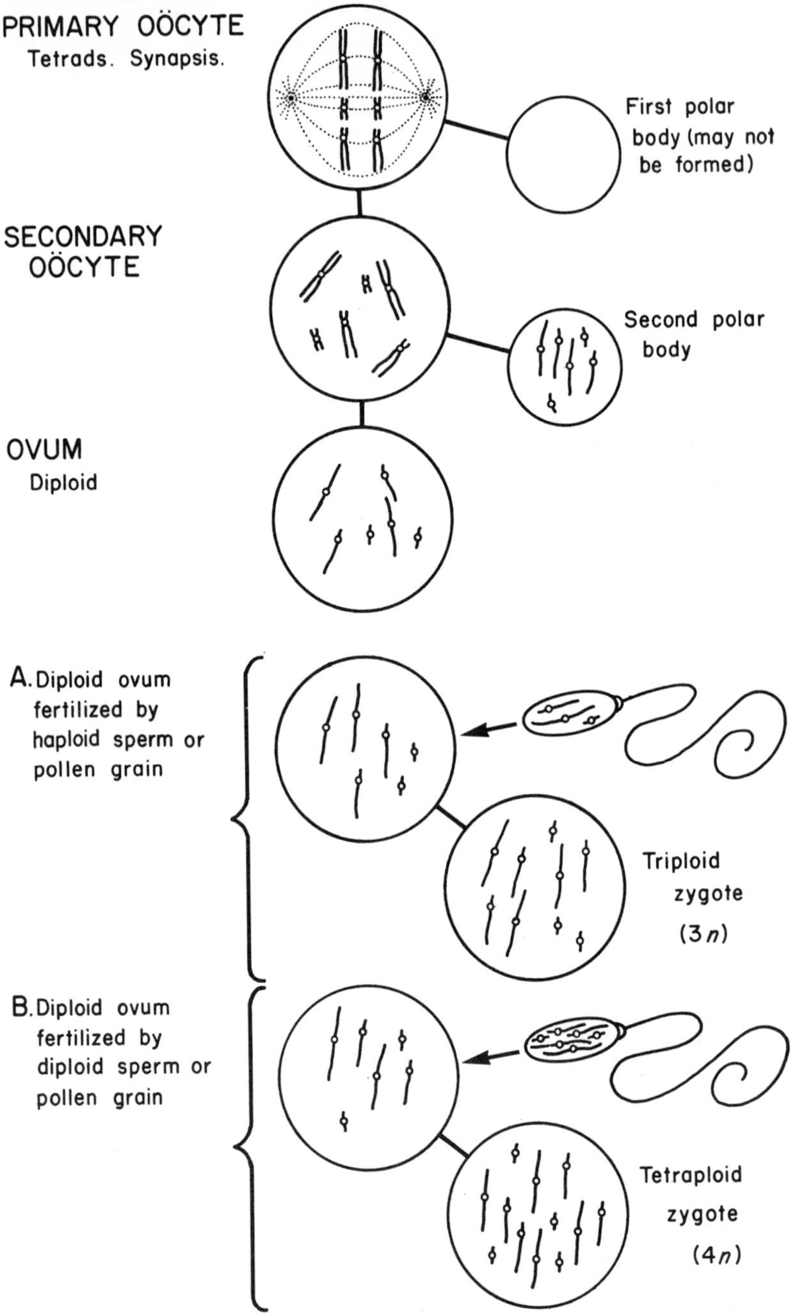

PRIMARY OÖCYTE
Tetrads. Synapsis.

First polar body (may not be formed)

SECONDARY OÖCYTE

Second polar body

OVUM
Diploid

A. Diploid ovum fertilized by haploid sperm or pollen grain

Triploid zygote (3*n*)

B. Diploid ovum fertilized by diploid sperm or pollen grain

Tetraploid zygote (4*n*)

FIG. 20.1 Polyploidy arising by production of diploid ova in a plant having three pairs of chromosomes. In meiosis the first polar body does not form, or if it forms does not contain chromosomes. A. Triploid zygote produced when a diploid ovum is fertilized by a haploid pollen grain. B. Tetraploid zygote formed when a diploid ovum is fertilized by a diploid pollen grain.

In Figure 20.2 we have used letters to represent genes but, of course, we cannot see the genes in actual chromosomes. We could probably arrange suitable breeding experiments in various organisms to determine which genes are missing, but man does not lend himself to such experimentation. If the missing piece contains important genes, (a) a gamete containing the deleted chromosome may not be viable, or (b) a zygote containing the deleted chromosome paired with a normal one from the other parent may not be viable, or (c) a zygote containing the deleted chromosome may not be viable *if* it happens to receive a deleted chromosome from the other parent also. In this latter case the zygote would be homozygous for the deletion (in terms of Fig. 20.2, it would be completely lacking in genes *H, I,* and *J,* and so, if these genes were important at all, the zygote would be likely to die). On the other hand, a heterozygote for the deletion (e.g., a zygote having *H, I,* and *J* in one chromosome but not in its partner) might survive if the genes "in single dose" could perform their functions.

2. **Duplication** or repeat. When a deletion occurs, what happens to the deleted portion? Usually it is lost, but it may become attached to

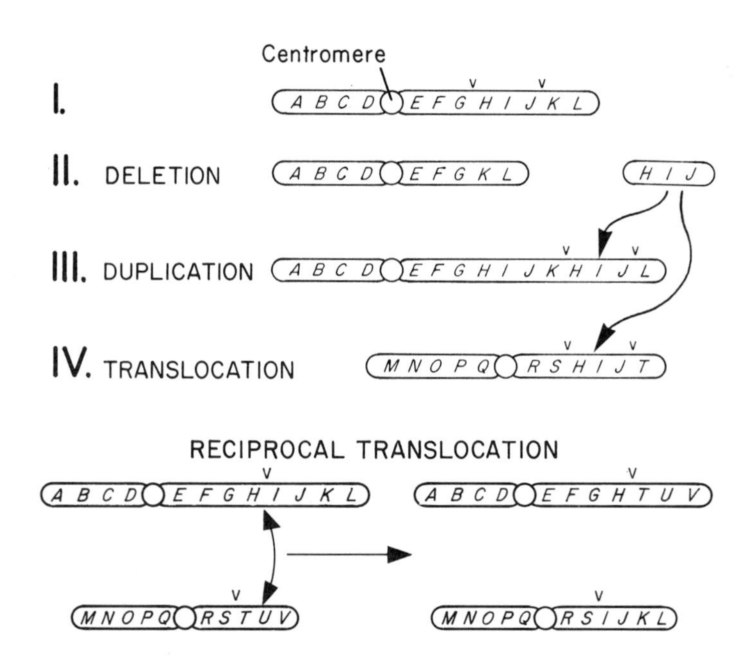

FIG. 20.2 Chromosomal aberrations. I. Normal chromosome, with letters representing gene loci. II. Deletion of the *HIJ* section. III. Duplication or repeat. The *HIJ* section is inserted into a homologous chromosome. IV. Translocation. In simple translocation the *HIJ* section is inserted into a nonhomologous chromosome (evidenced by the different letters it contains). In reciprocal translocation the ends of the arms of nonhomologous chromosomes are shown as being interchanged.

the homologous chromosome, causing a section of the chromosome to be repeated (Fig. 20.2, III, where there are two *HIJ* sections). The duplicated piece may or may not be placed next to the section that it duplicates. The visible result is a lengthening, usually slight, of the chromosome. Repeats in the banded salivary gland chromosomes of *Drosophila* can easily be identified. Because of the density of human chromosomes, demonstrating duplication has been difficult. But, as with deletions, development of techniques that show banding patterns in human chromosomes will change this.

3. **Translocation.** When part of one chromosome becomes attached to a chromosome not homologous to it (not a member of the same pair), translocation is said to have occurred. This may occur as shown first in Figure 20.2, IV, a piece of one chromosome simply becoming incorporated into another. But perhaps more usually there is a mutual exchange of pieces as shown in the second part of Figure 20.2, IV. This is called **reciprocal translocation.** In our discussion of Down's syndrome, we shall describe a human example of translocation.

4. **Inversion.** Sometimes a chromosome will divide into fragments and then the fragments will recombine, but one or more of them will be in a reversed relationship to the others. If the inversion is confined to one arm of the chromosome, it is called **paracentric** (Fig. 20.3, II). If the inverted portion includes the centromere, it is called **pericentric** (Fig. 20.3, III). As shown in the figure, pericentric inversions are more likely to change the shape of chromosomes than are paracentric ones, and hence have a greater chance of being seen in preparations of densely stained chromosomes, such as those with which most studies of human chromosomes are made.

We might expect that the last two chromosomal events, translocations and inversions, would have little genetic effect if no genes are missing or duplicated. This might be true if each gene performed its function all by itself, like a lone artisan in his home workshop, but in

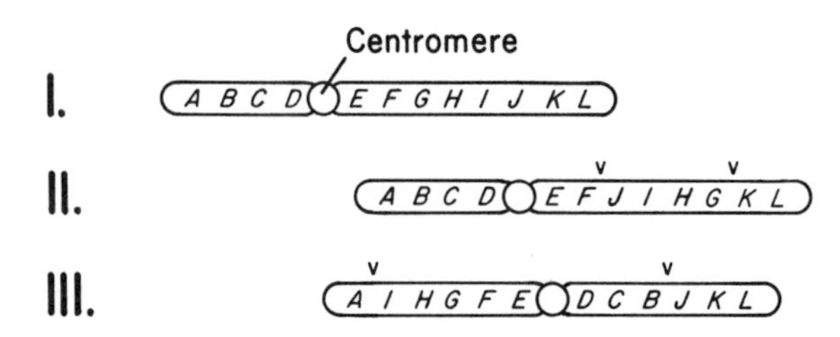

FIG. 20.3 Chromosomal aberrations: Inversions. I. Normal chromosome. II. Paracentric inversion. The *GHIJ* section in the right arm of the chromosome is reversed. III. Pericentric inversion. The *BCDEFGHI* section, including the centromere, is reversed. Note that this alters the relative lengths of the two arms.

point of fact, the function of a gene is frequently influenced by its position in relation to other genes: **position effect.** As we saw in Chapter 4, genes function by synthesizing proteins, which serve as enzymes. This synthesis is a chain reaction rather like an assembly line in a manufacturing plant. If one man in an assembly line inserts a bolt, and another man screws a nut onto it, this second man usually must be near the first one, not down at the other end of the line, especially if the bolt must be fastened before the next process in assembly can occur. It appears that similar assembly lines of genes function in much this way. Thus, changing the position of the genes may change their phenotypic effect.

5. **Formation of isochromosomes.** As we have noted, a typical chromosome consists of two "arms" separated by a centromere. The arms are not alike in genetic content, a fact we have indicated in our diagrams by using the letters of the alphabet in normal sequence (Fig. 20.4, I). An **isochromosome** is a chromosome in which the centromere is in the middle and the two arms *are* alike in genic content (although typically they are mirror images in arrangement). Figure 20.4, II, illustrates such chromosomes and the manner in which they may arise from normal chromosomes. When a daughter cell receives one of these isochromosomes, the genetic effect is that of deletion and duplication combined. For example, one cell will have genes *ABCDE* in a double dose and completely lack genes *FGHIJKLM,* whereas another cell will have genes *FGHIJKLM* in duplicate but will lack genes *ABCDE.* Both the lack of some genes and the presence of others in a double dose may have important consequences.

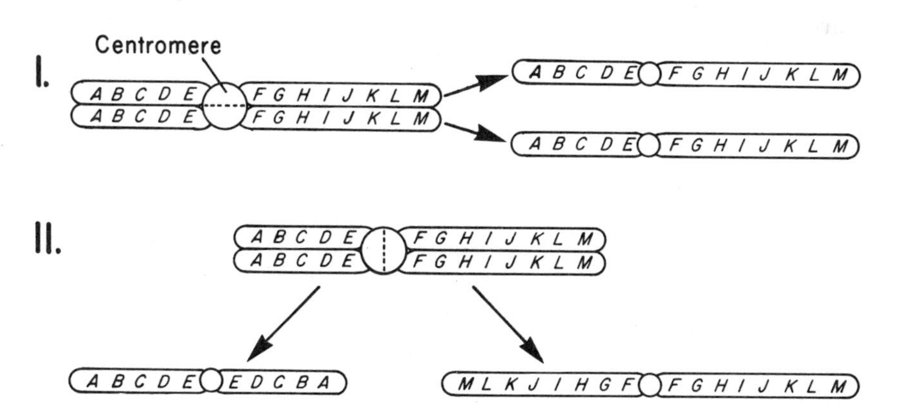

FIG. 20.4 Formation of isochromosomes compared to formation of normal chromosomes. I. Formation of normal daughter chromosomes from a metaphase chromosome consisting of two chromatids and one centromere. The centromere is shown dividing in a plane parallel to the long axis of the chromatids (dotted line). II. Formation of isochromosomes. The centromere divides abnormally, as indicated by the dotted line transverse to the long axis of the chromatids.

Knowledge of human isochromosomes and their significance is in its infancy. We may mention one suspected case involving the X chromosome. As we noted in Chapter 17, persons with Turner's syndrome (XO) usually do not have a Barr body in the nucleus (pp. 279–280). They are sometimes said to be *X-chromatin negative.* Occasionally, however, a person with Turner's syndrome does have one Barr body in the nucleus; she is *X-chromatin positive.* In some cases (Polani, 1962; Hamerton, 1971), one normal X chromosome was found to be present, as well as a large additional chromosome with a median centromere, resembling autosome No. 3 (Fig. 18.2, p. 293). It was suspected that this was an isochromosome consisting of the long arms from two X chromosomes. Radioactive tracers gave evidence that the two arms of the presumed isochromosome were indeed identical to each other. If this interpretation is correct, the corresponding two short arms were both missing. Thus, the only short arm present was that possessed by the single normal chromosome, as in the usual XO condition (Fig. 20.5). This would suggest that the genes for normal ovarian development (and for other bodily traits that Turner's syndrome patients exhibit) are located in the short arm of the X chromosome and must be present in duplicate if normal development is to occur, as in normal XX females.

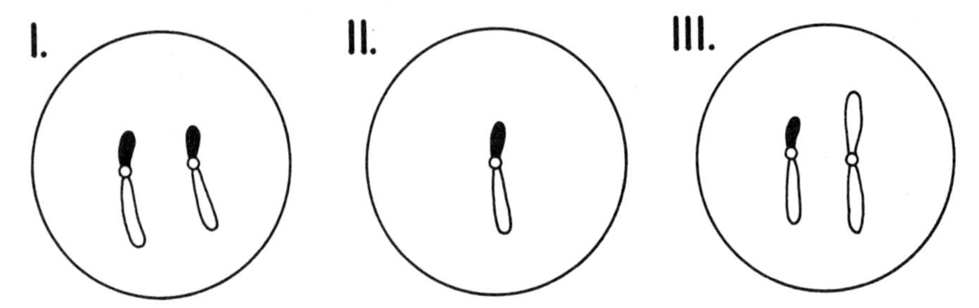

FIG. 20.5 Possible explanation of cases of Turner's syndrome having two X chromosomes, one an isochromosome. I. Normal female. Two X chromosomes, with a total of two short arms (black). II. Turner's syndrome with the usual XO constitution. One X chromosome, with one short arm. III. Turner's syndrome with one normal X chromosome and one X chromosome an isochromosome consisting of two long arms. Total: one short arm, as in the XO condition.

Down's Syndrome

Down's syndrome, a common form of mental defect, is also called mongolism because of an imagined physical resemblance to the Mongolian race, but because of possible racist implications, we shall not use the term. About one birth in 668 births is that of a child with Down's syndrome (Hamerton, 1971).

Children with Down's syndrome vary in their degree of mental deficiency, but in most cases it is severe. Such children vary in physical traits, but short stature, stubby fingers, a large, fissured tongue, and a round head are commonly found characteristics (Fig. 20.6). In Chapter 9 we noted characteristics of the dermal ridge pattern and of the flexion creases of the hands that aid in making a diagnosis in young babies (pp. 131–133). Life expectancy is low, various studies indicating average expectancies at birth of from 12 to 18 years (Hamerton, 1971). The most common causes of death are pneumonia and other infections of the respiratory system. Hence increased use of antibiotics is increasing life expectancy. Leukemia is much more common in children with Down's syndrome than it is in normal children.

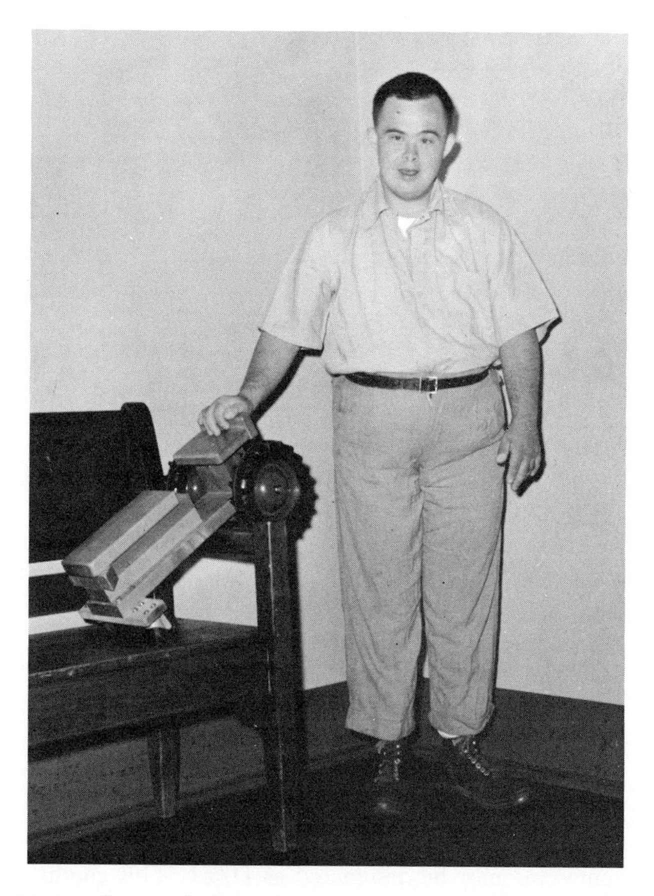

FIG. 20.6. Boy with Down's syndrome—age 25 years, Stanford-Binet IQ 26. The round, rather flat face and its expression are typical, as are the short stature and the stubby fingers. (Photo by Jim Reid. Courtesy of Dr. Raymond M. Mulcahy, Superintendent, Brandon Training School.)

Chromosomal Basis The syndrome seems always to be connected with the presence of the substance of three No. 21 chromosomes, a consitution called 21-trisomy. Most commonly, three separate No. 21 chromosomes are present, as in Figure 20.7. This is called **primary trisomy,** and using the standard nomenclature (Chapter 18), we designate such individuals as 47,XX,+21, or 47,XY,+21, depending on the sex. We recall from Chapter 18 that study of fluorescent banding patterns has confirmed the presence of three No. 21 chromosomes.

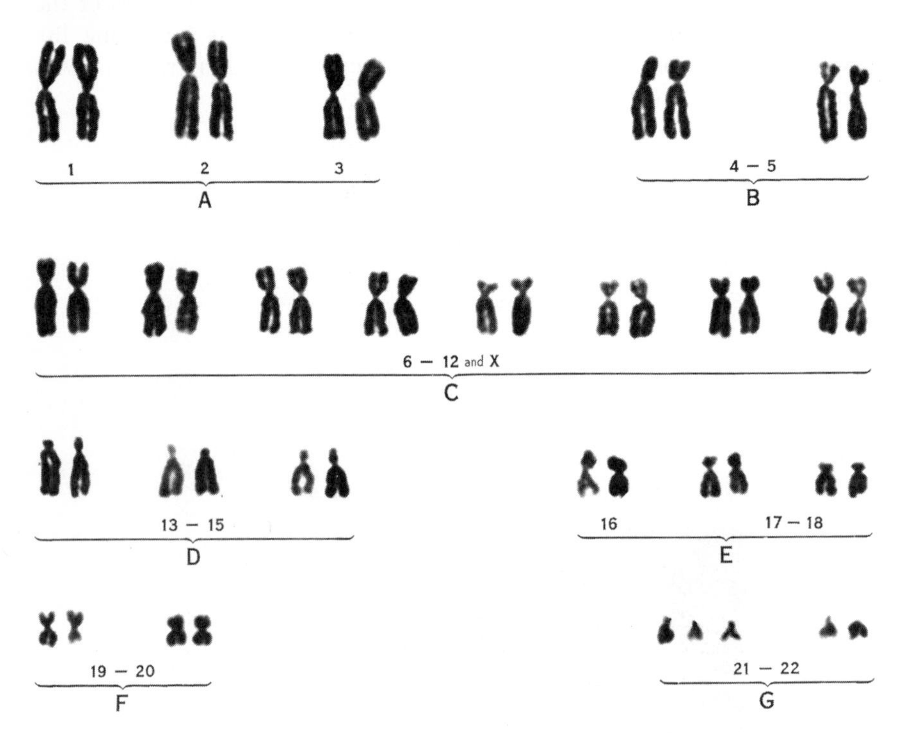

FIG. 20.7 Karyotype of a female with Down's syndrome (21- trisomy). Note the three No. 21 chromosomes, as well as the two X chromosomes. (Courtesy of Victor A. McKusick, M.D.)

Occasional individuals with Down's syndrome have only 46 chromosomes. In them it is found that one of the 46 is a double chromosome consisting of the long arm of a No. 21 chromosome joined to another chromosome, usually one in the D group (Fig. 18.2, p. 293). Such a double chromosome may be designated as D/21. This condition is called **secondary** or **translocation trisomy.** In addition to D/21, the two normal No. 21's are present, so the individual has the substance of three No. 21 chromosomes. In other cases the double chromosome consists of the long arms of two No. 21's joined together (21/21), or of a No. 21 joined to a No. 22 (21/22). Studies of fluorescence have confirmed the presence of a 21/21 chromosome (O'Riordan et al., 1971), and of the 21/22 condition (Mikkelsen, 1971a). O'Riordan et al. (1971) also demonstrated that in five cases of a D/21 chromosome, one case involved a No. 13, that is, was 13/21, and four involved No. 14, that is, were 14/21.

Primary 21-Trisomy How does this condition arise? *Nondisjunction* in the production of ovum or sperm seems to be the usual cause. Figure 20.8 shows nondisjunction in the first meiotic division of oögenesis, leading to production of an ovum containing two No. 21 chromosomes. Nondisjunction in the second meiotic division rather than the first would also give rise to such an ovum. As shown, the trisomic condition follows when such an ovum is fertilized by a normal sperm containing one No. 21. Conversely, nondisjunction in spermatogenesis may lead to production of a sperm containing two No. 21's. When such a sperm fertilizes a normal ovum, the trisomic condition results.

In this connection, we may note that although children with Down's syndrome may be born to mothers of any age, there is a marked tendency for them to be born to older mothers, the remainder of whose children have, usually, been normal. This tendency has been recognized for a long time, and its causes have been much debated.

In mammals, all of the oöcytes a female will ever have are present at birth or shortly thereafter. Evidence from mice indicates that as the oöcytes in the ovary grow older, they may exhibit an increased tendency to undergo nondisjunction. Perhaps this is also true of human oöcytes.

The increase in incidence with increasing age of the mother applies especially to primary trisomy. When young mothers *do* have Down's syndrome children, a disproportionately large number of them are of the translocation trisomy type.

Families having more than one Down's syndrome child with primary trisomy are rare, but occur frequently enough to suggest that some families may have an inherited predisposition for nondisjunction to occur. More evidence is needed before definite conclusions can be drawn, however.

Translocation (Secondary) 21-Trisomy One parent of a child with translocation trisomy usually has only 45 chromosomes, one of them being the D/21 chromosome. In addition, the parent has only one No. 21 chromosome, and so is normal. This parent may produce a germ cell containing both the D/21 chromosome and a No. 21. Such a germ cell, uniting with a normal one from the other parent, will produce a child

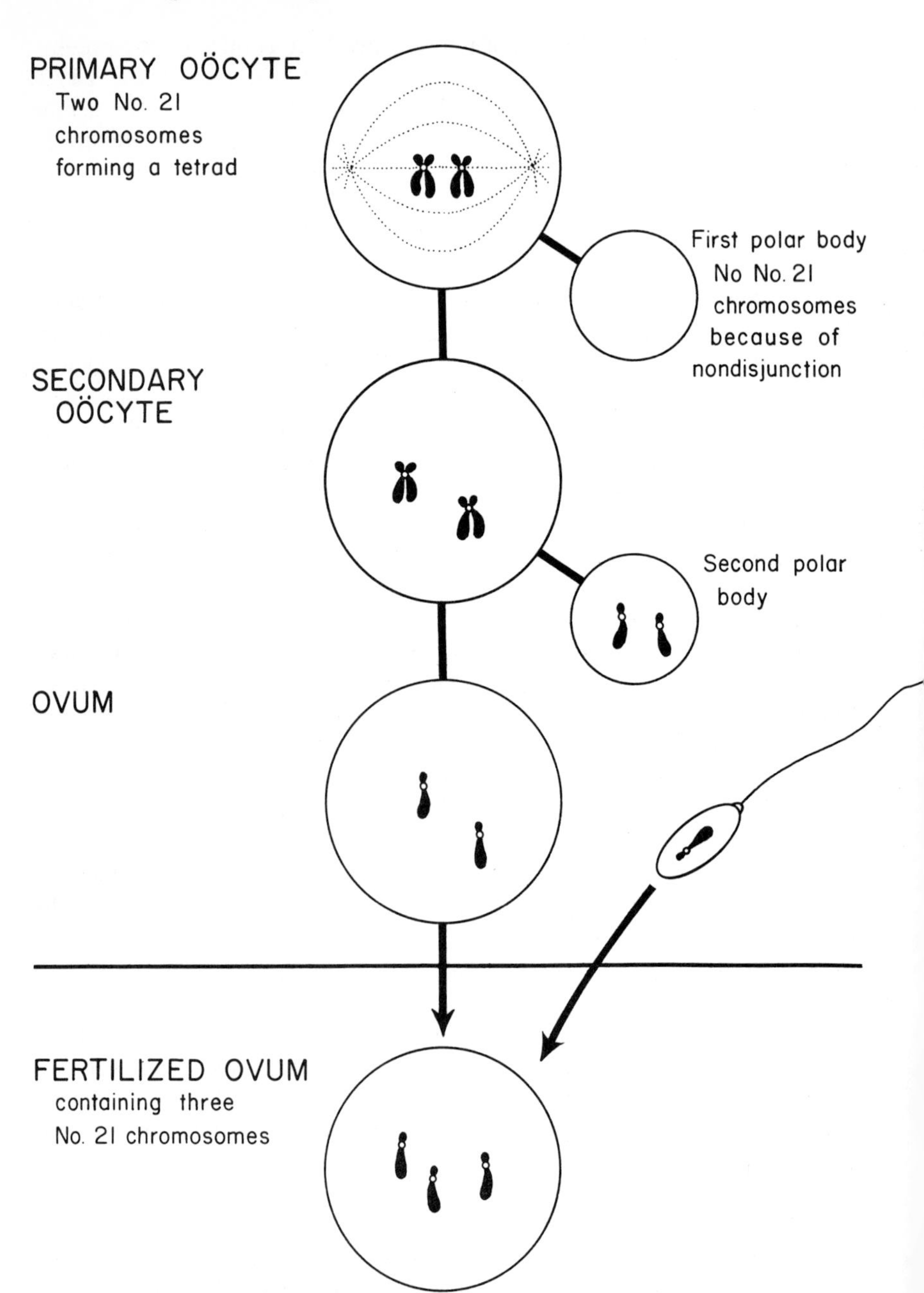

FIG. 20.8 Down's syndrome. Above the line: Origin of 21-trisomy through nondisjunction in production of an ovum. Below the line: Fertilization of an ovum containing two No. 21 chromosomes by a normal sperm containing one No. 21 chromosome.

with Down's syndrome (Fig. 20.9), where the D group chromosome is represented as being No. 14.

As Figure 20.9 indicates, a parent of this kind can also produce offspring of normal phenotype. Some of these, having the D/21 chromosome, can in their turn produce children with Down's syndrome. This affords a basis for the *inheritance* of the syndrome in families having the translocation chromosome. Such families are more likely to have a second affected child than are families in which Down's syndrome arises from having three separate No. 21's. Accordingly, as a basis for advising parents of an affected child concerning the chances that they will have another child with the syndrome, it is important to determine the type of trisomy present in that particular family.

The statement just made also applies to families in which the translocation chromosome is 21/22. But if the chromosome is 21/21, *all* the living children produced will have Down's syndrome. (One parent has only 45 chromosomes, one of them being this 21/21. Hence, the only germ cells produced will contain this chromosome *or* no No. 21 at all. So fertilization by a normal germ cell will result in (a) 21/21, 21 zygotes, or (b) zygotes having only one No. 21—monosomic—a lethal condition.) Fortunately, as noted above, studies of fluorescence make possible distinguishing between 21/21 and 21/22 chromosomes. (In some cases a 21/21 chromosome arises as an isochromosome, p. 317.)

Reproduction by Females Having Down's Syndrome Hamerton (1971) reported 13 known cases of parenthood by affected females (no affected males have been known to father children). In all 13 cases, the individuals had primary trisomy. When the primary oöcytes of an affected female contain three No. 21's, disjunction may be expected to produce equal numbers of ova containing (a) one No. 21 and (b) two No. 21's. Hence, following fertilization by normal sperm cells, half these ova will be expected to give rise to offspring with two No. 21's, half with three of them. The 13 females reported had ten normal children and four children with Down's syndrome. This departure from the expected 1:1 ratio may not be significant, but it might indicate that when an abnormal number of chromosomes is produced in meiosis, the abnormal number has an increased likelihood of being discarded in a polar body (Hamerton, 1971), or that abnormal zygotes are less viable than normal ones (Mikkelsen, 1971b). Mikkelsen (1971b) summarized data on four additional Down's syndrome females who had one child each; two children were normal, two had Down's syndrome.

For more extensive discussion of Down's syndrome, see Penrose (1966), Hamerton (1971), and Mikkelsen (1971b).

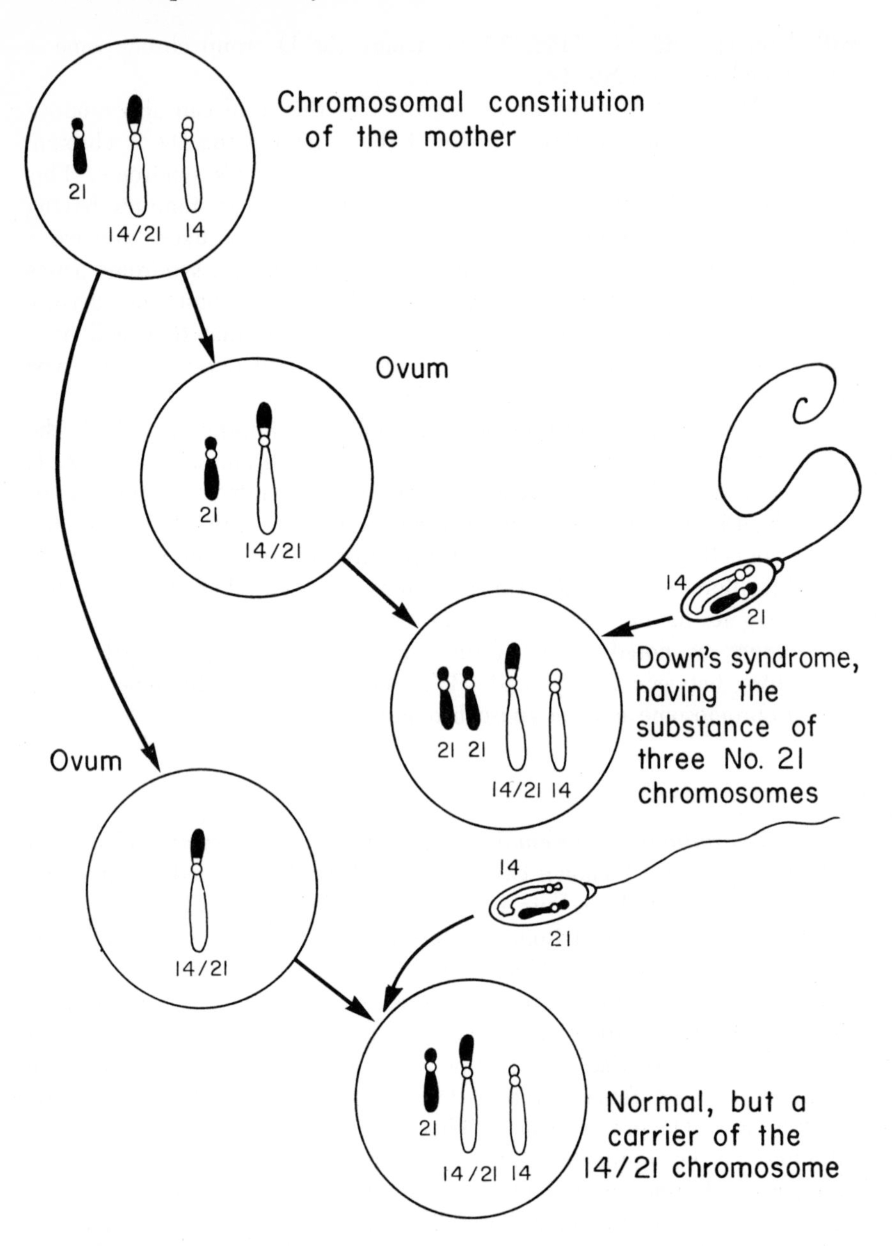

FIG. 20.9 Two offspring of a woman having a 14/21 translocation chromosome. In addition to the two types of ovum shown, such a woman can produce other types; those having one No. 21 and one No. 14 will give rise to normal offspring who will not be carriers of the 14/21 chromosome. In all cases, normal sperm are shown fertilizing the ova, although at times the 14/21 chromosome may be carried by sperm.

Other Chromosomal Anomalies

Many other defects connected with abnormal numbers or structures of chromosomes are known, and more are constantly being discovered. The literature on the subject is growing rapidly. For example, one defect that is receiving attention is the **cri du chat syndrome** (Lejeune et al., 1963). The name comes from the fact that the newborn babies have a peculiar, thin, mewing cry ("cat cry"). Affected children have abnormally small heads (microcephalic), various malformations of the face and body, and are mentally defective. The chromosomal basis of the syndrome is loss of a large portion of the short arm of chromosome No. 5 in the B group. The identification has been confirmed by studies of fluorescence (Caspersson, Lindsten, and Zech, 1970). Affected children have one normal and one abnormal No. 5 chromosome.

More detailed cataloguing of birth defects connected with abnormal chromosomes would be out of place here. Extensive discussions of them will be found in such books as Hamerton (1971, Vol. 2) and Levine (1971).

Many unanswered questions remain. For example, what are the frequencies of occurrence of chromosomal aberrations? This question is being answered by investigations of the chromosomal constitutions of all newborn babies in certain hospitals over a period of time, leading finally to accumulation of information concerning thousands of consecutive births.

Perhaps even more fundamental is the question of why and how chromosomal aberrations produce their unfortunate results. Why, for example, does having two of the tiny No. 21 chromosomes produce a normal phenotype, but having three of them produce the symptoms of Down's syndrome? Doubtless the presence of the abnormal number of the genes in these chromosomes upsets delicate enzyme balances during embryonic development. But *what* enzymes? Attempts to pinpoint exact enzyme lacks in Down's syndrome have met with but small success (Hamerton, 1971). We may be sure that such investigations will continue, spurred both by desire for knowledge and by the hope that increased knowledge may make possible remedial measures (as it did in the case of phenylketonuria, for example).

Problems

1. What would be the result if the tetraploid zygote shown in Figure 20.1B gave rise to a plant in which some of the primary oöcytes failed to form first polar bodies? How many haploid sets would the ova resulting from this failure contain? What would result if the ova were fertilized by haploid pollen grains? By diploid pollen grains?
2. Suppose that in human oögenesis nondisjunction of both the X chromosomes and the No. 21 chromosomes occurred in the first meiotic division,

the first polar body receiving neither type. If the resulting ovum were fertilized by a Y-bearing sperm, what would be the genotypic and phenotypic nature of the offspring produced?
3. How might nondisjunction in the production of the ovum result in offspring having both Turner's syndrome and Down's syndrome?
4. In Figure 20.9 we have shown two types of ovum that a woman of normal phenotype but having a 14/21 translocation chromosome might produce. What other types of ovum might she produce? Which of these would give rise to offspring that would be normal both in genotype and phenotype?
5. A woman with Down's syndrome may have 46 chromosomes, one of them being an isochromosome consisting of the long arms of two No. 21 chromosomes. If such a woman reproduces, and her husband is normal, what proportion of her children will be expected to have Down's syndrome?
6. A boy with Down's syndrome is found to have 46 chromosomes, one of which is a D/21 translocation chromosome. How many chromosomes does his phenotypically normal mother have?
7. A mother whose first child had Down's syndrome consulted a genetic counselor. She was found to have a G/G translocation chromosome. Banding patterns of the chromosome were determined using the quinacrine mustard technique. On the basis of the findings, the counselor concluded that the chance of a second child having Down's syndrome was about 33 percent. What did the analysis of banding patterns indicate about the nature of the G/G chromosome?

References

Böök, J. A., and B. Santesson, 1960. "Malformation syndrome in man associated with triploidy (69 chromosomes)," *Lancet,* 1960, **I**:858–859.

Caspersson, T., J. Lindsten, and L. Zech, 1970. "Identification of the abnormal B group chromosome in the 'cri du chat' syndrome by QM-fluorescence," *Experimental Cell Research,* **61**:475–476.

Hamerton, J. L., 1971. *Human Cytogenetics; Vol. 1, General Cytogenetics; Vol. 2, Clinical Cytogenetics.* New York: Academic Press.

Lejeune, J., J. Lafourcade, R. Berger, J. Vialatte, M. Boeswillwald, P. Seringe, and R. Turpin, 1963. "Trois cas de délétion partielle du bras court d'un chromosome 5," *Comptes Rendus des Séances de l'Académie des Sciences,* **257**:3098–3102.

Levine, H., 1971. *Clinical Cytogenetics.* Boston: Little, Brown & Co.

Mikkelsen, M., 1971a. "Identification of G group anomalies in Down's syndrome by quinacrine dihydrochloride fluoresence staining," *Humangenetik,* **12**:67–73.

Mikkelsen, M., 1971b. "Down's syndrome. Current stage of cytogenetic research," *Humangenetik,* **12**:1–28.

O'Riordan, M. L., J. A. Robinson, K. E. Buckton, and H. J. Evans, 1971. "Distinguishing between the chromosomes involved in Down's syndrome (Trisomy 21) and chronic myeloid leukemia (Ph^1) by fluorescence," *Nature,* **230**:167–168.

Penrose, L. S., 1966. *Down's Anomaly.* Boston: Little, Brown & Co.

Polani, P. E., 1962. "Sex chromosome anomalies in man." In J. L. Hamerton (ed.). *Chromosomes in Medicine.* London: National Spastics Society. Pp. 73–139.

21

Mutations, Chromosomal Aberrations, and Radiation

Gene Mutations

In our discussions in Chapters 5 and 20, we emphasized that gene or point mutations arise as changes in the DNA molecule. We saw that substitution of one nucleotide for another may suffice to cause significant change in the phenotype—to change hemoglobin A to hemoglobin S, for example (pp. 49–50). Such a change may be pictured as a *copying error.* Before the invention of printing, manuscripts were copied by hand, sometimes repeatedly. Occasionally the scribes made mistakes, so that the copy was not exactly like the original. In somewhat similar fashion, copies of the DNA molecule are made every time a chromosome duplicates itself in preparation for mitosis or meiosis. Usually the copying is precise, so that the replica is exactly like the original, but occasionally errors creep in. These errors are the mutations.

Copying errors may occur in the mitosis connected with replication of body (somatic) cells. Such **somatic mutations** are not passed on from one generation to the next, but are passed on from the somatic cell in which they occur to other somatic cells arising from that one by mitosis. Thus, in the embryo of a plant, a cell may undergo a mutation so that chlorophyll is not produced. If, in subsequent development, that cell gives rise to branches with leaves, those branches and leaves will be white. According to a widely recognized theory, somatic mutations may be involved in the production of cancer. A cancer may start from a cell that has undergone mutation in the genetic mechanism that controls cell division. The result would be that the cell would "lose control" and divide and subdivide in the unregulated fashion characteristic of malignant tumors.

Because our main concern here is genetics, we shall focus our attention on **germinal mutations**—copying errors in the production of germ cells. Such mutations *are* passed on from generation to generation.

What Causes Gene Mutations? In general, we may say that the factors concerned may affect the chromosomes directly, or do so indirectly by altering the microenvironment surrounding the chromosomes within the cell. One cause of mutations is believed to be **thermal agitation,** which we may picture as heat-induced movement of molecules (both inside and outside the chromosomes).

We also know of a variety of **chemicals** that can cause genes to mutate (mustard gas was perhaps the first to be recognized). For the most part, these chemicals are not normal constituents of cells. We can readily understand that it is much easier to add an unusual chemical and note its effect than it is to determine the mutagenic (mutation-causing) action of normal cell constituents. However, some of the observed mutation rate may be caused by chemicals that cells normally encounter.

Starting with the classic research of Muller (1927), we have much evidence that mutations are also induced by **radiations.** These radiations include ionizing radiations such as X-rays, neutrons, and cosmic rays, and nonionizing radiation such as ultraviolet light. Apparently, radiations may act in two general ways. They may strike a chromosome and its included DNA directly, producing a break in the chromosome or a change in the molecule, or they may alter the microenvironment (mentioned above), producing changes in the chemical compounds within the cell. These changes then affect the DNA of the chromosomes, producing mutations.

Mutation Rate

The effect of man-made radiations is to increase the rate at which mutations "normally" occur. This so-called **spontaneous rate** results from the action of thermal agitation, natural radiations such as cosmic rays and radiations from rocks containing radioactive minerals, and all other mutagenic forces that cells encounter naturally. Before we can measure the effect of man-made radiations, then, we need to have an idea of the magnitude of this spontaneous rate to serve as a basis for comparison.

Measuring the Spontaneous Mutation Rate How can this rate be determined in human beings, with whom experimentation is impossible? This is more easily done with normally occurring recessive genes that mutate to their dominant alleles than it is with dominant genes that mutate to their recessive alleles. When a new dominant gene is produced (by mutation of a recessive one), the mutated gene produces a phenotypic effect in every individual who inherits the gene. For example, if in a kindred lacking Huntington's chorea (Huntington's disease; pp. 66–69), an individual appears who has that abnormality, we may conclude that a new mutation has occurred in a germ cell of a parent and has been inherited by the individual offspring.

Thus, in theory, one could compute the mutation rate at which the gene for Huntington's chorea is produced by calculating the proportion of people who have the disease (are *Hh*) but who were born to parents neither of whom had it (they were *hh*). To obtain the mutation rate per gene (the rate at which *h* mutates to form *H*), this proportion of people would be divided by 2, because either the *h* gene from the mother or the one from the father might have mutated to produce the *H* gene. For example, if among 1,000,000 people, 40 were found to have Huntington's chorea although their parents did not have it, the mutation rate indicated would be

$$\frac{40}{1{,}000{,}000} \times \frac{1}{2} = 0.00002 \qquad \text{(or 2 mutations per 100,000 genes)}$$

(Check: 1,000,000 people possess 2,000,000 genes at the locus concerned. 2,000,000 $\times$ 0.00002 = 40).

The method sounds simple, but it is beset with difficulties. Errors of diagnosis, difficulty in determining true parentage, the fact that different genes may mutate to produce the same abnormality, the fact that nonhereditary traits (phenocopies) may appear and be indistinguishable from gene-determined traits—these are a few of the sources of uncertainty (Crow, 1956; UN, 1962).

Even greater difficulty is experienced in measuring the rate at which dominant genes mutate to form their recessive alleles. If the genes are in the autosomes, the mutant phenotype is shown only by homozygotes; in these one of the two recessive genes may be a newly formed one, but the other may have been inherited from distant ancestors. Thus rates of mutation cannot be determined by direct observation and counting. Methods of estimating such rates are based on principles of population genetics, utilizing certain assumptions, and they yield results of uncertain significance (Crow, 1956). If, on the other hand, the genes are in the X chromosomes, the problem is somewhat simpler, because *males* will show the phenotype produced by an X-linked gene even if the gene is recessive.

Spontaneous Mutation Rate Despite the difficulties, a consensus based on the results of many investigations is the estimate that the mutation rate averages around 1/100,000 (1×10^{-5}) per gene, per generation (e.g., UN, 1962). This means that only one gamete in 100,000 is likely to contain a mutation in any given gene in which we may be interested. This estimate is based in part on data from the fruit fly, *Drosophila,* and from the mammal whose genetics has been most thoroughly studied: the house mouse. Thus Schlager and Dickie (1971) reported results of examining more than 7 million mice for visible spontaneous mutations. The number of mutations observed was 249, of which 206 could be classified as either dominant or recessive. Despite large differences between genes, an average rate of 11×10^{-6} (11/1,000,000 or 1.1/100,000) was recorded for mutations of normal wild-type genes, and somewhat less (2.5×10^{-6}) for recessive genes mutating to their dominant alleles.

A rate of 1/100,000 is a very low rate, but it may be higher than the rates for many genes. The mutations we *observe* in man, mice, and flies may well be ones that oblige by occurring with greater frequencies than do most mutations. This is because huge numbers of individuals must be available if an investigator is to have a fair chance of observing even a few of the rare occurrences. In bacteria, where obtaining vast numbers of individuals is no problem, mutation rates as low as 10^{-9} per gene per cell have been measured (on the average, only one bacterial cell in a *billion* shows mutation of the gene being investigated).

Some of the advantages of investigations with bacteria are gained when human cells are grown in tissue culture. Large numbers of cells can be obtained and can be experimented with—thus overcoming the traditional disadvantage of human genetics of inability to perform experiments. Using this technique, Albertini and DeMars (1973) found that the spontaneous mutation rate of a particular locus under investigation seemed to approximate 1×10^{-6}. This would tend to confirm suspicions that the 1×10^{-5} we usually employ in human genetics may not be typical of all human genes.

What is the total frequency of spontaneous mutations? What proportion of gametes contain a new mutation of some one of the genes present? To answer these questions we need to know the number of genes in a gamete as well as the mutation rate. How many genes are there? McKusick (1971) estimated "not less than 100,000" (of which only some 1 percent have been identified so far). If these genes all mutated at a rate of 1/100,000, every gamete on the average would be expected to contain one new mutation. This estimate may be too low, however. Neel (1969) utilized the estimated rate of 1×10^{-5} and stated: "There is enough DNA in the human gamete to code for about 7,000,000 'messages' (i.e., polypeptides). If only 10 percent of this is functional, the total mutation rate per gamete would be $7 \times 10^5 \times 1 \times 10^{-5}$, or 7." All such estimates are rough approximations, of course, but they indicate the likelihood that most, if not all, of our gametes contain from one to several new mutations (plus mutant genes inherited from ancestors).

Radiation and the Mutation Rate

Of the physical and chemical factors that may cause genes to mutate, we focus attention on radiation as being one likely to increase in the modern world. What will be the genetic effect of such an increase?

Background Radiation In the first place, we should note that we are all subjected to some radiation throughout our lives. This background radiation arises from cosmic rays, radiation from radioactive minerals in rocks and soil, and radioactive isotopes in air, water, and food consumed.

How much background radiation does the average person receive? Only the most approximate of estimates can be given because environmental conditions vary greatly. For example, people who live in brick houses receive slightly more radiation than do people who live in wooden ones. In general, the amount is very small indeed. Estimates usually are of the order of 3 or 4 roentgen units spread over a 30-year period. This is about the amount of radiation one receives when two or three X-ray pictures are taken. Thus the international group of scientists that constitute the United Nations Scientific Committee on the Effects of Atomic Radiation estimated that, on the average, the dose received by the gonads amounts to 126 mrem* per year (UN, 1962) ($0.126 \times 30 = 3.78$ rem in a 30-year period). The committee estimated that those 126 mrem are comprised as follows: 50 mrem from cosmic rays, 50 from terrestrial radiation, 26 from radioactive isotopes entering the body in air, water, and food—elements of the radium and thorium series, potassium-40, and carbon-14, for the most part (UN, 1962, Table I, p. 21).

From results of experiments with plants and animals, we may infer that this low level of natural radiation may be responsible for some of our "spontaneous" mutations. As far as we know, there is no amount

*A word is in order about the units in which radiations are expressed, because different studies employ different terms. The most frequently employed unit is the roentgen (r), based on amount of ionization produced. The rad is a unit based on the amount of energy absorbed by the material being radiated. The rem is the biological equivalent of the rad, taking into account the difference in biological effectiveness of different sources of radiation—X-rays, gamma rays, etc. Mrem (millirem) is one-thousandth of a rem. The use of different units may seem confusing but need not concern us greatly because r, rad, and rem are of similar magnitude.

of radiation so small that it may not cause mutation. Yet experiments suggest that only a small proportion of the observed mutation rate would be caused by radiations totaling such a small amount. In this connection it would be of great interest to know whether or not the mutation rates are increased in people who live in regions where they are exposed to more radiation than most people receive. People who live at high altitudes receive greater cosmic radiation than do people at low altitudes—3 to 6 times the amount mentioned in the preceding estimate. People who live on certain types of soil receive more radiation than do people living on other soils. For example, in Travancore State in India, some 100,000 people live on a soil of monazite sand (containing radioactive thorium). It is estimated that they may receive from 50 to 150 units of radiation (roentgens or the equivalent) in a 30-year period (Gopal-Ayengar, 1957). Do they have a higher mutation rate than do people who receive only 3 or 4 units in a 30-year period? Here is a fertile field of investigation. At present, we can note with interest that apparently normal populations can live under widely varying intensities of background radiation. This may lead us to anticipate that people who receive 3 or 4 units in a 30-year period could have this amount increased somewhat without alarming consequences.

Man-Made Radiation How much is the radiation we receive from natural sources likely to be increased by radiation resulting from human activities? These latter radiations fall into two main categories: (1) medical radiations (such as diagnostic and therapeutic X-rays) and radiation exposure received by persons whose occupations place them in contact with radiation-producing equipment (e.g., X-ray technicians and radiologists); and (2) fall-out from the testing of nuclear weapons. The United Nations committee quoted above (UN, 1962) estimated that during the period 1954–1961, world population on the average experienced a genetic risk from medical and occupational exposures that was about one-third of that from natural sources. They concluded that "the comparative genetic risk from fall-out is about one-tenth of that from natural sources."

Hence a person born in 1931 may have received, by 1961, 3 or 4 r of natural radiation plus about 0.4 r from weapons testing. Although any increase is undesirable, such a small increase is hardly justification for grave concern, particularly when we recall the wide limits of natural radiation within which normal populations live. We should remember, however, that the estimates are world-wide averages and do not apply to populations living near testing sites. Radiation received by survivors of atomic warfare would be another matter entirely.

The probable genetic effect of fall-out from the testing of atomic weapons is overshadowed by the possibility of genetic damage from the medical and industrial use of X-rays and radioactive materials. We noted above the world-wide estimate that such sources may increase the genetic risk from natural radiations by one-third. In a country such as the United States, in which radiation diagnosis and therapy is widely used, the risk may well be greater, equalling if not exceeding the dosage received from natural sources (Glass, 1957). This would increase the 30-year total to at least 6 to 8 r. What would be the genetic effect of such an increase?

The Doubling Dose In estimating genetic risk from radiation, we often employ the term "doubling dose" to indicate that amount of radiation required to double the spontaneous mutation rate. Many investigations on various organisms suggest that usually the doubling dose is found in the range between 25 r and 100 r, 30 to 50 r being a common estimate (e.g., Freire-Maia, 1970; James and Newcombe, 1964; Lüning and Searle, 1971). Thus, the 6 to 8 r we mentioned falls far short of the doubling dose (but this is apparently not true of the induction of chromosomal aberrations; see below). Nonetheless, it remains true that *any* amount of radiation, no matter how small, *may* cause a mutation. Hence, because many mutations, at least, are detrimental, it is important to reduce radiation exposure of the gonads to a minimum. Modern radiologists are well aware of this, and improved methods are steadily reducing the amount of radiation received by the gonads when X-rays are employed for diagnosis. Obviously, protection of the gonads of children, before and after birth, and of young people before the end of their reproductive life, is especially important.

Acute Versus Chronic Irradiation Natural radiation is of low intensity and is received very slowly. Radiations from X-rays and other sources employed in medicine are ordinarily of relatively high intensity and are given in a short period of time. This may be called **acute irradiation** in contrast to the **chronic irradiation** mentioned first. Are the two types of irradiation equivalent in mutation-inducing potency? When the rate is slow, is there anything at all comparable to healing and repair?

A great body of information has been accumulated to show that when **sperm cells** (of fruit fly or mouse) are irradiated, the number of mutations depends on the total dosage of radiation and is independent of the rate at which that radiation is given. For example, if sperm cells receive 100 roentgen units of X-rays in 10 minutes, a certain average number of mutations will be induced. If they receive 100 units of radiation slowly, perhaps spread over 10 days, the same average number will be induced. This has been found true over a range of dosages from 5 r to

6000 r. Therefore, it is generally agreed that with mature sperm cells the induction of mutations varies with dosage and is independent of rate. In theory, at least, even the smallest amount of radiation will produce its proportionate effect, whether the amount is given slowly or rapidly. (In bacterial viruses, the direct proportionality between dosage and number of mutations has been found to extend down to 0.3 r; James and Newcombe, 1964.)

In **spermatogonia** and **oöcytes,** however, extensive experiments with mice have demonstrated that radiation given slowly produces fewer mutations than does the same amount of radiation given rapidly (Russell, Russell, and Kelly, 1958; Russell, 1963, a,b). In fact, a dosage given slowly produces only about one-fourth as many mutations as does the same dosage given rapidly. This reduction in number of mutations induced is called the **dose-rate effect.** It seems to be connected with the fact that, unlike sperm cells, spermatogonia and oöcytes are metabolically active and able to repair damage caused by radiation unless that radiation is received in too large amounts over a short period of time. At one time it was thought by some investigators that repair was possible only when chromosomal aberrations were produced by the radiation. But we now know that cells are capable of repairing damaged DNA. The damaged portion of the molecule is removed and a normal portion is synthesized in its place.

This dose-rate effect in spermatogonia and oöcytes is particularly important in view of the fact that these are the stages of meiosis in which much of mammalian germ plasm remains during the greater parts of the lives of individuals. That these cells are able to repair mutational damage should not surprise us. Man reached his present stage of evolution in a world in which he and his ancestors were constantly subjected to natural radiation of low intensity. It would be strange indeed if the body had not developed some ability to cope with this factor in the environment.

To what extent will this protective ability serve to keep the germ plasm buffered against harmful mutations in a world in which man himself is increasing the amount of radiation? Insofar as the increase is in chronic, low-level radiation, this ability to repair may be expected to reduce the amount of genetic damage, but there is no evidence that it will reduce the amount to zero. It still remains true that even the smallest amount of radiation *may* cause mutational damage. If, on the other hand, the radiation received by the gonads is of large amounts given quickly, as it may be in medical irradiation, the repair mechanisms are ineffective in reducing the damage caused.

Chromosomal Aberrations

A given amount of radiation seems to be even more potent in inducing chromosomal aberrations than it is in inducing gene mutations.

The doubling dose for aberrations may be as low as 1 roentgen (Freire-Maia, 1970).

If the radiation is given to germinal tissue undergoing meiosis, normal synapsing and separation (disjunction) of chromosomes may be interfered with, and abnormal or nonfunctional gametes may result. Abnormal gametes may result from irradiation of mature ova or sperm cells as well. Fertilized ova receiving these chromosomal aberrations in turn may be so abnormal that they fail to develop, or if they do develop, the resultant embryos may be abnormal, so abnormal in many cases that they undergo spontaneous abortion. We noted in Chapter 17 that chromosomal aberrations (e.g., XO) contribute significantly to the number of abortuses.

Because, therefore, radiations increase the frequency of chromosomal aberrations, we should expect that radiations would also increase the frequency of spontaneous abortion, even when it is the father who receives the radiation. Direct evidence was obtained by Freire-Maia (1970) in a study of Brazilian physicians some of whom worked with ionizing radiations and some of whom did not (the control group). Of 1449 pregnancies in the families of the control group, 7.87 percent (±0.71%) terminated in spontaneous abortions. But in the families of physicians who worked with radiations, 10.72 percent (±0.82%) of 1418 pregnancies terminated in such abortions.

When abortion does not occur, the child inheriting chromosomal aberrations may be abnormal in some way, as our earlier discussions made evident. Clearly, whether they cause gene mutations or chromosomal aberrations, ionizing radiations constitute a threat to the genetic well-being of future generations.

Significance of Mutations

Are Mutations "Good" or "Bad"? An organism, be it fly or man, is like a highly complex machine in that for perfect functioning all parts must work together smoothly. Both fly and man are highly adapted for a particular mode of existence. Each represents the product of age-long processes of evolution always tending to produce more and more perfect adaptation to the needs of existence. Although each may not be perfect in its own sphere, each is well up toward what is sometimes called its "adaptive peak." Each is a highly successful organism in its own way.

If we take a smoothly running machine and hit it with a hammer, we *may* improve the running of the machine, but the chances are against doing so. We are far more likely to injure the machine so that it runs poorly, or does not run at all.

Mutation is something like that—it is an accident to a chromosome or a gene in a chromosome. Because the machine is already running well, such an accident is more likely to do harm than it is to be of benefit (although we must never forget the possibility that it may be of benefit). Thus, we should expect that most mutations to modern, well-adapted organisms would be deleterious. This is what we do, in fact, observe.

However, we cannot leave the matter there. We have spoken of modern, well-adapted organisms. Well adapted to what? Obviously, to their environments. Adaptation has meaning only with reference to an environmental setting. Hence it is that in large measure whether a mutation is beneficial or harmful depends on the environment of the organism in question. Doubtless there are some mutations so harmful that they cannot be of benefit in any conceivable environment. In man, *Duchenne's muscular dystrophy,* which is dependent on an X-linked gene and always kills boys in their teens, would seem to be an example (p. 441). Yet great numbers of mutations are harmful in one environment and beneficial in another.

For example, a mutation that changed the physiology of an organism so that it required a food, vitamin, or chemical compound not available in its environment would certainly be detrimental. Indeed, if the requirement could not be met, the mutation would be lethal. Such a mutation occurs once in a billion times or so in the much-studied colon bacillus, *Escherichia coli.* Presumably from time immemorial, on rare occasions a mutation has occurred that results in its possessor's being unable to live without the antibiotic streptomycin. A cell with this mutation would promptly die unless streptomycin were present. Hence this is a lethal mutation in an environment lacking the antibiotic, but in a test-tube culture containing streptomycin the previously lethal mutation suddenly becomes most valuable (Demerec, 1950). Cells not possessing the mutation die by the million, but the cell with the mutation thrives and gives rise to a new streptomycin-dependent population. The mutation, lethal under usual conditions encountered by the bacillus, becomes highly beneficial in the new environment containing streptomycin.

In much the same way, mutations in house flies conferring resistance to the insecticide DDT would be of no value to the fly in an environment devoid of DDT, and they might even be harmful. However, as soon as man started using DDT, a great premium was placed on these very mutations. Flies died by the millions, but the fortunate possessors of these mutations survived and became the ancestors of the DDT-resistant strains that plague us today.

We might multiply the examples, but these two will suffice to emphasize the point that the harmfulness or beneficialness of many, if not most, mutations, depends on the environment. We cannot answer the question, "Is the mutation to streptomycin dependence in *E. coli* a harmful mutation?" until we know whether or not the environment contains streptomycin.

We have labored this point at some length to counteract the impression many people have that there is something essentially harmful about mutations, that they are always bad. As we have seen, mutations are genetic changes, and change in and of itself is neither intrinsically good nor intrinsically bad. We may have change for the better or change for the worse. To a great extent, environment determines which it shall be.

As a matter of fact, most biologists today regard mutations as the raw materials used in the evolution of organisms. Progressive evolution by this means would hardly be possible if mutations were always changes for the worse (see further discussion of evolution in Chapter 27).

Heterozygotes

So far we have focused attention on mutant genes as though they were isolated from other genes. In fact, each is a member of a pair of alleles, which may interact with each other in important ways.

Mutations Dominant Suppose that people normally have the genotype *aa*. What will be the effect when in some germ cell *a* mutates to form *A* (and this germ cell unites with an *a*-containing one from the other sex to form an *Aa* zygote)?

In the first place, gene *A* may be so abnormal that the resulting individual is not viable. The zygote, or the embryo, or the infant that develops from the zygote, may die. Such a gene is said to be **lethal,** or to be **semilethal** or **sublethal** if the individual is born but dies in infancy or childhood. Death may occur at any stage, and the effect is that of removing the gene before it can be passed on to a later generation.

In the second place, gene *A* may not be lethal, but it may cause lowered vitality or perhaps sterility. If *Aa* individuals are sterile, gene *A* is not passed on to the following generation and hence, the *genetic* effect is the same as it would be if the gene were lethal. On the other hand, *Aa* individuals may not be sterile, but they may have impaired health so that they produce fewer children on the average than *aa* people do. As a result, a reduced number of the next generation will have gene *A*, and if the trend continues in future generations, the gene may be lost entirely. Such loss is frequently called **genetic death** (Muller, 1950). The term means death of a gene: its loss from the gene pool of a population.

Mutations Recessive Suppose that most normal people are homozygous for a dominant gene—for example, they have the genotype *BB*. If gene *B* mutates to form its recessive allele, *b,* what will happen to this allele?

If *b* is completely recessive, *Bb* individuals are exactly like *BB* individuals, the only phenotypic effect of gene *b* being produced in *bb* individuals. If *bb* individuals suffer an important disadvantage compared to *BB* and *Bb* individuals, gene *b* may be expected to decline slowly in relative frequency in the gene pool in future generations. Individuals with the *bb* genotype may die without leaving children, or they may merely suffer some defect that reduces the number of children they produce. As we shall see in our discussions of selection (Chap. 25), the decline in the relative frequency of gene *b* in the gene pool will be slow, because most of these genes are possessed by heterozygotes (*Bb*), whom we are assuming to be completely normal.

When Heterozygotes Are Superior

What will be the effect when heterozygotes are superior to homozygotes—for example, when *Aa* (above) is superior to both *AA* and *aa,* and when *Bb* is superior to both *BB* and *bb*? In such a case genes *a* and *b* are not completely recessive, because they have some effect in heterozygotes.

The classic human example of this situation is that of sickle-cell hemoglobin, hemoglobin S (pp. 47–50). As we mentioned earlier, homozygous Hb^S Hb^S individuals suffer a severe, often fatal, anemia. Because they seldom if ever produce children, their genes are lost from the gene pool (suffer genetic death). Why, then, does gene Hb^S not disappear entirely? Much evidence indicates that in an environment in which severe malaria is present, heterozygous Hb^A Hb^S individuals survive the ravages of the disease better than do Hb^A Hb^A homozygotes (Allison, 1955). Hence, in an environment in which malaria is present, heterozygotes produce on an average an increased proportion of children as compared to Hb^A Hb^A homozygotes. Such an interaction of genes to produce a superior phenotype in a heterozygote is called **heterosis,** or **hybrid vigor,** or sometimes **overdominance.** (In this case, the advantage seems to lie in the fact that red blood cells containing both Hb S and Hb A afford a less suitable environment for the malarial parasite than do cells having Hb A only.)

In sum, because it is beneficial to heterozygotes *living in malarial environments,* gene Hb^S does not disappear from the gene pool entirely even though the gene is most deleterious to homozygotes. Here we have a perfect example of the importance of the *environment* in determining the value of a gene, be it a new mutation or one inherited from ancestors.

However, Hb^S Hb^A heterozygotes enjoy no advantage in such an environment as that of the United States.

Not All Heterozygotes Are Superior We do not wish to give the impression that being heterozygous is *always* beneficial. Sometimes a gene that is detrimental to homozygotes is also harmful in some degree to heterozygotes. For example, in the fruit fly *Drosophila,* the majority of genes that are *lethal* to homozygotes are also detrimental to heterozygotes. Genes that have less severe effects upon homozygotes are more likely to be heterotic (beneficial to heterozygotes). Genes having some degree of detrimental effect to homozygotes or heterozygotes, or both, are said to constitute a population's **genetic load.** These genes may have been inherited from ancestors or they may be newly arisen mutations. Muller (1950) calculated that in *Drosophila* one gamete in every 20 "contains a new spontaneously arisen lethal or detectable detrimental gene that arose within the span of the very last (parental) generation." He concluded that the rate is probably higher in man, with his greater number of genes and higher body temperature. These new detrimental genes, added to old ones received from ancestors, constitute our genetic load.

Despite the fact that heterozygotes are sometimes at a disadvantage, much experimental evidence indicates that populations, whether of lower animals or man, are largely composed of individuals who are heterozygous at many gene loci (see, for example, Dobzhansky, 1970). This means that heterozygosity for many genes and gene complexes underlies the production of the normal phenotype. Thus the value of a given gene *to a population* frequently seems to be determined by its effect on heterozygotes possessing it, rather than by its effect on homozygotes, who are likely to suffer some detrimental effect.

We now ask: If heterozygosity is normal, will it be beneficial to a population to increase that heterozygosity by production of additional new mutations, by irradiation, perhaps? Experiments on *Drosophila* have indicated that under some circumstances the viability of populations of these flies can be increased by causing new mutations to be produced by irradiation (e.g., Wallace, 1958). But we should note that such increase in viability of the population as a whole is purchased at a price. The price is the eventual production of more or less unfit homozygotes. The price may seem inconsequential when we are talking about fruit flies, but will our point of view be the same when we are talking about *people*?

Differing Value Judgments

Upon what basis shall we judge the value of a gene or mutation? Instead of speaking in abstractions, let us return to the example of the sickle-cell hemoglobin gene: Hb^S. Is this a "good gene" or a "bad gene"?

Unquestionably, from the standpoint of human evolution it is a "good gene." It has contributed significantly to enabling people to live in regions of the earth, especially Africa, where malaria is an environmental factor. Because much of early human evolution occurred in Africa, the advantage in viability and hence fertility enjoyed by heterozygotes has been highly valuable to ancestral populations as a whole striving to live in a hostile environment. The occasional children suffering from sickle-cell anemia may be thought of as "by-products" of the process—regretable, perhaps, but unavoidable, and relatively unimportant compared to the good of the population as a whole. But are we to imagine that the *parents* of a child with severe, often fatal, anemia regard the gene as "good"?

The differing value judgments concerning this gene may be contrasted by quoting the views of two authors. Fraser (1962) wrote: "It is even doubtful if the very considerable increase of resistance to malaria postulated for heterozygotes for the sickling gene is sufficient to balance the misery and early death of homozygotes with the full disease picture." While Brues (1969) commented on this statement: "This is a somewhat unimaginative point of view, for it overlooks the fact that the 'increased resistance,' translated into concrete terms, means that 2¼ individuals are spared from misery and early death by malaria for every one that succumbs to sickle-cell anemia." (The "2¼" is by way of example, assuming a certain frequency of Hb^S in the gene pool.)

Hence we see that value judgments vary with one's point of view.

Differing judgments are also inevitable concerning the subject of *genetic death,* mentioned earlier. Genes may be lost from the gene pool because gametes containing them are inviable, or because zygotes are inviable, or because early embryos are inviable, or because later fetuses are inviable, or because babies die young, or die later in childhood, or because the resulting individual is sterile. As we saw, also, genetic death may eventually befall a gene that is merely deleterious, reducing the fertility of individuals who possess it. Because all of these lead to genetic death, they are equivalent in the eyes of genetics.

But they are far from equivalent in social importance and in terms of human values. A *lethal* mutation may kill a zygote or embryo so early that the mother did not even know she was pregnant. In this case the lethal gene has no significance for her. But if miscarriage occurs later in pregnancy, the mother may find the experience traumatic.

Turning to mutations that are "merely" deleterious, we note that they may cause suffering and disabilities for many generations before they disappear. Hence their social impact is much greater than is that of lethal mutations (see also Newcombe, 1971). Thus we have a paradox: More drastic genetic changes suffered by genes or chromosomes may prove to be less serious in terms of human life and welfare than are less-damaging changes of the genetic material.

Problems

1. Suppose that a man is homozygous *AA*. If he produces five billion (5000 million) sperm cells, how many of them will contain the mutant allele, *a*, if the estimated mutation rate quoted from the UN report (p. 330) applies to this locus?
2. In a population of 2,000,000 people, 48 were found to show a certain dominant trait but to be the offspring of parents who did not show it. Estimate the mutation rate producing the dominant gene concerned.
3. If a 100 r dose of X-rays administered to sperm cells produces 3 mutations per 1000 cells when given within a span of 10 minutes, how many mutations will be produced when a 100 r dose is spread evenly over a 10-day period? How would your answer differ if the question specified spermatogonia or oöcytes instead of sperm?
4. To what extent is it true that the genetic risk arising from 3 r of X-rays received by the gonads in medical practice is greater than the risk from 3 r of cosmic rays?
5. Gene *a* is lethal in the homozygous state. Nevertheless, the gene pools of some populations contain fairly high percentages of this gene. Judging by analogy with known cases, what is likely to be the explanation for this fact?
6. Why may the frequency of the gene for sickle-cell anemia be expected to decline gradually in the United States (even in the absence of any conscious control)? What would be the probable effect on this trend of finding a "cure" for sickle-cell anemia so that persons with the Hb^S Hb^S genotype were phenotypically normal?
7. In one experiment Russell (p. 334) found that a dose of 400 r of radiation given mouse oöcytes induced 0.8 ($\times$ 10^5) mutations per locus per gamete. In another experiment with mouse oöcytes, 400 r induced 19.3 ($\times$ 10^5) mutations per locus per gamete. Judging from our discussion in the text, how did the experiments differ in the procedure he used?

References

Albertini, R. J., and R. DeMars, 1973. "Somatic cell mutation. Detection and quantification of X-ray-induced mutation in cultured, diploid human fibroblasts," *Mutation Research,* **18**:199–224.

Allison, A. C., 1955. "Aspects of polymorphism in man," *Cold Spring Harbor Symposia on Quantitative Biology,* **20**:239–255.

Brues, A. M., 1969. "Genetic load and its varieties," *Science,* **164**:1130–1136.

Crow, J. F., 1956. "The estimation of spontaneous and radiation-induced mutation rates in man," *Eugenics Quarterly,* **3**:201–208.

Demerec, M., 1950. "Reaction of populations of unicellular organisms to extreme changes in environment," *American Naturalist,* **84**:5–16.

Dobzhansky, T., 1970. *Genetics of the Evolutionary Process.* New York: Columbia University Press.

Fraser, G. R., 1962. "Our genetical 'load.' A review of some aspects of genetical variation," *Annals of Human Genetics,* **25**:387–415.

Freire-Maia, N., 1970. "Abortions, chromosomal aberrations, and radiation," *Social Biology,* **17**:102–106.

Glass, B., 1957. "The genetic hazards of nuclear radiations," *Science,* **126**: 241–246.
Gopal-Ayengar, A. R., 1957. "Possible areas with sufficiently different background-radiation levels to permit detection of differences in mutation rates of 'marker' genes." In *Effect of Radiation on Human Heredity.* Geneva: World Health Organization. Pp. 115–124.
James, A. P., and H. B. Newcombe, 1964. "The quantitative assessment of hereditary damage induced by radiation." In Steinberg, A. G., and A. G. Bearn, (eds.). *Progress in Medical Genetics,* **3**:217–259.
Lüning, K. G., and A. G. Searle, 1971. "Estimates of the genetic risks from ionizing irradiation," *Mutation Research,* **12**:291–304.
McKusick, V. A., 1971. *Mendelian Inheritance in Man,* 3rd ed. Baltimore: The Johns Hopkins Press.
Muller, H. J., 1927. "Artificial transmutation of the gene," *Science,* **66**:84–87. (Reprinted in J. A. Peters (ed.), 1959. *Classic Papers in Genetics.* Englewood Cliffs, N.J.: Prentice-Hall, Inc. Pp. 149–155.)
Muller, H. J., 1950. "Our load of mutations," *American Journal of Human Genetics,* **2**:111–176.
Neel, J. V., 1969. "Thoughts on the future of human genetics," *Medical Clinics of North America,* **53**:1001–1011.
Newcombe, H. B., 1971. "The genetic effects of ionizing radiations," *Advances in Genetics,* **16**:239–303.
Russell, W. L., 1963a. "The effect of radiation dose rate and fractionation on mutation in mice." In F. H. Sobels (ed.), *Repair from Genetic Radiation Damage.* New York: The Macmillan Company. Pp. 205–217.
Russell, W. L., 1963b. "Genetic hazards of radiation," *Proceedings of the American Philosophical Society,* **107**:11–17.
Russell, W. L., L. B. Russell, and E. M. Kelly, 1958. "Radiation dose rate and mutation frequency," *Science,* **128**:1546–1550.
Schlager, G., and M. M. Dickie, 1971. "Natural mutation rates in the house mouse. Estimates for five specific loci and dominant mutations," *Mutation Research,* **11**:89–96.
UN, 1962. *Report of the United Nations Scientific Committee on the Effects of Atomic Radiation.* General Assembly Official Records: 17th Session, Supplement 16 (A/5216). New York: United Nations.
Wallace, B., 1958. "The average effect of radiation-induced mutations on viability in *Drosophila melanogaster,*" *Evolution,* **12**:532–556.

22

Multiple Births

Twins have always been of special interest to students of human genetics. This arises from the fact that there are two kinds of twins, and that in one of these kinds we find the only human example of individuals with the same genotype. Hence we can study them with regard to a certain trait and attempt to answer the question: How much depends on heredity, how much on environment? The more we learn about such things, the more we realize that most traits are the result of the interaction of heredity *and* environment.

Dizygotic and Monozygotic Twins

The two kinds of twins just referred to are usually called "fraternal" and "identical." Identical, however, is too strong an adjective. No two individuals, not even the most similar of twins, are alike in every detail (i.e., identical). Most geneticists, therefore, prefer to use a term that refers to the differing origin of the two types of twins—the fact that in one type the members of the pair arise from separate ova, whereas in the other type the two members arise from a single ovum. So instead of speaking of fraternal twins, we call them **dizygotic twins** (two-egg twins) and use the initials **DZ** in referring to them. Similarly, identical twins (Fig. 22.1) are called **monozygotic** or **MZ** (one-egg twins).

Dizygotic Twinning In the production of DZ twins, two ova are ovulated simultaneously by the mother, and each is fertilized by a

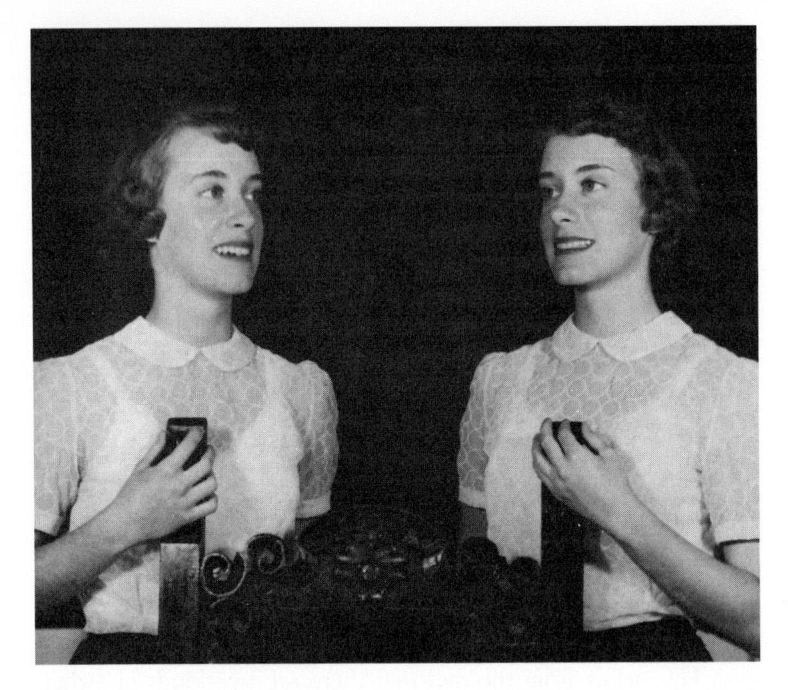

FIG. 22.1 Monozygotic twins Carolyn and Carol. (Photo by H. B. Eldred.)

different sperm cell. Each resultant embryo becomes separately implanted in the wall of the uterus and undergoes its own independent development. In general, DZ twins are no more alike than are other brothers and sisters in the same family, each of whom, of course, also arose from a separate ovum and a separate sperm. The DZ twins merely happen to be simultaneous instead of consecutive. Nevertheless, in the next chapter, we shall mention some embryonic processes that may cause members of a pair of DZ twins to be more alike than we should expect them to be (pp. 365–368).

Monozygotic Twinning Monozygotic twins, on the other hand, arise from a single fertilized ovum, and hence the genes they inherit are the same. Embryologists, like geneticists, are handicapped by their inability to experiment with human beings, and therefore we know little about the actual embryology of MZ twinning in man. Judging from observations on lower mammals, however, we may conclude that the single ovum, by a series of mitoses, produces an embryonic mass of cells. Ordinarily this mass would develop into a single embryo, but occasionally, as a result of influences as yet unknown, the mass may become subdivided into two masses, each of which will develop into an embryo. In the armadillo, which as we noted (p. 267) regularly produces monozygotic quadruplets, the embryonic mass divides into four masses by a sort of budding process (Patterson, 1913). Human monozygotic quadruplets also occur at times and doubtless originate in a comparable manner.

We should note that the *time* of separation of the embryonic masses may vary. It might occur as early as the two-celled stage (following the first mitosis of the fertilized ovum). Probably such early separation is not common in mammals; at least we know that it is not the program followed by armadillo embryos. Evidence suggesting that such early separation may occur in man, at least rarely, is afforded by a pair of twins consisting of a normal male and an individual with Turner's syndrome (Turpin et al., 1961). Blood groups and reciprocal skin grafting indicated the strong probability that this was a MZ pair. The chromosomal constitution was XY and XO, respectively. Suppose that an XY zygote underwent mitosis and in the process one daughter cell failed to receive a Y chromosome (through "anaphase lag," perhaps). If the two daughter cells then separated and each produced an embryo, one would be XY, the other XO, as in this case (Fig. 22.2). MZ twin pairs having a normal female the co-twin of an XO female have also been found (Turpin, 1970). The occurrence of such genetic accidents illustrates the point that members of a pair of MZ twins do not *always* have exactly the same genotype. Furthermore, gene mutations may occur in one twin that do not occur in the other.

As we noted earlier (p. 273), sibships in which XO individuals occur have an increased likelihood of also containing MZ twins. This suggests that mothers who have an increased tendency to produce ova with such genetic accidents as chromosomal aberrations also have an increased tendency to produce ova that will undergo the genetic accident of MZ twinning.

We do not know how late in embryonic development separation is still possible, but incomplete separation as in conjoined or so-called Siamese twins probably represents a belated attempt of an embryonic mass to form two embryos. In this connection, we may note the interesting but still mysterious fact that although separate MZ twins are usually very similar to each other, Siamese twins are frequently dissimilar to each other, although they are always of the same sex.

A Third Kind of Twinning? We may mention the *possibility* that there is a third kind of twin formation. A large polar body (Fig. 3.7, p. 33) might be fertilized and develop into an embryo along with the ovum itself. This happens in some invertebrates. Or the ovum might divide before fertilization and both daughter cells be fertilized (by different sperm cells, of course). In the latter case, the resulting twins would

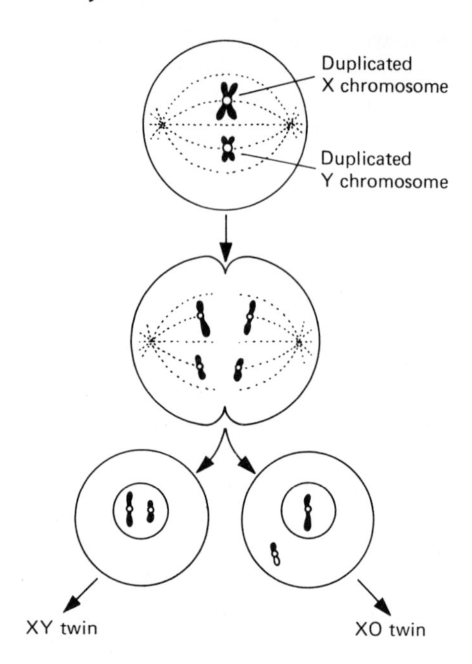

FIG. 22.2 Possible mode of origin of an XY-XO pair of MZ twins. When the fertilized ovum undergoes its first cleavage, one Y chromosome lags behind the other chromosomes and so is not included within the nuclear membrane of the corresponding daughter cell ("anaphase lag"). The daughter cells separate, each cell giving rise to a twin. The Y chromosome that lagged behind is eventually destroyed by enzymes. To simplify the diagram, autosomes are omitted.

have the same genes from their mother, but would differ in the genes received from their father. In the case of fertilized polar bodies, both members of the pair of twins would have different gene assortments from the father, but would vary in the similarities of gene assortments from the mother, depending on whether it were a second polar body or a cell produced by division of the first polar body. Second polar bodies are genetically like the accompanying ovum except as crossing over occurred in the first meiotic division. Following an extensive discussion of this kind of twinning, Bulmer (1970) concluded that there is no conclusive evidence that this kind of twinning does occur, although it remains a theoretical possibility.

Frequency of Multiple Births

United States Census data covering 50,000,000 births in a 22-year period indicated that among the white population one delivery in

92.4 gave rise to twins (Guttmacher, 1953). Among the portion of the population listed as "colored" (94 percent of whom were stated to have been Negroes), the rate of twin births was one in 73.8. An increased rate of triplet births also characterized the "colored" portion of the population. More recent studies have been in approximate agreement with this earlier one. Thus, Myrianthopoulos (1970), from data on over 50,000 births, found that the ratio of twin births to single births was 1:78.8 among Negroes, 1:100.3 among whites (1 percent or 10 per thousand; 1:92.4 is 10.8 per thousand).

In general, the rate of multiple births is highest among Negroes and lowest among Orientals; Caucasoids occupy an intermediate position. It is worthy of note that these racial differences are limited almost entirely to the amount of DZ twinning, not MZ twinning, which is nearly constant at about 3.5 per thousand births in all races (Bulmer, 1970). Bulmer stated as a broad generalization that the DZ rate is about 8 per thousand in Caucasoids, and that the rate in Negroes is about twice that, while the rate in Mongolian peoples is less than half that.

Hellin's Law Here we might appropriately mention a generalization usually called Hellin's law, although various people have had a hand in its formulation. According to this idea, if $1/n$ represents the fraction of births giving rise to twins, $1/n^2$ would represent the fraction of births giving rise to triplets, $1/n^3$ the fraction giving rise to quadruplets, and so on. That this may be a useful rule of thumb is indicated by data on 120,061,398 pregnancies occurring in a total of 21 countries over a 10-year period. The ratio of twin births to single births was 1:85.2. The ratio of triplet births to single births was 1:7628.7 or $1{:}(87.3)^2$. The ratio of quadruplet births to single births was 1:670,734 or $(1{:}87.5)^3$ (Greulich, 1930). Clearly the "law" is a mere approximation at best; not all data fit it as well as do these given as an example.

Hellin's law does not differentiate between the likelihood of the occurrence of the two types of twins. What proportion is dizygotic, what proportion monozygotic?

Proportion of Monozygotic and Dizygotic Twinning Weinberg (of Hardy-Weinberg fame) formulated a means of estimating the answer to this question. We consider first the frequencies of DZ twinning. For our purposes the approximation that the sex ratio at birth is 1:1 is sufficiently accurate. In other words, the chance for a male birth is ½, the chance for a female birth is ½. What are the chances that a pair of DZ twins will consist of two boys? The chance that the first-born twin will be a boy is ½, the chance that the second-born twin will be a boy is ½. Hence, the chance for a boy-boy pair is ½ $\times$ ½ $=$ ¼. In the same manner, the chance that both members of a DZ pair of twins will be girls is ½ × ½ = ¼. The total chance that the DZ pair will be of like sex (either boy-boy *or* girl-girl) is ¼ + ¼ = ½. Therefore, the other half of the DZ pairs must consist of unlike-sexed pairs (boy-girl pairs).*

According to calculation, then, one-half of the DZ twins consists of pairs differing in sex. Such pairs, of course, can be easily recognized. Suppose that we have 1000 pairs of twins of all kinds and that among them 335 are boy-girl pairs. How many of the 1000 pairs are monozygotic? According to our calculation, the 335 boy-girl pairs represent *half* of the *dizygotic* pairs. Thus the total number of DZ pairs must be twice 335 or 670. Then the number of *monozygotic* pairs must be 1000 — 670 = 330, or 33 percent, which is approximately the figure obtained from actual data (Myrianthopoulos, 1970). When more exact estimates are desired, the actual sex ratio in the population is used in the computation instead of the approximation that the chance of a male birth is one-half, and the chance of a female birth one-half.

Although twins are not uncommon, triplets are rare, and quadruplets and quintuplets are so unusual as to be always newsworthy. These multiple births of higher order may arise in a great variety of ways. They may all arise from one ovum, as the Dionne quintuplets are believed to have done; or each may arise from a separate ovum; or there may be a combination of the two types of origin. For example, Corner (1955) reported on four sets of quadruplets whose placentas and embryonic membranes had been studied carefully (see pp. 354–356). In one case, four ova were evidently involved. In a second case, three ova had participated, one giving rise to a MZ pair (in other words, the quadruplets consisted of one DZ pair and one MZ pair). In the third case, there were three girls, all of whom had developed within one amnion (an indication of monozygosity, p. 355) and one boy, who had a separate placenta and amnion; evidently only two ova were involved here. In the fourth case, all four quadruplets evidently arose from one ovum and were monozygotic. In other instances other combinations of monozygosity and the production of two or more ova are found.

Inheritance of Twinning

From our discussion so far, we can readily understand that monozygotic twinning and dizygotic twinning are two separate, and presumably

*This can be calculated directly. (a) Chance that first-born is a boy = ½; chance that second-born is a girl = ½; total chance for this combination = ½ × ½ = ¼. (b) Chance that first-born is a girl = ½; chance that second-born is a boy = ½; total chance for this combination = ½ × ½ = ¼. Total chance that one twin will be a boy, the other a girl = ¼ + ¼ = ½.

unrelated, phenomena. Hence we should not be surprised to find that they differ in many ways.

Repeat Frequency A woman who has had one pair of DZ twins is about four times as likely to have a second pair as are women in general (Bulmer, 1970). This is not true of a woman who has had one pair of MZ twins; her chances of having a second pair are about the same as they would have been if she had not had the first pair. Evidently some women have a tendency to DZ twinning. Evidence indicates that this tendency is probably hereditary. At any rate, a difference in repeat frequency is a respect in which DZ twinning and MZ twinning are not alike.

Incidence and the Age of the Mother Older mothers are more likely to have DZ twins than are younger mothers (Fig. 22.3). Note that the likelihood of DZ twinning rises with increasing age of the mother until about age 37. With further increase in age, the rate declines sharply, "probably a consequence of failing ovarian function as the menopause approaches" (Bulmer, 1970). By contrast, the frequency of MZ twinning increases only slightly with the age of the mother, although Myrianthopoulos (1970) found a sharp increase in mothers over 39 years old.

In dizygotic twinning two ovarian follicles mature simultaneously instead of the usual single ovulation per month. (A follicle is a tiny "bubble" containing liquid in which the ovum floats. This "bubble" gradu-

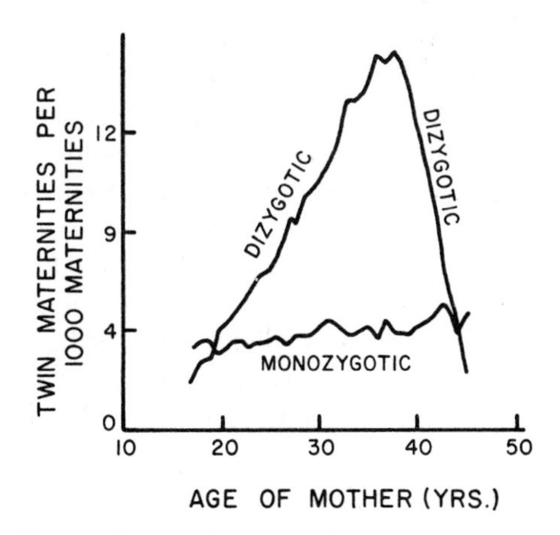

FIG. 22.3 Rates of monozygotic and dizygotic twinning related to the age of the mother. (Redrawn from Bulmer, M. G., 1959. "The effect of parental age, parity and duration of marriage on the twinning rate," *Annals of Human Genetics,* **23**:454–458.)

ally works its way to the surface of the ovary and ruptures, releasing its ovum near an opening into the oviduct or Fallopian tube). This process of ovulation is under the control of endocrine glands, the immediate stimulus being supplied by two so-called gonadotrophic hormones from the anterior lobe of the pituitary (the follicle-stimulating hormone and the luteinizing hormone). It is found that the amount of these gonadotrophic hormones increases as women grow older. This hormonal increase may explain why older women have more double ovulations (resulting in DZ twins) than do younger mothers (Bulmer, 1970). If, as evidence seems to suggest, tendency to produce DZ twins is inherited, the gene or genes concerned may well be operating through control of this hormonal mechanism.

Although DZ twinning and MZ twinning seem to be unrelated processes, the occurrence of both in the same woman sometimes makes us wonder (e.g., the quadruplets just discussed). Perhaps only coincidence is involved—two relatively rare events happening to occur simultaneously. Certainly we should need positive evidence before we concluded that something other than coincidence was involved. After all, we know of other rare conditions that happen to occur together in an occasional individual. Be that as it may, some women give evidence of having a remarkable tendency to multiple births (frequently of both types). For example, Davenport (1919) learned of a woman who was married three times and had multiple births by all three husbands. Her pregnancies totaled seven pairs of twins, five sets of triplets, and three sets of quadruplets. (Two sets of the triplets and two sets of the quadruplets terminated in miscarriages.)

Many pedigrees have been prepared showing the occurrence of several or many pairs of twins within a few generations. Here a word of caution is in order. As we have seen, twinning is not really a rare event. Consequently, even if there were no hereditary tendency involved, we should expect that by chance some pedigrees would include several sets of twins. We must be careful not to select pedigrees for their "curiosity value," that is, because they happen to include an unusual number of sets of twins. On the other hand, if no hereditary tendency were involved, the chance of occurrence of four successive generations like those shown in Figure 22.4 would be relatively slight. This pedigree is of particular interest in that four of the eight sets of twins are of unlike sex, thereby demonstrating that they are dizygotic. Note that I-1 had *only* twins in her family.

Incidence Among Relatives Evidence that tendency to DZ twinning has a hereditary component is afforded by study of the female relatives of mothers who have DZ twins. On the basis of extensive data

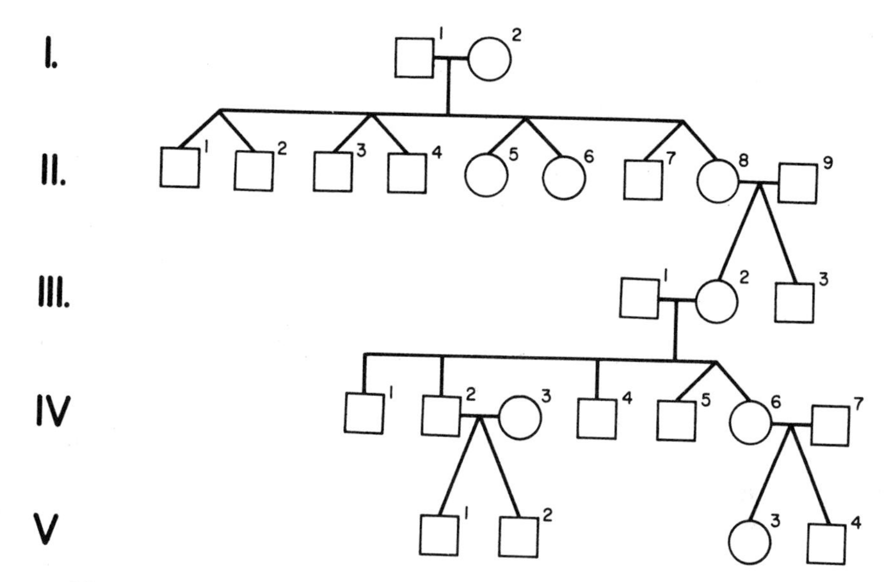

FIG. 22.4 Pedigree showing four successive generations of twinning: a total of eight pairs of twins. (Based on data in Taylor, C. E., 1931. "Four generations of heterosexual twins with prepartum amenorrhoea in two generations," *British Medical Journal,* **2**:384).

compiled by many workers, Bulmer (1970) concluded that "the dizygotic twinning rate among mothers and daughters of mothers of dizygotic twins is a little less than twice the rate in the general population, while the rate among sisters is increased about two and a half fold." Such an increase is not found among the relatives of mothers of MZ twins.

Some investigators have also reported an increase in rate among the female relatives of *fathers* of DZ twins, but the validity of such findings has been disputed (e.g., Bulmer, 1970). Certainly, it is difficult to imagine how a husband's genotype could cause double ovulation in his wife. If a man inherits genes for DZ twinning he can, of course, pass them on to his *daughters,* thereby increasing their tendency to produce twins. We recall (p. 220) that such genes are said to be **sex-limited** because, while both sexes may possess them, they find phenotypic expression in one sex only.

In summary: that MZ and DZ twinning are separate phenomena is indicated (a) by the fact that rate of DZ, but not MZ, twinning varies among different races, and (b) by the fact that incidence of DZ, but not MZ, twinning varies with the age of the mother. The probability that DZ, but not MZ, twinning has a hereditary basis is suggested by (a) the fact that a mother who has had one pair of DZ twins is more likely than women in general to have additional pairs, and (b) the fact of increased likelihood of having twins among the female relatives of mothers of DZ twins. Neither of these findings applies to MZ twinning.

In the present state of our knowledge, or lack of it, there is little to be gained by speculation on the nature and number of genes involved in tendency to DZ twinning. Bulmer (1970) proposed a hypothesis of monofactorial inheritance, mothers with a tendency to DZ twinning being homozygous for the recessive allele (i.e., *tt*), but he stated that not all findings can be explained on such a simple hypothesis. Other authors have suggested polygenic inheritance. For a phenomenon as complex as is tendency to DZ twinning, we are probably safest in suggesting that several or many genes are involved. Recalling earlier discussions, however, we note that our opinion on this point would change if someone were to demonstrate that tendency to double ovulation is based on action of a single enzyme.

For more complete discussions of multiple births, see Bulmer (1970), Nance (1959), and Scheinfeld (1967).

Problems

1. An investigation of the population of a certain city disclosed 500 pairs of twins, classifiable as follows: 166 boy-boy pairs, 170 girl-girl pairs, 164 boy-girl pairs. About how many of the pairs were probably monozygotic?
2. Judging from Figure 22.3, how much more likely are 30-year-old women to have DZ twins than they are to have MZ twins? 40-year-old women?
3. What would be the result if, in an XX zygote, the chromatids formed by one X chromosome failed to separate in mitosis, and then the two daughter cells formed a pair of twins? What would be the result if this happened in an XY zygote and it was the chromatids of the Y chromosome that underwent nondisjunction? Give the chromosomal constitutions of the twins in each case.
4. If in a certain population 111 of 10,000 births are twin births, how many of the births would Hellin's "law" lead us to expect to be triplet births?
5. If, as we stated (p. 347), the DZ twinning rate is about 8 per thousand among whites in general, what would we expect to be the rate among the daughters of mothers of DZ twins? Among the sisters of such mothers?

References

Bulmer, M. G., 1959. "The effect of parental age, parity and duration of marriage on the twinning rate," *Annals of Human Genetics,* **23**:454–458.

Bulmer, M. G., 1970. *The Biology of Twinning in Man.* Oxford: Clarendon Press.

Corner, G. W., 1955. "The observed embryology of human single-ovum twins and other multiple births," *American Journal of Obstetrics and Gynecology,* **70**:933–951.

Davenport, C. B., 1919. "A strain producing multiple births," *Journal of Heredity,* **10**:382–384.
Greulich, W. W., 1930. "The incidence of human multiple births," *American Naturalist,* **64**:142–153.
Guttmacher, A. F., 1953. "The incidence of multiple births in man and some of the other unipara," *Obstetrics and Gynecology,* **2**:22–35.
Myrianthopoulos, N. C., 1970. "A survey of twins in the population of a prospective collaborative study," *Acta Geneticae Medicae et Gemellologiae,* **19**:15–23.
Nance, W. E., 1959. "Twins: An introduction of gemellology," *Medicine,* **38**:403–414.
Patterson, J. T., 1913. "Polyembryonic development in *Tatusia novemcincta,*" *Journal of Morphology,* **24**:559–662.
Scheinfeld, A., 1967. *Twins and Supertwins.* Philadelphia: J. B. Lippincott.
Taylor, C. E., 1931. "Four generations of heterosexual twins with prepartum amenorrhoea in two generations," *British Medical Journal,* **2**:384.
Turpin, R., 1970. "Monozygotisme hétérocaryote," *Acta Geneticae Medicae et Gemellologiae,* **19**:188–198.
Turpin, R., J. Lejeune, J. La Fourcade, P. L. Chigot, and C. Salmon, 1961. "Présomption de monozygotisme en dépit d'un dimorphisme sexuel: sujet masculin XY et sujet neutre Haplo X," *Comptes Rendus de l'Académie des Sciences, Paris,* **252**:2945–2946.

23

The Twin Method in Human Genetics

Determination of Zygosity

Before twins can be used in studies of the role of heredity, accurate determination must be made as to whether each pair is monozygotic or dizygotic. In a sense, everything hinges on accuracy in this matter.

Twin pairs consisting of one boy and one girl are automatically identified as being dizygotic. In most studies this is the least interesting group because hereditary differences may be obscured by differences between the sexes. For example, if one were studying the hereditary component in stature, there would be little profit in comparing the difference in height between members of a pair of MZ boy twins with the difference in height between members of a DZ boy-girl pair, simply because in this latter case one twin is a boy, the other a girl.

For this reason, we are most interested in like-sexed pairs of twins. The problem then becomes: Among like-sexed pairs, how can we determine which ones are monozygotic, which ones dizygotic?

The Embryonic Membranes

Human embryos are enclosed within two membranes, an inner **amnion** and an outer **chorion.** Figure 23.1 shows a pair of twin embryos about 5 weeks of age. Note that in this case each embryo is enclosed within

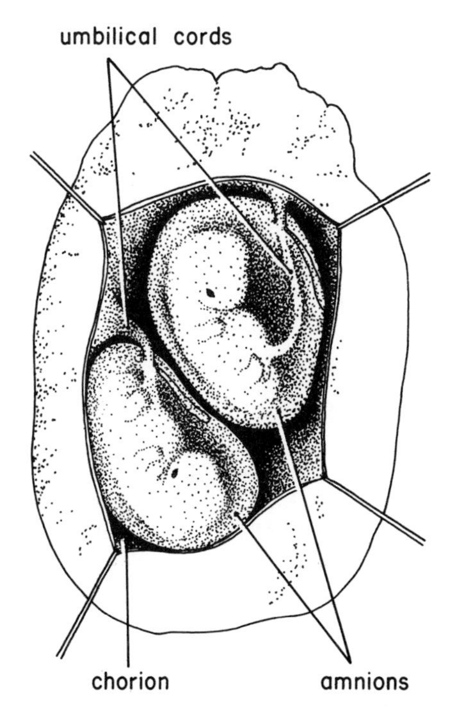

FIG. 23.1 Twin embryos of about 5 weeks of age. Each has its own amnion, but both are enclosed within one chorion. (Redrawn from Potter, J. C., 1927. "Human monochorial embryos in separate amnions." *Anatomical Record,* **34**:255, Fig. 2.)

its own amnion, but that the two amnions are enclosed within one chorion. Note, too, that the umbilical cords connected to the embryos run in different directions; this suggests that the placentas (p. 203) are separate. Sometimes the placentas may be so close together that they join to form one.

The situation shown in Figure 23.1 is typical of *monozygotic* twins. Usually MZ twins have separate amnions but a common chorion. Occasionally, MZ twins may even share a common amnion. This difference is presumably related to a difference in the time at which separation of the embryonic mass into two parts occurred (pp. 344–345). If separation occurred late, only after the chorion and amnion were both formed, then both twins would lie within one amnion. As far as determination of zygosity is concerned, this rule seems warranted: *If the twins are enclosed within one chorion* (and in some cases amnion), *they are monozygotic.* There is speculation about the possibility of two originally separate chorions merging into one, but if this ever occurs, it is so rare that it introduces little error into the general rule.

Dizygotic twins, on the other hand, have separate amnions and chorions. Figure 23.2 shows separate placentas, as well, but sometimes the placentas may lie close together and merge to form one. Placentas are of little aid in determining zygosity. Moreover, we cannot conclude that because twins have separate chorions they are dizygotic. About 33 percent

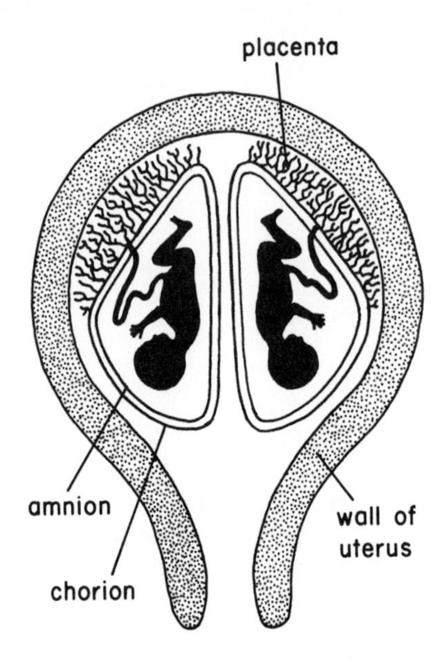

FIG. 23.2 Twin fetuses in separate chorions. (Redrawn from Potter, E. L., 1948. *Fundamentals of Human Reproduction.* Copyright © 1948, McGraw-Hill Book Co., Inc. Used by permission of McGraw-Hill Book Company.)

of MZ twin pairs have separate chorions. In these cases separation of the embryonic mass into two parts presumably occurred very early, perhaps as early as the two-celled stage (Nance, 1959).

Thus, in some cases, accurate investigation of the embryonic membranes at the time of birth will aid in determining zygosity, and in some cases it will not. If it can be determined definitely that there is only one chorion, then the twins are almost certainly monozygotic. However, if there are two chorions, the evidence is of little value. Furthermore, in many cases accurate study of the membranes was not made at the time of birth. As a result of all these factors, the zygosity of only a minority of an investigator's pairs of twins is determinable from data on the embryonic membranes.

The Similarity Method

For the most part, decisions concerning zygosity are based on an extension of the method employed by people generally in deciding whether a given pair of twins is identical or not. We ordinarily base our judgment on how much the twins look alike. Geneticists employ the same principle except that they utilize many traits besides similarity in physical appearance. Blood groups are particularly useful because their genetic basis is simple and well established.

The investigation by Osborne and De George (1959) may be

taken as a typical example of methods of diagnosing zygosity. In their investigation, the blood of the twins was tested for the following antigens: A_1, A_2, B, O; M, N, S; C, D, E, c of the Rh series; Kell, Duffy, and P. The mode of inheritance of most of these was discussed in earlier chapters. However, we did not mention Kell, Duffy, and P. Like the others, each of these is an antigen whose presence or absence from a red blood cell is demonstrable by using suitable antiserums. The presence or absence of each is dependent on a single pair of genes.

If the members of a pair of twins did not agree with regard to all 14 of these antigens, the pair was considered dizygotic. If the members of a pair of twins *did* agree with respect to all 14 antigens, did this fact prove that the twins were monozygotic? Unfortunately not. Osborne and De George found that five pairs of *unlike-sexed* twins agreed with respect to all these antigens. These five pairs were surely dizygotic, yet they had the same 14 antigens. How could this be? Evidently it reflected the genetic situation in the family. We can readily understand, for example, that if both parents belonged to blood group O, all the children would too, whether or not they were twins. If one parent were homozygous for blood group B and the other parent belonged either to B or O, all the children would belong to group B. The same principle applies to the other antigens. In general we can say then that although disagreement proves dizygosity, agreement does not prove monozygosity. The greater the number of traits employed, the greater will be the probability that dizygotic twins will differ in at least one of them.

What, then, did Osborne and De George do at this point? They added to the blood tests other traits having a fairly well-established mode of inheritance. The traits they found most useful were the following: hair color, eye color, eye detail pattern, ability or inability to roll the tongue lengthwise, ability or inability to taste PTC (phenylthiocarbamide), ear lobe form, chin form, and the presence or absence of hair on the middle phalanx of the fingers. Utilization of these traits increased the chance of detecting dizygosity. It is still possible, of course, that some dizygotic twins would agree on all these traits as well as on all blood tests, but the chance is relatively small. The error introduced would have the effect of causing investigators to include in their monozygotic group an occasional pair of DZ twins who were very similar to each other. This would in fact constitute a very small source of error.

The criteria employed by Osborne and De George have been cited as typical. Other investigators, employing the same principle, include a somewhat different list of traits, although emphasis is always placed on antigens. Sometimes investigators make a point of determining how frequently other members of the family, and friends, mistake the twins for each other. Indeed, investigators find that basing judgment on similarity of appearance usually results in the same identifications of MZ pairs as do the more elaborate methods we have been discussing (Gottesman and Shields, 1972).

In summary, we note that the similarity method of zygosity diagnosis cannot *prove* monozygosity, but it can make the conclusion that a given pair is monozygotic highly probable, and for practical purposes that is sufficient. As usual where probability is involved, mathematical expression of the probability is possible and useful. Interested readers will find concise treatments of the mathematics in Steinberg (1962) and Bulmer (1970).

Utilization of the Twin Method

General Principles When members of a pair of twins agree with regard to a certain trait, we say that they are **concordant;** when they disagree we say that they are **discordant.** Using these terms, we may make the following general proposition: *If a genetic component is concerned in producing a certain trait, concordance in MZ pairs should exceed that in DZ pairs.* The greater the genetic component, the greater the difference between MZ and DZ pairs is likely to be.

In most cases this "genetic component" will consist of chromosomes and genes inherited from ancestors. But at times it may include newly arisen mutations and chromosomal aberrations. For example, if one member of an MZ pair has Down's syndrome, the co-twin is highly likely to have it too (concordance about 89 percent). If the pair arose from an ovum that had three No. 21 chromosomes, both twins would be expected to have the three. (Rare exceptions could arise if nondisjunction occurred when a daughter cell destined to form one twin divided into two. One of the two might receive three 21's, the other cell only one. This second cell, being monosomic for No. 21, might well be inviable, leaving its trisomic partner to form the embryo. Or, alternatively, the monosomic cell might give rise to a minor cell line, the individual becoming a mosaic; Fig. 23.3).

By contrast, if one member of a DZ pair has Down's syndrome, the co-twin is highly *unlikely* to have it (concordance about 7 percent). This is because the two ova concerned are unlikely both to have been the result of nondisjunction, a rare event in itself, although occasionally they may both have been.

We mentioned earlier (pp. 321–323) that the tendency to 21-trisomy is not highly hereditary. Thus high concordance in MZ pairs does not *necessarily* indicate that a trait is highly hereditary, contrary to assumptions sometimes made. Because, however, gene mutations and chromosomal aberrations are rare events, most of the "genetic component" shared by MZ twins *has* been inherited from ancestors, rather than being

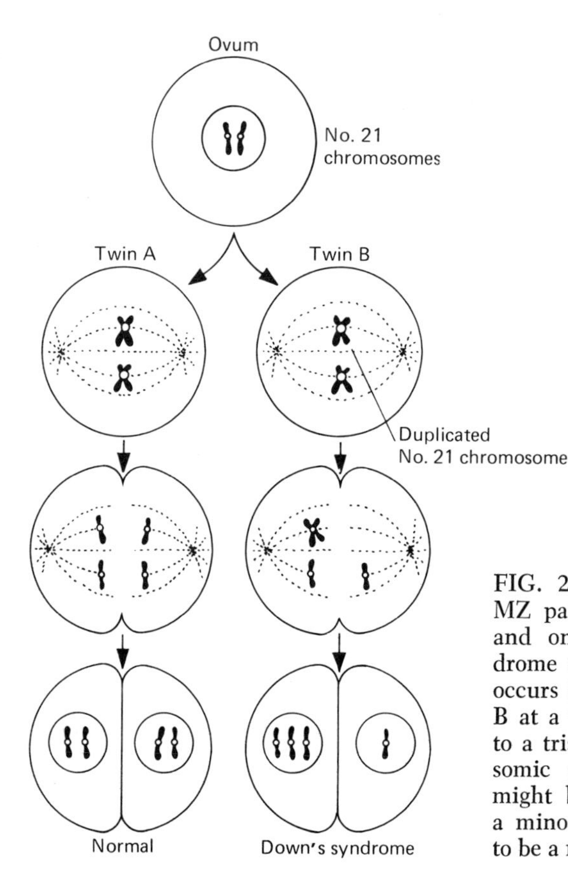

FIG. 23.3 One way in which an MZ pair having one normal twin and one twin with Down's syndrome could arise. Nondisjunction occurs in mitosis of a cell in twin B at a very early stage, giving rise to a trisomic cell line and a monosomic one. The monosomic line might be inviable, or might form a minor cell line, causing twin B to be a mosaic.

newly arisen. Consequently, in most cases we are safe in concluding that high MZ concordance reflects a hereditary basis for the trait being studied. The validity of the twin method in human genetics really rests on that assumption. Nevertheless, conclusions as to degree of heritability of a trait drawn from twin studies should, if possible, always be augmented by other methods of detecting heritability.

In the preceding chapter we discussed an MZ pair in which one twin was a normal boy, the co-twin being a girl with Turner's syndrome (XY/XO twins). We mentioned that such occurrences—and we might cite other examples—indicate that MZ twins do not always have identical genotypes, despite the fact that both came from one ovum. On balance, however, MZ twins average to be more alike genetically than are DZ twins. This difference between the two types of twins is sufficient to make the twin method a valuable tool.

High MZ and DZ Concordance We are interested, then in the extent to which these differences in genotypic similarity are reflected in differences in the phenotypes produced. Let us suppose in the first place that MZ and DZ twins are very similar with regard to some trait. For example, if one twin develops measles the other twin usually does, too, whether the twins are monozygotic or dizygotic (concordance about 95 and 87 percent, respectively). Thus we can conclude that the greater intrapair similarity of genotype of MZ pairs has little effect on the likelihood of having measles. The only clue that a slight genetic component may be involved in resistance to measles lies in the fact that concordance is slightly lower in DZ twins than it is in MZ twins. If the difference between 95 percent and 87 percent is statistically significant, we may suspect that a difference in genetic constitution may influence resistance to the disease. This would not seem to be improbable, of course, but we would still conclude that the principal factor is environmental: a matter of whether or not the measles virus enters one's environment.

Schizophrenia: Investigation of Gottesman and Shields

Instead of discussing the twin method in general terms, we select as an example what is probably the most thorough and carefully controlled twin study ever made (Gottesman and Shields, 1972).

Schizophrenia is the most common type of mental disorder or illness (as distinguished from mental retardation). Populations vary, but somewhere in the neighborhood of 1 percent of people are affected with the disorder. It is the form of mental illness most likely to occur in young people. Affected individuals experience a splitting of the personality, a tendency to withdraw from reality, emotional deterioration, and often hallucinations and delusions.

There has been prolonged, and sometimes bitter, disagreement as to whether liability to develop schizophrenia has a hereditary basis, or whether everyone in the population is equally likely to develop the symptoms if subjected to the necessary psychological, especially emotional, stresses and strains. In attempting to answer the question, many investigators have employed the twin method. For reasons mentioned above, we concentrate attention on the investigation by Gottesman and Shields.

Ascertainment In earlier chapters we have discussed the importance of obtaining an unbiased sample of the population. In twin studies this means that we must avoid selecting for study an atypical proportion of twin pairs that are concordant. If we were to rely on a search of the medical literature, for example, we should be likely to accumulate an undue number of reports on MZ pairs that were concordant. Less frequently would a physician or psychiatrist think it important to write a paper about an MZ pair only one member of which developed schizophrenia.

So the method of selection must be objective and without regard to concordance. Taking successive admissions to a psychiatric hospital as they chanced to come was the method employed by Gottesman and Shields. Their probands (propositi) were inpatients and outpatients consecutively admitted over a 16-year period (a) who were diagnosed as schizophrenic and (b) who were found to have a twin of the same sex (this latter provision was to avoid possible sex differences in liability to become schizophrenic). For various reasons, some potential probands ascertained in this manner could not be used, for example, the individual or his co-twin would not cooperate in the study; the co-twin died so young that the investigators had no way of knowing whether or not he would have developed the symptoms had he lived longer. The important point is that selection of the final probands had no reference to concordance. The number of MZ pairs finally included in the analysis was 22, the number of DZ pairs 33. Zygosity was determined in much the manner we have already discussed.

"Blind Judging" One of the strong points of this investigation was the fact that the investigators themselves did not make the diagnosis as to who was schizophrenic and who was not. If an investigator knows that an individual is the co-twin of a schizophrenic patient, that knowledge may well influence his judgment as to whether or not the co-twin is also schizophrenic, try as he may to be completely objective in his diagnosis. So the solution lies in having the diagnosis made by an expert who does not know anything about the twin of the individual he is diagnosing. Gottesman and Shields prepared detailed case histories on all of their twins and submitted the histories to a panel of six experts experienced in diagnosing mental disorders. The case histories, which included results of some psychological tests, did not indicate who was the twin of whom. Each judge's diagnosis was classified as: S, for schizophrenia; ?S, for uncertain schizophrenia; O, for other psychiatric abnormality; N, for normal.

The judges were unanimous in some of their diagnoses, but, understandably in a trait so complex and varied in its manifestations, there were differences of opinion in some cases. "In practice, all twins with three or more S votes and none with less than three were classified as S. Twins with an accumulation of at least three S or ?S votes were called ?S. All with four or more N votes were consensus N, and the remaining 18 twins were consensus O."

Using these criteria, the investigators found it necessary to omit two individuals, members of MZ pairs, who had originally been considered probands by reason of having been classified as schizophrenic by the hospital.

Pairwise Concordance Of the 22 pairs of MZ twins, in 11 pairs both twins were S or ?S by the criteria just mentioned, giving a concordance rate of 50 percent. In three of the 33 pairs of DZ twins, both twins were at least probably schizophrenic, a concordance rate of 9 percent.

If ?S individuals were omitted, the MZ concordance became 40 percent, the DZ concordance 10 percent.

On the average, probands in whom the schizophrenic illness was severe were more likely to have schizophrenic co-twins than were probands with less severe symptoms. This may in part explain why the MZ concordance of 50 percent is lower than the concordance reported by early investigators, who concentrated on severely affected, hospitalized patients.

Probandwise Concordance In four of the 22 MZ pairs, both twins were probands. This means that they had been independently ascertained because both had been admitted as schizophrenic patients during the 16-year period included in the study. This was also true of one DZ pair. Hence the number of probands in MZ pairs was 26, the number in DZ pairs was 34. We now ask: What proportion of the probands had an affected co-twin? The answer will give us the probandwise concordance. Of the 26 probands from MZ pairs, 15 had an affected co-twin, a concordance of 58 percent. Of the 34 probands from DZ pairs, 4 had an affected co-twin, a concordance of 12 percent.

The probandwise concordance rate is a more useful statistic than is the more traditional pairwise rate (see Bulmer, 1970). For one thing, it can be compared directly with rates of occurrence of the trait in populations as a whole. Gottesman and Shields presented the results of such concordance rates obtained in several recent investigations (Table 23.1).

TABLE 23.1 Probandwise concordance rates in recent twin studies of functional psychoses.

Investigation	MZ		Same-sex DZ	
Finland (1971)	7/20	35%	3/23	13%
Norway (1967)	31/69	45%	14/96	15%
Denmark (1969)	14/25	56%	12/45	26%
Gottesman and Shields (1972)	15/26	58%	4/34	12%
United States (1972)	52/121	43%	12/131	9%

From Gottesman, I. I., and J. Shields, 1972. *Schizophrenia and Genetics. A Twin Study Vantage Point.* New York: Academic Press.

They concluded: "The four European samples, in aggregate, yielded a probandwise rate approaching 50% in MZ and 17% in DZ co-twins." Recalling that about 1 percent of the population is schizophrenic, we see the basis for the statement: "Identical twins of schizophrenics have a risk about 50 times greater and fraternal twins one about 17 times greater than a member of the general population comparable in age." (Note that the comparable figures from the study in the United States are somewhat lower, especially for the DZ twins).

"A Necessary But Not Sufficient Cause" In line with the general proposition stated previously, we conclude that the greater concordance in MZ pairs than in DZ pairs indicates that there is a genetic basis for liability to become schizophrenic. It is noteworthy that in many cases concordance occurred even though proband and co-twin had little contact with each other during later life. A few pairs had even been reared in separate homes (Shields, 1962). But we also conclude that the genetic basis by itself is not sufficient to cause the illness. After all, the statement that about 50 percent of the probands in the studies just cited had co-twins who were also schizophrenic implies that in the other 50 percent of cases the co-twins did *not* develop the symptoms even though their genotypes were presumably identical with those of the schizophrenic probands. Clearly, environmental factors found in the stresses and strains of life determine in large measure whether or not a person with hereditary liability to become schizophrenic does in fact become so. (The question of whether a person without the hereditary liability may nevertheless become schizophrenic if given sufficient environmental stress lies outside our present discussion; see Kaplan, 1972).

Other Lines of Evidence We stated earlier that conclusions about heritability of a trait drawn from twin studies should be augmented by other methods. In the case of schizophrenia, we find an elevated incidence of the condition among the relatives of schizophrenics, the increased incidence being proportional to closeness of relationship to the probands (propositi). Such studies have been criticized, however, on the ground that there is much shared *environment* involved. This is especially true of close relatives, such as parents, sibs, and children of probands. This criticism has been countered by conducting studies of individuals who were *adopted* as children, and thus not raised in an environment shared by their biological relatives.

Detailed discussion of these studies would take us too far afield; we shall indicate the nature of some of them and the general findings. In one study, children of schizophrenic mothers were found to have an increased incidence of schizophrenia even though they had been separated from their mothers as early as the first 2 weeks after birth (Heston and Denney, 1968).

A different approach was to study schizophrenics who had been adopted or raised in foster homes as children and to compare the incidence of schizophrenia in their biological and adoptive relatives. An elevated incidence was found among the biological relatives but not among the adoptive relatives (Kety et al., 1968). From the standpoint of genetics, it is noteworthy that the increased incidence was found in half-sibs who shared a common *father* with the propositus. In these cases there could be no question of a possible effect of having developed within the same uterus.

Mode of Inheritance? While there is wide agreement that liability to become schizophrenic has a hereditary basis, there is no agreement as to the nature of that basis. By itself, the twin method does not indicate mode of inheritance. Hypotheses have been many and varied. Two among them have, probably, the most adherents.

One hypothesis postulates that "schizophrenia at least in the great majority of cases, is based on a single partially dominant gene with low penetrance" (Huxley et al., 1964). Such a gene would be dominant in the sense that an individual need have only one to show the trait, but because the gene has low penetrance (p. 70), many or even most people who have the gene would not show the trait.

The other leading hypothesis is that liability to develop schizophrenia is polygenic. This hypothesis, favored by Gottesman and Shields (1972), postulates an unknown number of genes working together to produce the trait. Almost certainly, they would interact in more complex ways than the simple additive action we used as an example in our original discussion of polygenes. In a trait as complex and varied as schizophrenia and its allied schizoid psychoses, action of many genes seems highly probable. Among them, one might be more prominent than the others, the arrangement amounting to one "main gene" with "modifying genes" (Chapter 8). Such a "main gene" would probably be almost indistinguishable from the gene postulated in the first hypothesis we mentioned.

This is not the place to enter into the arguments in support of the various hypotheses. Suffice it to say that the best support for the single "main" gene hypothesis would come from discovery of some unitary trait, "either biological or behavioral (psychometric pattern), which will not only discriminate schizophrenics from other psychotics, but will also be found in all identical co-twins of schizophrenics whether concordant or discordant" for schizophrenia (Gottesman and Shields, 1972). Best of all, in view of what we know of the way in which genes work, would be discovery of a single enzymatic defect underlying liability to become schizophrenic. We recall such discoveries in the cases of phenylketonuria and the Lesch-Nyhan syndrome. The defective enzyme might result in an abnormal product affecting the body as a whole, or acting on a particular area of the brain (Friedhoff, 1972; see also Kety, 1959). We may expect that future investigations of a biochemical nature will contribute much to our understanding of schizophrenia and its hereditary basis.

The literature on the genetics of schizophrenia is voluminous. Interested readers are referred especially to Gottesman and Shields (1972), Rosenthal and Kety (1968), and Kaplan (1972).

Possible Effects of Twinning on the Twins

Twin studies sometimes include comparisons of twins with their brothers and sisters who were born singly (singletons). Such comparisons may be useful, but it is well to bear in mind that twinning in itself may have consequences. This is reflected in the fact that the death rate of twins both before and shortly after birth is greater than is that of singletons. Also, on the average, twins are born about 3 weeks prematurely, and have a lighter birth weight—especially if they share the same chorion—than singletons have. After reviewing the evidence, Bulmer (1970) concluded that this retarded fetal growth is caused by insufficiency in the blood supply to the uterus, resulting in undernutrition and perhaps insufficient oxygen supply. When the birth weight is low, especially if it is much lower than 5 pounds, there is increased chance of mental retardation, and increased frequency of mental retardation is indeed characteristic of twins.

In this connection we might mention that while all the children of a certain mother develop in the same uterus, twins do so simultaneously. Thus they share whatever effects the state of the mother's health at the time, her hormonal balance, her use of drugs, and the like, may have upon fetuses developing in her uterus. At other times in her life, and so for other pregnancies, these conditions may differ.

In a large proportion of cases in which twins share the same chorion and also share a single placenta, there is some connection between the two fetal circulations. If this takes the form of an artery of one twin connecting to a vein of the other twin, the first twin may bleed into the second one. This has been called the "transfusion syndrome," and it results in one twin having too little blood, and so being anemic. In severe cases this may result in brain damage, or even death. The other twin has too much blood, a condition that is also detrimental to normal development. If both twins survive, one is frequently heavier than the other at birth.

Blood Chimeras While the fetal circulations of twins in the same chorion are frequently connected, such connection rarely occurs when twins have separate chorions, even when the twins have a fused placenta (Fig. 23.4). When connection of this kind does occur, however, hormones, and even hematopoietic tissues (blood-forming cells) may be exchanged. Such an exchange was first discovered in cattle, where joining of twin circulations is common. Lillie (1916) discovered that if one fetus is male, the other female, the sex hormones of the male enter the female fetus with harmful effect, usually causing her to be sterile (a "freemartin"). That hematopoietic tissue is also exchanged was demonstrated by the finding that DZ twin calves each have two kinds of red blood cells (Owen, 1945). One kind consists of cells having the antigens for which that twin has the genotype. The other kind is composed of cells for which the co-twin has the genotype.

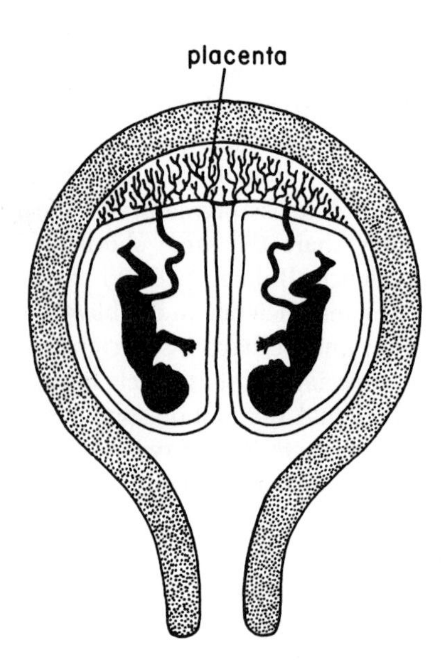

FIG. 23.4 Twin fetuses with separate amnions and chorions but with a single placenta. (Redrawn from Potter, E. L., 1948. *Fundamentals of Human Reproduction*. Copyright © 1948, McGraw-Hill Book Co., Inc. Used by permission of McGraw-Hill Book Company.)

Such blood chimeras have been found in man, although they are rare. For example, in one case of unlike-sexed twins (Booth et al., 1957), 86 percent of the brother's cells were of group A_1, 14 percent of group O. He was a secretor, secreting antigens A_1 and H in his saliva (pp. 192–193). His sister had 99 percent of group O cells and 1 percent of group A_1 cells. She secreted only antigen H in her saliva. Evidently, then, she was genetically of group O, while her brother was genetically of group A_1.

Further evidence of chimerism in this case was afforded by the white blood cells. Some of these cells in the brother showed "drumsticks" on their nuclei, a characteristic of leucocytes of *females* (Fig. 17.7, p.

280). This suggests that he had received ancestral white blood cells from his co-twin.

In this case, also, skin grafts were transferred from one twin to the other and survived successfully. Evidently the embryonic exchange of blood-forming tissues had included creation of an antigenic similarity of the kind necessary for successful transplantation. Usually such successful transplantation is possible only between MZ twins, and, as we have seen, it is sometimes used to distinguish MZ from DZ twins. Because of the rarity of fusion of fetal circulations in man when the twins have separate chorions, Bulmer (1970) concluded that the success of skin grafts can only very rarely be explained as the result of fusion. Thus, the test remains generally useful for distinguishing the two types of twins.

Incidentally, no human case is known in which the female co-twin of a male was rendered sterile by the association; that is, there are no human "freemartins."

Effect on Concordance and Discordance Some of the aspects of twinning we have mentioned may cause DZ twins to be more similar than their differing genotypes would cause them to be. Insofar as this occurs, it has the effect of increasing the amount of concordance between members of DZ pairs, and hence of decreasing the difference in concordance between MZ and DZ pairs. As a result, heredity might seem to be less important than it actually is. But the error introduced in this way is doubtless small. We have mentioned the fact that MZ twins do not always have exactly the same genotype. But even when they do, there are factors in development that may cause them to differ. We have mentioned the "transfusion syndrome," which may produce various effects, including brain damage, intrapair differences in birth weight, or even death of one twin. Even in the absence of this syndrome, however, the twins may differ in the adequacy of their contacts with the mother's blood in the placenta. This may result in differences in available nutrition and oxygen supply, and so affect the respective embryonic and fetal developments of the twins. Some of the resultant differences between the twins disappear in time, but some have lasting effects (see Price, 1950, for further discussion).

Interesting light on these matters was furnished by an investigation on intraset differences in newborn, MZ, armadillo quadruplets (Storrs and Williams, 1968). Twenty traits were investigated, including weights of the body as a whole, of adrenals, brain, heart, kidneys, liver, and spleen, as well as various hormonal and biochemical levels. The investigators found substantial differences within the quadruplet sets. No doubt these differences arose during the separate embryonic developments of the four, and, as the authors suggested, they may be of a type to cause important differences in such characteristics as strength, speed, intelligence, fertility, and so on.

So we see that there are factors tending to cause DZ twins to be more alike than they "should be," and causing MZ twins to be less alike than they "should be." In comparisons of concordance of MZ pairs to that of DZ pairs, both tendencies would cause heredity to seem to be less important than it actually is. Nevertheless, the twin method of analysis has made substantial contributions to human genetics.

Problems

1. A husband and wife have the following genotypes relative to ABO blood groups, MN blood types, Rh, the secretor trait (pp. 193–195), and the tasting of PTC (the gene for tasting, *T*, is dominant to the gene for non-tasting, *t*).

   $$I^A\ I^O\ L^N\ L^N\ R\ r\ Se\ Se\ T\ t \times I^B\ I^O\ L^M\ L^M\ R\ r\ se\ se\ T\ t$$

   They have a pair of male twins both of whom belong to group O, type MN, are Rh-positive, are secretors and nontasters. Does this agreement form strong evidence that the twins are monozygotic? In answering, compute the chances that if one member of a *dizygotic* pair of twins born to these parents is a male having the traits listed, the other member of the dizygotic pair will be a male having them also.
2. With regard to trait No. 1, concordance of MZ twins is 90 percent, and concordance of DZ twins is 80 percent. With regard to trait No. 2, concordance of MZ twins is 90 percent, and concordance of DZ twins is 30 percent. With regard to which trait would genetic factors seem to be the more important?
3. A husband belongs to blood group O, his wife to group B. They have a pair of unlike-sexed twins. Most of the red blood cells of the girl react with anti-B antibodies, but a small percentage of the cells fail to do so. Most of the red blood cells of the boy do not react with anti-B, although some of them do. What is the probable cause of this situation? If the girl eventually marries a man of group O, and the boy marries a woman of group O, what types of offspring may each expect to have?
4. If Gottesman and Shields had been able to study 150 schizophrenic MZ twins, instead of 26, how many of the co-twins would one expect to have been schizophrenic (probandwise concordance)? If they had been able to study 150 schizophrenic DZ twins, how many of the co-twins would one expect to have been schizophrenic?
5. In the results of Gottesman and Shields (our Table 23.1), the MZ concordance rate is how many times greater than the DZ concordance rate? Compute this also for the data from the United States. How well do the two sets of data agree in this respect?
6. It has been found that in 53 percent of MZ twin pairs, if one twin has tuberculosis the other twin does too, and that in 22 percent of DZ twin pairs, if one twin has tuberculosis the other twin does too. (Both members of all pairs were reared together.) What does this difference in percentage indicate?

References

Booth, P. B., G. Plaut, J. D. James, E. W. Ikin, P. Moores, R. Sanger, and R. R. Race, 1957. "Blood chimerism in a pair of twins," *British Medical Journal,* **1**:1456–1458.

Bulmer, M. G., 1970. *The Biology of Twinning in Man.* Oxford: Clarendon Press.

Friedhoff, A. J., 1972. "Biochemical aspects of schizophrenia." In Kaplan (1972), pp. 339–350.

Gottesman, I. I., and J. Shields, 1972. *Schizophrenia and Genetics. A Twin Study Vantage Point.* New York: Academic Press.

Heston, L. L., and D. Denney, 1968. "Interactions between early life experience and biological factors in schizophrenia." In Rosenthal and Kety (1968), pp. 363–376.

Huxley, J., E. Mayr, H. Osmond, and A. Hoffer, 1964. "Schizophrenia as a genetic morphism," *Nature,* **204**:220–221.

Kaplan, A. R. (ed.), 1972. *Genetic Factors in "Schizophrenia."* Springfield, Ill.: Charles C Thomas.

Kety, S. S., 1959. "Biochemical theories of schizophrenia," *Science,* **129**: 1528–1532; 1590–1596.

Kety, S. S., D. Rosenthal, P. H. Wender, and F. Schulsinger, 1968. "The types and prevalence of mental illness in the biological and adoptive families of adopted schizophrenics." In Rosenthal and Kety (1968), pp. 345–362.

Lillie, F. R., 1916. "The theory of the free-martin," *Science,* **43**:611–613.

Nance, W. E., 1959. "Twins: An introduction to gemellology," *Medicine,* **38**:403–414.

Osborne, R. H., and F. V. De George, 1959. *Genetic Basis of Morphological Variation.* Cambridge, Mass.: Harvard University Press.

Owen, R. D., 1945. "Immunogenetic consequences of vascular anastomoses between bovine twins," *Science,* **102**:400–401.

Potter, E. L., 1948. *Fundamentals of Human Reproduction.* New York: McGraw-Hill Book Company.

Potter, J. C., 1927. "Human monochorial twin embryos in separate amnions." *Anatomical Record,* **34**:255.

Price, B., 1950. "Primary biases in twin studies," *American Journal of Human Genetics,* **2**:293–352.

Rosenthal, D., and S. S. Kety (eds.), 1968. *The Transmission of Schizophrenia.* New York: Pergamon Press.

Shields, J., 1962. *Monozygotic Twins.* London: Oxford University Press.

Steinberg, A. G., 1962. "Population genetics: Special cases," In W. J. Burdette (ed.). *Methodology in Human Genetics.* San Francisco: Holden-Day, Inc. Pp. 76–91.

Storrs, E. E., and R. J. Williams, 1968. "Study of monozygous quadruplet armadillos in relation to mammalian inheritance," *Science,* **160**:443.

24

The Twin Method: Quantitative Traits

Thus far in our discussion of the twin method, we have dealt with alternative traits—either/or situations. One either develops a certain disorder or one does not. But many of the investigations using the twin method have been concerned with traits involving differences in the amount of something: height, weight, intelligence, and so on. Here we do not speak of concordance but of intrapair difference or variance: the amount by which members of a pair differ from each other.

The general principle here is: *If genetic components are involved in the production of quantitative traits, the intrapair difference between MZ twins should be less than the intrapair difference between DZ twins* (or, more concisely, the intrapair variance in MZ twins should be less than it is in DZ twins).

Total Finger Ridge Count

In Chapter 9 we discussed finger and palm prints (pp. 128–133). We mentioned that the total number of ridges one possesses on all ten fingers is a highly hereditary trait, dependent upon polygenes that act in a simple additive manner.

We mentioned twins in this connection. If, on the average, both twins of a pair had exactly the same total number of ridges, we could express the fact by saying that the coefficient of correlation is 1. Analysis of 80 pairs of MZ twins yielded a correlation of 0.95 ± 0.01 (Holt,

1968). Obviously this is very close to 1, indicating that the virtually identical genotypes of the twins are of primary importance in determining the number of ridges. Such intrapair difference as does exist is probably attributable to influences upon the fetus encountered during the first 4 months of fetal life, because by the end of that time the finger print patterns have developed.

The coefficient of correlation for 92 DZ pairs was found to be 0.49 ± 0.08 (Holt, 1968). This is an almost perfect fit to the theoretical expectation of 0.5 for pairs of sibs who are not twins (recall that each child receives the same number of polygenes from each parent).

In this connection we should mention that there has long been interest in the use of finger prints as a means of *diagnosing* zygosity, a problem discussed in the preceding chapter. Gottesman and Shields (1972), for example, included ridge counts in the data used to diagnose zygosity in 21 of their pairs of twins. Occasionally members of a DZ pair, so identified by other means, are found to have very similar ridge counts, similar enough to suggest monozygosity. So the trait does not establish the zygosity of a given pair beyond doubt. Yet it is useful when one is computing the mathematical probability that a certain pair is monozygotic.

PTC Tasting

In Chapter 9 we discussed the genetic basis for differences in sensitivity to tasting the bitter thioureas (pp. 138–141). In the present connection, there is interest in an investigation of taste thresholds using the twin method (Kaplan et al., 1967). The investigators studied 75 MZ pairs and 70 DZ pairs. They found significant difference in intrapair variance in taste threshold between MZ pairs and same-sex DZ pairs. (They used the related compound PROP: propylthiouracil.) There was no significant difference between DZ pairs and nontwin sibling pairs.

Stature

Stature was one of the first human traits to interest students of human genetics. Statistical studies of it were made by Pearson and others in the nineteenth century, before the importance of the gene theory of inheritance, based on the work of Mendel, was realized. Subsequently, data were accumulated which indicated that, insofar as differences in height have a hereditary basis, inheritance is polygenic.

The twin method has been employed to analyze the importance of inheritance in determining how tall one becomes. Of the various investigations, we select that of Osborne and De George (1959), whose method of determining zygosity was described in the preceding chapter (p. 357). In this investigation all twins were over 18 years of age. The twins were voluntary participants, selected in various ways that had nothing to do with the traits being studied.

The results of the study were expressed in terms of "mean intrapair variance," a statistic indicating the average amount of difference between co-twins. This variance was calculated by the formula: $\Sigma \chi^2/2n$. In this formula the capital sigma indicates summation; "n is the number of twin pairs, and χ the difference between the two members of a twin pair for a given measurement."

TABLE 24.1 Mean intrapair variances in stature between like-sexed pairs of twins.

	Number of twin pairs	Variance
Male MZ pairs	25	1.604
Male DZ pairs	10	7.581
Female MZ pairs	34	1.387
Female DZ pairs	27	18.329

Reprinted by permission of the publishers from R. H. Osborne and F. V. De George, *Genetic Basis of Morphological Variation.* Cambridge, Mass.: Harvard University Press, Copyright © 1959 by The Commonwealth Fund.

Results of the study of stature are summarized in Table 24.1, without including the statistical tabulations which demonstrated that differences are significant. We note immediately that the intrapair difference is much greater for DZ pairs than it is for MZ (almost 5 times as great for male twins, and 14 times as great for female twins). From this we may conclude that there is a strong genetic component in stature. We may find surprising the fact that males differ from females in the difference between MZ pairs and DZ pairs. The small number of DZ male pairs may have contributed to this, but the authors concluded that it was due largely to the somewhat amusing circumstance that women were more cooperative than men about having their measurements taken. Thus, when the same individual was given the same measurement at different times, the results of the two measurings were more nearly the same if the subject was a woman than they were if the subject was a man.

Table 24.1 includes only like-sexed pairs. As we should expect, the variance in unlike-sexed pairs (114.789) was much greater than the variance in either MZ or DZ like-sexed pairs.

As we saw in connection with studies on finger-print ridge counts, one way of expressing degree of similarity is by means of the coefficient of correlation. The correlation between the height of DZ twins is about the same as it is for nontwin siblings, about 0.45. But the correlation between MZ twins is about 0.95 (Bulmer, 1970). This difference indicates a large genetic component in variability in height.

Weight

Another trait studied by Osborne and De George was body weight. Table 24.2 presents the intrapair variances. We note that they are much

TABLE 24.2 Mean intrapair variances in weight between like-sexed pairs of twins.

	Number of twin pairs	Variance
Male MZ pairs	25	62.400
Male DZ pairs	10	65.450
Female MZ pairs	34	38.515
Female DZ pairs	27	66.722

Reprinted by permission of the publishers from R. H. Osborne and F. V. De George, *Genetic Basis of Morphological Variation*. Cambridge, Mass.: Harvard University Press, Copyright © 1959 by The Commonwealth Fund.

greater than are those for stature. This indicates in general that an MZ twin differs from his co-twin in weight more than he does in stature. We also see that male MZ twins are closely similar to male DZ twins in variance. Even in the case of the female twins, the difference in variance between MZ and DZ pairs is not statistically significant. Accordingly, the authors drew the conclusion "that after the end of the growth period, weight, in essentially healthy people, is predominantly under environmental influence." (Roughly, this is a way of saying that it depends in large part on how much we eat and how much physical work we do.)

Earlier studies with twins had indicated a genetic component in weight. However, many of the twins in these studies were children. In the investigation by Newman, Freeman, and Holzinger (1937), for example, mean intrapair weight difference was 10.0 pounds for like-sexed DZ twins, 4.1 pounds for MZ pairs, but the average age of the subjects was 13 years. Accordingly, Osborne and De George concluded that "in the twin studies of growing children, the hereditary influence measured by intrapair weight differences is principally associated with growth rate." As we all know, children have periods when they grow rapidly, other periods when they grow slowly. If, as may well be, such rates are under genetic control, we might expect MZ twins to resemble each other more in pattern of growth than do DZ twins. If so, measurements taken at any given time would show less intrapair difference between MZ twins than between DZ twins, on the average.

Interested readers will find in the book of Osborne and De George (1959) similar data concerning inheritance of other body measurements (e.g., arm length, trunk length, and chest circumference). The results on stature and weight have sufficiently illustrated the principles of the twin method applied to such traits.

Intelligence

Nature versus Nurture Do "normal people" differ genetically with respect to intelligence? We cannot enter into an extensive discussion of this large and much-debated question, which has been argued from every conceivable point of view. One extreme view is that everyone comes into the world with the *same* intellectual endowment and that differences in intellectual attainment depend solely on differences in environment (including differences in training and education). Probably few people would subscribe to so bold a statement as this; after all, we do seem to sense that there was some difference between Einstein's mind and ours that was not entirely dependent on a difference in upbringing and education. Yet many people conclude that the effects of environmental differences are more important than genetic differences in determining intelligence. In no other field has the "nature-nurture" question been so hotly debated. What evidence is obtainable by application of the twin method?

Intelligence Tests We must note at the outset one real handicap in attempting to answer this question. Aside from the fact that intelligence is difficult to define, there is no exact and objective means of measuring the native intellectual endowment with which a person is born. There are accurate scales for measuring weight, but nothing comparable for measuring intelligence. Tremendous ingenuity has entered into the devising of intelligence tests, and serious attempts have been made to devise "culture-free" tests. Even the best of them, however, are only partially successful in separating what is innate from what is acquired. Most tests employ language; this immediately introduces the effects of different environments and education. Nonverbal tests attempt to avoid the difficulty, but one can go only so far in measuring intelligence without the use of words. What kind of test can measure with reasonable objectivity and completeness the native intelligence of both a college graduate and a person with no schooling? Or that of a child and that of an adult? Surprisingly, some tests are partially successful in meeting such demands, and the best of them are the best means we have available. In the discussion that follows, intelligence is of necessity defined as "that which intelligence tests measure." We hope that there is reasonable correlation between what intelligence tests measure and what we may think of as native intellectual endowment: "the brains one is born with."

Intrapair Variance The underlying theory in applying the twin method to the study of intelligence is the same as that applied to the study of stature and weight. If there is a hereditary component in intelligence, intrapair variance in MZ twins should be less than it is in DZ twins. Many studies have indicated that this is in fact the case. A review of the extensive literature would be out of place here, but we offer as a typical example the results obtained by Newman et al. (1937). These investigators gave the Stanford-Binet test to 50 MZ pairs and 52 like-sexed DZ pairs. The average intrapair difference for the MZ pairs was 5.9 IQ points; for the DZ pairs it was 9.9 points. Woodworth (1941) applied a correction to compensate for the fact that "retest of the same person within a week usually shows some shift of score up or down, averaging about 5 points of IQ." This correction reduced the MZ difference to 3.1 points, the DZ difference to 8.5 points. Thus, the DZ pairs differed about 2.7 times as much as did the MZ pairs (Woodworth, 1941). If we may assume that the environment of DZ pairs raised together is not significantly more different than the environment of MZ pairs raised together, then the increased intrapair variance of DZ twins may be ascribed to the fact that the co-twins differ genetically.

The assumption just mentioned has been challenged on the ground that parents frequently treat "identical" twins differently than they do "fraternal" twins, even when the latter are of the same sex. Parents may wish either to accentuate the similarity (by dressing the MZ twins alike, for example) or to deemphasize the similarity so that each member of the pair may become an independent, self-sufficient individual. When such differential treatment of MZ and DZ twins occurs, will it affect the degree of similarity the twins will show in intelligence test scores?

Coefficient of Correlation Applying the correction mentioned above to the data for Newman et al., Woodworth (1941) found that the correlation coefficient for MZ pairs was 0.93, whereas that for DZ pairs was 0.66. If there is a genetic component in intelligence, we should expect the correlation of DZ pairs to be fairly high because the co-twins share the same parents. (Only unrelated pairs of persons in differing environments would be expected to show correlation coefficients approaching 0; see Figure 24.1).

Many subsequent studies have yielded similar results. Bulmer (1970) summarized the composite findings of Burt (1966) and earlier studies cited by Burt (Table 24.3). We note that the correlation for MZ twins reared together is 0.88, that for DZ twins 0.55.

TABLE 24.3 Typical correlations between relatives for intelligence.

Type of relationship	Correlation coefficient
Monozygotic twins reared together	0.88
Monozygotic twins reared apart	0.78
Dizygotic twins reared together	0.55
Sibs reared together	0.55
Sibs reared apart	0.45
Unrelated children reared together	0.25
Parents–child (parent tested as child)	0.56
Husband–wife	0.50

From Bulmer, M. G., 1970. *The Biology of Twinning in Man.* Oxford: Clarendon Press. Modified from Burt, 1966.

"Dissecting" Intelligence Although the evidence seems conclusive that there is a genetic, as well as an environmental, component in determining one's performance on intelligence tests, we have little basis for drawing conclusions as to the mode of inheritance. It is safe to conclude that many genes must be involved in such a complex trait. Actually, intelligence is not a single trait at all; it is comprised of many component parts. "It is, for instance, rather common to find someone who is highly gifted verbally but who is poor in numerical ability or in the ability to understand mechanical principles and vice versa" (Vandenberg, 1962). Investigators have attempted to break down intelligence into these component parts and to identify the ones that seem to be most strongly gene-controlled. Thus Blewett (1954) employed Thurstone's Primary Mental Abilities Test (PMA) in testing 52 pairs of like-sexed twins aged 12 to 15 years; half were boy-boy pairs, half were girl-girl; half were monozygotic pairs, half were dizygotic. He found the usual difference in correlation coefficients in the PMA test as a *whole,* but observed that this difference was not maintained within the various parts of the test. Hence, he concluded that there is a larger hereditary component in verbal comprehension and word fluency than there is in such abilities as numbers and spatial factors.

Vandenberg (1962), also employing the PMA test, obtained results indicative of a hereditary component in verbal completion, addition of numbers, and word fluency. A great variety of tests was given to

82 pairs of like-sexed twins of high school age (45 MZ and 47 DZ). Aside from the PMA tests, there were various cognitive and achievement tests, tests for motor skills, tests of perceptual skills, sensory tests and measures of musical ability and interest, and tests concerning personality. As might be expected, evidence of a genetic component was obtained in some instances but not in others.

Investigators in the Psychometric Laboratory of the University of North Carolina studied 142 pairs of twins, with an average age of 14 years (Thurstone, Thurstone, and Strandskov, 1953). They employed 34 tests, which yielded 53 test scores. They found that large intrapair differences were more frequent in DZ pairs than they were in MZ pairs. They found that tests in which there were significant differences between DZ and MZ twins involved visual, verbal, and motor functions.

Studies on Adults Most of the investigations we have described utilized children and young people in school. It is relatively easy to ascertain twins through school administrators and teachers, and to secure cooperation in the testing program. Schoolchildren are used to being tested, whereas many adults find it an annoyance and an interference with the day's work. Nevertheless, there is interest in older twins. It has been suggested that differences between young DZ twins may disappear with the passage of years. According to this view, genetic elements may be important in determining intellectual development during childhood, but cease to be important in later life. Jarvik (1962) gave a battery of tests to 134 like-sexed pairs of twins 60 years of age and older. Two years later most of them were given the same tests a second time. A smaller number survived to take the tests six years later, and 17 individuals were available to take the tests a fourth time, after a two-and-a-half-year lapse. The initial testing showed that "the scores of one-egg twin partners were more similar than those of two-egg twins even after the age of 60, indicating the persistence during adult life of gene-specific differences in mental functioning." Subsequent tests demonstrated no significant change in intrapair correlations.

Monozygotic Twins Reared Apart

As attempts to determine the relative contributions to intelligence of heredity and environment, the investigations we have discussed so far have the shortcoming that members of a pair shared the same home and school environment as children. Furthermore, as we intimated, to some extent it is likely that members of an MZ pair share a more similar environment than do members of a DZ pair.

Faced with this problem, a geneticist working with plants or lower animals would devise two types of experiment: (a) experiments in which genetic constitution is held constant, but the environment varied, and (b) experiments in which the genetic constitution is varied, but the environment is held constant. Workers in human genetics are not free to conduct such experiments, of course, but they can take advantage of a situation that approximates the first type: cases in which MZ twins were separated in early life and reared in different homes.

Such situations only approximate the ideal experiment, because members of a pair shared the same intra-uterine environment, and in some cases were not separated until they were several years old, whereas an investigator would like his subjects to be separated at birth. Also in many cases the homes in which the twins were reared were those of relatives, or of people of the same socioeconomic group, and the schooling of the separated children was not very different. Nevertheless, significant results have been obtained by this type of study.

The Classic Investigation Nineteen pairs of separated twins were studied by Newman, Freeman, and Holzinger (1937), and a twentieth was added by Gardner and Newman (1940). The team of investigators brought to the study the differing backgrounds of zoology, psychology, and statistics.

In 11 of these 20 pairs, separation had occurred not later than 1 year of age; 8 pairs were separated between their first and second birthdays; one pair was 3 years of age, and another 6 years old when separation occurred. Despite the fact that they were being reared apart, and in many cases without knowledge of the co-twin's existence, the twins continued to be remarkably alike in many psychological traits.

The average difference in IQ was 7.5 points, which we may compare with the 5.9 points of difference shown by the 50 pairs of MZ twins reared together (p. 375). If the corrections recommended by Woodworth (1941) are applied, these figures become 6 points and 3 points, respectively. Evidently being reared apart did make a little difference in average attainment on intelligence tests, but not enough difference to mask the underlying similarity determined by the identical inheritance.

The distribution of differences is of interest. Eight of the 20 pairs differed by 4 points or less (4 pairs by only 1 point). At the other extreme, only 4 pairs differed by 15 points or more (the greatest difference was 24 points). A study of sibling pairs (not twins) reared apart showed 17 percent of the pairs differing by *more* than 24 points; the average difference was 15.5 points.

Attainment on intelligence tests is known to be affected by education. Newman and his colleagues therefore established criteria for estimating the similarity of educational opportunities enjoyed by the twins. As we might expect, the investigators found a high correlation between IQ and the education received, the coefficient being 0.791. In general, the greater the difference in educational opportunity experienced by the separated twins the greater the difference in IQ. Thus, in the pair showing the 24 points of difference, one twin had a college degree and was a teacher, whereas her sister's education consisted of only three years of grade school. (The data correlating educational opportunity with IQ are nicely summarized for the original 19 pairs in Table 15 of Osborn, 1951. The twentieth pair was studied when the twins were in the same college; their IQ difference was 3 points [Gardner and Newman, 1940]).

The conclusion reached was that monozygotic twins have an inherited tendency to similarity of response in testing. Small environmental differences cause little or no deviation in this inherited similarity, but larger differences may produce detectable effects. Newman concluded that in only 4 of the 20 pairs "were there any very striking differences, exceeding those of some identical twins reared together. The remaining sixteen cases show no greater mean differences than did the fifty pairs of identical twins reared together. It might be concluded then that in these sixteen cases, although we know of many differences in their environments, these differences were not sufficiently large to reach the threshold of effectiveness" (Gardner and Newman, 1940).

These statistics only dimly reveal a wealth of interesting information concerning the 20 pairs of twins. This information is presented in the form of case histories in the references cited and also in Newman (1940). As samples we shall summarize two cases briefly.

Two Case Histories Gladys and Helen were the given names of the twins who differed by 24 IQ points. They were separated at 18 months of age and did not meet again until they were 28 years old. Helen was adopted twice, the second time by a childless farmer and his wife who were determined to give her every advantage. She graduated from a good college and taught school for 12 years. Gladys was adopted by a Canadian railroad conductor. Her schooling was interrupted in the third grade when the family moved to an isolated part of the Canadian Rockies so that her foster father could recuperate from tuberculosis. Although the family later returned to Ontario, Gladys did not resume her schooling. She worked in a knitting mill and then as a salesgirl in various stores. Finally, she got a job in a printing establishment and worked up to a position as assistant to the proprietor. Both Helen and Gladys married; Helen had one child, Gladys two.

Newman's comment concerning Helen was that "she had acquired considerable social polish, showed much facility in social intercourse and possessed a good deal of feminine charm." Concerning Gladys he wrote: "She had always been a hard worker and had not acquired any of the social ease and polish possessed by her sister. She was a business woman without airs or coquetry" (Newman, 1940).

Because the subject under discussion is intelligence, we have said nothing of the personality tests also given the 20 pairs of separated twins. As we might expect, Gladys and Helen scored very differently in these tests, the differences no doubt arising from the great differences in their lives. In the matter of intelligence, Helen had an IQ score of 116 on the Stanford-Binet test, whereas Gladys scored 92. Thus Helen was "high normal" and Gladys was "low normal." Thirteen years' difference in the amount of schooling showed its effect in ability to cope with an intelligence test although innate ability was probably the same for both.

We choose the second pair of twins because of the curious similarities of their lives. Such similarities are often encountered by students of twins—too often, it would seem, to be matters of mere coincidence.

Edwin and Fred were separated at about one year of age and were reared as only children in their respective homes. When they were small they attended the same school, noticed how much they looked alike but did not suspect that they were brothers. Later their families moved to different cities, and the twins did not meet again until they were 25 years old. Newman stated: "The oddest feature about these twins is that, although neither one knew of the other's existence, they lived strangely parallel lives. They were both reared as only children of childless foster parents, both under the impression that they were own children. They had about the same amount of education, were both interested in electricity and both became expert repair men in different branches of the same large telephone company. They were married the same year to young women of about the same age and type. Each had a baby son and each owned a fox terrier dog named Trixie. Believe it or not!"

The twins differed slightly in the amount of schooling, Edwin having had one year of high school, Fred three years. On the Stanford-Binet test Edwin scored 91 and Fred 90. In other tests, including those of personality, the twins showed themselves to be strikingly similar. Newman commented: "From this case one might be led to infer that the identical heredity and closely similar environments had interacted to produce almost identical results" (Newman, 1940).

Later Investigations Shields (1962) studied 44 pairs of separated twins, most of whom had been parted in the first six months of life. Juel-Nielson (1965) investigated 12 pairs; all were separated in early childhood, the age varying from the day of birth to almost six years. The investigation of Burt (1966) included 53 pairs reared apart, all separated at birth or during their first six months.

Jensen (1970) combined the data from these three investigations with the data from the study by Newman et al. (1937). This yielded a total of 122 pairs of separated twins for whom scores on intelligence tests were available. He found that the IQ's of these 244 people showed a normal distribution similar to that of people in general, although the mean was 96.82 rather than the usual 100. This was to be expected in view of previous evidence that the average IQ of MZ twins is a little lower than that of the general population.

Because the IQ scores of these 244 people were "quite typical and representative of the distribution of intelligence in the general population," Jensen analyzed the combined data statistically as though they had all arisen from one investigation. He found that the mean absolute intrapair difference was 6.6 IQ points, the largest difference being that between Gladys and Helen (see above). We note how closely this corresponds to the finding that the average difference was 6 points in the study by Newman et al. (p. 378).

Jensen found the correlation coefficient to be 0.824, slightly higher than the 0.78 calculated by Bulmer (1970) using most of the same data (Table 24.3). But both figures are of the same magnitude and indicate that separation has relatively little effect on IQ scores of MZ twins. (Incidentally, such separation was found to have much greater effect on scholastic attainment.) Hence studies of MZ twins reared apart provides additional evidence of an important genetic component in intelligence.

We have concentrated solely on the parts of these four investigations dealing with intelligence tests. Actually, the investigators studied much more than that (e.g., physical traits, personality traits, scholastic attainment, normal and abnormal psychological traits), and the original papers should be consulted by readers interested in these matters.

Heredity Varied, Environment Constant We mentioned earlier (p. 378) that an experimental geneticist faced with the "nature-nurture" problem would wish to design two types of experiments. We saw that one of these is approximated by study of MZ twins reared apart. The second is approximated when unrelated children are reared in the same home or institution. To what extent will the common environment cause increased similarity of attainment on intelligence tests? In degree of intelligence, will foster children bear closer resemblance to their biological or to their foster parents? Obtaining unequivocal answers is difficult for many reasons (see Woodworth, 1941). Without detailing any of the investigations, which are not really germane to a discussion of the twin method, we note from Figure 24.1 that rearing together does raise the correlation coefficient slightly as compared to that of unrelated persons reared apart. But the correlation in intelligence test scores remains low despite being reared together (Table 24.3). Here again we see evidence of the importance of genetic differences.

A Composite Picture

Erlenmeyer-Kimling and Jarvik (1963) summarized evidence for genetic contribution to intelligence in Figure 24.1. The chart summarizes data from 52 studies, carried on in eight countries on four continents over a period of 50 years. Each dark circle on the chart represents a correlation coefficient obtained from these studies. Over two-thirds of the coefficients relate to IQ scores, the remainder to special tests, such as the Primary Mental Abilities Test mentioned earlier. Despite the diversity and heterogeneity of the tests included, the general trend is obvious: the greater the degree of relationship, the more similar the individuals are in intelligence. Note the low foster-parent-to-child correlation and the larger own-parent-to-child correlation (the median here being almost exactly 0.50, as would be expected on genetical grounds). Note the similarity of siblings and of DZ twins in median and in range of coefficients, and, as previously, the difference in degree of similarity between DZ twins

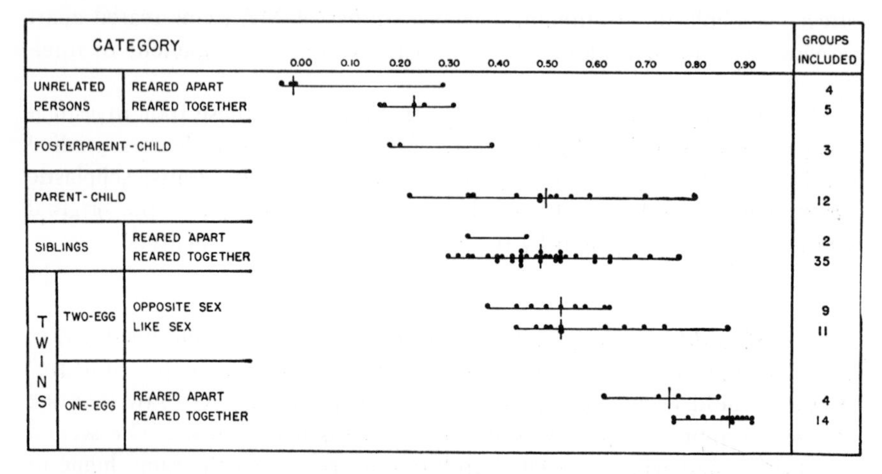

FIG. 24.1 Correlation coefficients for intelligence test scores from 52 studies. Over two-thirds of the coefficients were derived from IQ's, the remainder from special tests (e.g., Primary Mental Abilities). Mid-parent-child correlation was used when available, otherwise mother-child correlation. Correlation coefficients obtained in each study are indicated by dark circles; medians are shown by vertical lines intersecting the horizontal lines, which represent the ranges. (From Erlenmeyer-Kimling, L., and L. F. Jarvik, 1964. "Genetics and intelligence: A review," *Science,* **142**:1477–1479. Copyright © 1964 by the American Association for the Advancement of Science.)

and MZ twins. The survey "shows that mean intrapair differences on tests of mental abilities for dizygotic twins generally are between 1½ to 2 times as great as those between monozygotic twins reared together." Finally, we note again the great degree of similarity of members of MZ pairs, and the fact that rearing separately does, in many but not all cases, make some difference in attainment on tests.

Problem

1. In their study of morphological variation, Osborne and De George obtained the following data on within-pair variance in male twins. The numbers represent mean variance in each case.
   (1) upper arm length: MZ = 0.316; DZ = 0.846
   (2) upper arm circumference: MZ = 1.120; DZ = 1.528
   What conclusions are warranted concerning relative genetic components in these two traits?

References

Blewett, D. B., 1954. "An experimental study of the inheritance of intelligence," *Journal of Mental Science,* **100**:922–933.

Bulmer, M. G., 1970. *The Biology of Twinning in Man.* Oxford: Clarendon Press.

Burt, C., 1966. "The genetic determination of differences in intelligence: A study of monzygotic twins reared together and apart," *British Journal of Psychology,* **57**:137–153.

Erlenmeyer-Kimling, L., and L. F. Jarvik, 1963. "Genetics and intelligence: A review," *Science,* **142**:1477–1479.

Gardner, I. C., and H. H. Newman, 1940. "Mental and physical traits of identical twins reared apart. Case XX. Twins Lois and Louise," *Journal of Heredity,* **40**:119–126.

Gottesman, I. I., and J. Shields, 1972. *Schizophrenia and Genetics. A Twin Study Vantage Point.* New York: Academic Press.

Holt, S. B., 1968. *The Genetics of Dermal Ridges.* Springfield, Ill.: Charles C Thomas.

Jarvik, L. F., 1962. "Biological differences in intellectual functioning," *Vita humana,* **5**:195–203.

Jensen, A. R., 1970. "IQ's of identical twins reared apart," *Behavior Genetics,* **1**:133–148.

Juel-Nielsen, N., 1965. "Individual and environment: A psychiatric-psychological investigation of monozygotic twins reared apart," *Acta Psychiatrica Scandinavica,* Supplement 183.

Kaplan, A. R., R. Fischer, A. Karras, F. Griffin, W. Powell, R. W. Marsters, and E. V. Glanville, 1967. "Taste thresholds in twins and siblings," *Acta Geneticae et Gemellologiae,* **16**:229–243.
Newman, H. H., 1940. *Multiple Human Births.* New York: Doubleday, Doran & Company, Inc.
Newman, H. H., F. N. Freeman, and K. J. Holzinger, 1937. *Twins: A Study of Heredity and Environment.* Chicago: University of Chicago Press.
Osborn, F., 1951. *Preface to Eugenics,* rev. ed. New York: Harper & Brothers.
Osborne, R. H., and F. V. De George, 1959. *Genetic Basis of Morphological Variation.* Cambridge, Mass.: Harvard University Press.
Shields, J., 1962. *Monozygotic Twins.* London: Oxford University Press.
Thurstone, T. G., L. L. Thurstone, and H. H. Strandskov, 1953. "A psychological study of twins. 1. Distributions of absolute twin differences for identical and fraternal twins," *Research Report No. 4, Psychometric Laboratory, University of North Carolina,* pp. 1–9.
Vandenberg, S. G., 1962. "The Hereditary Abilities Study: Hereditary components in a psychological test battery," *American Journal of Human Genetics,* **14**:220–237.
Woodworth, R. S., 1941. "Heredity and environment," *Bulletin No. 47, Social Science Research Council, New York.*

25

Eugenics, and the Nature of Selection

The eugenics movement was started by Sir Francis Galton in the latter part of the nineteenth century. A cousin of the more widely known Charles Darwin, Galton was a pioneer in the study of human genetics, specializing in statistical studies of such quantitative traits as stature. As we noted in Chapter 8, these traits are now explained as the action of polygenes, but Galton, a contemporary of Mendel, yet unaware of his work, did not think in terms of genes.

Galton coined the name **eugenics**, which may be dissected into "eu-genics" and translated as "true-born" or "well-born." Galton defined the term as follows: "Eugenics is the study of agencies under social control that may improve or impair the racial [i.e., hereditary] qualities of future generations, either physically or mentally" (Popenoe and Johnson, 1933). It is worthy of note that Galton referred to eugenics as a "study," and, indeed, investigation into human genetics has always been a primary factor in the movement. Following such studies, however, it is the idealistic hope of eugenists to *do* something about improving the human race. Man has worked wonders in improving cultivated plants and domestic animals. Could he not apply to his own improvement some of the methods that have been so successful with these lower forms?

Plant and animal breeders rely heavily on the complementary practices of hybridization and selection. Hybridization implies control of mating. This is not practicable in human relationships. True, there have been communities in which marriages have been arranged by a leader or by a committee of elders; this was the case in the early years of the Oneida

Community in New York State. However, the transitoriness of such practices testifies to their unacceptability to most people in Western cultures.

Might not the other procedure, that of selection, be applied with profit to man himself? Mankind presents great heterogeneity, even without controlled hybridization. Might not selection, therefore, be employed for human improvement? That was the eugenic dream.

Positive Eugenics Traditionally, eugenics was divided into the aspects, *positive eugenics* and *negative eugenics*. Without going into detail, we may simply state that the goal of positive eugenics was to encourage people having superior genetic traits to have more children. The goal of negative eugenics was to discourage the production of children on the part of people having inferior genetic traits.

Positive eugenics as a separate entity always was a bit vague and uncertain. *Who* should be urged to have large families? What qualities should one select as the ones to be encouraged? Because most eugenists were intellectuals, intelligence was strongly emphasized. However, life is not intellect alone. Who should decide? And how could "desirable" parents be encouraged to have more children than they naturally were inclined to have? (We shall ignore the perversion of eugenics by Hitler and his followers. That was not part of the mainstream of eugenics.)

Negative Eugenics In point of fact, most of the emphasis was placed on negative eugenics. Actually, positive and negative eugenics are simply two sides of the same coin, for if one discourages reproduction among people having undesirable genetic traits, then *ipso facto* one is encouraging reproduction among people *lacking* these undesirable traits. In a sense, absence of undesirable traits is equivalent to presence of desirable traits.

Can we agree on what traits are undesirable? To some extent we are in the same dilemma as before, but only partly so, for there are some traits that are generally agreed to be undesirable. The eugenics movement concentrated heavily on insanity and feeblemindedness (to use the older and more picturesque names for mental illness and mental deficiency or retardation).

Hence an important aspect of the eugenic dream was this hope: If we can prevent, or reduce, reproduction on the part of people who are insane or feebleminded, can we not succeed in reducing greatly the number of such individuals in future generations? Clearly the ideal is a worthy one; from the standpoints of society as a whole and of individuals with their personal and family problems it *would* be a boon to have fewer people suffering from these sad defects.

The "Old" Eugenics In the early days of the eugenics movement, two methods of preventing reproduction were emphasized. The first, termed **institutionalization,** simply meant keeping the sexes separate in institutions. This is necessary for people who are unable to be self-supporting or to care for themselves. But it is expensive for society, and it is also undesirable for the people themselves if they are capable of living independently, at least to some degree.

For mentally ill or retarded persons who were not institutionalized, **eugenic sterilization** was proposed as a means of preventing reproduction. In those days sterilization was much less familiar to the general public than it is today, when it is gaining increased use as a means of birth control. During the 30 years beginning with 1907, many states in the United States passed laws providing for the voluntary or compulsory sterilization of certain types of defectives.

Whatever the methods employed, the key question remains, can the proportion of people suffering from mental deficiency and mental illness in future generations be reduced substantially by preventing individuals of these types from reproducing? Earlier in the present century, this seemed to be a reasonable hope. What lessened that hope? To a considerable extent, hopes were dimmed when the principles of population genetics became generally understood. We turn now to consideration of what selection can and cannot be expected to accomplish. We refer readers to Haller (1963) for more details of the history of the eugenics movement, and to Osborn (1951) for a summary of the principles of eugenics.

Total Negative Selection

Gene Dominant Total negative selection means that possessors of a certain gene, or genotype, produce no offspring. Let us consider, first, total negative selection against a dominant gene, *A;* that is, possessors of gene *A* produce no offspring. Let us suppose that a population in one generation has the following constitution: 1 percent *AA,* 18 percent *Aa,* 81 percent *aa.* If the 19 percent who have an *A* gene either as homozygotes or as heterozygotes do not produce any offspring at all, the entire generation of offspring will have the genotype *aa,* having inherited solely from the *aa* parents. This statement is subject to one qualification: We may expect some mutation to occur—an occasional *a* gene mutating to its dominant allele, *A.* In general, we may conclude that *total negative selection against a dominant gene will, in one generation, reduce the frequency of that gene to the low level maintained by recurrent mutation.*

In the case of a dominant gene, negative selection is very effective. When dominant genes are involved, cannot the eugenic dream be realized? Unfortunately, many human abnormalities are not determined by dominant genes. There is one notable exception: **Huntington's chorea** (Huntington's disease), which, as noted earlier (pp. 66–69), gives evidence of simple dominant inheritance. In this case, it is safe to say that if every person who has the gene for Huntington's chorea did not produce children, Huntington's chorea would virtually disappear in one generation. However, the matter is not as simple as it sounds. Many potential parents who have the gene do not know it, because they do not develop choreic symptoms until after they have married and produced children. In the future, it may be possible through tests to identify which child in a family having a choreic parent will later become choreic himself. If all these children were prevented from having children when they reached reproductive age, Huntington's chorea could be eliminated (except for that originating from a fresh mutation).

Even then it would be necessary to bear several things in mind. In the first place, practically every person who has Huntington's chorea is heterozygous, having had only one choreic parent. Thus, on the average, half of the children of choreic people are normal. If we prevent a choreic person from reproducing, we are preventing the birth of normal offspring as well as of choreic ones. Would this be right? There is a difficult value judgment here—and only one of several. Would a program that prevented the birth of both normal and abnormal children be worthwhile, or just? Other questions also arise, but we shall defer consideration of them until after our discussion of selection against recessive genes.

We should note here that although our subject is *total* negative selection, selection need not be total to contribute to the realization of eugenic ideals. This is not an all-or-none situation. Every time an individual with the gene for Huntington's chorea fails to pass on the gene to offspring, the proportion of choreic individuals in the next generation is decreased. Such a decrease, even though it is fractional rather than total, contributes to the genetic improvement of mankind, and hence is eugenic.

In sum, we see that prevention of reproduction by all people who possess a trait determined by a dominant gene would quickly reduce the frequency of the gene to the level maintained by recurrent mutation. What would be the result if people who had a phenotype determined by a *recessive* gene did not reproduce?

Gene Recessive Let us consider a gene that is completely recessive; the only people showing the phenotype are the ones homozygous for the gene.

Let us suppose that we have a population with the following constitution: $\frac{1}{4}$ *AA*, $\frac{2}{4}$ *Aa*, $\frac{1}{4}$ *aa*. As noted previously (pp. 145–149), in such a population p (frequency of the dominant allele) is $\frac{1}{2}$, and q (frequency of the recessive allele) is $\frac{1}{2}$. If there is no selection, the proportions as stated will remain indefinitely. However, let us suppose that there is selection, of such nature that none of the *aa* individuals produce offspring. What will be the nature of the next generation?

First, we must calculate the changed values of p and q. Let us call them p_1 and q_1 to distinguish them from the p and q of the first (original) generation. Now only the *AA* and *Aa* individuals reproduce, and we note that there are twice as many *Aa*'s as there are *AA*'s. Therefore, the effective breeding population is: ⅓ *AA*, ⅔ *Aa*. In this effective breeding population, the frequency, p_1, of the dominant gene is comprised of the entire contribution of the *AA* parents plus half the contribution of the *Aa* parents. q_1 is comprised entirely of the other half of the contribution of the *Aa* parents.

Thus:

$$p_1 = 1/3 + 1/2(2/3) = 1/3 + 1/3 = 2/3$$
$$q_1 = 1/2(2/3) = 1/3$$

Then:

$$p_1^2 + 2p_1q_1 + q_1^2$$
$$(2/3)^2 + 2(2/3)(1/3) + (1/3)^2$$
$$4/9\,AA + 4/9\,Aa + 1/9\,aa$$

Thus, one generation of total negative selection against *aa* individuals has reduced the frequency of such individuals from ¼ to 1/9 (from 25 percent to 11.1 percent), and the frequency of gene *a* in the gene pool from ½ to ⅓. In other words, the number of *aa* individuals has been reduced by 13.9 percent. This might be considered a substantial accomplishment even though it does not compare with the effectiveness of total negative selection against a dominant gene.

What will be the frequency of *aa* individuals in the third generation if *aa* members of the second generation do not reproduce? Because *AA* and *Aa* members of the second generation occur in equal numbers, the effective breeding population may be written: 2/4 *AA*, 2/4 *Aa*.

Then:

$$p_2 = 2/4 + 1/2(2/4) = 3/4$$
$$q_2 = 1/2(2/4) = 1/4$$
$$p_2^2 + 2p_2q_2 + q_2^2$$
$$(3/4)^2 + 2(3/4)(1/4) + (1/4)^2$$
$$9/16\,AA + 6/16\,Aa + 1/16\,aa$$

Thus, two successive generations of total negative selection have reduced the frequency of *aa* individuals to 1/16 or 6.2 percent. The reduction from the second to the third generation was from 11.1 percent to 6.2

percent, or 4.9 percent, considerably less than the reduction between the first and second generations (13.9 percent).

If total negative selection continues in the third generation, what will be the relative frequency of *aa* individuals? We already have a simple formula for solving this problem, the formula we derived in Chapter 10 to answer the question: What proportion of recessive-phenotype children is to be expected from matings in which both parents show the dominant phenotype? The formula is

$$\left(\frac{q}{1+q}\right)^2$$

(p. 156). Here we need only change the formula to recognize that the frequency of the recessive gene is changing from generation to generation. So we write:

$$q_3^2 = \left(\frac{q_2}{1+q_2}\right)^2$$

Then, because $q_2 = 1/4$ (above):

$$q_3^2 = \left(\frac{1/4}{1+1/4}\right)^2 = \left(\frac{1/4}{5/4}\right)^2 = (1/4 \times 4/5)^2 = (1/5)^2 = 1/25$$

In the fourth generation the frequency of *aa* individuals is $1/25$ or 4.0 percent, the frequency of the recessive gene being $1/5$. Between the third and fourth generations, selection succeeded in achieving only a 2.2 percent reduction in the frequency of *aa* individuals.

What would be accomplished by total negative selection in the fourth generation? Because $q_3^2 = (1/5)^2$; $q_3 = 1/5$.

$$q_4^2 = \left(\frac{q_3}{1+q_3}\right)^2 = \left(\frac{1/5}{1+1/5}\right)^2 = (1/5 \times 5/6)^2 = (1/6)^2 = 1/36$$

In the fifth generation, then, *aa* individuals will constitute $1/36$ or 2.8 percent of the population, a further reduction of only 1.2 percent. Apparently as generation follows generation, continued selection accomplishes less and less. Why is this?

Table 25.1 summarizes expectations for six generations. Note the simple progressions in the fractions, for example, in the column headed "Frequency of recessive gene." What would be the gene frequency in the seventh generation? What then would be the frequency of *aa* individuals (fifth column)? The table shows the declining effectiveness of selection, and also the reason for it. Look at columns 4 and 5. In the first generation

TABLE 25.1 Total negative selection against a recessive genotype (*aa*).

Generation	Frequency of recessive gene	Proportions of genotypes			Reduction in
		AA	*Aa*	*aa*	*aa*
1	1/2	1/4	2/4	1/4 = 25.0%	
2	1/3	4/9	4/9	1/9 = 11.1%	> 13.9%
3	1/4	9/16	6/16	1/16 = 6.2%	> 4.9%
4	1/5	16/25	8/25	1/25 = 4.0%	> 2.2%
5	1/6	25/36	10/36	1/36 = 2.8%	> 1.2%
6	1/7	36/49	12/49	1/49 = 2.0%	> 0.8%

the ratio of *Aa* individuals to *aa* individuals is 2:1, in the second generation it is 4:1, in the third generation 6:1, in the fourth generation 8:1, and so on. As the gene decreases in frequency in the gene pool, *more and more of the genes that remain are found in heterozygotes.* This is the reason that selection against homozygous recessives declines in effectiveness.

The point is illustrated graphically in Figure 25.1, which begins with the same population as does Table 25.1, continues for ten generations, and then, in ten generation jumps, continues on up to the 100th generation. Note here, as in the table, that the relative frequencies of both *Aa* and *aa* individuals decline, although at a steadily decreasing rate, but that the *Aa* individuals outnumber the *aa* individuals more and more as generations pass. Hence, selection applied against *aa* individuals alone becomes less and less effective as the gene becomes more and more rare (see Dahlberg, 1948).

Total negative selection against a dominant gene will eliminate the gene entirely (down to the level of replacement by mutation) in one generation. How many generations will be required for elimination of a recessive gene if total negative selection is applied against homozygous recessives only? In theory the gene is *never* eliminated; it approaches zero asymptotically. However, as shown in Figure 25.1, it becomes vanishingly small by the 100th generation (0.01 percent). How long is 100 generations? At least 2500 years. Can we imagine any program for human betterment being continued consistently for so long?

A Theoretical Example Perhaps we should be satisfied with less than nearly complete elimination of the trait in question. Might not prevention of reproduction accomplish *something* worthwhile?

Let us imagine an undesirable trait possessed by 1 percent of our population. We shall suppose that the trait has some of the hallmarks of

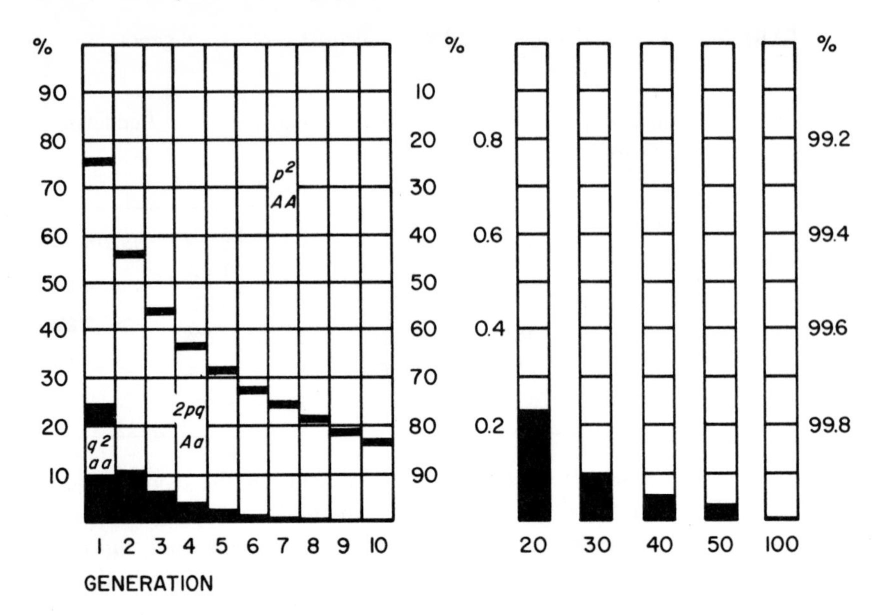

FIG. 25.1 Effect of total negative selection against homozygous recessives (*aa*). For the first ten generations the vertical scale is in percentage, for later generations in fractions of 1 percent. *AA*: homozygous dominants; *Aa*: heterozygotes. In each vertical column the height of the black portion at the base indicates the percentage of *aa* individuals, the portion above this up to the heavy horizontal line represents the percentage of *Aa* individuals, and the portion at the top (above the heavy horizontal line) represents the percentage of *AA* individuals. (Redrawn from Dahlberg, G., 1948. *Mathematical Methods for Population Genetics*. New York: Interscience Publishers. Used by permission of the Institute for Medical Genetics, University of Uppsala, Sweden.)

recessive inheritance, such as "skipping" of generations. We shall assume that persons who show the trait have the genotype *aa*. Using the Hardy-Weinberg formula, we equate the proportion of *aa* individuals to q^2 (p. 153):

$$q^2 = 0.01$$

$$\text{Then } q = \sqrt{0.01} = 0.1$$

$$q_1^2 = \left(\frac{q}{1+q}\right)^2 = \left(\frac{0.1}{1.1}\right)^2 = \left(\frac{1}{11}\right)^2 = \frac{1}{121} = 0.0083 = 0.83\%$$

Thus, complete negative selection against *aa* individuals in one generation has reduced the proportion produced in the following generation from 1.0 percent to 0.83 percent. Is this worthwhile? If our population is 200,000,000, then 1 percent of that number or 2,000,000, now show the trait in question. If they were prevented from reproducing, the number of persons showing the trait in the next generation (assuming population size remained stationary) would be $0.0083 \times 200{,}000{,}000$ or 1,660,000. In other words, there would be 340,000 *fewer* persons with the undesirable trait. Now, 340,000 is a large number when we think of such things as needed medical services and hospital beds. Were the eugenists of a former day right, after all? Should not society embark upon a program that would seem to accomplish so much?

Would Theory Work in Practice? What are some of the flies in the ointment? In the first place, when we used the Hardy-Weinberg formula we were tacitly assuming that *aa* individuals have the same fertility as *AA* and *Aa* individuals. With many undesirable human traits, persons who show the trait have reduced fertility. As a result, if they are permitted to reproduce, they produce less than their predicted "share" of the children with the trait. This is the equivalent of saying that more of the affected children than the formula would predict are the offspring of carrier (heterozygous) parents. As a result, preventing *aa* individuals from reproducing would have less than the predicted effect, because they already fail to produce "their share" of offspring.

When we use the Hardy-Weinberg formula we also imply that mating is *random*—determined only by the relative frequencies of the different genotypes. If, on the other hand, possessors of the trait had an increased tendency to marry each other, such assortative mating would increase the frequency of *aa* × *aa* matings over the frequency predicted by the formula. Because all children from *aa* × *aa* matings are *aa,* increasing the number of *aa* parents who marry each other—instead of *AA* or *Aa* individuals—would increase the effectiveness of selection in reducing the number of people in the *next* generation who show the trait. (When *aa* persons mate with *AA* and *Aa* individuals, most of the children are heterozygous and so presumably normal. Hence, preventing *aa* × *aa* and *Aa* × *aa* matings has less *immediate* effect in reducing the number of *aa* children than does prevention of *aa* × *aa* matings.)

In this connection, we note that if we could identify carriers by some test—and the number of traits for which this is possible is increasing rapidly—we could in theory prevent them from having children, and so greatly increase the effectiveness of negative selection. But could we ever gain the support of society for a program that prevented two phenotypically normal people from having children, particularly when three-fourths of those children would be expected to be normal, anyway?

Another difficulty lies in the fact that our theoretical calculation implies complete success: *total* negative selection. Actually, any program involving millions of people, many of whom would have no interest in cooperating, would be only partly successful at best.

Turning to genetic aspects of the problem, we ask: How sure are we that the trait is really monofactorial, that only one pair of genes is involved? Most of the traits in which eugenists are interested are polygenic. In such cases selection is less effective than it is represented as being in our simple example.

And how about penetrance (p. 70)? Do all persons with the *aa* genotype show the trait? If some do not, they would be missed by the selection process, and so pass on *a* genes to children.

A closely allied problem concerns traits that are the combined product of heredity and environment, a very common situation, as we have seen. In some environments *aa* individuals might not develop the trait. Hence they would not be included in the selection process.

Perhaps, also, the trait is correctable by medical means, or by improvement of the environment. If so, we should find little popular support for a program that prevented such "cured" individuals from having children. Even if no "cure" existed at present, other similar medical successes might lead people to hope that a "cure" would be found by the time the next generation needed it.

We should remember, however, that the "cure" has not affected the genes. If these "cured" people reproduce, gene *a* will be passed on to the next generation, which will then have more *aa* individuals to be "cured" than it would have had if the "cured" individuals in the preceding generation had not had children. Many eugenists worry about this piling up of the "genetic load" from generation to generation as more and more "cures" for genetic defects are found.

In times past, programs of the kind we are discussing were advocated on the basis of saving society the expense of providing additional medical services, hospitals, and the like. But how about the expense involved in putting on such a massive program? Imagine the number of eugenic sterilizations that would have to be performed!

And this brings us to the point that we could never gain public support for such a program. It is contrary to the temper of our times. It would require a bureaucratic organization on a national scale, and be in danger of suffering from the abuses that plague such organizations. Also, many people are opposed to sterilization for religious and other reasons. Other people would denounce the program as discriminatory and unfair. We can almost read the headlines now: "Desecrating the Sanctity of the Home"—"Invasion of Personal Rights"—"the God-Given Right to Have Children"!

The "New" Eugenics We have indicated some of the reasons why the eugenics movement has changed. Modern eugenists retain their predecessors' concern for the genetic welfare of future generations. But they recognize that both improvement of environment and improvement of genotype are necessary ingredients in programs for human betterment. And they realize that such programs, to be successful, must be of a nature acceptable to society. They are concerned, not only with future generations, but also with the welfare and best interest of those members of the present generation who have the misfortune to suffer from defects of various kinds.

Before we consider the "new" eugenics in more detail, however, we must discuss the matter of differential fertility (Chap. 26).

Problems

1. A population has the following constitution: 3000 *AA*, 2000 *Aa*, 500 *aa*,
   (a) If no possessors of the gene *A* reproduce, how many members of the next generation will be *AA*? *Aa*? *aa*?
   (b) If the *aa* individuals do not reproduce, how many members of the next generation will be *AA*? *Aa*? *aa*? Assume that the population remains stationary in total size (5500).
2. The gene *e* is semilethal, homozygotes dying before puberty. In a certain community the child population has the following constitution: 98,010 *EE*, 1980 *Ee*, 10 *ee*. Thus, one child in 10,000 has the genotype *ee*. What fraction of the next generation will be expected to have the genotype *ee*?
3. In a certain panmictic population, 4 percent of the people are homozygous for a certain recessive gene. If they die before reaching reproductive age, what percentage of the next generation will be expected to be homozygous recessive?
4. In population I the frequency, q, of a recessive gene is 0.01. In population II the frequency of the gene is 0.1. What is the ratio of homozygous recessives to heterozygotes in each population? In which population will negative selection against homozygous recessives prove most effective in reducing the proportion of homozygous recessives in the next generation? Why?

References

Dahlberg, G., 1948. *Mathematical Methods for Population Genetics.* New York: Interscience Publishers.

Haller, M. H., 1963. *Eugenics.* New Brunswick, N.J.: Rutgers University Press.

Osborn, F., 1951. *Preface to Eugenics,* rev. ed. New York: Harper & Brothers.

Popenoe, P., and R. H. Johnson, 1933. *Applied Eugenics.* New York: The Macmillan Company.

26

Differential Fertility, and Modern Eugenics

The central idea of differential fertility is nothing more remarkable than the obvious fact that some people have more children than others. Our point of interest is whether or not there are, on the average, significant genetic differences between people who have many children and those who have few, or none.

It will be convenient to divide our discussion into two parts: (1) differential fertility among groups of people, and (2) individual differential fertility. The distinction is somewhat artificial because, of course, groups are composed of individuals, but it is convenient for emphasis.

Group Differences in Fertility

The Inverse Ratio For generations, even for centuries, students of mankind have been classifying people into groups and then observing differences among these groups in the number of children produced. The groupings may be based on income, social position, the occupation of the husband, or some other means of distinction. To have a general term for such groupings we call them **socioeconomic groups,** and at the risk of some snobbishness we classify them as "higher" and "lower." For generations an inverse ratio, or negative correlation, existed, between the "height" of the group and number of children produced. The larger the income the fewer the children. The families of professional men were smaller than those of businessmen, which, in turn, were smaller than those of skilled artisans, which were smaller than those of unskilled laborers, and so on. The greater the amount of education of the parents, the smaller the families they produced.

This was the traditional picture for generations. We can see why it was a matter of concern to persons interested in the genetic future of mankind. *If* the differences between socioeconomic groups rest to a considerable extent on differences in ability, and *if* these differences in ability are at least in part determined by heredity, then the inverse ratio is **dysgenic.** Dysgenic is the opposite of eugenic; it means "detrimental to the future of mankind," especially to man's genetically determined nature. For example, *if* a man who earns $20,000 a year is genetically superior to a man who earns $5000 a year, then the fact that this second man was producing more children than the first would be dysgenic—more of the second man's than the first man's genes would be present in the gene pool of the next generation. The prospect was not one a eugenist could regard with equanimity.

The "ifs" we have used in the preceding sentences suggest that many questions arise concerning the *genetic* significance of differences among socioeconomic groups. Without doubt, environment and what we usually think of as "luck" play large roles in determining whether a child becomes an adult who is classed as "lower" or "higher" on the socioeconomic scale. One unskilled laborer is unskilled because he lacks the ability to be anything else; another is unskilled because he had no opportunity for education and training as a child, not because he could not have profited from them if he had had the opportunity. This second man's genes are likely to be all right, and therefore his producing of a large family is not dysgenic. Whether or not his doing so is *socially* desirable depends on whether he, and society, can provide the children with education and opportunities commensurate with their abilities. That, however, is another question.

Because of the great variability within each socioeconomic group, therefore, the major *genetic* interest centers on differences between people *within* the groups rather than on differences *between* the groups themselves (see Osborn, 1968). What we desire, of course, is some means of measuring each person's genetically determined endowment. Such knowledge would enable us to answer the question: Does the inverse ratio apply when people are classified by their genetically determined qualities? Or more simply: Are genetically superior people having fewer children on the average than genetically inferior people are? Unfortunately, attempts to answer this question are handicapped by lack of objective means of measuring one's genetically determined endowment. The best means presently available are intelligence tests, despite their shortcomings (Chap. 24). Consequently, attempts to answer the question have concentrated on studies of fertility with relation to intelligence. Before we consider the relationship between fertility and IQ, however, we shall discuss briefly the relationship between fertility and amount of education received. Amount of education is far from perfect as an indicator of amount of intelligence, but it has the advantage that it, as well as number of children produced, is included in information gathered by the United States Census.

Amount of Education and Fertility Through the years many studies have indicated an inverse ratio between amount of education and number of children produced. This relationship is strikingly shown in Figure 26.1, which is based on three successive censuses. The figure represents the number of children produced by women who were from 35 to 39 years old at the time of each respective census. "Wives 30 to 39 years old have already completed about 96 percent of their expected lifetime fertility" (U.S. Bureau of the Census, 1973). The figure includes two sets of graphs: one for all white women aged 35 to 39 whether married or not, and one for white women aged 35 to 39 who, at some time in their lives, at least, were married. Women who were never married are assumed to have been childless.

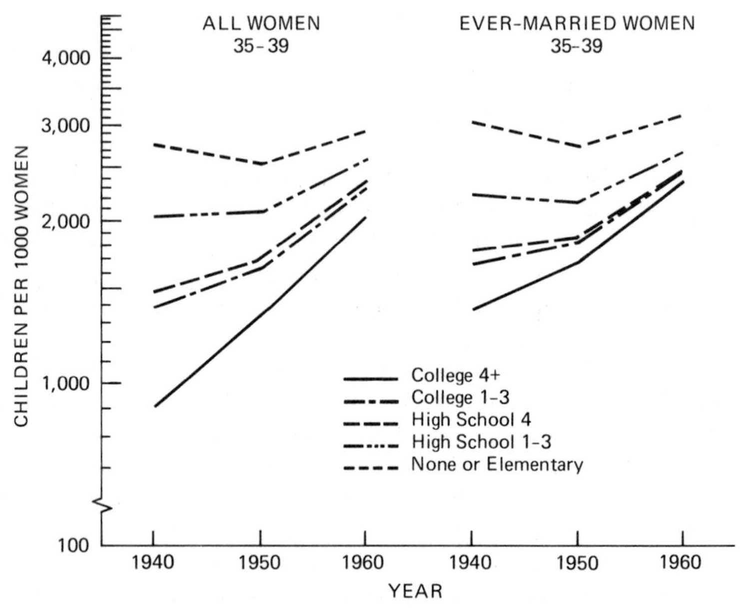

FIG. 26.1 Number of children ever born per 1000 white women 35–39 years old classified by educational attainment of the woman. Based on data from the U.S. censuses of 1940, 1950, and 1960 (data for 1940 relate to native white women). (Redrawn from Kiser, C. V., 1970. "Changing patterns of fertility in the United States," *Social Biology*, **17**:302–315.)

Note the strong differential fertility shown by the 1940 census. The women with four or more years of college averaged less than one child apiece, while women with no or little education averaged well toward three children apiece. Other groups fell in between. Turning to the "ever-married" group, we note that such women with four or more years of college had substantially more children than did "all women" with that amount of education. This difference reflects the fact that in those days many women with college educations remained unmarried.

The most striking aspect of Figure 26.1 is the indication that in later censuses the differences in fertility between groups have declined sharply (shown by convergence of the lines). The change lies mostly in the fact that women with high school and college educations were having more children than such women formerly had. This increasing fertility since World War II reflected "a trend toward more marriages, earlier age at marriage and. . . reduction in interval between marriage and first and successive births" (Kiser, 1970).

Further increase in fertility was shown by women aged 35 to 39 at the time of the 1970 census, and the differential by amount of education was about the same as it was in 1960 (Kiser, *personal communication*). Among women aged 25 to 29 in 1970, however, the census data indicated decreased fertility as compared to women of comparable age in 1960 (see below). Furthermore, Kiser noted that the differential between educational groups in these younger women was somewhat increased, mainly because the fertility of women who had attended college had decreased more than had that of women with less education. If in the future the fertility of less-educated women declines as much as has that of college women, the rates will again converge, though at a somewhat lower level than in 1960 (Fig. 26.1). Time will tell.

Trends Since 1960 "Between 1960 and 1972, the average number of children ever born declined sharply among women in the 15 to 24 age group, reflecting the sharp decline in fertility that occurred during the 1960's" (U.S. Bureau of the Census, 1973). As we have seen, this trend did not characterize older women (35–39 years) at the time. The decline in fertility among young women was due partly to a change in marriage patterns: slight increase in age of first marriage, and increase in the proportion of young women who did not marry (the proportion rose from 28 percent to 36 percent between 1960 and 1972). No doubt increased use of effective birth control was also important in the decline.

There is interest in the number of children these young wives said they *expected* to have. The average number of births expected by wives 18 to 24 years old declined from 2.9 in 1967 to 2.3 in 1972 (this applies to wives of all races; Negro wives and white wives did not differ significantly). In these days when **zero population growth** is being urged to prevent overpopulation, there is interest in whether or not a birth rate of 2.3 will approximate stability in population size. The Bureau of the Census (1973) suggested, on the basis of past experience, that young women who expect to have 2.3 children on the average will actually have about 2.1 births each. "This average approximates 'replacement level fertility,' which is the fertility level required for the population to eventually reach zero growth under projected mortality rates and in the absence of immigration." "Replacement level fertility of about 2.1 births per woman is comparable to about 2.2 births per ever-married woman."

Data at present available indicate that young wives of different educational levels expect to have about the same number of children. Hence our total evidence indicates that insofar as amount of education is an indicator of genetic differences in intelligence, having large amounts of education is not so dysgenic as it formerly was.

We should bear in mind, however, that use of amount of education as a measure of intelligence suffers from the same limitations we mentioned for socioeconomic groups in general: There are many reasons aside from amount of native intelligence why one does or does not attend, or complete, high school or college. And within any high school or college population, we find great differences in intelligence.

IQ and Fertility Earlier studies indicated that people with higher IQ's had, on the average, smaller families than people with lower IQ's (studies summarized by Falek, 1971). On the basis of these findings, the prediction was made that the intelligence of the population is declining slowly from generation to generation.

Other studies, however, were failing to find evidence that intelligence as measured in IQ scores is in fact declining. For example, in 1932, over 87,000 11-year-old Scottish children were given an intelligence test; they achieved a mean score of 34.5 points. In 1947, nearly 71,000 11-year-old Scottish children were given the same test; their mean score was 36.7 points. A random sample of the children in the two surveys was given the Binet test, with no appreciable change in scores being noted (Scottish Council for Research in Education, 1949).

Again, investigation of intelligence test scores of military personnel in World War I and World War II showed better performance by the personnel of the second war. This paralleled an increase in the amount of education the men had received (Tuddenham, 1948).

A third investigation was an analysis of mental test scores achieved by some 130,000 American high school pupils over a period of 26 years (Finch, 1946). The results indicated that the average mental ability of the pupils in more recent years was equal, or slightly superior, to the mental ability of the pupils who had been in high school during the earlier years. This was true despite the fact that the proportion of children going to high school had increased substantially in the meantime.

Thus we had a paradox (sometimes called "Cattell's paradox" after one of the early investigators.) Studies of differential fertility indicated that intelligence should be declining; other studies indicated that it was not doing so.

The contradiction was found to exist in large measure because the earlier investigations concentrated entirely on those members of each IQ grade who did in fact have children. The investigators neglected the single and nonreproductive siblings of the parents they studied. When these individuals were included, the picture changed.

The largest investigation of this kind utilized data from a study of more than 82,000 individuals who were ascertained by reason of descent from the grandparents of 289 retardates in an institution (Higgins, Reed, and Reed, 1962; Reed and Reed, 1965; Waller, 1971; Reed, 1971). Six generations were included in the study, and the relationship of many of these people to the 289 probands (propositi) was so distant that they could fairly be said to constitute an unselected sample of the general population. Retardates constituted 2.7 percent of the population, about the same percentage we find in our general population.

In this population there were 1016 families in which the IQ scores of the parents (tested as children) and one or more children were known. Data from these families were "revised to exclude spouses (who are not in the line of descent) but include tested married siblings" (Reed, 1971). This yielded 1900 tested parents. These were ranked by IQ scores. Table 26.1 summarizes the findings, giving the average number of children for each rank (second column; parents and their siblings are all entitled "Siblings" in this table). We note that among the 1900 people, those with lowest IQ had the largest families—a finding directly in line with results of the earlier studies we mentioned.

But column 3 of the table indicates the result of including the 66 unmarried siblings of the parents. This changes the result entirely: Now people in the lowest IQ rank are seen to have the smallest average number of children. The change results from the fact that the mean IQ of the unmarried siblings was 80.46, because many of them were so severely retarded as to be unmarriageable. Also, none of the siblings of parents having IQ scores of 131 or above failed to reproduce, whereas more than 40 percent of the siblings of retarded parents did not reproduce (Reed, 1971). "Thus by including the childless members of a generation as well as the fertile we find that our paradox has been resolved; the lowest IQ group produces the fewest children and the highest IQ group produces the most" (Reed and Reed, 1965).

Other investigators have recognized the fact that in estimating

TABLE 26.1 Reproductive rate of (a) married siblings, and (b) all siblings, married and unmarried.

	Average number of children per sibling	
IQ of sibling	Married siblings only (number: 1900)	All siblings, married and unmarried (number: 1966)
55 and below	3.64	1.38
56–70	2.84	2.46
71–85	2.47	2.39
86–100	2.20	2.16
101–115	2.30	2.26
116–130	2.50	2.45
131 and above	2.96	2.96

(Data from tables 51 and 52 in Reed, E. W., and S. C. Reed, 1965. *Mental Retardation: A Family Study.* Philadelphia: W. B. Saunders Company.)

the relative reproductive success of a group, other factors must be taken into account in addition to number of children produced (Bajema, 1963; Waller, 1971). Aside from (a) number of children per fertile individual, and (b) proportion of nonreproductive individuals, Bajema stressed: (c) mortality rates up to the end of the child-bearing period, and (d) generation length. He made use of a statistic called the *intrinsic rate of natural increase,* which takes all four of these variables into account. The parents he studied had received intelligence tests when in the sixth grade of a public school system. Thus, none of the parents were severely retarded, although three (of 1144) had scores between 70 and 56. These three produced no children. Seventy-five parents with scores between 85 and 71 averaged 2.05 children, whereas, at the other extreme, 23 parents with IQ's of over 130 averaged 3.00 children apiece. Computation of the intrinsic rate of natural increase substantiated the conclusion evident from this fact. Bajema concluded that "the population under study has probably been in equilibrium with respect to the genetic factors favoring high intelligence or, more likely, has experienced a slight increase in the frequency of the genetic factors favoring high intelligence."

A statistical analysis made by Waller (1971) on part of the individuals included in the Reed and Reed (1965) study yielded a similar general conclusion: "It seems that natural selection is favoring an increase in mean IQ score in the sample of the present study, or at least there is no evidence of a decrease."

These studies all yield, therefore, no evidence that people of low intelligence are outbreeding people of high intelligence. The studies were all made on white Americans living in the Midwest. Similar studies of people in other regions and nations and in other ethnic groups would be highly desirable as a basis for drawing general conclusions.

Parentage and Children of Retardates Of the many facets of the investigation by Reed and Reed (1965), two more are of especial interest in the present connection. What proportion of the retardates in one generation are the offspring of retardates in the preceding generation? Reed (1971) concluded that about 17 percent of retarded persons had one or both parents retarded, the remaining 83 percent being offspring of parents both of whom had IQ's of 70 or over.

Eugenists have long been interested in the levels of intelligence of children in the families produced by retarded parents. The investigation of Reed and Reed (1965) yielded the results shown in Figure 26.2, which shows that when the IQ's of both parents were below 70, about 40 percent of the children were retarded (shaded columns). When only one parent was retarded, about 12 percent of the children were retarded. When neither parent was retarded (and both had normal siblings), 0.5 percent of the children were retarded (Reed, 1971).

Modern Eugenics

At the end of the preceding chapter we mentioned that the new eugenics is concerned with the welfare of people now living as well as with those yet to be born, and that improvement of environment and of genotype are both necessary ingredients in programs for human betterment. "Man and his culture are inextricably interrelated; the culture cannot develop beyond his genetic potentials, and his genetic potentials do not evolve beyond the possibilities offered by his environment" (Osborn, 1973). At the outset we ask: For what sort of world must programs for human betterment be planned ?

The Setting: Overpopulation Because of the great amount of publicity being given the subject, every reader of this book is undoubtedly aware that the threat of overpopulation—the "population explosion"—is a most important problem facing mankind. Most of our other problems—exhaustion of natural resources and of sources of energy, destruction of the environment by pollution, and the like—stem directly from the large number of people now making demands on our resources and environment. And the problems will become more severe.

At the time this is being written, the United States is in a period of declining birth rates as compared to the high rate characterizing

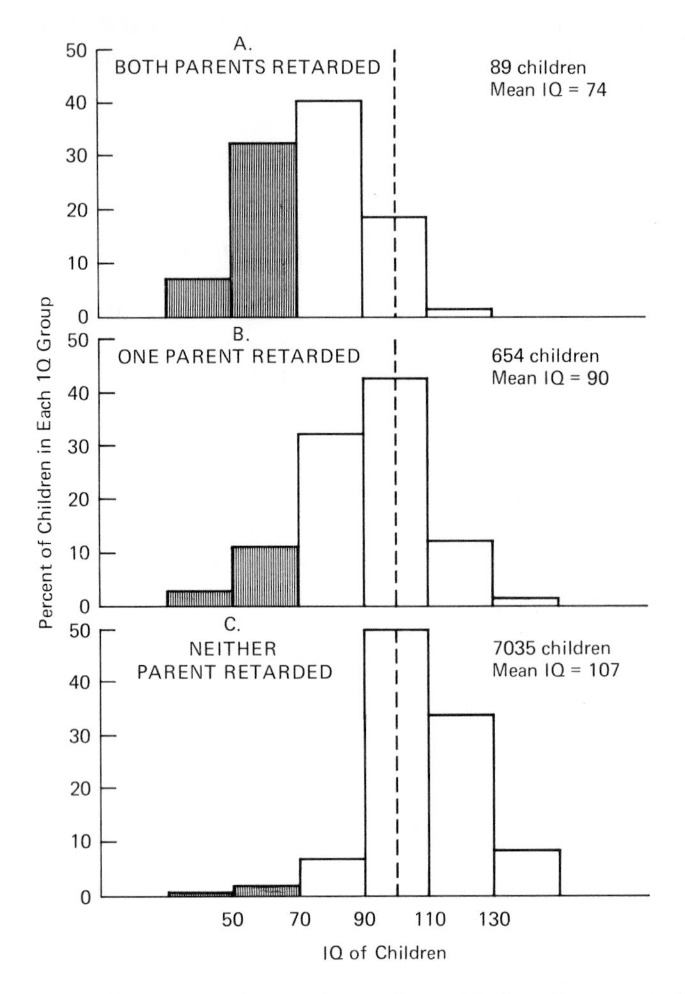

FIG. 26.2 Parentage of mental retardates. IQ distribution of children from different parental combinations (based on data from Reed and Reed, 1965). (Redrawn from Reed, E. W., 1971. "Mental retardation and fertility," *Social Biology,* 18 (Supplement): S42–S49.)

the years following World War II. As we noted earlier (p. 400), young wives now expect to have families of a size that will result in simple replacement of the present childbearing generation by an equivalent number of offspring who will live to reproductive age in their turn: "replacement-level fertility." This will not lead immediately to "zero population growth," however. If the average number of children per family is 2.1, the population will continue to grow for some 70 years, increasing from the 1970 census total of around 209 million to 320 million. This is because of the age structure of the present population, which means mainly the large number of girls born during our period of higher birth rates who will be in their child-bearing years.

Some amount of optimism will lead us to conclude that the United States of the future can provide for 320 million people without completely exhausting our resources and destroying our environment. But the problems entailed will be enormous. And where will these 320 million people live? If present trends are any indication, more and more of them will live in urban communities, especially those concentrated along the Atlantic and the Pacific seaboards. If so, problems connected with overcrowding—transportation difficulties, air pollution, provision of adequate water supplies, disposal of wastes, crime in the streets, and so on—will increase. It is our earnest hope, of course, that they will be solved. They *must* be if society as we know it is to survive.

Such, then, is the setting in which eugenic programs will have to operate.

What Kind of People Will Be Needed? The answer to this question depends upon how well we solve the problems just mentioned. Can we solve them and still maintain the standard of living and the quality of life now characterizing the United States (to say nothing of improving them)?

Perhaps playing the Devil's Advocate, Simpson (1967) visualized that people who will succeed best in surviving and raising families in this future world will be people who *prefer* to live in overcrowded ghettos surrounded by pollution and crime, and who will actually find obnoxious much that we prize: space, fresh air, creativity, gentility, and so on. "If that is true, human nature will indeed change, and the change will entail loss of what is generally considered most refined, most civilized, most humane, and most admirably human among us now."

Simpson would be the first to hope that such gloomy prophesies will not prove to be correct. But if they are not to prove correct, we and our descendants must solve enormous problems of improving the *environment* in the face of demands placed upon it by increasing numbers of people (Fig. 26.3). No wonder Frederick Osborn—who deserves to be called the father of modern eugenics—stated (1968): "The improvement of the environment is as important as the improvement of the hereditary base, for the two are closely related in the evolutionary process." And again: "The aim of eugenic proposals is an increase in the proportion of children born to the individuals who are most successful in their particular environment, and a decrease in the proportion of children born to the least successful in their environment."

This last statement reminds us that, contrary to ideas some people have had, eugenists do not seek production of a uniform type of person—

FIG. 26.3 Heredity has done its part. What will be the effect of environment? (Photo by Tony Rollo. From *Newsweek*.)

standardized men and women, so to speak. In the world of the future there will continue to be many environments, many jobs to be done, and many services to be performed. These will call for people of as many types as does the world today. Imagine, for a moment, a nation composed entirely of college professors (appalling thought!). Who would produce, process, and distribute the food, construct the buildings, manufacture and service the automobiles, airplanes, household appliances, and so on, and in fact do all the countless other tasks that must be done when large populations of people live together in a society?

In this connection we must remember also that high intelligence is not the only desirable human quality. We are all too familiar with the great harm that has been done throughout history by highly intelligent but unscrupulous people. High IQ is no guarantee of social worth. For this reason, Gottesman and Erlenmeyer-Kimling (1971) urged "would-be eugenists to abandon their fixation on IQ as *the* trait to be maximized in our species in favor of an Index of Social Value." Despite its limitations, IQ is easier to measure with some objectivity than is "social value," but that fact does not excuse us from recognizing the importance of social value as we plan for the future of mankind.

Emphasis: Voluntary Methods Despite the urgency of the environmental problems, a textbook of human genetics is perforce more concerned with the "hereditary base" mentioned by Osborn. Within the framework of modern society, how can we ensure that the genetic endowment of our population will not deteriorate, but will, if possible, improve?

Birth Control "The most urgent eugenic policy at this time is to see that birth control is made equally available to all individuals in every class of society" (Osborn, 1968). It is generally believed that much if not most of the traditional differential fertility among socioeconomic groups and educational levels discussed earlier in the chapter was the result of differences in use of the means of birth control. Poorer and less educated people used these means less, and less effectively, than did people of higher financial and educational levels. And it is thought that the reason the differential fertility between groups has been decreasing (Fig. 26.1) is because less-advantaged people are now using contraception more, and more effectively, than their predecessors did. This being the case, making birth control "equally available to all individuals in every class of society" will reduce the differences between groups still further, and thus reduce whatever dysgenic effect such group differences have had.

Here we should emphasize that the gain will not relate entirely to a decrease in proportion of less desirable genes to be inherited by future generations. Equally important will be the gain to the children themselves when every child is a "wanted child," and parents produce only as many children as can be provided with a good home environment, adequate educational opportunities, and the like. Hence, whether one views it from a genetic or an environmental point of view, making effective birth control equally available to everyone is an important goal.

To be of maximum value to people of all classes, birth control methods should be not only effective but also easy to use. As everyone knows, great improvements have been made, and more will be made. Apparently the most widely used of the new methods are "the Pill" (hormones suppressing ovulation in women), and intrauterine devices (IUD), which work in ways not clearly understood. But constant research being done is certain to produce still more effective methods and ones simpler to use (Marx, 1973).

We should remember, also, that voluntary *sterilization* is the most effective means of all, and is being used with increasing frequency. Simplest is the severing of the sperm ducts in a male (vasectomy), no hospitalization being required. But operations of severing or closing the oviducts of a female have been simplified to a point where hospitalization is not necessary and will be further simplified in the near future. These operations have been given so much publicity that it should not be necessary to remind readers that they do not "unsex" people. Sex hormones continue to be produced and secreted into the bloodstream.

While we are mentioning means of birth control, we must mention legalized abortion. Highly controversial in the United States, abortion is an accepted and widely used means of birth control in nations with different cultural background, for example, Japan. We certainly do not regard it as the method of choice, but it is available as a last resort when other methods have not been used, or have proved ineffective.

The Mentally Retarded When we speak of making birth control equally available to everyone, we include people who are mentally retarded. In listing voluntary and socially acceptable means of reducing the incidence of deleterious genes, Osborn (1968) included "training and constant help in the use of the new methods of contraception for the mentally defective and borderline cases not in institutions." Such individuals may have some capability of being useful members of society and of being self-supporting, although often they are the wards of welfare agencies. In either case, their income is not large and their standard of living is likely to be low, and to be made lower if they have children to rear. Furthermore, the intellectual and cultural home environment provided by such people is not such as to be a stimulus for the best development of children, even when the latter do not inherit their parents' deficiencies. All of which means that it is best for the people themselves, and for society, if they do not have children, or at most do not have many of them. Many times they can be led to understand the advantages to themselves of limiting their families.

For such people, voluntary sterilization is the surest method, but experience has shown that even people with very limited mentality can be taught to use other methods (Morgan, 1965). New, simpler methods that are sure to be developed in the future will be a boon for them.

Parents with Little Education and Income Moving on to parents who are not mentally retarded but who for various reasons can provide children with only an inferior home environment, we note that such people often have more children than they wish to have. Many surveys have indicated that people with little income and/or with little education have larger families than they desired. For example, Bumpass and Westkoff (1970) analyzed the results of the 1965 National Fertility Study in the United States, an interview survey of a probability sample of 5600 married women. They found that "in general, the proportion of unwanted births is approximately twice as high among wives with less than a high school education as among wives who have attended college," and is "more than twice as high for families with incomes of less than $3,000 as for those with incomes of over $10,000."

From the standpoint of total population, Bumpass and Westkoff estimated that if there had been no unwanted births between 1960 and 1965, the total number of births in the nation would have been reduced by 4.7 million, which represents one-fifth of all births during the period. Thus we see that making birth control equally available to everyone will (a) serve to reduce population size, (b) help to limit the number of children reared in substandard environments, and (c) serve eugenic ends insofar as less-advantaged people are that for genetic reasons.

Heredity Clinics Another important means of reducing the number of deleterious genes in future generations is by means of heredity clinics where people who desire it may receive genetic counseling. We shall discuss this subject in Chapter 28; here we note that people do not really want to have defective and deformed children and will seek advice when it is made available to them. Oftentimes a counselor can advise them only in terms of statistical probabilities, but answers will become more positive and definite as more tests are developed for identifying individuals who are carriers of genes that cause defects in the homozygous state.

World Population We have confined our attention mainly to the United States, but as everyone knows, the "population explosion" is a world-wide phenomenon. Indeed in some regions, such as Latin America and India, problems arising from it are more acute than those faced by the United States.

Under primitive conditions populations maintain approximate equilibrium because high birth rates are balanced by high death rates. But with the advent of modern medicine, death rates in most countries have been greatly reduced. If birth rates continue high, as they have in many countries, population size increases at a much faster rate than ever before. Obviously, the solution to the problem is to decrease the birth rates. This has been done effectively in Japan, and progress is being made in other countries. In India, for example, there are active programs promoting use of birth control, especially intrauterine devices, and vasectomies. But with huge populations having little education, will such programs succeed in controlling total population size? Only the future will tell.

How Effective Will Voluntary Methods Be? In the first place we note evidence that reducing average family size does have a eugenic effect in reducing the number of defective children born. In Japan, "as a direct result of limiting family size, the deaths from congenital malformation were reduced by 20%, Down's syndrome (mongolism) was reduced by 40%, and erythroblastosis by 50%" (Gottesman and Erlenmeyer-Kimling, 1971). Here we recall that older mothers have a greater tendency than do younger mothers to have children with Down's syndrome, and that when there is Rh incompatibility, later pregnancies are more likely than earlier ones to result in erythroblastosis. Smaller families born to younger mothers will tend to reduce the proportion of any defects the incidence of which increases with the age of the mother or with the number of pregnancies.

But will voluntary methods prove sufficient to control total population size? There is no consensus on an answer to this question. An affirmative answer is given by advocates of family planning, who point out that people in many countries state a preference for a small family even though they do not now have the means of limiting their families to that small number. To the extent that their excess fertility depends upon not having means of birth control, providing those means will be effective. Will this be sufficient to lead to population equilibrium?

Such writers as Spengler (1969) and Hardin (1968, 1970, 1972) concluded that voluntary cooperation must be reinforced with some system of incentives, or if that fails, of coercion.

Under the heading "Conscience is Self-Eliminating," Hardin (1968) wrote: "Confronted with appeals to limit breeding, some people will undoubtedly respond to the plea more than others. Those who have more children will produce a larger fraction of the next generation than those with more susceptible consciences. The difference will be accentuated, generation by generation. . . . To make such an appeal is to set up a selective system that works toward the elimination of conscience from the race."

For people with this point of view, what is the solution? Proposals have been many. The less drastic ones suggest incentives—bonuses, tax rebates, and the like—for *not* having more than a specified number of children. If incentives do not prove sufficient, involuntary measures such as governmental licenses to have children, compulsory sterilization of parents following birth of a certain number of children, and the like, have been proposed. See Berelson (1969) for more details. We may hope that such governmental meddling with the private lives of our descendants will not prove necessary.

Returning to the thesis advanced by Hardin, we note that while it is based on impeccable Darwinian reasoning (Chap. 27), the genetic basis of conscience is completely unknown. Insofar as it has a genetic basis, this basis is unlikely to be a simple one. Furthermore, as Hardin clearly understood, mankind is not divided neatly into two types: those with conscience versus those without conscience. There are all degrees of sensitivity of response to appeals for cooperation. Until it is proven otherwise, we may reasonably suspect that the portion of the population having some degree of sensitivity and willingness to cooperate is far larger than the portion that will not cooperate at all. If so, the fact that people who "couldn't care less" have large families would not be sufficient to make great change in the total composition of subsequent generations. Note also that at best the visualized elimination of conscience would require many generations, that is, hundreds or thousands of years. By that time we may be sure that our successors will be faced with problems, *and* means of solving them, that we cannot foresee at present.

Obviously, also, the observed trend toward smaller family size noted above is a hopeful portent for the future of population control.

We conclude with the statement of Pilpel (1971): "I personally believe that if all have the means as well as the right to free choice in matters of family planning (and that includes statutory sterilization and abortion) there will probably be no need for any of the more coercive approaches prematurely being advanced today. In other words, give 'voluntarism' a chance before deciding that it doesn't work."

Individual Differential Fertility

The gene pool of one generation is not necessarily a random sample of the gene pool of the parents. Not all the genes in the parental generation are passed on to offspring. Some members of the parental generation do not marry. Some who do marry do not have children. Some who have children have fewer of them than do others. This is what we mean by individual differential fertility.

A striking example is afforded by data collected many years ago by Powys on the fertility of married women, primarily of British ancestry, in New South Wales, Australia (Powys, 1905). These women had died after the age of 46, that is, after their child-bearing years were over. Hence the data represent completed families. From Table 26.2 we note that of the 10,276 women, the more fertile 50 percent produced 76.4 percent of the children. The data are noteworthy for a few families of rather spectacular size; even if we omitted these, a strong differential fertility would still be evident.

The data illustrate a trend common to both advanced and primitive cultures. Neel (1958) summarized data showing that in a certain township on the Gold Coast of Africa the more fertile 50 percent of the women produced 69.7 percent of the children. A survey in Liberia showed

TABLE 26.2 Differential fertility of married women in New South Wales, Australia. (The women had died after the completion of the reproductive period, i.e., after 46 years of age.)

	Size of family	No. of married women	No. of children	
5138 women = 50%	0	1110	0	15,030 children = 23.6%
	1	533	533	
	2	581	1162	
	3	644	1932	
	4	702	2808	
	5	813	4065	
	6	755	4530	
5138 women = 50%	6	100	600	48,623 children = 76.4%
	7	976	6832	
	8	963	7704	
	9	847	7623	
	10	786	7860	
	11	568	6248	
	12	422	5064	
	13	226	2938	
	14	129	1806	
	15	57	855	
	16	39	624	
	17	12	204	
	18	5	90	
	19	2	38	
	20	2	40	
	21	1	21	
	22	1	22	
	24	1	24	
	30	1	30	
		10,276	63,653	

Data from Powys, A. O., 1905. "Data for the problem of evolution in man. On fertility, duration of life and reproductive selection," *Biometrika,* 4:233–285.

that 50 percent of the women produced 78.5 percent of the children. In two villages in Pakistan, 50 percent of the women produced 69 percent of the children. In the United States, as revealed by the 1940 Census, 50 percent of the women produced 88 percent of the children (Neel, 1958).

This tendency for a descendant generation to be produced by only a portion of the parental generation was noted long ago by Karl Pearson, the eminent British biometrician. He wrote: "It is a point which seems to me of the utmost significance that, allowing for the proportion of unmarried in the population, about ⅕ to ⅙ only of the adults produce quite one-half of the next generation, and any correlation between inheritable (physical or social) characteristics and fertility must thus sensibly influence that next generation" (Neel, 1958).

What is the significance of such differential fertility? Pearson and others since his time have been concerned that the more fertile portion of the population might be inferior to the less fertile portion. Insofar as the superiority had a genetic basis, lessened fertility on the part of superior individuals would be dysgenic. We have discussed this question as it relates to *group* differences in fertility, and as groups are only assemblages of individuals we need not repeat the discussion here. At any rate, differentials of this type may well be decreasing, as we have noted.

Turning the question around, we ask: Why do some parents have fewer children than do others, or have no children at all? In part, reduced fertility is a matter of voluntary control, as we have seen, but to an important extent it is involuntary. Sterility and lowered fecundity may have physical bases, or the difficulties may be primarily psychological. Penrose (1962) has concluded that the extremely low birth rate of low-grade mental defectives has a psychological basis rather than being a matter of physical inability to bear children.

Multiple neurofibromatosis is a disfiguring defect in which the skin is covered with unsightly nodules formed of benign tumors growing at the ends of peripheral nerves. The trait seems to be inherited as a simple autosomal dominant. Studies at the University of Michigan reveal that the fertility of affected individuals is only 52.7 percent that of their normal siblings (Neel, 1958). Only 32 percent of affected males marry, whereas 58 percent of affected females do so; but among those who do marry, the fertility is only 79.5 percent that of unaffected siblings. Here we have an example of differential fertility that is *eugenic* in the sense that reduced fertility decreases the number of members of the next generation who will have the deleterious gene.

Differential fertility in multiple neurofibromatosis is also an example of **natural selection** in action. When possessors of one genotype produce fewer offspring on the average than do possessors of another genotype, natural selection is operating. Because the reasons that some people have more children than others are determined at least in part by genes (manifesting themselves physically or psychologically, or both), individual differential fertility results in a change in the gene pool of the subsequent generation. Some genes present in the gene pool of the parental generation are not represented at all in the gene pool of the offspring (e.g., if the gene is dominant and lethal, or produces sterility). More commonly, deleterious genes of the parents are transmitted to offspring, but with decreased frequency. This is natural selection working for the good of future generations. In the next chapter we shall discuss its role in producing human diversity.

References

Bajema, C. J., 1963. "Estimation of the direction and intensity of natural selection in relation to human intelligence by means of the intrinsic rate of natural increase," *Eugenics Quarterly,* **10**:175–187.
Berelson, B., 1969. "Beyond family planning," *Science,* **163**:533–543.
Bumpass, L., and C. F. Westoff, 1970. "The 'perfect contraceptive' population," *Science,* **169**:1177–1182.
Falek, A., 1971. "Differential fertility and intelligence: Current status of the problem," *Social Biology,* **18** (Supplement):S50–S59.
Finch, F. H., 1946. "Enrollment increases and changes in the mental level of the high school population," *Applied Psychology Monographs,* **10**:1–75.
Gottesman, I. I., and L. Erlenmeyer-Kimling, 1971. "A foundation for informed eugenics," *Social Biology,* **18** (Supplement):S1–S8.
Hardin, G., 1968. "The tragedy of the commons," *Science,* **162**:1243–1248.
Hardin, G., 1970. "Parenthood: Right or privilege?", *Science,* **169**:427.
Hardin, G., 1972. *Exploring New Ethics for Survival.* New York: Viking Press.
Higgins, J. V., E. W. Reed, and S. C. Reed, 1962. "Intelligence and family size: A paradox resolved," *Eugenics Quarterly,* **9**:84–90.
Kiser, C. V., 1970. "Changing patterns of fertility in the United States," *Social Biology,* **17**:302–315.
Marx, J. L., 1973. "Birth control: Current technology, future prospects," *Science,* **179**:1222–1224.
Morgan, D., 1965. "The acceptance by problem parents in Southampton of a domiciliary birth control service." In Meade, J. E., and A. S. Parkes (eds.). *Biological Aspects of Social Problems.* New York: Plenum Press. Pp. 199–204.
Neel, J. V., 1958. "The study of natural selection in primitive and civilized human populations," *Human Biology,* **30**:43–72.
Osborn, F., 1968. *The Future of Human Heredity, An Introduction to Eugenics in Modern Society.* New York: Weybright and Talley.
Osborn, F., 1973. "The emergence of a valid eugenics," *American Scientist,* **61**:425–429.
Penrose, L. S., 1962. *The Biology of Mental Defect,* 3rd ed. London: Sidgwick & Jackson Ltd.
Pilpel, H., 1971. "Discussion: Family planning and the law," *Social Biology,* **18** (Supplement): S127–S133.
Powys, A. O., 1905. "Data for the problem of evolution in man. On fertility, duration of life and reproductive selection," *Biometrika,* **4**:233–285.
Reed, E. W., 1971. "Mental retardation and fertility," *Social Biology,* **18** (Supplement):S42–S49.
Reed, E. W., and S. C. Reed, 1965. *Mental Retardation: A Family Study.* Philadelphia: W. B. Saunders Company.
Scottish Council for Research in Education, 1949. *The Trend of Scottish Intelligence.* London: University of London Press.
Simpson, G. G., 1967. "Biology and the public good," *American Scientist,* **55**:161–175.
Spengler, J. J., 1969. "Population problem: In search of a solution." *Science,* **166**:1234–1238.
Tuddenham, R. D., 1948. "Soldier intelligence in World Wars I and II," *American Psychologist,* **3**:54–56.
U.S. Bureau of the Census, 1973. *Current Population Reports,* Series P-20

No 248, "Birth Expectations and Fertility: June 1972." Washington, D.C.: U.S. Government Printing Office.

Waller, J. H., 1971. "Differential reproduction: Its relation to IQ test score, education, and occupation," *Social Biology*, 18:122–136.

27

Human Diversity and Its Origin

No one needs to be convinced that human beings show great diversity. We differ from each other in eye color, hair color, stature, blood groups, and in almost countless other ways. We say that man is **polymorphic** (many-formed). When we think of the diversity among races, we sometimes use the term **polytypic** (many-typed), but the distinction is principally one of degree.

The polymorphism that we see or detect by serological and biochemical tests is, of course, phenotypic. It is the outward expression of the *genetic* polymorphism. In saying so, we must never forget that most traits are the product of the interaction of heredity and environment. Blood groups may be an exception, being wholly gene-determined, but environment contributes to most phenotypic traits. We may have genes for the production of sufficient melanin in our skins to cause us to be brunets, but the actual shade of our skins will depend in part on how much time we spend sunbathing.

The question we are interested in now is: How did human diversity arise? How does the gene pool of a population attain such variety? How do the gene pools of different populations come to differ from one another?

As we recall our previous discussions, we readily recognize that the basic process responsible for genetic diversity is the process of mutation, using the term in its most general sense to include changes in chromosomal material, both those that are, and those that are not, visible with the microscope (p. 311). Mutations are the raw materials of change.

Even if a gene pool were uniform at one stage in its existence, mutations would occur to introduce diversity.

What happens to mutations once they occur? In the first place, we recall from our discussion in Chapter 10 the underlying tendency of populations to remain in equilibrium. Suppose that with respect to a certain chromosomal locus, a gene pool contains 90 percent *A* genes and 10 percent *a*. We shall also suppose that the *a* genes arose from the *A* genes by mutation, although this need not have occurred in the immediate past. Then if random mating occurs among the possessors of these genes, we shall expect that 81 percent of the offspring will be *AA*, 18 percent will be *Aa*, and 1 percent will be *aa* (see Hardy-Weinberg formula, p. 149). If the offspring when mature also mate at random, their offspring in turn will also occur in approximately the same proportions. This Hardy-Weinberg equilibrium is predicated on the assumption that the gene pool does not change. Obviously every time an *A* gene mutates to form an *a* gene, the pool *is* changed by just that much. This tendency of genes to mutate, at a low but rather constant rate, is called **mutation pressure.** In itself, it forms one means by which diversity is introduced into a gene pool. There is, however, a tendency to *reverse* mutation (of gene *a* to mutate back to gene *A*), so that these opposing mutation pressures will be expected to balance each other and produce an equilibrium of their own. Without recourse to the customary mathematics, we can understand that an equilibrium will be attained when the number of newly formed *A* genes in each generation equals the number of newly formed *a* genes. The point we wish to make here is that although mutations are the raw materials for evolutionary change, mutation pressure by itself is not a very potent force for making such change.

Indeed, the term "polymorphism" is frequently restricted to the diversity present in a population in excess of the amount introduced by mutation pressure. A widely accepted definition is that of Ford (1964): "Genetic polymorphism is the occurrence together in the same locality of two or more discontinuous forms of a species in such proportions that the rarest of them cannot be maintained merely by recurrent mutation." What, then, are the forces that lead to polymorphism by producing changes in populations and their gene pools? We shall emphasize two: genetic drift and natural selection.

Genetic Drift

The term genetic drift refers to chance occurrences in small populations. The Hardy-Weinberg formula assumes large populations in which breeding is random. However, the equilibrium predicted by the formula may be upset if the population is small. By way of illustration, let us take the smallest possible "population," one male and one female. Let us suppose that one is *Aa,* the other *aa.* Here the frequency, p, of *A* is ¼ or 25 percent, the frequency, q, of *a* is ¾ or 75 percent. We cannot use the Hardy-Weinberg formula because we cannot have random mating. However, because this formula is only one aspect of Mendelian thinking, we state the simpler expectation that in such a mating half the offspring are expected to be *Aa,* half to be *aa.* As we have emphasized repeatedly, such a 1:1 ratio is only expected on the average. Suppose that the two parents produce only two offspring. What is the chance that one child will be *Aa,* the other *aa*? Because such a family can arise in two ways (older child *Aa,* younger child *aa, or* older child *aa,* younger child *Aa*), the chance is $2 \times \frac{1}{2} \times \frac{1}{2} = \frac{2}{4}$. Only two-fourths of the time will the family of two be expected to fit the 1:1 ratio exactly. What will happen the remaining two-fourths of the time? One-fourth of such families of two will be *Aa* and *Aa.* If this occurs, the frequency of the gene *A* has been doubled; p is now 50 percent. Alternatively, one-fourth of the families of two will be expected to be *aa* and *aa.* If this occurs, the gene *A* is lost, $p = 0$. So merely by chance, with no regard for usefulness, a gene may increase in frequency in the gene pool, or it may decrease in frequency, down to the point of being completely lost. These chance fluctuations in gene frequencies are what we mean by drift.

Our example of two parents replacing themselves by producing two children is hardly a population, although it does illustrate the fact of chance fluctuations. Similar fluctuations may occur when numbers of individuals are large enough to be considered populations, even though they are small ones. Starting with the gene frequencies just employed ($p = 0.25$; $q = 0.75$), we may imagine a population of 16 people constituted as follows: 1 *AA;* 6 *Aa;* 9 *aa.* In theory this is a population in Hardy-Weinberg equilibrium, as a moment with pencil and paper will convince us. If the population were 100 *AA,* 600 *Aa,* and 900 *aa,* we could feel reasonably confident that the offspring would occur in about the same proportions as the parents, the gene pool remaining unchanged. However, we have only 16 people, not 1600. How will these 16 people marry each other (we will assume that they are equally divided as to sex)? We can imagine many combinations, but let us suppose that the matings actually occur as follows: (1) *AA* × *aa;* (2) *Aa* × *Aa;* (3) *Aa* × *aa;* (4) *Aa* × *aa;* (5) *aa* × *aa;* (6) *aa* × *aa;* (7) *Aa* × *Aa;* (8) *aa* × *aa.* We shall next suppose that this population replaces itself; the 16 parents being represented in the next generation by 16 offspring who live to reproductive age in their turn. Each mating produces two such offspring. What will be the nature of these offspring? Here the laws of chances will enter in, but let us suppose that the outcome is as follows:

	Mating	**Offspring**	
1.	*AA* × *aa*	*Aa, Aa*	(this is necessarily so)
2.	*Aa* × *Aa*	*aa, aa*	(other combinations are possible)
3.	*Aa* × *aa*	*aa, aa*	(other combinations are possible)
4.	*Aa* × *aa*	*Aa, Aa*	(other combinations are possible)
5.	*aa* × *aa*	*aa, aa*	(this is necessarily so)
6.	*aa* × *aa*	*aa, aa*	(this is necessarily so)
7.	*Aa* × *Aa*	*Aa, aa*	(other combinations are possible)
8.	*aa* × *aa*	*aa, aa*	(this is necessarily so)

Although the outcome might have been different, this is what actually did happen. As a result there has been a change in the population:

Parents: 1 *AA*, 6 *Aa*, 9 *aa*
Offspring: O *AA*, 5 *Aa*, 11 *aa*

What is the nature of the gene pool of these offspring?

$$p = 5/32 = 15.6\%$$
$$q = 27/32 = 84.4\%$$

Thus we see that the frequency of the dominant gene has "drifted" downward, whereas the frequency of the recessive gene has "drifted" upward correspondingly (from the 25 percent and 75 percent, respectively, of the parental generation). In a third generation this drift might continue, or it might be reversed. If it continued, gene *A* might disappear from the population completely in a generation or two. In that event we would say that gene *a* had attained "fixation." If the direction of drift were reversed, gene *A* would increase in frequency and this trend might continue until gene *a* was eliminated and gene *A* had reached fixation.

Anything could happen. The point we wish to make is that genetic drift is a matter of random fluctuations in small populations, entirely without regard to the usefulness of the genes concerned.

It is possible to simulate the conditions leading to genetic drift by employing simple models composed of marbles, or of colored beads, representing the genes. For a description of such models see Moody (1970).

Can we demonstrate instances of genetic drift in actual human populations? An apparent instance of it has been studied by Glass and his students (1952, 1953, 1956). The population consisted of an Old Order Dunker (Old German Baptist Brethren) community in Pennsylvania. The community has existed for over two centuries, ever since the ancestors migrated from Germany. The average size has remained between 250 and 300 persons, with less than 100 parents present in any given generation. Fecundity is high, but the total size remains fairly constant because of emigration from the group. "The maintenance of the group as a genetic isolate has thus depended not so much on the failure of marriages with outsiders to occur, as rather upon the characteristic exclusion from the group of those who do marry outside" (Glass et al., 1952).

Glass and his associates studied a variety of genetic traits, comparing their frequency of occurrence with frequencies in (a) West Germany, whence the ancestors came, and (b) the United States in general. For our purposes we shall consider only one example: the MN blood groups (pp. 81–84). The findings were as follows:

	M	MN	N
West Germany	29.85%	49.9%	20.2%
Dunker isolate	44.5%	41.9%	13.5%
United States	29.16%	49.58%	21.26%

(From B. Glass et al., 1952.)

We note that the figures for West Germany and for the general population of the United States agree closely in the frequencies of the three types, but in the Dunker community the proportion of type M people is considerably elevated. There is apparently no advantage to be gained by being of type M (certainly no advantage that would apply to Dunkers more than to other people), so that the higher frequency of type M Dunkers may be ascribed to chance, that is, to genetic drift.

Similar evidence of drift was found in the distribution of the ABO blood groups and of some, but not all, of the other traits studied.

The MN groups are of interest, in that drift seems to be demonstrable within the lifetime of people still living. If we let p represent the frequency of gene L^M, and q the frequency of gene L^N, in the United States in general the frequencies are approximately as follows: $p = 0.54$, $q = 0.46$.

The investigators divided their Dunker population into three age groups, representing roughly three generations: (a) 56 years old or older, (b) 28 to 55 years old, (c) 2 to 27 years of age (Glass 1953, 1956). Among people of the oldest generation the gene frequencies were approximately those of the surrounding population, just given; among people of the second generation, $p = 0.66$, $q = 0.34$; and among the young people of the third generation, $p = 0.74$, $q = 0.26$. Glass commented: "These genes were apparently caught in the act of drifting." We note at once the similarity to our hypothetical example above.

Another example of genetic drift may be provided by the high percentage of blood group A people in the Blackfoot and Blood tribes of Indians in Montana. Approximately 80 percent of these people belong to group A, in contrast to the 2 to 22 percent found in other tribes of North American Indians. Because there seems to be no reason why being of group A should be of more advantage to Blackfeet and to Bloods than it is to other tribes of Indians, we ascribe the occurrence of this high percentage to genetic drift.

Founder Principle A consideration related to genetic drift has been called the founder principle (Mayr, 1942). Suppose that the population of a country has this distribution of the MN blood groups: 29 percent M; 50 percent MN; 21 percent N. We shall also suppose that a group of 100 people emigrate from this country to some previously uninhabited region, to form a colony. The emigrants may consist of a random sample of the original population, or they may not. As an extreme example we may take the possible, albeit improbable, case in which all 100 emigrants belong to type M. Then in future generations all the people in the colony will be found to be of type M, instead of having the distribution of groups that characterized the ancestral population "back in the Old Country." Less extreme deviations from the original proportions are more likely, of course. The point is that by chance the founders of a new colony may differ in gene frequencies from the population from which the founders come. Thus, by chance, the colony may differ from the ancestral population.

Importance of Drift We see, then, that chance may cause populations to differ from one another. By itself chance seems to operate most effectively when populations are small and isolated. Without doubt some of the differences between small, isolated groups of people can be attributed to this cause. Of how much importance is genetic drift in producing the diversity found in human populations on a larger scale? What, if any, role has it played in human evolution?

Here we encounter considerable differences of opinion. Theoretical considerations and observations and experiments with plants and animals indicate that most rapid evolutionary change occurs in small populations, although not in populations so small as to be in danger of extinction. It is in such small populations that drift is likely to occur. Hence it is argued that in the past a small population may by chance have become characterized by a high frequency of a certain gene-determined trait. Perhaps that trait proved to be a valuable one, better fitting its possessors to live in their environment, or permitting them to enter a new environment and exploit its possibilities. When this occurred, the "lucky" population would flourish, and perhaps expand at the expense of less well-endowed contemporaries. Thus evolutionary progress toward more perfect adaptation to the old, or a new, environment would be made.

Note that this progress involved two steps: (a) a small population becoming possessed of a trait by chance, that is, through genetic drift; (b) natural selection operating on the trait, in terms of its value to the population. The first step is important only insofar as it leads to the second. "Nonadaptive differentiation is obviously significant only as it ultimately creates adaptive differences" (Wright, 1948). The words are those of Sewall Wright, long identified with the idea of genetic drift—so much so that it is sometimes termed "the Sewall Wright effect." Even he sees it only as a first step, providing a small population with something that natural selection can then utilize.

Is genetic drift really necessary as a first step? A number of students of the subject answer in the negative (see Ford, 1949). They argue that genetic drift occurs only in populations having less than 1000 breeding individuals in any given generation, and that such small populations have no evolutionary future. They may continue to exist as they are for a long time, but they are not "going anywhere." Furthermore, proponents of this view maintain that natural selection itself operates from the start on every significant genetic change that occurs. Hence there is no need to invoke a first step in the form of a random occurrence such as genetic drift. Probably, as usual, both views have validity. In some instances, random drift may be a contributing factor, in other instances not. At any rate, all agree that the ultimate arbiter is natural selection.

Neutral or Nonadaptive Traits Before leaving the subject of genetic drift, we may note that it has sometimes been invoked to explain the occurrence in a population of what are called **nonadaptive traits.** Nonadaptive traits are *neutral;* they confer no benefit, and neither are they disadvantageous. Because natural selection can act only on traits that are not neutral, how can we explain nonadaptive traits? Why is it that various populations have an MN blood group distribution of around 29 percent M, 50 percent MN, 21 percent N? Here we are regarding the MN blood groups as nonadaptive traits. We are free to do so until someone finds that they have some significance in human life. Frankly, we cannot answer the question; but chance may have been involved. Perhaps as a result of genetic drift in a small ancestral population the 29:50:21 ratio came into existence, and it has persisted ever since in the multitudinous descendants of those ancestors. Perhaps the founder principle was operative. Under stringent conditions, populations go through "bottlenecks." Famine, natural catastrophes, and epidemics cause great loss of life, leaving small numbers of survivors to become the parents of the next generation. The collection of genes that the survivors chance to have will then determine the genetic constitution of future generations.

The foregoing comment about regarding the MN blood groups as nonadaptive traits is indicative of the fact that our list of such traits is steadily dwindling with advancing knowledge. Had this book been written a few years ago, the ABO blood groups would have been mentioned as being nonadaptive. However, in our discussion of these groups (pp. 209–210), we noted that they may be connected to susceptibility to disease and so may not be neutral at all. Many, if not most, genes are *pleiotropic,* having two or more effects on their possessors. Some of these effects may be noticeable but insignificant; others, perhaps not so noticeable, may be of real significance in the lives of their possessors. The noticeable effects may be anatomical; the significant effects may be biochemical or physiological. If so, natural selection will act on the significant effects, and the insignificant ones will simply "tag along."

Imagine, for example, a red-eyed insect living in the Temperate Zone. In this insect a dominant mutation occurs, which changes the eye color to black *and* alters the physiology so that the insect can live in extreme cold. Eventually biologists might find a black-eyed race of this insect living in the Arctic. This would occur, not because black eyes were advantageous for arctic living but because of the physiological accompaniment that made it possible for the insect to survive in extreme cold. The example is hypothetical but undoubtedly true to life, and it explains why, as our knowledge of biochemistry, physiology, and pathology increases, our list of nonadaptive traits becomes smaller and smaller. In other words, there remain fewer and fewer traits to be explained merely on the basis of chance. More and more of what we had previously regarded as nonadaptive traits we now see to be subject to the action of natural selection.

Natural Selection

Natural Selection Is Differential Fertility It is appropriate that our discussion of natural selection follows that of differential fertility, because both terms have essentially the same meaning. Both are operative whenever possessors of one genotype leave fewer offspring than do possessors of another genotype. As a result, the gene pool of the next generation contains a disproportionate number of the genes from the more fertile parents.

This equating of natural selection to differential fertility may seem strange to some readers who are more used to thinking of natural selection as "the survival of the fittest," a term that Darwin himself used in his classic (1859) entitled *On the Origin of Species by Means of Natural Selection, or the Preservation of Favoured Races in the Struggle for Life,* to quote the full title as we seldom do. Darwin emphasized survival although he recognized full well that unless the individuals who survived also reproduced, nothing significant for future generations would ensue.

And who are the "fittest"? They are the individuals who leave the largest numbers of offspring. "The 'fittest' is nothing more remarkable than the producers of the greatest number of children and grandchildren" (Sinnott, Dunn, and Dobzhansky, 1958). This concept of Darwinian fitness has not always been recognized and, indeed, will seem strange to some readers today. In times past, the fittest have been thought of as the strongest and most belligerent—the best fighters. Sadly, this misconception has been used to justify all manner of social injustices: "to the victor belongs the spoils," "the devil take the hindmost," "might makes right," and the like. Cut-throat competition and the exploitation of less fortunate peoples have been justified on grounds connected with this misconception. Some of the saddest events of world history during the past century have evolved from mistaken ideas concerning what constitutes fitness and what are the privileges and prerogatives of those who can demonstrate their fitness by becoming "top dog."

This perversion of Darwin's theory received the name "Social Darwinism" (interested readers will find writings by its proponents collected in Appleman, 1970). But we are wisest to reserve the term "Darwinism" for the central idea of natural selection itself, employing the term "Darwinisticism" for the various, often unfortunate, philosophical theories that have been built upon it (Peckham, 1959).

We see, therefore, that fitness in the Darwinian sense is contributed to by any aspect of body and mind that leads to the production of large numbers of offspring. Obviously, individuals must survive, at least to reproductive age, to do this, but survival in itself is not enough. A strong, healthy, intelligent person may survive, but if he does not pass on genes to offspring, the *genetical* constitution of the future is not improved by his having lived.

Please note that we did not say society is not improved by his having lived. Man has two kinds of inheritance: *genetic* and *cultural.* Cultural inheritance is the accumulated wisdom of generations, transmitted by the spoken and written word, by art, music, and so on. People who have no children may make great contributions to this type of inheritance. However, cultural inheritance is characteristically a human phenomenon. Among plants and animals, and even in at least the earlier stages of the evolution of man, an individual's greatest contribution to posterity was his genes. Consequently, those who contributed their genes most copiously had the most influence on the future.

Origin of Diversity Through Natural Selection How does natural selection give rise to diversity? When mutations occur, they will increase in frequency in the gene pool if possessors of them have more offspring on the average than do the individuals who do not possess them.

Let us consider first the situation in which a normally occurring recessive gene, *a,* mutates to form a dominant allele, *A.* At first all individuals possessing the gene will be heterozygous, *Aa*. If the gene contributes to survival and fertility, such *Aa* individuals will leave a disproportionate share of offspring, and hence the genes *A* will increase in frequency in the gene pool. Eventually, homozygous *AA* individuals will arise (offspring in the first instance of *Aa* × *Aa* matings). If they are as viable and fertile as *Aa* individuals, they also will contribute to the increase of gene *A* in the gene pool.

On the other hand, *AA* individuals may be less viable or fertile than *Aa* individuals. In this case, every time homozygotes are produced, genes will be lost through failure of the homozygotes to produce "their share" of offspring. Such superiority of heterozygotes is termed *heterosis,* a term we employed in Chapter 21 in our discussion of the sickle-cell gene, Hb^S. We noted that in an environment containing malaria, heterozygotes for this gene have an advantage. Hence, although homozygotes for the gene suffer severe anemia and have few children, the gene is kept in the gene pool at a substantial frequency.

A third possibility is that the new allele *A decreases* the viability or fertility of its possessors. If so, the gene will be eliminated or at least reduced to the low frequency maintained by recurrent mutation.

Recessive Mutations We turn next to the situation in which a normally occurring dominant gene, *A*, mutates to form a recessive allele, *a*. If recessivity is complete, *Aa* individuals will be exactly like *AA* individuals in viability and fertility. Hence the gene will accumulate in the gene pool until heterozygotes become common enough so that they chance to mate together and produce the homozygous recessive, *aa*. If these individuals enjoy some advantage involving viability or fertility, the frequency of the gene will increase in the gene pool. If, however, they suffer some disadvantage, the frequency of the gene will decrease. However, as we noted in Chapter 25, as long as selection operates on homozygous recessives only, change in gene frequency, be it upward or downward, will occur very slowly. Much more rapid change will occur if recessivity is not complete, if *Aa* individuals differ from *AA* individuals. If both *Aa* and *aa* individuals are *inferior* in viability or fertility, gene *a* will be eliminated much more rapidly than it will be if *aa* individuals alone are deficient. If both *Aa* and *aa* individuals are superior to *AA* individuals, gene *A* will tend to disappear from the gene pool.

If, however, *Aa* individuals are superior to both *aa* and *AA* individuals, we have the condition of heterosis mentioned above. This leads to a *genetic polymorphism* in that the gene pool tends to reach equilibrium as a result of the two opposing forces. Although detrimental to homozygotes, a gene (Hb^S, for example) does not disappear entirely if it is beneficial to heterozygotes.

In sum, we see that the fate of any given new mutation will depend on the answer to several questions. Does it produce an effect in heterozygotes or does it not? If it does, are homozygotes superior or inferior to heterozygotes? Are homozygotes for the new mutation superior or inferior to homozygotes for the "old" (unmutated) gene? For the sake of simplicity, we have confined our discussion and questions to a single pair of alleles, *A* and *a*. When multiple alleles, polygenes, or other complexes of genes are involved, as they frequently are, the genetic situation becomes much more complex, but, of course, natural selection acts on the phenotypes produced by gene complexes as well as on the phenotypes produced by single genes.

In stating the above questions we have used the terms "superior" and "inferior." In the present context these terms refer to all traits (anatomical, physiological, psychological) that in one way or another are concerned with relative ability to contribute genes to offspring. It is important to emphasize again that natural selection always operates in an *environmental* setting, as we noted in connection with the sickle-cell gene. Indeed, in a very real sense it is the environment that *does* the selecting. It is the arbiter that determines which phenotypes shall leave the greatest number of offspring.

Modern Medicine and Natural Selection Modern medicine tends to alleviate the harsh "judgments" of the arbiter just mentioned. As we mentioned in Chapter 14, for example, in times past males with hemophilia characteristically died young, and so seldom married and produced offspring. This was natural selection in action. Now modern medicine enables hemophiliacs to live, marry, and produce children. Does this mean that natural selection no longer operates against hemophilia? As our discussion above should have made evident, natural selection will continue to operate against hemophilia unless and until "cured" hemophiliacs *average to produce as many* children as do nonhemophiliacs. In the preceding chapter we noted another example of this as we discussed the differential fertility of persons with multiple neurofibromatosis. We might multiply examples, but these will suffice.

While modern medicine does not eliminate natural selection, it does reduce the effect of natural selection. This reduction tends to increase the number of deleterious genes that accumulate in the gene pool—in other words, to increase the "genetic load" (p. 339). For example, with hemophiliacs now having children, we may expect that the percentage of hemophilia genes in the gene pool will increase. This will lead to an increase in the number of hemophiliacs in future generations. Granted that these future hemophiliacs can all have their symptoms alleviated by medical means, is it good for mankind to have an ever-greater proportion of individuals dependent upon medical treatment in order to live normal lives? In view of the great number of deleterious genes whose frequencies may be increased by this relaxation of natural selection, grave problems are sure to arise. Some geneticists are pessimistic about our genetic future. Thus Muller (1950) visualized a future in which people's time and energy "would be devoted chiefly to the effort to live carefully, to spare and to prop up their own feeblenesses, to soothe their inner disharmonies and, in general, to doctor themselves as effectively as possible. For everyone would be an invalid, with his own special familial twists."

Such pessimism may be premature, however. We still have much to learn about genes and their interactions in producing phenotypes. And as mentioned in Chapter 1, there is even the possibility that we may eventually be able to repair or replace harmful genes through "genetic engineering." "But even the most optimistic geneticists agree that the probable increase of genetic defects and deficiencies poses a serious threat to man's future" (Osborn, 1968).

Returning to the subject of the origin of diversity by natural selection, we recall that natural selection is constantly tending to cause organisms to become better adapted to the environments in which they live. To the extent that two populations live in differing environments, that place different demands on their inhabitants, natural selection will cause those populations to differ from each other. In this way, natural selection promotes diversity.

Origin of Races

Microgeographic Races Let us imagine a large population of animals that is thriving and growing in size so that it spreads over more and more territory. As it extends its range in this manner, portions of the population become separated from each other by distance and perhaps also by geographic barriers. Perhaps the animals spread from the lowlands up into neighboring mountain valleys, so that the inhabitants of each valley are no longer in contact with the inhabitants of other valleys. Such isolated portions of a large population we frequently call subpopulations. What will happen in each of these subpopulations?

If a subpopulation is small, genetic drift may operate. By chance some genes may become common in one subpopulation, not in another. This will introduce diversity, so that in time one subpopulation will come to differ from another in various ways.

In addition, natural selection will operate on each subpopulation to cause it to become adapted to its particular mountain valley. Perhaps one valley is wooded, another covered with sagebrush. One valley may be well watered, another dry and barren. All manner of differences may be found. In each valley a premium will be placed on ability to adapt to specific conditions. As time goes by, the inhabitants of the wooded valley will come to differ in various ways from the inhabitants of the unwooded valley; the inhabitants of the well-watered valley will have developed some traits not possessed by the inhabitants of the barren valley, and vice versa.

As a result of the combined action of genetic drift and natural selection, the subpopulations have become different from each other and from their ancestors. Now if an inquiring biologist comes along and studies them, he will probably wish to recognize the differences by designating each subpopulation a **microgeographic race.** By this he will mean that each subpopulation has some distinguishing features and occupies a small unit of territory. Examples are numerous in biological literature. As one instance, Dice (1937) studied small populations of whitefooted mice (*Peromyscus leucopus*) inhabiting woodlots in the neighborhood of Ann Arbor, Michigan. The woodlots were separated from each other by open fields, which the woodland inhabitants seldom crossed. Thus they formed subpopulations of the sort we have been discussing. Although separated from each other by only three or four miles, the subpopulations showed statistically significant differences in a variety of bodily and skeletal measurements and in hair color.

Geographic Races If the territory inhabited by a race is larger, the "micro-" is dropped and the term **geographic race** or **subspecies** is employed. Usually, the physical differences between one geographic race and another are more marked than are the differences between microgeographic races. In many if not most instances the geographic races probably arose from microgeographic races that, although small and isolated, developed traits that fitted them for a larger territory. In cases in which this larger territory was available and unoccupied by animals of similar kind, the microgeographic race could expand to become a geographic race.

Aside from its territoriality, what are the attributes of a geographic race? In general, two are important. Geographic races usually differ from each other in a variety of bodily traits, although the differences are small compared to the basic similarities. Members of different geographic races are usually entirely interfertile if and when they do come into contact. They have not developed what we call **reproductive isolation;** they will interbreed freely when the opportunity is offered.

Races usually differ from each other in the *frequencies* with which genes occur rather than in the presence of a gene in one race and its absence from another. For example, at one time I studied the distribution of blood group antigens in some populations of deer mice (*Peromyscus*) from the Columbia River Valley (Moody, 1948). These antigens were similar in many ways to the human A, B, M, N, etc., we have discussed. The different populations, which were in part representatives of different geographic races, differed from each other in the percentages of mice possessing each of the antigens. This situation exactly parallels that found in human races. The four blood groups, A, B, AB, and O, are represented in every race, but the percentages of people who belong to each one differ from race to race. In Chapter 12 we discussed the differing percentage of antigen B possessed by inhabitants of England and China. A high percentage of antigen B is characteristic of Mongolian peoples, exemplified by the Chinese. Interestingly enough, some peoples who are not Mongolian also have a high proportion of B, for example, Abyssinians, and Pygmies in the Congo. Eskimos, Portuguese, and Australian aborigines, members of three different races, resemble one another in blood group distributions.

This illustrates a very important point about races: *The differences between races are of the same kind as the differences between groups of people within races.* Race differs from race in frequencies of genes concerned with blood antigens, ability to taste PTC and other chemicals, production of melanin in the skin, and all manner of anatomical, physiological, and biochemical traits. Populations *within* one race differ in these ways, too; members within one family may differ in them. Thus, races are merely "constellations of characters" (Boyd, 1950). The "characters" are traits by which people within races also differ. The genetic bases of some of the characters have been analyzed. We are still ignorant of the genetics involved in the others, although we have no reason to doubt that principles learned from the genetically analyzed traits apply to them as well.

It is no accident that anthropologists have difficulty in classifying races. To a degree, like students of races in plants and lower animals, they are attempting to draw lines where no lines exist. We find no difficulty in distinguishing Scandinavians from Japanese, or either of these from Congolese, and we may say that they belong to white (Caucasoid), Mongolian, and Negroid races, respectively (although some anthropologists will quarrel with the names chosen). So far everything has been easy. "It will, however, be far from easy to delimit these races if one observes also the inhabitants of the countries geographically intermediate between Scandinavia, Japan, and Congo, respectively. Intermediate countries have intermediate populations, or populations which differ in some characteristic from all previously outlined races" (Dobzhansky, 1963). Thus one "constellation of characters" differs from another, but only in part, and this other constellation differs to some extent, but not completely, from a third, and so on. No wonder anthropologists have difficulty in deciding where one race begins and another leaves off. Biologists are faced with the same problem. As we are now in a position to understand, the problem arises from the very nature of the origin of races: small isolated populations acted on by drift and natural selection, with interbreeding (hybridization) whenever the populations come into contact through migration, conquest, or other means of expanding the originally small territories.

Origin of Human Races We shall never know the history of the races of man, but from our knowledge of race formation in general we can draw some probable conclusions.

The evidence suggests that before the development of civilization, even before the advent of agriculture, our remote ancestors were hunters and food gatherers. Judging from modern peoples who have retained this means of livelihood, they were joined together into rather small family groups or tribes. Because a hunting tribe must have a considerable area available to it as a source of food, the population cannot be of the density possible to an agricultural society. Hence each tribe defends its own territory from encroachment and so keeps itself comparatively isolated from its neighbors. "The requirements of hunting and collecting keep the number of people who live near enough to one another to breed as a unit within about 500 or 600 individuals" (Coon, 1962). Coon estimated that of this number only about a third would at any one time be of childbearing age. Hence each such group would constitute a subpopulation of the type we discussed above. To a considerable extent, also, each would be a genetic isolate although the isolation would break down at times through migration, conquest, wife-stealing, and so forth.

Insofar as each of these small groups retained its identity, it constituted a subpopulation within which genetic drift and natural selection could operate, as we have discussed. In part by chance, in part in response to differing environmental needs, the subpopulations came to differ from one another in a manner comparable to the formation of the microgeographic races we discussed above. Most of these groups existed for a time and then disappeared, or were incorporated into other groups. Some of them, especially ones well separated from each other for thousands of years, persisted and became the ancestors of today's major races.

If we are correct, then, the differences between races may be explained largely as a result of the action of genetic drift and natural selection. Ancestral populations would have been small enough so that drift could have occurred. This would have been especially true at times when a tribe was small as an aftermath of famine, war, or disease. The survivors, and hence "founders" (p. 421) of the future population, might by chance have had gene frequencies quite unlike those that prevailed before the tribal catastrophe. Consequently, future generations might have differed quite strikingly from previous ones, and the differences might have been transmitted down to the present time. In this manner, genetic drift in small ancestral populations can account for racial differences that are neutral or nonadaptive. At present we do not know what proportion of racial traits falls into this category. It is often assumed that most of them do. Yet, as we noted previously, the more our knowledge of the body's physiology and biochemistry increases, the more we realize that many traits previously thought to have no significance are, in fact, of importance to their possessors, and thus subject to the action of natural selection.

Skin Color The analysis of racial differences in physiology and biochemistry is in its infancy, and therefore any conclusions drawn at the present time must be extremely tentative. Let us consider, as our example, the trait that everyone thinks of first when race is mentioned: skin color. The distribution of people with much melanin in their skin suggests that a dark skin is an adaptation for living in open country with strong sunlight. Such people are found in open regions of Africa south of the Sahara Desert, in Melanesia, New Guinea, southern India, and Australia—thus this is a trait common to different races. It seems likely that the dark skin is a protection against damage by ultraviolet radiation, although Blum (1961) raised doubts as to its actual importance in evolution. Recalling that vitamin D, the "sunshine vitamin," is synthesized in the skin, Loomis (1967) pointed out that dark skin is a protection against overproduction of vitamin D, while white skin is an adaptation for increased production of that vitamin in regions where people are exposed to reduced amounts of sunlight. In northern latitudes, for example, white skin would thus tend to protect children from developing rickets.

There is much we do not know about the importance of differences in skin color, however. Melanin, the principal pigment, is contained in cells called melanocytes located between the two layers of the skin, the dermis and epidermis. These cells are produced by embryonic nervous tissue. People of all races have the same number of these cells; the difference is in the amount of melanin the cells contain. The amount of pigmentation is controlled by at least three hormones, two from the pituitary gland, one from the pineal gland (Lerner, 1961). Hence we see here a complex contributed to by the nervous system and the endocrine glands. Lerner suggested (1961): "The color-regulating hormones may have even more important functions as members of the mysterious group of neurohormones, which mediate the interplay between the endocrine and nervous systems in the regulation of the body." If this proves to be correct, color differences may be a more or less unimportant by-product of physiological differences that were important at the time the races originated. As in so many aspects of human genetics, we await the results of future research with great interest.

Adaptations to Heat and Cold World War II stimulated interest in physiological differences between races because "the global nature of modern warfare stimulated the interest of several nations in man's ability to live in all climates, particularly the arctic" (Coon, 1962). Of the researches summarized by Coon, we shall mention only adaptation for living in the cold. Investigations demonstrated that Mongolian peoples, such as Indians in Alaska and in Tierra del Fuego, compensate for heat lost from their bodies by having an elevated basal metabolism. In the case of the Canoe Indians of Tierra del Fuego, the "basal metabolism is 160 per cent higher than the norm for whites of the same age and stature" (Coon, 1962). When their hands are placed in cold water, the blood flow to the hands is increased to twice that of white men tested under the same conditions. By contrast, such increased flow of blood to hands placed in cold water is an adaptation not possessed by Lapp reindeer hunters, who live under similar stringent conditions, or by white Norwegian fishermen living above the Arctic Circle.

On the other hand, Australian aborigines who sleep naked on the desert where temperatures may go below freezing have a different adaptation. The hands and feet become chilled (with temperature as low as 54° to 59° F), but the internal body temperature remains normal. A complex arrangement of blood vessels makes this possible.

Going from cold to heat, we may mention that Negroes have been found to have a greater tolerance for humid, hot conditions than have whites (Dobzhansky, 1962).

Various anatomical characteristics have their physiological concomitants. Peoples who live in a dry atmosphere tend to have noses with narrower openings than do peoples who live in a damp atmosphere. In the narrower nose the air inhaled is humidified more efficiently. People who live in cold climates tend to be stockier, with shorter limbs, than do people who live in warm climates. The nearer the body form approaches a sphere, the less the heat lost through the surface. These are only a few of the adaptive differences between peoples. (For more complete discussion see Coon, 1962, and Dobzhansky, 1962.)

Unfortunately, in examples like those cited we do not know to what extent the differences are genetic, to what extent they depend on the ability of the body to become acclimatized to extreme conditions. Yet even this ability to adjust to a variety of environmental conditions has a hereditary basis. One's genotype produces a "norm of reaction," an ability to adjust to a range of conditions. The range may be broad or it may be narrow. One genotype may make possible accommodation to a range of high temperatures, another to a range of lower ones. Hence, in one form or another, genetic differences are certainly involved.

"Identity Versus Equality" This heading, borrowed from Mayr (1963), suggests that equality and identicalness or sameness are two different things. Some people are disturbed by the thought that there *are* any differences between races. In their devotion to race equality they may even deny the evidences that such differences exist. In our discussion we have tried to introduce the matter in such a way as to emphasize the fact that race formation in man is part and parcel with race formation throughout the living world. Drift and natural selection in virtually all organisms cause populations to develop differences in the frequencies of genes, and such differing populations we call races (in plant, insect, bird, lower mammal, or man). These differences have nothing to do with *equality.* This is an ethical concept, not a biological one. When the framers of our Declaration of Independence stated that "all men were created equal" they did not mean that all men were created identical. They knew better than that. In the words of Mayr: "Simply stated, equality means equal status before the law and equal status in human social relations in spite of genetic difference." It means that skin color, form of hair, shape of nose, blood groups, and all manner of other racial differences are irrelevant in determining legal and social status.

Intelligence Do races differ in intelligence? This question has been answered in various ways.

One point of view is that of those who deny that intelligence is to any degree gene-determined. If that were true, any genetic differences between the races would have no effect on intelligence, which presumably would be determined entirely by environment. Such a point of view totally ignores the results of twin studies and other evidence of a genetic component in intelligence (Chap. 24).

Another viewpoint is that of those who recognize that genes are important in determining intelligence, and who recognize that genetic differences exist between races. If races differ in the frequencies of genes controlling such things as skin color and blood groups, should we not also expect them to differ in the frequencies of genes controlling intelligence? The small ancestral groups from which the races sprang differed in physical and social environments. Might not these differences have led to differences in the distribution of genes controlling intelligence?

At least two answers have been given to these questions. Investigators who have answered in the affirmative have presented evidences that such differences do exist. The evidence usually consists of attainment in school and results of intelligence tests. We noted in Chapter 24 that attainment on intelligence tests varies with the amount of education one has had. It also varies with the socioeconomic group to which one belongs. Studies indicate that the subject matter used in constructing the tests is more likely to be familiar to children of the "upper" groups than it is to children of the "lower" groups. If this is true *within* one race, how much more bias will be introduced by using one and the same test for children from different racial and cultural backgrounds, who frequently differ in socioeconomic rank as well. Until it is possible to devise a truly culture-free intelligence test, conclusions as to racial differences will be in doubt.

Readers interested in the pros and cons of the question of racial differences in intelligence are referred to Scarr-Salapatek (1971a, b) and the references she cited. Our purpose at present is merely to view the question from the perspective of evolution.

Let us suppose for the moment that the difficulties are overcome, and that an investigation does afford conclusive evidence that two races differ in average intelligence. As every schoolchild knows, when large numbers of people are tested, the scores tend to approximate the familiar bell-shaped normal frequency curve. The largest numbers of individuals group around a midpoint or median, smaller numbers attaining higher or lower scores, respectively. Figure 27.1 presents a hypothetical situation of this kind. Here we have two overlapping bell-curves with slightly differing medians. The median of race A differs from the median of race B by a few IQ points, but note the huge area of overlap. Large numbers of people in race B are more intelligent than are large numbers of people in race A. The curves illustrate a most important point about racial differences: the differences found *within* races are greater than the differences *between* races. This is observably true and can be documented abundantly. Applied to our hypothetical races, it means that great contributions may be expected from members of both races, and it means that programs for education and environmental improvement in general must consider each individual as an individual without regard to race. Gifted children of both races must be accorded similar opportunities. Mental defectives of both races must be similarly cared for. What we are saying, then, is that even if an ideal investigation does reveal average differences in intelligence between races, that finding will have no real significance for the managing of human affairs.

Figure 27.1 oversimplifies in representing "intelligence" as a unit. As we noted in Chapter 24, there are many component parts to intelligence, and heredity seems not to be equally important in all of them. It may well be that race B will be superior to race A in some components, and vice versa. In our present state of knowledge concerning intelligence we can merely suggest a not-improbable possibility.

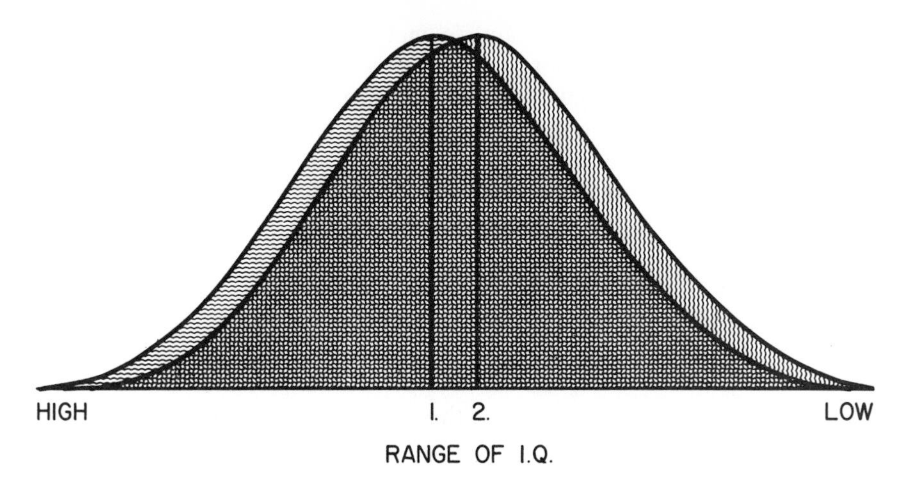

FIG. 27.1 A suppositious case: two hypothetical races differing from each other in average IQ. Each race is represented by a normal frequency curve, one curve shaded vertically, the other horizontally.

Educability Thus far we have been discussing the first answer to our question as to whether racial differences in intelligence are to be expected—the affirmative answer. Other writers consider that differences of this kind are unlikely to have arisen in human evolution (Dobzhansky and Montagu, 1947; Montagu, 1959; Dobzhansky, 1962). Man's greatest superiority over lower animals is his ability to control his environment instead of being controlled by it. Other organisms become adapted to live in one environment, or at most a restricted number of environments. Man can and does live in all environments, from the Tropics to the Arctic, from the desert to the rain forest, from sea level to the high Andes and Himalayas. This plasticity or adaptability (as opposed to narrow adaptation) is man's most distinctive attribute. At its base lies the ability to learn by experience, educability. At all times and in all societies, so the argument runs, natural selection will have placed a high premium on educability, which undoubtedly is a gene-controlled trait, at least in part. We have a tendency to underestimate the complexity of the lives and social structures of people we regard as "primitives." We do this largely from ignorance. As we learn more of the demands placed on them for survival, the more we respect their abilities, much as they may differ from ours. An American capable of contributing to the space program might well die of thirst in the Australian desert, where an aborigine could easily locate a subterranean water supply. "Life at any level of social development in human societies is a pretty complex business, and is met and handled most efficiently by those who exhibit the greatest capacity for adaptability, plasticity" (Dobzhansky and Montagu, 1947). Hence it may well be that natural selection has produced the same degree of educability in all races. If there are differences in averages, they are likely to be small and of no real practical significance, as noted above.

Race Mixture We mentioned earlier the fact that when previously isolated populations come into contact, through migration, conquest, or expansion of territory, they interbreed. In this way, gene frequencies acquired by the populations during their period of isolation are altered, and traits are combined through hybridization to form new "constellations." This interfertility of all peoples has contributed to the problems of classification faced by anthropologists.

At this point we may appropriately note the death of an old idea. This was the theory that in the beginning there were "pure" races (e.g., pure Caucasoids, pure Mongolians, pure Negroids), and that the present great diversity of peoples arose through intermarriage between these original pure races. The more we learn about prehistoric peoples, including the "fossil men," the more we realize that mankind has always been characterized by great diversity. The pure races are a myth.

As far as evidence is available, it indicates that race mixture has always been characteristic of mankind. Biologically, then, it is entirely normal. All peoples share so many fundamental similarities, despite superficial dissimilarities, that hybrids are normal and healthy. This, of course, assumes that in each case the specific parents and ancestries were such as to provide the genes for health. Indeed, we might well have mentioned hybridization, along with natural selection and drift, as a force in the production of new races. For example, some anthropologists consider that a new race is now arising in Hawaii. Called "Neo-Hawaiian" by Coon, Garn, and Birdsell (1950), the new race represents an amalgam of genes from many other peoples, notably Polynesian, European, Japanese, and Chinese (see also Garn, 1961; Dobzhansky, 1962). These writers also suggest that a new race is emerging in our own country: "North American Colored." Here the genes are derived mainly from African, European, and American Indian sources. Just so, new races with their internal variabilities and fluctuating gene frequencies have doubtless been arising from time immemorial.

Because race mixture is biologically normal, is it therefore desirable? Biologically, yes. The problems involved are *sociological*. However they are very real and difficult problems. They stem principally from the fact that "half-castes have been more often than not treated as outcasts by society" (Montagu, 1959). In those societies in which this is not true, as apparently in Hawaii, race mixture is normal and without disadvantage. When will this attitude be true of all societies? Prejudices die hard, and usually only with the deaths of the holders thereof, but I think we are not being too optimistic if we foresee a future in which race mixture will seem normal and matter-of-course.

References

Appleman, P. (ed.), 1970. *Darwin.* New York: W. W. Norton & Company.

Blum, H. F., 1961. "Does the melanin pigment of human skin have adaptive value?" *Quarterly Review of Biology,* **36**: 50–63.

Boyd, W. C., 1950. *Genetics and the Races of Man.* Boston: Little, Brown & Company.

Coon, C. S., 1962. *The Origin of Races.* New York: Alfred A. Knopf.

Coon, C. S., S. M. Garn, and J. B. Birdsell, 1950. *Races.* Springfield, Ill.: Charles C Thomas.

Darwin, C., 1859. *The Origin of Species by Means of Natural Selection.* (Available in many reprint editions, e.g., Collier Books series, The Crowell-Collier Publishing Company, New York; Mentor Books series, New American Library, New York; Modern Library series, Random House, New York.)

Dice, L. R., 1937. "Variation in the wood-mouse, *Peromyscus leucopus noveboracensis,* in the northeastern United States," *Occasional Papers of the Museum of Zoology, University of Michigan,* **352**: 1–32.

Dobzhansky, T., 1962. *Mankind Evolving.* New Haven, Conn.: Yale University Press.

Dobzhansky, T., 1963. "Genetics of race equality," *Eugenics Quarterly,* **10**: 151–160.

Dobzhansky, T., and M. F. Ashley Montagu, 1947. "Natural selection and the mental capacities of mankind," *Science,* **105**: 587–590.

Ford, E. B., 1949. "Early stages in allopatric speciation." In G. L. Jepsen, E. Mayr, and G. G. Simpson (eds.). *Genetics, Paleontology, and Evolution.* Princeton, N.J.: Princeton University Press. Pp. 309–314.

Ford, E. B., 1964. *Ecological Genetics.* New York: John Wiley & Sons, Inc.

Garn, S. M., 1961. *Human Races.* Springfield, Ill.: Charles C Thomas.

Glass, H. B., 1953. "The genetics of the Dunkers," *Scientific American,* **189**: 76–81.

Glass, H. B., 1956. "On the evidence of random genetic drift in human populations," *American Journal of Physical Anthropology,* **14**: 541–555.

Glass, H. B., M. S. Sacks, E. F. Jahn, and C. Hess, 1952. "Genetic drift in a religious isolate: An analysis of the causes of variation in blood group and other gene frequencies in a small population," *American Naturalist,* **86**: 145–159.

Lerner, A. B., 1961. "Hormones and skin color," *Scientific American,* **205**: 98–108.

Loomis, W. F., 1967. "Skin-pigment regulation of vitamin-D biosynthesis in man," *Science,* **157**: 501–506.

Mayr, E., 1942. *Systematics and the Origin of Species.* New York: Columbia University Press.

Mayr, E., 1963. *Animal Species and Evolution.* Cambridge, Mass.: Harvard University Press.

Montagu, Ashley, 1959. *Human Heredity.* Cleveland, Ohio: The World Publishing Company.

Moody, P. A., 1948. "Cellular antigens in three stocks of *Peromyscus maniculatus* from the Columbia River valley," *Contributions from the Laboratory of Vertebrate Biology, University of Michigan,* **39**:1–16.

Moody, P. A., 1970. *Introduction to Evolution,* 3rd ed. New York: Harper & Row.

Muller, H. J., 1950. "Our load of mutations," *American Journal of Human Genetics,* **2**:111–176.

Osborn, F., 1968. *The Future of Human Heredity.* New York: Weybright and Talley.

Peckham, M., 1959. "Darwin and Darwinisticism," *Victorian Studies,* **3**:3–40. (Also in Appleman, 1970).

Scarr-Salapatek, S., 1971a. "Unknowns in the IQ equation," *Science,* **174**: 1223–1228.

Scarr-Salapatek, S., 1971b. "Race, social class, and IQ," *Science,* **174**: 1285–1295.

Sinnott, E. W., L. C. Dunn, and T. Dobzhansky, 1958. *Principles of Genetics,* 5th ed. New York: McGraw-Hill Book Company, Inc.

Wright, S., 1948. "On the roles of directed and random changes in gene frequency in the genetics of populations," *Evolution,* **2**:279–294.

28

Genetic Counseling

In this chapter we shall discuss some of the objectives, methods, problems, and difficulties of genetic counseling. There will be no thought of equipping readers to do such counseling themselves. This is a specialized field of medical science and practice, as the following discussion will make abundantly evident, and hence no more appropriate for amateurism than is neurosurgery, for example. On the other hand, we hope some readers will be stimulated to enter this fascinating field. If so, they should acquire extensive knowledge of both genetics and medicine, including if possible the earning of doctor's degrees in both fields. If only the M.D. degree is earned, the potential counselor should supplement the medical training with extensive graduate study in genetics, including practical experience in genetical research. If the future counselor holds only the Ph.D. in genetics, he will be wisest to become a member of a team the other members of which are medical specialists in appropriate fields, such as pediatrics.

Our viewpoint in this chapter will be more that of the potential "consumer" of genetic counseling. If at some time we find it desirable to consult a genetic counselor, what may we expect in the way of advice, and what resources will be available to the counselor upon which to base that advice?

We begin with the most common situation presented to genetic counselors. A married couple have produced a deformed or defective child. They come to the counselor with the question: "If we have another child, will he also have this abnormality?"

Necessity of Exact Diagnosis

Before the counselor can answer the couple's question, he must be certain of the exact nature of the abnormality. This seems so obvious as not to require mentioning, but in many cases exactness in diagnosis requires the services of highly trained and experienced specialists. This is because of the occurrence of "look-alikes," disorders that have similar appearance or symptoms—phenotypes, in genetic terms—but are inherited differently.

As an example, we mention two types of **dwarfism:** diastrophic dwarfism, which is inherited as a recessive, and achondroplastic dwarfism, which is inherited as a dominant. The signs and symptoms are so similar that physicians who are not specialists in birth defects can easily confuse them. A case is on record of a normal couple who had a child diagnosed as an achondroplastic dwarf. Because the trait is dominant, normal persons do not serve as "carriers," and most occurrences represent new mutations. As we know, mutations of a given gene are rare, hence parents of an achondroplastic dwarf are very unlikely to have a second one. Encouraged by this information, the couple had a second child, and this child was also a dwarf. Nevertheless, they persisted and produced a third child; this child was a dwarf too. Then the children came to the attention of a specialist in birth defects. He found that they were really diastrophic dwarfs. Both parents were evidently heterozygous for the recessive gene concerned. Thus the chance that any given child would have the condition was ¼, rather than being extremely small as it would have been if the original diagnosis had been correct. Here we have dramatic evidence of the crucial importance of exact diagnosis (I am indebted to William E. Hodgkin, M.D., for this example).

No good purpose would be served by cataloging all the "look-alikes," but we shall mention two more pairs. We mention first two types of **muscular dystrophy:** a rare type inherited as an autosomal recessive, and a more common X-linked type, Duchenne's muscular dystrophy, having an incidence of about 2.8 per 10,000 births (Porter, 1970). Obviously a counselor must be certain which type a child has before he can advise the parents as to the probability of recurrence.

Another pair of "look-alikes" is the **Hurler syndrome** and the **Hunter syndrome.** Both are characterized by retarded growth and mental development, and by various abnormalities of skeleton and biochemistry (Porter, 1970). Distinguishing between them is difficult, yet the Hunter syndrome is an X-linked recessive, the Hurler syndrome an autosomal recessive.

Pedigree Study

Having made certain of the diagnosis, the counselor next needs to gain as much information as he can about the relatives and the ancestors of the couple who have come to him for advice.

To give focus to our discussion, we picture the mother of a boy with the Duchenne type of muscular dystrophy coming to a counselor for advice. First we ask: "What does it mean to parents to have a child with Duchenne's muscular dystrophy?" It means seeing a normally active little boy gradually become weaker and less coordinated in his movements, so that typically by the time he is 12 years old he is confined to a wheelchair, and before he reaches the age of 20 he has died. This indicates how strong is the desire of the parents of such a boy not to have another child who will go through the same tragic sequence. The question they ask the counselor is no mere matter of academic interest.

A New Mutation? The first question facing the counselor is: "Is the mother heterozygous for the X-linked gene, or did the gene arise as a new mutation in the ovum from which the affected boy developed?" To answer this question, study of the family is important. If such study shows that the mother had an affected brother, or an affected uncle on her mother's side of the family, the counselor has evidence that the mother is probably a carrier for a gene that is "in the family." On the other hand, if the mother has no known affected relatives, the question remains open. She may be heterozygous, or the affected boy may have arisen as a new mutation. Because mutations are rare events—of the order of 1/100,000 (p. 330)—the probability is that she is heterozygous. If she is, the risk that the next *boy* will be affected is 50 percent. [Murphy (1973) pointed out that the total risk should be reduced from ½ somewhat by reason of the fact that the first son *might* have been a new mutation, in which case the chance of recurrence is very slight. His calculations led him to conclude "that the risk to the next son is for practical purposes ⅓." Such computations would be difficult to explain to parents, however, and we anticipate that the ½ will continue to be regarded by counselors as sufficiently accurate.] Because the chance that the next child will be a girl is ½, the chance that the next child will be affected is $\frac{1}{2} \times \frac{1}{2} = \frac{1}{4}$ (25 percent).

Expected Burden With as much skill and tact as possible, the counselor explains the statistical probabilities to the parents: The chance that their next child will have the dread syndrome is ¼. Or, to turn the matter around: The odds that their next child will be *normal* are 3:1. The parents themselves must then make the decision. Are the odds for normality good enough? Shall they take the risk?

As pointed out by Murphy (1973), in reaching a decision the parents will be influenced greatly by the *burden* that the phenotype imposes. If, for example, the mother had been heterozygous for deutan color blindness, the odds that the next child would not be color-blind would also be 3:1. But clearly, the burden imposed by being color-blind is of an entirely different order of magnitude from the burden imposed by having Duchenne's muscular dystrophy. With color blindness the parents would almost certainly accept the odds; with muscular dystrophy they probably would not, and should not.

Murphy (1973) used arbitrary units in estimating severity of burden, and plotted this against the risk of recurrence: the genetic probabilities (Fig. 28.1). In this diagram we see that the "risk" of Rh+ parents having an Rh+ child is very high, but the "burden" is zero (upper left-hand corner). On the other hand, the risk in the cases of hemophilia and of Duchenne's muscular dystrophy is smaller but the burden is incomparably greater. At the bottom of the diagram we note that recurrence risk of achondroplasia (achondroplastic dwarfism) is very low—the mutation rate of 10^{-5}—but the burden is large. Of the other traits

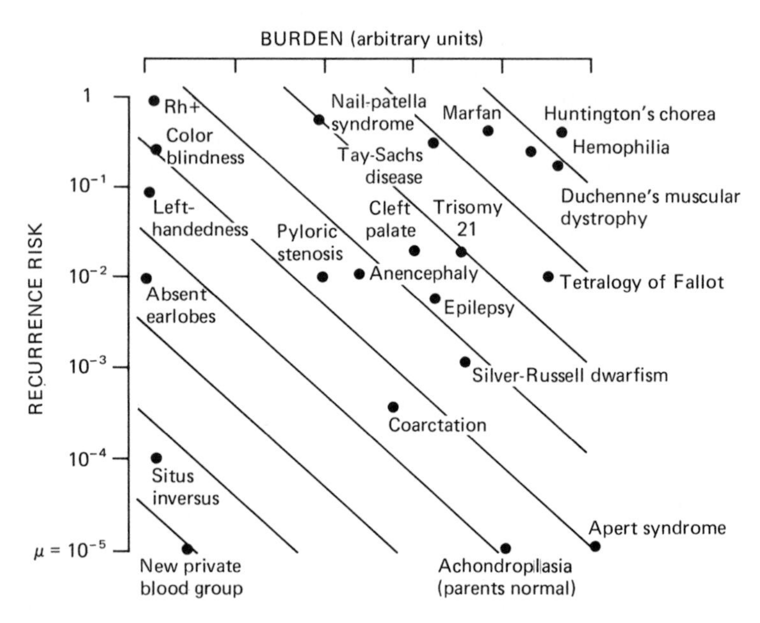

FIG. 28.1 Expected burdens imposed by various inherited traits. The risk of occurrence of the condition is plotted on the vertical axis; on the horizontal axis an attempt has been made to represent the burden imposed when the condition occurs. (From Murphy, E. A., 1973. "Probabilities in genetic counseling," in Bergsma, D., ed., 1973. *Contemporary Genetic Counseling*. National Foundation, Birth Defects: Original Article Series, Vol. IX, No. 4, pp. 19–33.)

shown in the diagram, note the position of 21-trisomy. As Murphy put it, the parents will ask both "What is the risk?" and "What is the risk of?"

Aid to Diagnosis At times the study of the family may aid in diagnosing the phenotype of the child who caused the parents to seek counseling. If it is found, for example, that a *female* relative has the same symptoms of muscular dystrophy, the counselor knows that the X-linked Duchenne type is not the type involved. Or in the case of autosomal genes, if the trait "skips" generations (Chap. 6), the gene involved is most likely recessive, the parents of the affected child being heterozygous. Finding that the parents were first cousins would increase the probability that the counselor was dealing with a recessive autosomal trait.

One Phenotype, Varied Modes of Inheritance The counselor will search the scientific literature to learn what previous investigators have found about the inheritance of the trait with which the parents are concerned. He may find that in some families the trait shows autosomal inheritance, in other families X-linked inheritance.

For example, aside from Duchenne's muscular dystrophy, there are other types of muscular dystrophy and "heterogeneity exists (1) because of mutations involving *different* genes, one in an X-chromosome and another in an autosome and (2) because of different mutations of presumably the *same* gene, namely, the one situated in the X-chromosome" (Porter, 1970).

Such differences are to be expected in traits involving more than one enzyme. If two enzymes are concerned, one ("enzyme *a*") may be determined by an autosomal gene, the other ("enzyme *x*") by an X-linked gene. So in one family it may be "enzyme *a*" that is abnormal, in another family "enzyme *x*." Consequently, the counselor must study the family of the inquiring parents to determine whether the autosomal gene or the X-linked gene is defective in their family.

Possible Value of Linkage Relationships Study of the family may reveal linkage between certain traits, and in some cases this linkage may be of value in counseling. Here we recall our mention of "marker genes" in Chapter 15 (pp. 247–249), where we discussed their possible usefulness in genetic counseling.

In addition, linkage between genes that are not usually considered markers may be useful to the counselor. In Figure 28.2 we present an imaginary but true-to-life family in which the X-linked traits of deutan

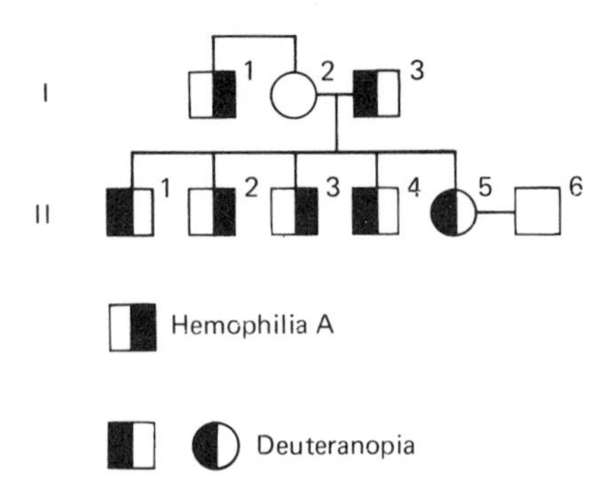

FIG. 28.2 Linkage as an aid to counseling in an imaginary but realistic family situation. See text.

color blindness and hemophilia A are both present (Chap. 14). The couple who come to the counselor are II-5 and II-6. They wish to know the chance that they may have a hemophilic child. The wife (II-5) is color-blind, having inherited from a color-blind father and a mother who was a carrier, as evidenced by her color-blind sons (II-1 and II-4). The wife's mother (I-2) was evidently a carrier for hemophilia also, because she had two hemophilic sons (II-2 and II-3). She also had a hemophilic brother. Now the daughter (II-5) wonders what the chance is that she is also a carrier for hemophilia, like her mother.

Because the color-blind sons of I-2 are not hemophilic, and the hemophilic sons are not color-blind, I-2 evidently had the defective genes in different X chromosomes (repulsion linkage). Thus, using the symbols we employed in Chapter 15, we give her genotype as ($D\ a^H$) ($d\ A^H$). Because II-5 is color-blind, she evidently inherited gene d from her mother, and the probability is that she inherited gene A^H (for normal blood clotting) along with it. She could inherit the gene for hemophilia only if crossing over had occurred in her mother, producing a ($d\ a^H$) chromosome [and a ($D\ A^H$) one]. As we noted in Chapter 15, crossing over between these two loci occurs about 12 percent of the time. So the counselor can tell her that the chance that she is a carrier is 12 percent. If her serum is tested for level of antihemophilic globulin (factor VIII), however, the counselor may be able to determine much more precisely whether or not she is a carrier (see "Detection of Carriers" below). If the 12 percent probability is used, the chance that any child of hers will receive the ($d\ a^H$) chromosome will be half that or 6 percent. Of course, if the child were a girl she would not be hemophilic, although she might be a carrier in turn. In sum, the chance that the first child of II-5 and II-6 will be a hemophilic boy is about 3 percent. The couple would probably find quite reassuring the advice that their child has a 97 percent chance of being nonhemophilic.

Of course, in most families in which hemophilia is present, color blindness is *not* also present, and linkage cannot be of help. In such families the sister of a hemophilic male has a 50 percent chance of being a carrier (assuming, of course, that her father is not hemophilic or the brother's hemophilia the result of mutation). Obviously, such a sister would wish more information. Her probable carrier status could be determined by use of the antihemophilic globulin test mentioned above, although the test has not been perfected so that it will identify all carriers.

Some Sources of Confusion We mentioned above the fact that a given abnormal phenotype may be produced by different abnormal genotypes in different families. Conversely, the same abnormal genotype may sometimes produce different phenotypes or no visible abnormality at all. These are examples of the *variable expressivity* and *partial penetrance* we discussed previously (pp. 70–74). Obviously, they are a source of confusion for a counselor. If, for example, an individual has a certain dominant gene but does not show the phenotype usually associated with it, the counselor will have no evidence that he or she has the gene, and hence may advise incorrectly about the probability that children will inherit the trait usually produced by the gene.

Among the other complicating factors listed in a World Health Organization report on genetic counseling (1969) is the fact that some genetic traits show their effects only when a particular environmental influence is present. The report gives as an example acute intermittent porphyria. This is an inborn error of metabolism with various pathological symptoms including neurological ones. Persons with the genotype develop the symptoms if they take certain drugs, especially barbiturates. (At a somewhat different level, schizophrenia is also an example, as mentioned previously; p. 363.)

Another source of confusion is the occurrence of **phenocopies.** At times environmental factors can combine to produce a phenotype in one individual that closely resembles a phenotype usually produced by genes. In this connection, the embryonic environment is of particular importance. Such an environmentally induced "copy" is not inherited, nor is the trait passed on to offspring. The production of phenocopies is a well-known phenomenon in experimental organisms, and it may well occur in man, too, although we have little actual information on the subject. It is suspected, for example, that retinoblastoma sometimes appears as a phenocopy, in the absence of the dominant gene discussed previously (p. 69; Fraser, 1956). If the survivor of retinoblastoma has the condition because of possession of the dominant gene, half of his children will be expected to be affected. However, if he has the condition because it arose in him as a phenocopy, no children he produces will be affected. This poses a real problem for the counselor. Obviously we need means of distinguishing phenocopies from genetic cases.

Onset After Reproductive Age In some cases, particularly with dominant genes, the abnormal symptoms may not appear until after the individual may have married and produced children. The most prominent example of this is Huntington's chorea (Huntington's disease), in which possessors of the dominant gene concerned do not develop the tragic symptoms until they are in their 30's or 40's (pp. 68–69). By this age many affected people have produced their families, some members of which are likely to have inherited the gene.

If a parent develops the symptoms of Huntington's chorea, his children as they reach marriageable age are faced with two important questions: (a) Will I become insane when I grow older? (b) If I marry, will my children have the disease? As we saw earlier, affected individuals are heterozygous, Hh. If, as almost always happens, they marry individuals who do not have the gene, the mating is $Hh \times hh$. Hence, half of the children are expected to be Hh. This means that what a counselor can tell an individual who has a choreic parent is this: There is a 50 percent chance that you will eventually become choreic; if you do not become choreic, none of your children ever will either; if you do become choreic, 50 percent of any children you have produced in the meantime will be likely to become choreic in their turn. Not very comforting information. But Fraser (1956) wrote of a physician who was so skillful in giving counsel that he managed to convey to the son of a choreic mother the dangers of his having choreic children "without revealing to him the fact that he was a candidate for the disease." In this particular case the man and his wife decided to adopt children instead of having any of their own.

Obviously, what is needed for counseling in Huntington's chorea is some means of determining which young people in choreic kindreds have the gene. A closely linked marker gene might be helpful. Better would be a biochemical test indicating some way in which a young person who has the gene differs from persons who do not. Klawans et al. (1972) tested young people who had a choreic parent, giving them the compound L-DOPA by mouth. This caused some of the group to manifest choreic symptoms temporarily. Others were not affected, and none of the normal controls were affected. Will those who reacted positively to the L-DOPA test develop the disease in later life, while none of the others do? Only time will tell how valuable this test may be in identifying possessors of the gene before the disease develops. It is also possible that abnormal brainwave patterns, electroencephalograms, might be a means of identifying young people who will eventually become choreic (Patterson, Bagchi, and Test, 1948). Other tests may be developed; for example, Falek and Glanville (1962) employed a test that records the muscular tremors made by the hand.

When we eventually have a test enabling us to tell which *children* have the gene, genetic counseling will be greatly facilitated. When the presence of the gene is thus identifiable, a young person found to have it can be told that there is a 50 percent chance that any child he produces will become choreic. Conversely, if he is found not to have the gene, he need have no worries of producing choreic children.

But what of the counselee himself? Unless the counselor is as skillfull as the physician mentioned above, a young person told that he has the gene will realize that as he grows older he is certain to develop increasing muscular incoordination accompanied by severe mental deterioration. What would be the effect on a young person's emotional and psychological health of knowing this? Truly, if knowledge brings power, it also creates problems. The hope is, of course, that remedial measures may be discovered to prevent the tragic and fatal symptoms, as they have been in the case of phenylketonuria (PKU), for example.

Other Requests for Counseling

Parents of a Child Having a Recessive Autosomal Trait The parents discussed at the outset had a child with an X-linked trait (Duchenne's muscular dystrophy being our principal example). When the gene involved is autosomal and recessive, the affected child is homozygous, may be of either sex, and both parents are heterozygous if both have the normal phenotype. In this situation, the chance that the next child will have the same defect is 25 percent. Earlier in the book we discussed various traits of this type, such as albinism, sickle-cell anemia, galactosemia, and PKU. Sly (1973), an experienced counselor, mentioned that he also told parents that their children were unlikely to have affected offspring in their turn *unless* (a) the gene is common, as in sickle-cell anemia (the more common the gene, the greater the chance that two heterozygotes will marry each other), or (b) the children marry a close relative, such as a first cousin. As will have been evident from our earlier discussion (Chap. 11), first-cousin marriage has its dangers.

Siblings of an Individual Having a Recessive Autosomal Trait What is the chance that a normal brother or sister of a child with PKU, for example, is a carrier? Because the parents both have the genotype *Pp* (p. 54), the expectation for children is ¼ *PP*, $\frac{2}{4}$ *Pp*, ¼ *pp*. Hence among the *normal* children twice as many are *Pp* as are *PP*. In other words, the chance that the *normal* sibling of a *pp* individual is heterozygous is ⅔. Such a sibling will, of course, run the risk of having a child with PKU only if he or she marries another heterozygote (or a recovered PKU patient).

In this connection we may appropriately mention that Reed (1963) emphasized the point that ideally genetic counseling should be *premarital,* especially when serious disorders are involved. This is particularly true for conditions in which the carrier state can be identified with some certainty. In Chapter 5 we mentioned that carriers for the sickle-cell gene have the "sickle trait," which can be identified by a simple blood test. Similarly carriers of the gene that produces thalassemia (Cooley's anemia) can be identified. Homozygotes suffer a severe anemia called thalassemia major, whereas heterozygotes have a milder condition called thalassemia minor. If they have no outward symptoms, they can be identified by biochemical tests. In the province of Ferrara, Italy, one person in ten is heterozygous (in some townships, one in five) (Roberts, 1962). Roberts pointed out that with random mating in these townships, one child in every hundred has the severe anemia. In some regions, tests are made and registers posted so that heterozygotes may avoid marrying each other. He commented: "Of course, it is a little late when the banns have been published and the wedding arrangements made, but this knowledge may well prevent those early approaches which ultimately lead to courtship and marriage."

At least, this would be the *hope* of counselors who are consulted by young people before marriage. Another possibility is that a couple who know they are both carriers for some serious defect might marry, but not produce children of their own, acquiring children by adoption (or by artificial insemination of the wife using sperm from a male who is not a carrier?). Obviously, such decisions are often accompanied by emotional and psychological problems.

As we shall mention below, the number of genetic defects for which the carrier state can be identified is now large and steadily increasing.

Parents of Children with Common Congenital Malformations

Under this heading Sly (1973) listed such conditions as cleft palate, neural tube defects (e.g., spina bifida), and congenital heart disease. These conditions are not inherited in any simple Mendelian manner. In each case the counselor must rely on empirical findings as to the experience of others on how frequently the trait in question is likely to recur in another child. Sly stated that while the risk is greater than it is for parents who have not produced such a child, it generally lies below 10 percent.

Parents of Children Having Chromosomal Aberrations The most common example is Down's syndrome. As noted in Chapter 20, the risk that a couple who have one child with the syndrome will have another varies with the chromosomal basis. Accordingly, for accurate counseling it is important that karyotypes of the affected child, and if possible, of his parents, be prepared.

In some 95 percent of the cases, ordinary 21-trisomy is found. When it is, the risk of recurrence varies with the age of the mother, being greatest in older mothers (Chap. 20). With mothers under 30 years of age, recurrence is rare, although it is more likely than is birth of a child with Down's syndrome to women in general. But the risk rises to about 2 percent for mothers over 45 (Hamerton, 1971).

The picture is entirely different when translocation trisomy is present in the family. With D/21 or 21/22 trisomy, the risk may exceed 30 percent (WHO report, 1969), and with 21/21 trisomy it rises to 100 percent (p. 323). Reed (1968) stated: "None of the numerous carriers with whom I have talked wished to take the risk of further children."

Increasing the Precision of Counseling

Detection of Carriers As will be evident from our discussion so far, counseling is much more precise when the counselor can say "You are a carrier" than it is when the best he can do is state the statistical probability that the inquirer is a carrier. For this reason, much research is devoted to methods for detection of the carrier state.

Our earlier discussions have mentioned examples of successful identification. We noted above that carriers for the sickle-cell gene and for thalassemia can be identified by blood tests. Detection of heterozygotes for PKU was discussed on p. 56, and for galactosemia on p. 62 (Fig. 5.7).

Tests for heterozygosity vary with the trait being studied. Sometimes they involve giving the person being tested a large dose of the substance that affected persons do not metabolize normally, as in the "phenylalanine loading" test for persons suspected of being carriers of PKU. Sometimes the enzyme levels in parents of an affected child can be measured directly and compared to the normal level. At other times, tests are made on leucocytes or fibroblasts grown in tissue culture. Some conditions produce changes that can be detected by microscopic study.

As the result of application of these various methods, Hsia (1968) was able to state that the carrier state could be detected in 67 of the known 99 inborn errors of metabolism. And the number steadily increases. Hsia (1968) and Nadler (1970a; 1972) published lists of hereditary disorders for which detection of carriers was possible. Among the disorders we have mentioned previously were (in addition to the ones mentioned above): afibrinogenemia (p. 173), Hurler's syndrome, Hunter's syndrome, hemophilia A, hemophilia B, G6PD deficiency, crystinuria and pentosuria (p. 53), Duchenne's muscular dystrophy.

In some of the tests there is overlap between some heterozygotes and some homozygous normals. But the tests are constantly being improved and "sharpened" to attain the ideal of distinguishing every carrier individual from every homozygous normal individual. Such improvement is evident in detection of carriers for Duchenne's muscular dystrophy, for example. Formerly the sister of an affected boy could be told that her chance of being a carrier was 50 percent. By the time the WHO report (1969) was published, determination of the level of creatine phosphokinase (an enzyme) in the blood made possible identification of about 80 percent of the carriers. No doubt in time 100 percent detection will be attained or at least closely approached.

This goal has apparently been attained in the case of Tay-Sachs disease. As Figure 28.1 indicates, this disease involves a heavy burden in that it leads to death by about the third year of life, and death is preceded by loss of muscular function, paralysis, progressive deafness, blindness, and other tragic symptoms. As noted in Chapter 11, the gene concerned is an autosomal recessive, and in some populations (e.g., Ashkenazi Jews) its frequency is such as to result in some 2.6 percent of the population being heterozygotes. Fortunately, there is a relatively simple serum test that distinguishes heterozygotes from homozygous normal individuals (O'Brien, 1972). The test is based on level of activity of an enzyme: hexosaminidase A. No overlap of values between heterozygotes and normal controls has been found. Hence screening programs for identification of heterozygotes are practicable, and would be particularly desirable among the brothers and sisters in families in which the disease has occurred. O'Brien noted that among them the incidence of heterozygotes is ten times higher than it is for Ashkenazi Jews as a whole.

Furthermore, the hexosaminidase A test can be used for prenatal diagnosis (see below). Thus when two heterozygotes are married and the wife becomes pregnant, the parents can be told whether or not the fetus will develop into a child with Tay-Sachs disease. And this can be done in time so that the fetus can be aborted, thereby forestalling the trauma of a birth that is sure to result in early death in any case.

Karyotyping We have mentioned the importance of chromosomal analysis in counseling for Down's syndrome. Karyotypes should be prepared whenever the symptoms suggest that abnormalities of the chromosomes may be present. At present, karyotyping is time-consuming and hence expensive. But introduction of automated methods will undoubtedly change this situation in the near future.

Prenatal Diagnosis Identification of abnormal fetuses has been made possible by use of **amniocentesis.** This is the process of obtaining some of the amniotic fluid surrounding a fetus. It is done for various medical purposes, but our interest is in the use of the fluid and the fetal cells it contains as means for detecting fetuses with genetically determined abnormalities. A sterile hypodermic needle is inserted through the mother's abdominal wall and the uterine wall into the amniotic cavity (Fig. 28.3). Amniotic fluid containing cells produced by the fetus is withdrawn with a syringe. The cells are separated from the fluid. Biochemical and enzymatic analyses can be made on the fluid and on the cells, and the latter can be examined for the presence of Barr bodies, as an indication of the sex of the fetus. The cells may also be grown in tissue culture for further biochemical and enzymatic studies, and for chromosomal analysis by karyotyping.

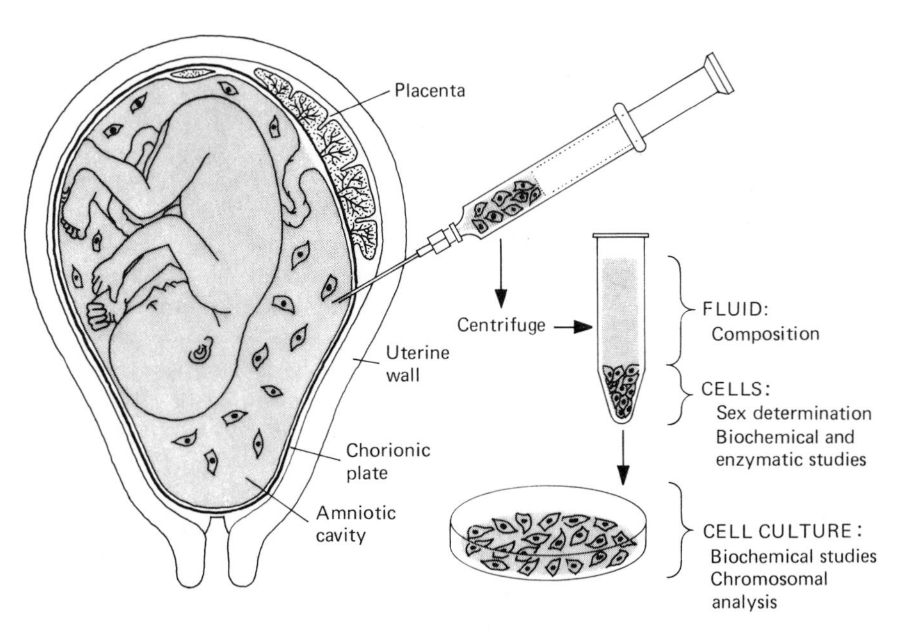

FIG. 28.3 Amniocentesis, with indications of the types of tests performed on the amniotic fluid and the cells it contains. (Redrawn from Friedmann, T., 1971. "Prenatal diagnosis of genetic disease," *Scientific American,* **225**:34–42.)

Thus amniocentesis is a useful tool when there is reason to believe that there is substantial risk that the fetus may have an important disorder. Two other provisos are that (1) the disorder must be detectable in the fluid or in the cells it contains, and (2) the parents must be willing to consider seriously having the fetus aborted if it is found to have the abnormality in question. Otherwise there is little point in determining whether or not a fetus is abnormal. (This means that people who, for religious or other reasons, object to abortion under any and all circumstances will not find prenatal diagnosis helpful.) The process involves some small risk of injuring the fetus or of causing the fetus to abort, and consequently it should be undertaken only when there are serious reasons for doing so.

As a first example we cite cases in which the abnormality of the fetus is not certain. A mother has a hemophilic son or a son affected with Duchenne's muscular dystrophy. She becomes pregnant again. Study of the amniotic fluid cells reveals that the fetus is a male. There is, therefore, 50 percent chance that the child will have the disorder. This poses a difficult problem for the parents. Shall they take the chance that the boy will be normal? Or is the chance that he is abnormal so great that an abortion is best? We sense that important emotional and even psychological stresses and strains accompany the making of such a decision. As we should expect, faced with this problem, some parents decide one way, some the other.

Although they are not without emotional accompaniments, cases in which diagnosis of the fetal disorder is positive are easier to decide. For example, a mother has a son with the Lesch-Nyhan syndrome (pp. 282–283). She becomes pregnant again, and enzyme studies show that the fetus is deficient in HGPRT activity. Abortion is clearly indicated. As noted above (p. 450), the same situation is faced when the fetus is found to be deficient in hexosaminidase A activity, thereby heralding the birth of a child with Tay-Sachs disease.

In other cases a mother has had a child with Down's syndrome, and chromosome studies have indicated that this is translocational trisomy, a D/21 chromosome being present, for example. In a second pregnancy, karyotyping of the fetal cells reveals that the fetus has the D/21 chromosome together with two normal 21's. Although this syndrome is not so severe as is the Lesch-Nyhan syndrome, most parents would conclude that abortion is wise. Nadler (1970b) reported the results of prenatal diagnosis of 38 fetuses of translocation-carrier mothers. Of the 38, 10 were diagnosed as affected, and in all but one case the parents decided to have the fetus aborted.

If a mother has had a child with Down's syndrome of the ordinary 21-trisomy type, the chance that a second child will have the syndrome is slight if the mother is young. If the mother is over 40 years of age, however, determination of the chromosomal constitution of the fetus is highly desirable with a view to possible abortion. Counselors differ in judgment as to whether the risk of having a second affected child warrants amniocentesis if the mother is younger.

The number of disorders in which prenatal diagnosis can be made is increasing steadily. In addition to chromosomal aberrations, some 30 disorders were listed by Friedmann (1971). Of the disorders we have mentioned, the list included the Lesch-Nyhan syndrome, galactosemia, Hunter's syndrome, and Hurler's syndrome. Undoubtedly, future advances will make possible more and more fetal diagnosis. Will increasing capability be used wisely? Many people are concerned about ethical, as well as legal, problems that may arise.

New Capabilities Bring New Responsibilities

Every time advance in science enables us to do something we had not been able to do before, our responsibilities are increased. Think, for instance, of the added responsibilities placed on our shoulders by learning to split the atom. The same principle applies to medical advances. Every advance in this field increases our ability to control our own destinies, our own evolution. And so it is our responsibility to ensure that each advance is used for the maximum good of the people concerned in both this generation and succeeding ones.

Guidance As we indicated, the decision as to whether a counselor's advice is or is not followed rests with the parents. But we, the "potential consumers" mentioned at the outset of this chapter, are going to expect from genetic counselors more than mere statements of statistical probabilities. We are going to expect a counselor to counsel, with wisdom and with compassion for the dilemma in which we find ourselves when we consult him (or her). We shall not expect him to say: "If I were in your place I should do so-and-so." But we shall expect him to share his wisdom as to the physiological, psychological, and emotional consequences of the various courses of action open to us.

This being the case, how much responsibility are we going to place upon the counselor? If, following his advice, an abortion is not performed and an abnormal baby is nevertheless born, are we going to blame the counselor? Or, alternatively, if a pregnancy is terminated on his advice and the fetus is subsequently found to have been normal, are we going to consider bringing charges of malpractice? Let us hope not. We must always remember that biochemical tests, and even chromosomal analyses, are subject to error (Macintyre, 1972). Such tests are being constantly improved, but they can cause a thoroughly competent counselor to reach the wrong conclusion.

With regard to competence, we may mention that professionals in the field recognize that as the number of genetic counselors increases along with the demand for their services, professional standards will be established. With this will come certification and licensing, to ensure that only competent people offer professional advice.

Some Vexing Problems One such problem may arise when the husband and wife cannot agree on whether an abortion should be performed.

The examples we have used so far have concerned such extremely serious abnormalities as Down's syndrome and Duchenne's muscular dystrophy, where disastrous outcome is inevitable. But how serious must an abnormality be to justify termination of pregnancy? If a fetus is found to be XO or XXY, for example, should it be aborted? And how about a fetus found to be XYY? As discussed in Chapter 17, we have only the *possibility* that the extra Y may result in abnormality, not the certainty that it will.

And what if the abnormality is one, like hemophilia, in which the symptoms can be treated after birth? (There is also active research in the possibility of intra-uterine treatment for some conditions.) Most of the abnormalities caused by chromosomal aberrations are not now amenable to remedial measures, however.

And how late in pregnancy can an abortion be performed without being considered infanticide? Neel (1970) stated: "Unfortunately for our peace of mind, it is clear that many pregnancies in which the question of interruption can be raised will not come to [our] attention until the third or even fourth month, and the necessary studies may well require three or four weeks. It is very easy to find ourselves facing the fifth month of pregnancy with the knowledge that a birth only a month later would be treated as a premature infant, to whose salvage special nurseries are devoted."

Another rather frightening possibility is that eventually prenatal screening for genetic defects, and the abortion of defective fetuses when found, may become *compulsory*. The cost to society of rearing and caring for an individual with Down's syndrome, for example, is great (lifetime institutional care for each one is estimated to cost $250,000; Friedmann, 1971). Will the time come when all pregnancies must be monitored and all trisomic-21 fetuses aborted (Miller, 1970)? Here we should point out that the cost of such a massive screening program on a national scale would be enormous. Nevertheless, professionals in this field are concerned about the possibility that parents' decisions may not always be permitted to remain voluntary.

Change in the Gene Pool? In the preceding chapter we mentioned that modern medicine tends to increase the number of deleterious genes that are passed on to future generations, and so to alter the gene pool. Some people are concerned that prenatal diagnosis followed by abortion of defective fetuses is one aspect of medical practice that will have this effect (see Neel, 1970).

Without recourse to mathematics, we can visualize the reasoning behind this point of view quite simply. We picture gene *a* as being a recessive gene that does not kill homozygotes but that in some way renders reproduction by homozygotes highly unlikely. A couple, both carriers, have decided to limit their family to a certain number, two for example. The mating is thus $Aa \times Aa$. It happens that the first child is normal, though a carrier (*Aa*). We shall suppose that the second child is defective (*aa*). Hence, in this sibship, although three *a* genes are present, only *one* will be passed on to future generations. There was no genetic counseling in this case.

Now suppose that the couple, knowing they were carriers, *did* seek genetic counseling, and that amniocentesis was used successfully. When the wife became pregnant the first time, the fetus was diagnosed as normal, and so was not aborted. With the second pregnancy, the fetus was diagnosed as defective, and was aborted. With the goal of having two children, the wife became pregnant a third time, and, we shall suppose, this fetus was normal and hence not aborted. It might or might not have been a carrier, although the odds were 2:1 that it was. For illustrative purposes we shall picture it as being a carrier. Thus the couple achieved their goal of two children: *Aa* and *Aa*. The result is that there are two *a* genes to be passed on to future generations instead of only the one that would be passed on had there been no counseling. If this type of happening were repeated thousands of times over many generations, we see how gene *a* would increase in frequency in the gene pool. But the change would occur very slowly, and in most cases it is only rare genes whose increase would be of concern. Carriers would slowly become more common, and their marriage to each other more frequent. Hence there would be more defective fetuses to be aborted in the future, and more defective children born to those future parents who did not seek counseling. But, as mentioned in our previous discussion, we may be optimistic that future advances in medical science may enable our successors to cope with the defect better than we can today, conceivably even to the extent of *correcting* the genotype so that the unfavorable trait is no longer produced by it.

Also, our discussions earlier in this chapter will have made evident that much might be accomplished by identification of carriers followed by premarital counseling, advising carriers not to marry each other.

Encouragement of Parents Sometimes parents who know there is a serious genetic defect "in the family" are afraid to have children. If they know they can depend on prenatal diagnosis followed by abortion of any defective fetuses, they may be encouraged to have children (Macintyre, 1972). Furthermore, Macintyre pointed out that prenatal diagnosis may actually *save* lives. When the wife in a family of the type pictured unintentionally becomes pregnant, she is very likely to seek an abortion for fear the baby *may* be abnormal. Prenatal diagnosis often shows that the fetus is not abnormal, however, and that there is no need to terminate pregnancy. On the basis of his experience, Macintyre concluded: "There are in fact, a number of healthy, desired youngsters living today whose lives literally have been saved by this type of genetic engineering, and I think that's great!"

Much is being written about ethical and legal aspects of genetic counseling. We have been able to mention only a few of them. For further introduction into this important field, see Fraser, 1970, Friedmann, 1971, and the following books (listed in References under the names of their editors): Bergsma and Abramson, 1970; Harris, 1970, Bergsma, 1972, 1973; Hilton et al., 1973.

Problems

1. On p. 440 we mentioned a pair of parents who, wrongly informed concerning the nature of their first child's affliction, eventually produced three children with diastrophic dwarfism. What is the numerical probability that three children born to parents having these parents' genotypes will have this affliction?
2. We mentioned the "look-alikes": Hurler syndrome and Hunter syndrome, both recessive, but the first autosomal, the second X-linked. The first child of a pair of phenotypically normal parents had the Hurler syndrome. What is the chance that a second child would also have that syndrome? Answer the same question for the Hunter syndrome.
3. In the kindred shown in the accompanying pedigree chart, III-2 had the Hunter syndrome. All other members of the kindred were normal. Contemplating marriage to a normal man, II-3 came to a genetic counselor asking whether she would be likely to have a child with the syndrome. What advice should she receive? Give reasons.

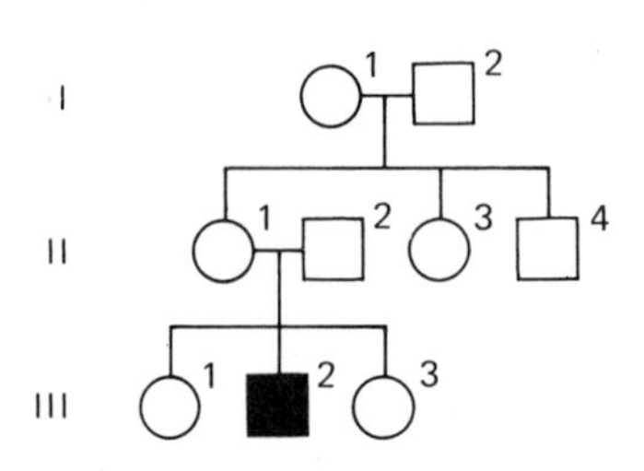

4. What is the chance that *if* the first child of II-5 × II-6 (Fig. 28.2) is a daughter, she will be a carrier of *both* deuteranopia and hemophilia? That she will be a carrier of deuteranopia only? That she will be a carrier of hemophilia only?
5. Figure 28.1 indicates that parents who have a child with pyloric stenosis, parents who have a child with anencephaly, and parents who have a child with teratology of Fallot run the same risk that a second child will be similarly afflicted. Judging from the diagram, will these three pairs of parents be likely to differ in their decision as to whether to have a second child? If so, in what way, and why?
6. A pair of parents who have a child with Down's syndrome come to a genetic counselor for advice. The counselor has the child's karyotype determined. What is the risk that a second child would have the syndrome if the chromosome study indicates (a) that the child has three No. 21 chromosomes (and the mother is 25 years old), (b) that the child has three No. 21 chromosomes (and the mother is 45 years old), (c) that the child has a D/21 or 21/22 translocation chromosome, or (d) that the child has a 21/21 translocation chromosome?
7. Utilizing information given on page 450 concerning Ashkenazi Jewish people, compute the probability that the sister of a boy with the Tay-Sachs syndrome will produce a child with the syndrome if (a) she marries a man whose brother had the syndrome, or (b) she marries a man with no known relatives who had the syndrome.

References

Bergsma, D., (ed.), 1968. *Human Genetics.* National Foundation–March of Dimes, Birth Defects: Original Article Series, Vol. IV, No. 6.

Bergsma, D., (ed.), 1972. *Advances in Human Genetics and Their Impact on Society.* National Foundation–March of Dimes, Birth Defects: Original Article Series, Vol. VIII, No. 4.

Bergsma, D., (ed.), 1973. *Contemporary Genetic Counseling.* National Foundation–March of Dimes, Birth Defects: Original Article Series, Vol. IX, No. 4.

Bergsma, D., and H. Abramson (eds.), 1970. *Genetic Counseling.* National Foundation–March of Dimes, Birth Defects: Original Article Series, Vol. VI, No. 1. Baltimore: Williams & Wilkins Company.

Falek, A., and E. V. Glanville, 1962. "Investigation of genetic carriers." In Kallmann, F. J. (ed.). *Expanding Goals of Genetics in Psychiatry.* New York: Grune & Stratton, Inc. Pp. 136–148.

Fraser, F. C., 1956. "Heredity counseling. The darker side," *Eugenics Quarterly,* **3**:45–51.

Fraser, F. C., 1970. "Counseling in genetics: Its intent and scope," in Bergsma and Abramson, 1970. Pp. 7–12.

Friedmann, T., 1971. "Prenatal diagnosis of genetic disease," *Scientific American,* **225**:34–42.

Hamerton, J. L., 1971. *Human Cytogenetics,* Vol. II. New York: Academic Press.

Harris, M., (ed.), 1970. *Early Diagnosis of Human Genetic Defects.* Fogarty International Center Proceedings No. 6; Washington: U.S. Government Printing Office, H.E.W. Publication No. (NIH) 72–25.

Hilton, B., D. Callahan, M. Harris, P. Condliffe, and B. Berkley (eds.), 1973. *Ethical Issues in Human Genetics.* Fogarty International Proceedings No. 13. New York: Plenum Press.

Hsia, D. Y.-Y., 1968. "Recent advances in the detection of heterozygous carriers," in Bergsma, 1968. Pp. 75–87.

Klawans, H. L., Jr., G. W. Paulson, S. P. Ringel, and A. Barbeau, 1972. "Use of L-Dopa in the detection of presymptomatic Huntington's chorea," *New England Journal of Medicine,* **286**: 1332–1334.

Macintyre, M. N., 1972. "Professional responsibility in prenatal genetic evaluation." In Bergsma, 1972. Pp. 31–35.

Miller, O. J., 1970. "An overview of problems arising from amniocentesis." In Harris, 1970. Pp. 23–30.

Murphy, E. A., 1973. "Probabilities in genetic counseling." In Bergsma, 1973. Pp. 19–33.

Nadler, H. L., 1970a. "Newer procedures in the preconceptional, prenatal and early postnatal diagnosis of birth defects." In Bergsma and Abramson, 1970. Pp. 26–33.

Nadler, H., 1970b. "Risks in amniocentesis." In Harris, 1970. Pp. 129–137.

Nadler, H. L., 1972. "Prenatal detection of genetic disorders." In Harris, H., and K. Hirschhorn (eds.), *Advances in Human Genetics, Vol.* 3. New York: Plenum Press. Pp. 1–37.

Neel, J. V., 1970. "Ethical issues resulting from prenatal diagnosis." In Harris, 1970. Pp. 219–229.

O'Brien, J. S., 1972. "Ganglioside storage diseases." In Harris, H., and K. Hirschhorn (eds.), *Advances in Human Genetics, Vol.* 3. New York: Plenum Press. Pp. 39–98.

Patterson, R. M., B. K. Bagchi, and A. Test, 1948. "The prediction of Huntington's chorea: An electroencephalographic and genetic study," *American Journal of Psychiatry,* **104**: 786–797.

Porter, I. H., 1970. "The detection of carriers and the problem of heterogeneity in genetic counseling." In Bergsma and Abramson, 1970. Pp. 13–25.

Reed, S. C., 1963. *Counseling in Medical Genetics,* 2nd ed. Philadelphia: W. B. Saunders Company. Also available in paperback under the title: *Parenthood and Heredity.* New York: John Wiley & Sons, Inc. (Science Editions, 1964).

Reed, S. C., 1968. "Human factors in genetic counseling." In Bergsma, 1968. Pp. 105–109.

Roberts, J. A. F., 1962. "Genetic prognosis," *British Medical Journal,* **1**: 587–592.

Sly, W. S., 1973. "What is genetic counseling?" In Bergsma, 1973. Pp. 5–18.

WHO Report, 1969. "Genetic counseling," *World Health Organization Technical Report Series,* No. 416.

Appendices

Appendix A

Derivation of the formula expressing the probability that a child of first cousins will be homozygous for a recessive gene at a certain locus in an autosome.

In Figure A.1 the fact that II-2, II-3, III-2, and IV-1 are represented as female is arbitrary and immaterial, as sex linkage is not involved.

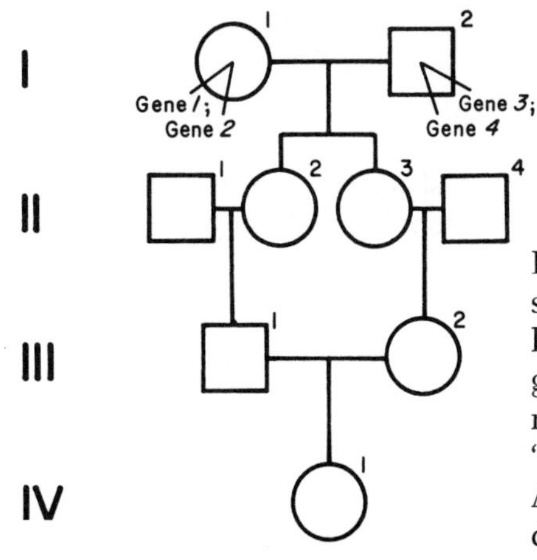

FIG. A.1 Marriage of first cousins.

What are the chances that IV-1 will be homozygous for a recessive gene at a certain locus? The allelic pairs at this locus in the great grandparents, I-1 and I-2, are designated as "gene 1" and "gene 2," and "gene 3" and "gene 4," respectively. Any one of the four genes may be dominant or recessive.

What are the chances that IV-1 will inherit "gene 1" *from her father* (III-1)? The chance that I-1 will pass "gene 1" on to her daughter, II-2, is ½. The chance that if II-2 has the gene she will pass it on to her son III-1 is ½. The chance that if III-1 has the gene he will pass it on to

IV-1 is $1/2$. Thus the chance that IV-1 will receive "gene 1" *from her father* is $1/2 \times 1/2 \times 1/2 = 1/8$.

Similarly, what are the chances that IV-1 will inherit "gene 1" *from her mother* (III-2)? The chance that I-1 will pass the gene on to II-3 is $1/2$. The chance that if II-3 has the gene she will pass it on to III-2 is $1/2$. The chance that if III-2 has the gene she will pass it on to IV-1 is $1/2$. So the chance that IV-1 will receive "gene 1" *from her mother* is also $1/2 \times 1/2 \times 1/2 = 1/8$.

Hence, the chance that IV-1 will inherit "gene 1" from *both* parents (i.e., be homozygous for it) is $1/8 \times 1/8 = 1/64$.

By the same reasoning, we see that the chance that IV-1 will be homozygous for "gene 2" is also $1/64$. And the same reasoning applies to "gene 3" and "gene 4."

Thus the chance that IV-1 will be homozygous for *any one* of the four genes from her great grandparents, I-1 and I-2, is $1/64 + 1/64 + 1/64 + 1/64 = 4/64$ or $1/16$. Note that we *add* the fractions because they represent mutually exclusive events. By this, we mean that IV-1 cannot be homozygous for more than one of the four genes. If she is a "gene 1–gene 1" homozygote she cannot also be a "gene 2–gene 2" homozygote, and so on.

What are the chances that the gene IV-1 is homozygous for is a recessive gene? The answer depends on the frequency of the gene in the population of which this kindred is a part. As usual we designate this frequency of the recessive gene by q. Therefore, the chance that IV-1 will be homozygous for a recessive gene inherited from the great grandparents shown in the pedigree is $1/16q$. It follows that the chance that IV-1 will *not* be homozygous for a recessive gene inherited from the great grandparents shown in the pedigree is $15/16q$ ($1/16q$ subtracted from unity, the totality of cases).

Thus far we have neglected the husbands of II-2 and II-3. These husbands, II-1 and II-4, of course, had parents in their turn although the parents are not shown in the diagram. There is some chance that the recessive gene in which we are interested may have come from them. This chance is represented by the frequency of the gene, q. Hence the chance that homozygosity for the gene may arise in IV-1 from these "outside sources" (*not* from the great grandparents shown in the diagram) is

$$15/16q \times q = 15/16q^2$$

Thus the total chance that IV-1 will be homozygous for a recessive gene is the chance that she will be homozygous because of inheritance from the great grandparents shown ($1/16q$) plus the chance that she will be homozygous because of inheritance from other ancestors ($15/16q^2$).

$$\tfrac{1}{16}q + \tfrac{15}{16}q^2 = q/16\,(1 + 15q)$$

See Dahlberg, G., 1948. *Mathematical Methods for Population Genetics.* New York: Interscience Publishers. Pp. 53–54. Stern, C., 1973. *Principles of Human Genetics,* 3rd ed. San Francisco: W. H. Freeman & Co. Pp. 473–474.

Appendix B

The von Dungren-Hirszfeld hypothesis of ABO inheritance. The basis of the statement by Bernstein (1925) that if the hypothesis were valid the following equation should be correct:

$$(\overline{\mathrm{A}} + \overline{\mathrm{AB}}) \cdot (\overline{\mathrm{B}} + \overline{\mathrm{AB}}) = \overline{\mathrm{AB}} \qquad \text{(p. 182)}$$

Let

$$\begin{aligned} s &= \text{relative frequency of gene } a \\ (1 - s) &= \text{relative frequency of gene } A \qquad \text{(see p. 181)} \\ t &= \text{relative frequency of gene } b \\ (1 - t) &= \text{relative frequency of gene } B \end{aligned}$$

Then taking the gene pairs separately, we find that

$$\begin{aligned} \text{frequency of } AA &= (1 - s)^2 \\ \text{frequency of } Aa &= 2(1 - s)s \\ \text{frequency of } aa &= s^2 \end{aligned}$$

(Note here the similarity to the Hardy-Weinberg formulation.)

$$\begin{aligned} \text{frequency of } BB &= (1 - t)^2 \\ \text{frequency of } Bb &= 2(1 - t)t \\ \text{frequency of } bb &= t^2 \end{aligned}$$

GROUP O. Persons in this group would have the genotype *aabb*. Their expected frequency would be s^2t^2.

GROUP A. Persons in this group would have one of two genotypes:

AAbb, with a frequency of $(1 - s)^2 \times t^2$
Aabb, with a frequency of $2(1 - s)s \times t^2$

Total frequency for the group would be

$$(1 - s)^2t^2 + 2(1 - s)s \times t^2$$

which may be simplified as follows:

$$\begin{aligned} &t^2[(1 - s)^2 + 2(1 - s)s] \\ &t^2[(1 - s)^2 + 2s - 2s^2] \\ &t^2(1 - 2s + s^2 + 2s - 2s^2) \\ &t^2(1 - s^2) \end{aligned}$$

In sum, the expected frequency of group A people would be $t^2\ (1 - s^2)$.

GROUP B. The two genotypes would be

aaBB, with a frequency of $s^2 \times (1 - t)^2$
aaBb, with a frequency of $s^2 \times 2(1 - t)t$

The total, $s^2\ [(1 - t)^2 + 2(1 - t)t]$, simplifies to $s^2\ (1 - t^2)$ in the manner indicated for group A.

GROUP AB.

Genotypes	**Frequencies**
AABB	$(1 - s)^2 \times (1 - t)^2$
AaBB	$2(1 - s)s \times (1 - t)^2$
AABb	$(1 - s)^2 \times 2(1 - t)t$
AaBb	$2(1 - s)s \times 2(1 - t)t$

The total of these frequencies simplifies to $(1 - s^2)\ (1 - t^2)$.

IN SUMMARY:

Group O, $\overline{\mathrm{O}}$, would have a frequency of s^2t^2
Group A, $\overline{\mathrm{A}}$, would have a frequency of $t^2\ (1 - s^2)$
Group B, $\overline{\mathrm{B}}$, would have a frequency of $s^2\ (1 - t^2)$
Group AB, $\overline{\mathrm{AB}}$, would have a frequency of $(1 - s^2)(1 - t^2)$

It follows that $\overline{\mathrm{A}} + \overline{\mathrm{AB}} = 1 - s^2$, calculated as follows:

$$t^2\ (1 - s^2) + (1 - s^2)(1 - t^2) = t^2 - t^2s^2 + 1 - s^2 - t^2 + t^2s^2 = 1 - s^2$$

In the same manner:

$$\overline{\mathrm{B}} + \overline{\mathrm{AB}} = 1 - t^2$$

Therefore

$$(\overline{\mathrm{A}} + \overline{\mathrm{AB}}) \cdot (\overline{\mathrm{B}} + \overline{\mathrm{AB}}) = \overline{\mathrm{AB}}$$

because

$$(1 - s^2)(1 - t^2) = (1 - s^2)(1 - t^2)$$

(Recall that this equation does not in fact fit the data, p. 182.)

Bernstein, F., 1925. "Zusammenfassende Betrachtungen über die erblichen Blutstrukturen des Menschen," *Zeitschrift für induktive Abstammungs- und Vererbungslehre,* 37:237–270.

Appendix C

Chi-square test and its application to data concerning offspring of parents (pp. 194–195).

The chi-square test is sometimes called a test of "goodness of fit." How well do actual data fit expectation based on theory? In the specific example (p. 193), 105 families in which both parents were secretors produced 241 secretor children and 33 nonsecretor children. Use of the formula

$$\left(\frac{q}{1+q}\right)^2$$

indicates that 28.6 nonsecretor children were to be expected. In using this formula, we are advancing the hypothesis that nonsecretion depends on a recessive gene, secretion on its dominant allele. Is the observed deviation in numbers significant evidence against the hypothesis?

Table C.1 shows the application of the chi-square test to this

TABLE C.1 An example of the chi-square test.

	o	e	o − e	$\frac{(o-e)^2}{e}$	
	241	245.4	−4.4	$\frac{19.36}{245.4}$	= 0.079
	33	28.6	+4.4	$\frac{19.36}{28.6}$	= 0.677
Totals	274	274.0	0		$0.756 = \chi^2$

question. In the first column, headed *o* for "observed," we find the number of secretor and nonsecretor children counted in the investigation. In the second column, headed *e* for "expected," we find the numbers expected on the basis of the formula. Note that columns *o* and *e* must have the same total number. (The 245.4 is obtained by subtracting 28.6 from 274, the total number of children.)

In the column headed $o - e$ we find the *deviations,* obtained by subtracting the expected values from the observed ones. Note that the plus deviations must equal the minus deviations and hence the total of the $o - e$ column must be zero.

The deviations are then squared and the squares are divided by the respective expected values: column headed $\frac{(o-e)^2}{e}$. Each fraction is then divided out to form a decimal, and these decimals are added to give a total, which is called the *chi-square value,* χ^2.

We may summarize this computation of chi-square by the formula

$$\chi^2 = \sum \frac{(o - e)^2}{e}$$

(The chi-square value equals the sum of the squared deviations divided by the expected values.)

Reference is then made to the table of chi-square (Table C.2).

TABLE C.2 Table of chi-square.

Degrees of freedom	Probability, *P* 0.90	0.70	0.50	0.20	0.10	0.05	0.01
1	0.0158	0.148	0.455	1.642	2.706	3.841	6.635
2	0.211	0.713	1.386	3.219	4.605	5.991	9.210
3	0.584	1.424	2.366	4.642	6.251	7.815	11.341
4	1.064	2.195	3.357	5.989	7.779	9.488	13.277
5	1.610	3.000	4.351	7.289	9.236	11.070	15.086

Abridged from Table IV of Fisher and Yates, *Statistical Tables for Biological, Agricultural, and Medical Research,* published by Oliver & Boyd Ltd., Edinburgh, and by permission of the authors and publishers. See the book for a more complete table.

The body of the table is composed of chi-square values. The column headings are percentages expressed as decimal fractions, the values of *P*. The percentage in each case is the probability that the amounts of deviation represented by the chi-square values below will be exceeded *by chance.* If the probability is great that chance alone will produce a certain amount of deviation, the occurrence of that amount of deviation is not considered significant evidence against the hypothesis being tested.

The horizontal rows in the table are numbered at the left to indicate the respective "degrees of freedom." To illustrate the idea of degrees of freedom, let us imagine a bag containing three marbles—one red, one white, one black. If, without looking, we reach into the bag and pick up a marble, we may pick up any one of the three; freedom of chance is complete. Suppose that by chance we remove the white marble on the first trial. On a second trial chance can still operate, because the marble we pick up may be the red one or the black one. Suppose that by chance we remove the black marble on this second trial. Now if we reach in our hand a third time, chance can no longer operate; the marble we pick up *must* be the red one. So with three marbles we have *two* "degrees of freedom," two trials in which chance can operate. In general, there is one less degree of freedom than there are classes or categories. In the present example, when we first encounter a child we do not know whether he is a secretor or a nonsecretor. However, if by tests we find that he is a secretor, we have removed the chance that he might be a nonsecretor, and vice versa. Thus with the two categories, "secretor" and "nonsecretor," there is only one degree of freedom.

Accordingly, we enter the table on the first row and look for the chi-square value closest to the one we have obtained (0.756). We see that the chi-square value in the 50 percent column is 0.455, that in the 20 percent column is 1.642. The 0.756 is between these two, nearest to 0.455. This means that approximately 40 percent of the time the amount of deviation represented by our chi-square value would be expected by random chance. Because this probability is so great, we conclude that obtaining the deviation expressed by a chi-square value of 0.756 is not significant evidence *against* the hypothesis that the nonsecretor trait depends on a recessive gene (p. 193).

Where shall the line of significance be drawn? Usage differs, but many biologists draw the line at 5 percent. This means that if the probability of obtaining the observed deviation by random chance is less than 5 percent, the data probably do not fit the hypothesis being tested. The investigator should then obtain more data or formulate a hypothesis that fits the present data more closely.

Appendix D

Haldane's formula for computing the frequencies of X-linked genes, applied to the Xg blood groups.

Let

$$p = \text{frequency of the dominant allele}$$
$$q = \text{frequency of the recessive allele}$$

$$q = \frac{[4(2f + m)(b + 2d) + a^2]^{1/2} - a}{2(2f + m)}$$

where

$$m = \text{number of males}$$
$$a = \text{number of } Xg^a \text{ Y males}$$
$$b = \text{number of } Xg \text{ Y males}$$
$$f = \text{number of females}$$
$$d = \text{number of } Xg\ Xg \text{ females}$$
$$p = 1 - q$$

Derivation of the formula is given in Haldane (1963); see references for Chapter 14. Application of the formula to the Xg blood groups is found in Chown, Lewis, and Kaita, 1964; Noades et al., 1966; Race and Sanger, 1968.

How much difference does computation with Haldane's formula make, as compared to equating q to the frequency of males who show the phenotype of a recessive X-linked gene? In the discussion of the Xg blood groups, we noted that males who have the recessive gene *Xg* are classified as Xg(a—). Accumulated data on 3418 persons indicated that 35.4 percent of the males were Xg(a—). With Haldane's formula it was computed that the frequency of the gene *Xg* was 0.341 (Noades et al., 1966; Race and Sanger, 1968). Thus the comparison is between 0.354 and 0.341. Both are estimates. Although one estimate is doubtless superior to the other, equating to q the frequency of males who show a recessive X-linked trait may frequently give a *useful* approximation.

Glossary

ABO blood groups. Classification of people on the basis of the presence or absence of the red blood cell antigens A and B.

ABO incompatibility. Refers to a marriage in which the husband has antigen A and/or B, which his wife lacks.

Acrocentric. Refers to chromosomes having the centromere located very near one end.

Acute radiation. A given amount of radiation received quickly within a brief span of time.

Adenine. One of the organic bases (purines) found in DNA and RNA.

Adenylate kinase (AK). An enzyme concerned with release of energy within cells. A pair of alleles determine the inheritance of two isozymes of it (p. 261).

Afibrinogenemia. An abnormality in which the blood does not clot normally due to lack of the plasma constituent called fibrinogen.

Agglutinins. Antibodies that react with, for example, red blood cells containing the corresponding antigen (agglutinogen), causing the cells to clump together (agglutinate).

Agglutinogen. An antigen in, for example, red blood cells that will react with the corresponding antibodies (agglutinins), causing the cells to clump together (agglutinate).

Albinism, ocular. An X-linked trait characterized by loss of pigment from the eyes but not from skin, hair, etc.

Albinism, oculocutaneous. Absence of pigment (melanin) from eyes, skin, hair, etc. Occurs in two forms, both inherited as autosomal recessives: tyrosinase-negative, and tyrosinase-positive (pp. 58–61).

Alkaptonuria. A metabolic abnormality in the chain of reactions by which amino acids are broken down into simpler compounds for excretion. An intermediate product, homogentisic acid, collects in the urine, causing it to darken.

Allele (allelomorph). One of two or more alternative genes that may be present at a given locus in a chromosome; e.g., *A* is the allele of *a,* and vice versa.

Allozymes. Two enzymes are said to be allozymes of each other if they are essentially similar (differing only as isozymes do) but are derived from different species (e.g., thymidine kinase from man, and thymidine kinase from mouse). *see* Isozymes.

Amaurotic idiocy. *see* Tay-Sachs disease.

Amino acids. Complex organic compounds from which proteins are formed (see Table 4.1).

Amniocentesis. Sampling of the amniotic fluid surrounding a fetus by inserting a hypodermic needle through the mother's abdominal wall and the uterine wall into the amniotic cavity (see Fig. 28.3).

Amnion. The inner membrane that encloses the embryo and its surrounding fluid (amniotic fluid) (see Fig. 23.1).

Anaphase. A stage in mitosis (see Fig. 3.1).

Anaphase lag. Said to occur when, following metaphase in mitosis, a chromosome fails to move normally toward the pole of the spindle (see Fig. 3.1). As a result, the chromosome is not included in either daughter nucleus; it is "lost" and later resorbed.

Anastomosis. Joining together of blood vessels so that blood flows from one into the other.

Androgen. General term for male sex hormones.

Anencephaly. Absence of brain, or of all but a rudimentary brainstem and basal ganglia.

Aneuploidy. Chromosomal aberration in which the number of chromosomes is increased by less than the haploid number.

Angiokeratoma (Fabry's disease). X-linked defect in which affected individuals have wart-like thickenings of the skin overlying vascular tumors, abnormalities of lipid metabolism, and kidney failure.

Angstrom unit. 1/10,000,000 of a millimeter.

Anthocyanin. A pigment found in flowers.

Antibodies. Proteins synthesized by the human or animal body in response to antigens such as viruses, bacteria, or proteins foreign to the body. The antibodies can react with the antigens that stimulated their formation.

Anticodon. Triplet of transfer-RNA nucleotides complementary to a codon in the messenger-RNA molecule (see Fig. 4.5).

Antigen. A substance that will stimulate the production of antibodies and react with them.

Antihemophilic globulin. A plasma protein forming part of the blood-clotting mechanism; people having hemophilia A are deficient in this globulin.

Antimutagenic agent. Any material or technique that might protect cells and their contents against the mutation-inducing effects of radiation.

Antiserum. Blood serum containing antibodies.

***a priori* method** (of correcting pooled data). A means of compensating for the fact that when families of two heterozygous parents are identified only by virtue of having at least one child with the presumed recessive trait, more than one-fourth of the children will be expected to show that trait.

Artificial insemination. Use of instruments to introduce sperm cells into the uterus of a female.

Ascertainment. In this context, the method by which an investigator learns of individuals or matings—the method of sampling.

Assortative mating. Mating based on choice or preference relative to the genetic trait being investigated, as opposed to random mating.
Autosomes. Chromosomes other than the sex chromosomes, X and Y.
Autozygous. Homozygous for a particular gene (not merely for a particular type of gene) received from an ancestor. It results from consanguinity.

Back cross. Mating of a heterozygote to a homozygote, especially to the homozygote for a recessive gene; e.g., $Aa \times aa$.
Balanced polymorphism. A condition of equilibrium in a population having individuals of two or more forms (phenotypes). The relative frequencies of the two phenotypes remain constant from generation to generation.
Barr bodies. Sex-chromatin (X-chromatin) bodies; deeply staining particles found attached to the inner surfaces of the nuclear membranes of the cells of females (see Fig. 17.7A).
Bilirubin. Red bile pigment formed in the liver from hemoglobin of destroyed red blood cells.
"Bleeder." A person who has hemophilia.
Blending inheritance. Inheritance in which F_1 offspring are intermediate between their parents with respect to some quantitative trait, and F_2 offspring vary widely, usually approximating a normal frequency curve. *see* Polygenes.
Brachydactyly. A hereditary trait characterized by markedly shortened fingers.
Buccal smear. A stained preparation of cells scraped from the lining of the mouth.

Carbon-14. Radioactive isotope of carbon.
Carrier. A heterozygote; a person who has a certain gene but does not show the effects of it himself.
Castration. Surgical removal of ovaries or testes.
Cataract. An abnormality of the eye in which the lens becomes opaque.
Cattell's paradox. The apparent contradiction between two ideas: (a) that people of lower intelligence average to produce more children than do people of higher intelligence, and (b) that the average intelligence of the population does not seem to be declining with successive generations, as one might expect it to do as a consequence of (a).
Centriole. A tiny body outside the nucleus of a cell, concerned with formation of the spindle in mitosis and meiosis (see Fig. 3.1).
Centromere. A nonstaining body forming part of a chromosome. Chromosomes vary as to its location. It is important in causing separation of chromatids following metaphase in mitosis and meiosis (see Fig. 3.1).
Chiasma (*plural: **chiasmata***). Cross-like pattern formed when two chromatids that have undergone crossing over start to separate following synapsis (see Fig. 3.5).
Chimera. Individual to whose composition two different zygotes contributed, hence comprised of two different cell lines. *see* Mixoploid; Mosaic.
Chi-square test. A test of "goodness of fit"—of the likelihood that deviation between observed data and values expected on the basis of some hypothesis is due to chance and hence does not constitute significant evidence against the hypothesis (see Appendix C).
Chorion. The outer embryonic membrane surrounding the amnion and its contents (see Figs. 23.1, 23.2).
Christmas disease. Hemophilia B.
Chromatid. When a chromosome has duplicated itself, but the original and

newly formed elements are still united by a centromere, each element is called a chromatid (see Fig. 3.5).

Chromatin. A term for the material of which chromosomes are composed, rich in nucleoproteins and especially the nucleic acid DNA.

Chromatin negative. Having no Barr bodies in the cell nuclei.

Chromatin positive. Having Barr bodies in the cell nuclei.

Chromatography. In this context, a means of separating into its component parts a mixture of peptide fragments obtained by digestion of a protein with trysin. It depends on the fact that different peptides migrate along a strip of filter paper to different distances from the starting point (paper chromatography).

Chromomeres. Irregularly spaced, nodule-like enlargements on chromosomes (see Fig. 3.3).

Chromosomal aberration or mutation. Alteration in number or structure of chromosomes, visible with a microscope. *see:* Aneuploidy; Deletion; Duplication; Inversion; Isochromosome; Polyploidy; Translocation.

Chromosomes. Deeply staining bodies in the nuclei of cells, rich in nucleoproteins and especially the nucleic acid DNA.

Chronic radiation. A given amount of radiation received slowly over an extended period of time.

cis phase (of linkage). *see* Coupling linkage.

Cistron. The gene as a functional unit of heredity—having a significant effect on the phenotype.

Clone. In somatic cell genetics: a colony of cells all arising from a single somatic cell by repeated mitoses.

Codominant genes. Alleles each of which produces an independent phenotypic effect in heterozygotes.

Codon Triplet of messenger-RNA nucleotides (see Fig. 4.5, Table 4.2).

Coefficient of consanguinity or inbreeding. Probability that an individual is homozygous and that both genes of the pair were derived from one gene possessed by one ancestor. *see* Autozygous.

Coefficient of relationship. The probability that two individuals have the same gene derived from a common ancestor.

Colchicine. A drug used on living cells to cause the arrest of mitosis at the metaphase stage.

Color blindness. (1) **Complete.** Virtually total inability to distinguish colors; rare. (2) **Partial.** *see* Deuteranopia; Protanopia.

Complementary genes. Two (or more) dominant genes, not alleles of each other, that interact to produce a phenotype although neither gene produces a phenotypic effect by itself.

Concordance (of twins). Members of a pair of twins are alike with respect to a certain trait.

Consanguinity. Marriage between relatives.

Cooley's anemia. *see* Thalassemia.

Copying error. Production of a mutation by failure of a gene to duplicate itself precisely in preparation for mitosis or meiosis.

Correlation coefficient. A statistic expressing the degree of similarity between one individual and another.

Cosmic rays. Radiations originating in outer space.

Cotyledon. Embryonic leaf within a plant seed.

Coupling linkage. With reference to two pairs of genes: the dominant members of both pairs are located in the same chromosome, and the recessive members of both pairs are located in the same chromosome; e.g., (*AB*) (*ab*).

Cretinism. Mental deficiency arising from lack of adequate thyroid secretion during infancy and early childhood.
Cri du chat syndrome. Affected newborn babies have a peculiar thin, mewing cry. Abnormally small head (microcephalic), various malformations of face and body, and mental deficiency characterize the syndrome.
Crossing over. Exchange of material between one chromatid and another at the time of synapsis in meiosis (see Figs. 3.5, 15.1).
Cultural or social inheritance. The accumulated wisdom of mankind, transmitted by the written and spoken word, by art, music, etc.
Cystic fibrosis. An autosomal recessive defect in which the glands do not function normally: absence of pancreatic enzymes; sweat with elevated salt concentration; excess secretion of thick mucus in the lungs, obstructing air passages.
Cystinuria. An inborn error of metabolism in which large amounts of the amino acids cystine, lysine, arginine, and ornithine are excreted by the kidneys; kidney stones formed of cystine are characteristic.
Cytogenetics. Study of the contributions to heredity of structures and phenomena within the cells, especially of changes in chromosomes.
Cytoplasm. The portion of a cell lying outside the nucleus.
Cytosine. One of the organic bases (pyrimidines) found in DNA and RNA.

Daltonism. Partial color blindness. *see* Deuteranopia; Protanopia.
Darwinism. *see* Natural selection.
Deficiency. *see* Deletion; Mental deficiency.
Deletion. Chromosomal aberration in which a portion is lost from the chromosome (see Fig. 20.2).
Deuteranopia. A type of partial color blindness in which there is no marked reduction of sensitivity to any color but there is confusion in distinguishing red, yellow, and green (cf. Protanopia).
Diabetes mellitus. A disease in which the body lacks the normal ability to utilize sugar; the excretion of sugar in the urine is one symptom.
Differential fertility. Differences among individuals or groups in the number of offspring produced.
Diploid number (2*n*). The number of chromosomes in a somatic cell or in a germ cell before meiosis. The chromosomes occur in homologous pairs.
Discordance (of twins). Members of a pair of twins are not alike with respect to a certain trait.
Disjunction. Normal separation of chromosomes in mitosis and meiosis.
Distal hyperextensibility (of the thumb). Ability to bend the thumb sharply backward (dorsally) (see Fig. 10.1).
Dizygotic or **DZ.** Literally, two zygotes; refers to twins that develop from separate ova; "fraternal twins."
DNA (deoxyribonucleic acid). Nucleic acid characteristic of chromosomes; its sugar component is deoxyribose.
Dominant. (1) A gene that produces a phenotypic effect in a heterozygote; (2) A trait that is shown phenotypically by both homozygotes and heterozygotes.
DOPA. Dihydroxyphenylalanine, formed from tyrosine by enzyme action (see Fig. 5.3). **l-DOPA** rotates the plane of polarized light to the left.
Double crossing over. Simultaneous breakage and recombination of synapsing chromatids at two points (see Fig. 15.3).
Down's syndrome. Type of mental deficiency accompanied by a variable syndrome of physical traits, such as short stature, stubby fingers, a large, fis-

sured tongue, simian crease in the palm of the hand (see Fig. 20.6).

Drift. *see* Genetic drift.

Drosophila. The genus of fruit flies or "vinegar flies" widely used in genetical research.

"Drumstick." A lobe-like appendage to the nucleus in white blood cells of females (see Fig. 17.7B).

Duchenne's muscular dystrophy. *see* Muscular dystrophy.

Duodenum. The portion of the small intestine attached to the stomach.

Duplication. Chromosomal aberration in which a portion of a chromosome is repeated within the same chromosome (see Fig. 20.2).

Dyad. A chromosome composed of two chromatids united by a centromere; usually applied to chromosomes in the secondary spermatocyte or secondary oöcyte stage of meiosis (see Figs. 3.4, 3.7).

Dysgenic. Adjective applied to any force or trend that contributes to genetic deterioration of mankind; opposite of eugenic.

DZ. Dizygotic.

EEG (electroencephalogram). "Brain waves"; recordings of fluctuations in the electrical potentials generated by the brain.

Effective allele. A term applied to those polygenes that produce an increase in a quantitative trait.

Electrophoresis. In this context, separation of proteins or their constituents by differential rate of migration of the particles when placed in an electrical field.

Elliptocytosis. A hereditary condition in which the red blood cells are elliptical instead of circular.

Ellis-van Creveld syndrome. Characterized by dwarfism, striking shortening of hands and fingers, polydactyly, abnormal teeth and fingernails, and other malformations.

Enzyme. An organic catalyst; an organic compound, usually protein, that increases the rate at which chemical reactions occur in the body. Usually the reaction will not occur to an effective extent without the enzyme.

Equatorial plate. *see* Metaphase plate.

Equilibrium, genetic. The state manifested by a population that remains the same, genotypically and phenotypically, generation after generation.

Erythroblastosis fetalis. A pathological condition in which red blood cells are destroyed in the fetus or newborn with resultant anemia and other severe symptoms, including jaundice.

Erythroblasts. Nucleated cells from which red blood cells are formed.

Erythrocytes. Red blood cells.

Escherichia coli. A common bacterium found in the intestines; extensively used for genetical research.

Estrogen. A female sex hormone.

Eugenic. Adjective applied to any force or trend that contributes to genetic improvement of mankind.

Eugenics. "The study of agencies under social control that may improve or impair the racial [i.e., hereditary] qualities of future generations, either physically or mentally" (Francis Galton).

(1) **Negative eugenics.** Programs for decreasing the number of children produced by people with inferior genetic traits.

(2) **Positive eugenics.** Programs for increasing the production of children by people with superior genetic traits.

Eugenic sterilization. Surgical operation on the reproductive organs to prevent (1) ova from reaching the uterus in females, or (2) sperm cells from being included in the semen of males.

Eugenist. A person interested in the possibility of genetic improvement of mankind and prevention of genetic deterioration.

Expressivity, variable. A situation in which individuals with the same genotype show different phenotypes.

F_1. First filial generation; offspring of homozygous dominants mated to homozygous recessives and therefore necessarily heterozygous.

F_2. Second filial generation; offspring of F_1 individuals mated together, or, more broadly, offspring of any heterozygotes mated to each other, as $Aa \times Aa$.

Factor VIII deficiency. Hemophilia A.

Factor IX deficiency. Hemophilia B.

Fecund. Capable of producing offspring.

Fertilization. Union of sperm and ovum to form a zygote (fertilized ovum).

Fetus. Unborn young in the final six months before birth.

Fibrin. Elastic, fibrous material present in blood clots; formed from fibrinogen.

Fibrinogen. A blood plasma protein necessary for normal clotting of the blood; it is converted to fibrin.

Fixation. In this context, a condition within a population in which all the genes at a certain locus are the same (e.g., all *A*), the allele (e.g., *a*) having been lost from the gene pool.

Follicle (in the ovary). A "bubble-like" structure within the tissue of the ovary; it contains an ovum surrounded by fluid.

Founder principle. When a new colony is founded by emigration from a population, the emigrants may not possess a typical (random) sample of the genes present in the population. Thus by chance the colony may come to differ from the ancestral population.

Freemartin. A female calf, usually sterile, born twin to a male.

Galactosemia. An inborn error of metabolism characterized by inability to digest galactose, which with its harmful derivatives accumulates in the body. If not corrected by diet the condition leads to mental retardation, cataracts in the eyes, and malfunction of the liver.

Gamete. A mature germ cell, sperm or ovum.

Gene. The unit of inheritance; a portion of a DNA molecule within a chromosome. *see* Cistron.

Gene mutation. Alteration of that portion of a DNA molecule comprising a gene.

Gene pool. The totality of genes possessed by a population. The term may be used to refer to a single pair of alleles, or to two or more pairs.

"Genetic death." Removal of a gene from the gene pool of a population or from a given line of descent.

Genetic drift. Chance fluctuations in gene frequencies; more characteristic of small populations than of large ones.

Genetic homeostasis. Self-regulating capacity of populations, enabling them to adapt to varied or changing environments and conditions of life.

Genetic isolate. *see* Isolate.

"Genetic load." The total of the mutations in an individual's germ plasm, consisting of new mutations plus old ones inherited from ancestors.

Genetic polymorphism. Genetically based diversity in a population, e.g., blood groups, variant hemoglobins, and the like.

Genotype. The assemblage of genes found in the chromosomes of an individual or a selected portion thereof; one's "genetic formula."

Geographic race. A population occupying a geographic territory and having certain traits that, on the average, distinguish its members from members of other populations. Differences usually reflect differences in gene *frequencies* in the gene pools of the various geographic races.

Germ cell. A reproductive cell, sperm, or ovum, or one of the precursor stages in meiosis of sperm or ovum.

Germinal mutations. Genetic changes in germ cells.

Germ plasm. A general term for germ cells in all stages and the tissues from which they arise.

Glucose-6-phosphate dehydrogenase (G6PD). An enzyme; a deficiency of this enzyme causes hemolysis if antimalarial drugs such as primaquine are taken or if the fava bean is eaten.

Glutamic acid. One of the amino acids entering into the structure of proteins, e.g., of hemoglobin.

Gonadal dysgenesis. *see* Turner's syndrome.

G6PD. *see* Glucose-6-phosphate dehydrogenase.

Guanine. One of the organic bases (purines) found in DNA and RNA.

H antigen. An antigen found in the red blood cells of all group O people (and in varying amounts in the cells of people in other blood groups).

Haploid number (*n*). The number of chromosomes in a germ cell, gamete, following meiosis.

Hardy-Weinberg formula. In a population characterized by random mating and Mendelian inheritance, the ratio of individuals homozygous for the dominant gene to heterozygous individuals to individuals homozygous for the recessive gene is as $p^2:2pq:q^2$, where p is the relative frequency of the dominant allele and q is the frequency of the recessive allele.

Hellin's law. If $1/n$ represents the fraction of births giving rise to twins, $1/n^2$ represents the fraction of births giving rise to triplets, and $1/n^3$ the fraction of births giving rise to quadruplets.

Hematopoietic tissue. Embryonic tissue from which blood cells are formed.

Hemizygous. Adjective applied to an individual who has a single gene at a given locus instead of a pair of alleles. It is true of males for genes carried in the X chromosome and not homologous to genes in the Y chromosome.

Hemoglobin. The iron-containing compound in red blood cells, giving them their color. Important in the transportation of oxygen.

Hemolysis. Destruction of red blood cells.

Hemophilia. One of several diseases in which the blood does not clot normally.

Hemophilia A. The deficiency is lack of antihemophilic globulin (factor VIII) in the serum.

Hemophilia B. The deficiency is lack of plasma thromboplastin component (factor IX). Also called PTC deficiency or Christmas disease.

Hermaphrodite. An intersex.

1. **"True" hermaphrodite.** An individual having both male and female reproductive organs or, at least, both ovarian and testicular tissue.
2. **Pseudohermaphrodite.** An individual having only ovarian or testicular tissue but having some of the reproductive organs of the other sex.

Heterochromia iridis. The condition of having eyes differing in color, usually one brown eye and one blue eye.

Heterosis. A condition in which hybrids are superior to both parents in some respect: hybrid vigor. In this context it refers to a situation in which heterozygotes (*Aa,* for example) are superior in some way to both homozygotes (*AA* and *aa*).

Heterozygote (heterozygous). An individual in which the members of a pair of genes are unlike, e.g., *Aa.*

Hexosaminidase A. An enzyme concerned in Tay-Sachs disease.

HGPRT. Hypoxanthine-guanine phosphoribosyl transferase. *see* Lesch-Nyhan syndrome.

Holandric gene. *see* Y-linked genes.

Homeostasis. Self-regulating ability of the body, or of a population (*see* Genetic homeostasis).

Homologous chromosomes. Chromosomes that belong to the "same pair," as evidenced by the fact that they pair together in synapsis. They contain the same gene loci, except for the XY pair.

Homozygote (homozygous). An individual in which the members of a pair of genes are alike, e.g., *AA,* or *aa.*

Hormones. "Chemical messengers"—substances secreted into the bloodstream by the endocrine glands; they exert control over bodily development and metabolism.

Hunter syndrome. An X-linked defect characterized by markedly retarded growth and mental development, and by abnormalities of skeleton and biochemistry.

Huntington's chorea (disease). A nervous disorder characterized by spasmodic, involuntary movements of the face and limbs, with gradual loss of mental faculties and the development of dementia.

Hurler syndrome. A defect inherited as an autosomal recessive; characterized by markedly retarded growth and mental development, and by abnormalities of skeleton and biochemistry.

Hybrid. Individual of mixed ancestry. The term is variously used: (1) to apply to heterozygotes, (2) to apply to offspring produced when members of one race (or species) mate with members of another one.

Hybrid cell (somatic cell genetics). A cell formed by the fusion of two somatic cells of different type (e.g., derived from different species).

Hybrid vigor. *see* Heterosis.

Hypertrichosis. In this context, growth of hair on the surface of the pinna (auricle) of the ear and along the rim (see Fig. 14.5).

Ichthyosis congenita. A severe abnormality characterized by deeply fissured, leathery skin.

Idiogram. *see* Karyotype.

Inbreeding. *see* Consanguinity.

Independent assortment. In the formation of gametes—the distribution of the members of one pair of genes does not influence the distribution of the members of other pairs (see Fig. 3.8). True when the different pairs of genes are in different chromosomes.

Individual differential fertility. Differences among individuals in number of children produced; as a result of it the gene pool of one generation is not necessarily a random sample of the gene pool of the preceding generation.

Institutionalization. As employed in eugenics, prevention of reproduction by keeping males and females separated in institutions.

Interphase. The state of a cell not undergoing mitosis (see Fig. 3.1).

Inversion. Chromosomal aberration in which one portion of a chromosome becomes reversed in relation to the original sequence of contained genes (see Fig. 20.3).

Paracentric inversion. The reversed portion is confined to one arm of the chromosome, i.e., it does not include the centromere.

Pericentric inversion. The reversed portion includes the centromere.

IQ (intelligence quotient). Mental age divided by chronological age, multiplied by 100. (This is the traditional computation; another method of computation is being used increasingly.)

Isoagglutinogen. An antigen (agglutinogen) that occurs normally in an individual, i.e., without artificial stimulation.

Isoantibodies. Antibodies that people normally possess.

Isochromosome. A chromosome in which the centromere is in the middle and the two arms are alike in genic content, typically being the mirror images of each other (see Fig. 20.4).

Isolate, genetic. The group of individuals among whom (random) mating may be considered to occur. Large populations are more or less completely subdivided into isolates by such factors as distance, geographical barriers, race, religion, social status, etc.

Isozymes. Chemically different forms of an enzyme. The different forms have the same function but differ structurally in some way so that they differ in properties (e.g., in rate of migration in an electrical field). Isozymes derived from different species are called allozymes.

Karyotype. The chromosomes contained in a cell. The term is especially applied to diagrams made by lining up the chromosomes in homologous pairs (see Fig. 17.5).

Kernicterus. Destruction of nerve centers by an elevated level of bilirubin in newborn infants.

Kindred. Group of relatives in one or more generations.

Klinefelter's syndrome. Abnormality caused by possession of two X chromosomes and one Y chromosome (XXY). Characteristics are those of a male, but the testes are not functional; the individual may have feminine breast development.

Lesch-Nyhan syndrome. An X-linked trait characterized by elevated level of uric acid, mental retardation, spastic cerebral palsy, and compulsive self-mutilation. It results from inactivity of the enzyme hypoxanthine-guanine phosphoribosyl transferase (HGPRT).

Lethal gene. A gene that kills its possessor, usually before birth. The ones that can be studied are recessive in the sense that they are carried by heterozygotes and kill only homozygotes.

Leucocytes. White blood cells.

Leukemia. Malignancy of blood-forming tissues leading to production of excessive numbers of white blood cells.

Linkage. Location of genes in the same chromosome so that they do not assort independently in meiosis.

Locus (*plural:* **loci**). The location in a chromosome occupied by a gene or, alternatively, by one of its alleles.

Lutheran gene. Gene determining the presence of the Lutheran antigen in the red blood cells of some people.

Lyon hypothesis. The hypothesis that in any given cell in a female one X chromosome is active, the other inactive. The inactive X chromosome is believed to form the Barr body.

Lysine. One of the amino acids entering into the structure of proteins, e.g., of hemoglobin.

Macromolecule. A structured aggregation of smaller molecules functioning as a unit, e.g., DNA.

Marital fertility. Number of children relative to the number of wives in a population.

Marker gene. A gene utilized in investigations to indicate the presence of a certain chromosome with its genetic contents.

Maximum-likelihood methods of detecting autosomal linkage. Methods that depend on the probability that an observed relationship of traits in the parents to traits in the children would arise if there were linkage, compared to the probability that the relationship would arise if the traits were not linked.

Meiosis. The process by which diploid precursor cells give rise to haploid gametes (ova or sperm) (see Figs. 3.4, 3.7).

Melanin. Brown pigment, found, for example, in cells of skin, hair, and iris of the eye.

Melanocytes. Cells containing pigment (melanin).

Mendelian laws. First law: Genes occur in pairs in the cells of individuals; when gametes are produced the members of a pair separate so that each gamete receives only one member of the pair (law of segregation). Second law: The manner in which the members of one pair of genes are distributed to gametes does not influence the manner in which other pairs of genes are distributed (law of independent assortment).

Mental deficiency. Intelligence below the normal level. Frequently the line is drawn at an IQ of 70, persons below that score being considered mentally deficient or retarded.

Mental retardation. *see* Mental deficiency.

Messenger RNA. RNA that is synthesized in the nucleus on a DNA template (pattern), then passes into the cytoplasm, attaching to one or more ribosomes and serving as a template on which amino acids are assembled to form a polypeptide chain (see Fig. 4.5).

Metabolism. The total of the chemical processes carried on in the body, including digestion, respiration, secretion, excretion, release and utilization of energy, and so on.

Metacentric. Referring to chromosomes that have the centromere in the middle or close to it.

Metaphase. The stage in mitosis or meiosis in which the chromosomes are lined up at the equator of the spindle (see Fig. 3.1).

Metaphase plate. Arrangement of chromosomes in the metaphase of mitosis or meiosis (see Fig. 3.1).

Microcephaly. Type of mental deficiency in which the brain is abnormally small.

Microcythemia. A defect characterized by abnormally small red blood cells.

Microgeographic race. A population occupying a small unit of territory and characterized by some, usually small, differences from other such populations.

Micron (μ). 1/1000 of a millimeter.

Mitosis. The process of nuclear division in which the chromosomes are duplicated and distributed equally to the daughter cells, so that each of the latter has exactly the same chromosomal content as the other.

Mixoploid. A mosaic characterized by two or more cell lines that differ in chromosome number or structure. *see* Mosaic.

Modifying gene. A gene that alters the effect of another gene not its own allele.

Mongolism. *see* Down's syndrome.

Monosomic. Having one less chromosome than the normal number.

Monozygotic, MZ. Literally, one zygote; referring to twins, triplets, and so on, that develop from a single ovum; "identical" twins or triplets.

Moron. Person with an IQ between 70 and 50; "educable."

Mosaic. Individual developed from a single zygote but having two or more kinds of cells with respect to their genetic constitution, the antigens they contain, the enzymes they produce, and the like. *see* Mixoploid; Chimera.

mrem. Millirem; unit employed in measuring the amount of radiation (see footnote on p. 331).

Multiple alleles. A series of genes, any one of which can occupy a given locus in a chromosome.

Multiple genes. *see* Polygenes.

Multiple neurofibromatosis. A hereditary trait characterized by the growth of benign tumors at nerve endings in the skin.

Muscular dystrophy (Duchenne type). An X-linked pathological condition in which the muscles degenerate progressively; the boy becomes more and more crippled and dies before reaching maturity.

Mutagenic agents. Chemicals or radiations that induce the formation of mutations.

Mutation. Alteration of the genetic material; it may involve microscopically visible changes in the chromosomes (chromosomal aberrations), or changes in the structure of the DNA molecule (gene or point mutations).

Mutation pressure. The tendency of genes to mutate at a rather constant, though low, rate.

Myotonic dystrophy. A defect characterized by progressive weakness and atrophy of muscles, cataract, and abnormalities of heart function and reproductive organs. Apparently inherited as an autosomal dominant.

MZ. Monozygotic.

Nail-patella syndrome. A hereditary anomaly in which the fingernails are abnormal, the patella (kneecap) is small or missing, and other skeletal abnormalities occur.

Natural selection. Literally, selection by Nature; determining which individuals or groups shall live and reproduce and which shall not. It results from the cumulative action of all forces that tend to cause individuals of one genetic constitution to leave larger numbers of offspring than do individuals of another genetic constitution. *see* Differential fertility.

Negative eugenics. *see* Eugenics.

Negative selection. Prevention of reproduction of persons having a certain genotype and/or phenotype.

Neurohormones. Hormones concerned with the interrelationships of the nervous system and the endocrine glands in regulating bodily functions.

Nonadaptive traits. Characteristics that are neutral: neither beneficial nor detrimental.

Nondisjunction. Failure of chromosomes to separate normally. Homologous chromosomes may fail to separate from each other following synapsis in meiosis (see Figs. 17.1, 17.2). Chromatids (daughter chromosomes) may fail to separate from each other in the anaphase of mitosis.

Nonsecretor. A person who has antigens A or B in red blood cells but not in such secretions as saliva.

Normal frequency curve. The bell-shaped curve usually approximated when large populations are measured with respect to some quantitative trait that varies continuously, e.g., stature, weight, IQ scores (see Fig. 8.2).

Norm of reaction. The limits within which an individual can adapt to living under different environmental conditions.

Nucleic acids. Organic compounds composed of a phosphoric acid radical, sugar, two purines, and two pyrimidines. *see* DNA; RNA.

Nucleolus. A conspicuous, deeply staining body often found within the nucleus of a cell.

Nucleoproteins. Organic compounds composed of proteins and nucleic acids.

Nucleotide. Organic compound composed of a molecule of sugar, a phosphoric acid radical, and an organic base (purine or pyrimidine) (see Fig. 4.2).

Nucleus. A relatively large body found inside a cell in the interphase stage; it contains the chromosomes (see Fig. 3.1).

Oöcytes, primary and secondary. Stages in gamete formation (meiosis) in the female (see Fig. 3.7).

Oögenesis. Production of gametes (meiosis) in the female (see Fig. 3.7).

Oögonium. Primordial germ cell in a female (see Fig. 3.7).

Overdominance. *see* Heterosis.

Ovum. The mature female germ cell or gamete; the egg.

p. Relative frequency of a dominant gene. *see* Hardy-Weinberg formula.

Panmictic population. One in which matings occur according to the laws of probability, insofar as the genetic trait being studied is concerned.

Panmixis. *See* Random mating.

Paracentric inversion. *see* Inversion.

Parental selection. *see* Preadoption.

Parthenogenesis. Development of an individual from an unfertilized ovum.

Partial penetrance. *see* Penetrance.

Pattern baldness. The common type of baldness in which the sides and the lower portions of the back of the head retain hair.

Pedigree. A record of inheritance, frequently presented as a diagram, for two or more generations of a kindred.

Penetrance, partial. When a dominant gene does not produce the expected phenotype in every possessor of the gene, or recessive genes do not produce the expected phenotype in every individual homozygous for them.

Pentosuria. An inborn error of metabolism in which increased amounts of a derivative of pentose (sugar having five carbon atoms) are excreted in the urine.

Peptide. A compound containing two or more amino acids. Peptides join to form polypeptides, which in turn join to form proteins.

Pericentric inversion. *see* Inversion.

Phenocopy. A characteristic or trait that is not hereditary nor the product of genetic change, but closely resembles a phenotype that in other individuals is produced as a result of genetic change (mutation).

Phenotype. The appearance or observable nature of an individual.

Phenylalanine. One of the amino acids; concerned in phenylketonuria.

Phenylalanine hydroxylase. Deficiency of this enzyme causes phenylketonuria (PKU).

Phenylketonuria, PKU. An inborn error of metabolism characterized by inability to oxidize phenylalanine to tyrosine; it is accompanied by severe mental retardation unless remedial measures are taken.

Pituitary gland. A small gland of internal secretion attached to the brain.

Placenta. Structure formed in the uterus by a combination of tissue of the uterine wall with tissue from the embryonic membranes. Within it the blood vessels of the embryo come into close contact with the blood vessels of the mother (see Figs. 23.2, 28.3).

Plasma. The fluid portion of the blood.

Plasma thromboplastin component, PTC. A portion of the blood-clotting mechanism that is deficient in people who have hemophilia B.

Pleiotropic gene. Any gene that has more than one phenotypic effect.

Point mutation. *see* Gene mutation.

Polar bodies, first and second. In meiosis in the female—tiny cells almost devoid of cytoplasm, homologous to secondary oöcyte and ovum, respectively (see Fig. 3.7).

Pollen. Spore containing the male gamete in flowering plants.

Polydactyly. The condition of having more than the normal number of digits. Usually characterized by having six fingers and/or toes.

Polygenes. Genes that are independent in mode of inheritance but combine to produce a trait. In many cases the genes are cumulative in their effect and the trait is a quantitative one, e.g., stature, weight, degree of pigmentation. More complex interaction between the genes may occur, however.

Polymorphism. Literally, having many forms; frequently applied to differences among individuals within one race.

Polyovular follicle. An ovarian follicle containing more than one ovum.

Polypeptide chain. A group of peptides linked together. *see* Peptide.

Polyploidy. Chromosomal aberration in which the number of chromosomes is increased by the full haploid number, or multiples thereof.

Polysomy. *see* Aneuploidy.

Polytypic. Literally, of many types; a term sometimes applied to the differences between races.

Population genetics. Genetical principles applied to groups of individuals. *see* Hardy-Weinberg formula.

Porphyria. In this context, an inborn error of metabolism in which abnormal neurological symptoms appear if barbiturates are used.

Position effect. The fact that the functioning of a gene may be influenced by its position relative to other genes in the chromosome.

Positive eugenics. *see* Eugenics.

Potassium-40. Radioactive isotope of potassium.

Pre-adoption. The idea that in the future parents may elect to have germ cells other than their own, derived from individuals having known superior traits, used in production of their children.

Prenatal adoption. *see* Pre-adoption.

Primordial germ cells. Embryonic cells that, through meiosis, will give rise to gametes; spermatogonium and oögonium (see Figs. 3.4, 3.7).

Proband. *see* Propositus.

Prophase. Early stage of mitosis (see Fig. 3.1).

Propositus. Index case: the individual possessing a certain trait who attracted the attention of the investigator to the family.

Protanopia. A type of partial color blindness in which the eye is relatively insensitive to red, and has difficulty distinguishing red from dim yellow or green.

Proteins. Nitrogenous compounds of high molecular weight forming the most important structural and enzymatic constituents of the body. Composed of amino acids.

Pseudohypertrophic muscular dystrophy (Duchenne type). *see* Muscular dystrophy.

PTC. (1) Phenylthiocarbamide. A chemical that has a taste (usually bitter) to some people but is tasted by other people only when the solution is very concentrated. People who are relatively insensitive to it are called nontasters. **(2) Plasma thromboplastin component.** *see* Hemophilia B.

Purine. One class of organic bases found in nucleic acids.

Pyloric stenosis. Hypertrophy (enlargement) of the sphincter muscle of the pylorus, resulting in obstruction of the passage of food from stomach to intestine.

Pyrimidine. One class of organic bases found in nucleic acids.

q. Relative frequency of a recessive gene. *see* Hardy-Weinberg formula.

Races. Populations that differ in the frequencies of certain of the genes in their gene pools.

rad. Unit employed in measuring the amount of radiation (see footnote on p. 331).

Random mating. Mating determined by chance (i.e., by the laws of probability) insofar as the genetic trait being studied is concerned, as opposed to assortative mating.

Random sample. In this context, a sample of a population so selected that it has the same gene frequencies as does the entire population.

Recessive. (1) A gene that does not produce a phenotypic effect in a heterozygote. (2) A trait that is shown phenotypically only by homozygotes.

Recombination. When linked genes do not stay together in inheritance but are reassorted. It is observable when crossing over occurs in heterozygotes.

Red blood cell mosaic. *see* Mosaic.

rem. Unit employed in measuring amount of radiation (see footnote on p. 331).

Repeat. *see* Duplication.

Repeat frequency. In this context, the frequency with which mothers who have had one pair of twins subsequently have additional pairs.

Replacement level of a population. The number of births necessary to ensure that the number of adults of reproductive age shall remain constant from generation to generation.

Reproductive isolation. A situation in which the gene pools of two or more populations are kept separate because the members of the different populations do not interbreed even though they come into contact with each other.

Repulsion linkage. With reference to two pairs of genes, the dominant gene of one pair is located in the same chromosome as the recessive gene of the other pair, and vice versa; e.g., (Ab) (aB).

Retinoblastoma. Cancer of the retina of the eye.

Rh, "rhesus factor." A group of red blood cell antigens discovered by the finding that some human cells react with antibodies formed against rhesus monkey cells.

Rh incompatibility. A marriage in which the wife is Rh-negative and the husband Rh-positive.

Rh-negative. People whose red blood cells do not react with anti-**Rh**$_0$ (anti-D) antibodies (see Table 13.1).

Rh-positive. People whose red blood cells react with anti-**Rh**$_0$ (anti-D) antibodies (see Table 13.1).

Ribosomes. Small particles in the cytoplasm of a cell forming the sites of protein synthesis. The messenger RNA attaches to them.

RNA (ribonucleic acid). Nucleic acid found in both the nucleus and cytoplasm of cells; its sugar component is ribose.

roentgen, r. Unit commonly employed in measuring the amount of radiation, based on the amount of ionization produced (see footnote on p. 331).

Satellites. Tiny masses of chromatin found attached to the short arms of some chromosomes (see Fig. 18.2, No.'s 13, 14, 15, 21, 22).

Schizophrenia. A form of mental disorder or psychosis in which there is a splitting of the personality and a withdrawal from normal human relationships.

Secondary nondisjunction. Production of gametes with abnormal chromosome numbers as a result of nondisjunction in individuals who already have an abnormal chromosomal constitution, e.g., production of XX gametes by an XXX female.

Secondary sex characteristics. Those bodily traits, other than the reproductive organs themselves, by which members of one sex differ from members of the other.

Secretor. A person possessing a water-soluble form of antigens A and/or B; these antigens are found in such secretions as saliva.

Seminiferous tubule dysgenesis. *see* Klinefelter's syndrome.

Sendai virus. A virus used in somatic cell genetics to increase the frequency of fusion of somatic cells.

Serum. The fluid portion of the blood remaining after blood has clotted (with removal of the cells and of fibrin).

Sewall Wright effect. *see* Genetic drift.

Sex chromatin bodies. *see* Barr bodies.

Sex chromosome mosaic. *see* Mosaic.

Sex chromosomes. The X and Y chromosomes.

Sex-controlled genes. *see* Sex-influenced genes.

Sex hormones. Substances secreted into the blood by, especially, the ovary and testis; they control many phases of bodily development and functioning, including the development of the secondary sex characteristics.

Sex-influenced genes. Autosomal genes that in one sex produce a phenotypic effect in both homozygotes and heterozygotes, but in the other sex produce a phenotypic effect in homozygotes only.

Sex limitation. A term applied to traits that appear phenotypically in only one sex, although the genes for them may be carried by both sexes.

Sex linkage. Having the gene carried in the sex chromosomes, usually the X chromosome.

Siamese twins. Conjoined twins, connected by some portion of their bodies.

Siblings (sibs). Brothers and sisters.

Sib-pair method of detecting autosomal linkage. The method involves comparing sibs taken two at a time as to agreement or lack of agreement in possession of two traits under investigation.

Sibship. A family of brothers and/or sisters.

Sickle-cell anemia. A severe disease in which the red blood cells become sickle- or crescent-shaped at low oxygen tensions and are destroyed in large numbers (see Fig. 5.1). Genotype: $Hb^S\ Hb^S$.

Sickle trait. Characteristics of people whose red blood cells assume bizarre shapes when deprived of oxygen, but in whom symptoms of anemia do not develop (see Fig. 5.1). Genotype: $Hb^S\ Hb^A$.

Simian crease. A crease in the hand extending completely across the palm (see Fig. 9.5).

Similarity method of demonstrating monozygosity. A method whereby the similarities of twins are checked with respect to a large number of genetic traits.

Simple sib method of correcting pooled data. A means of correcting for the fact that when heterozygotes cannot be identified directly, only parents who have at least one child showing a presumed recessive trait can be included in an investigtaion. The sibs of the propositi, but not the propositi themselves, are counted in computing the normal-to-affected ratio.

Socioeconomic groups. Classifications of people on such bases as income, means of livelihood, education, and the like.

Somatic cell. A body cell, as opposed to a germ cell.

Somatic mutations. Genetic changes in somatic cells.

Sperm, Spermatozoan. The mature male germ cell or gamete.

Spermatid. A stage in the production of gametes, meiosis, in the male (see Fig. 3.4).

Spermatocytes, primary and secondary. Stages in gamete formation, meiosis, in the male (see Fig. 3.4).

Spermatogenesis. Production of gametes, meiosis, in the male (see Fig. 3.4).

Spermatogonium. Primordial germ cell in a male (see Fig. 3.4).

Spina bifida. Defect in the spinal column consisting of absence of vertebral arches. Spinal membranes and cord may protrude. In the simplest form only membranes protrude.

Sterilization. *see* Eugenic sterilization.

Stratification. In this context, a concurrence of traits because of uniformity of the gene pool of a population, rather than because of genetic linkage.

Sublethal gene. A gene that kills its possessor in infancy or childhood. *see* Lethal gene.

Submetacentric. Referring to chromosomes having the centromere somewhat nearer one end than the other but with both arms quite long.

Subspecies. *see* Geographic race.

Survival of the fittest. The idea that under conditions of competition for existence some individuals are superior to others and hence survive and reproduce, whereas the inferior individuals fail to do so, at least to as great an extent. *see* Natural selection.

Synapsis. Pairing of chromosomes in primary spermatocytes and primary oöcytes (see Figs. 3.4, 3.7).

Syndactyly. Characterized by varying amounts of fusion to each other of fingers, and sometimes toes.

Syndrome. A term for the group of symptoms characterizing a certain disease or abnormality.

Tay-Sachs disease. Birth defect in which the child seems nearly normal at birth but develops loss of muscular function, paralysis, deafness, blindness, and other symptoms, and dies by about the third year.

Telophase. Final stage in mitosis (see Fig. 3.1).

Test cross. *see* Back cross.

Testicular feminization. An intersexual condition in which individuals with the XY chromosomal constitution have the secondary sex characteristics of females.

Tetrad. Cluster of four chromatids formed when two chromosomes pair in synapsis in the metaphase stage of primary spermatocytes and primary oöcytes (see Figs. 3.4, 3.5, 3.7).

Tetraploid. Having twice the usual number of chromosomes, i.e., four haploid sets of chromosomes.

Tetrasomic. Having two more chromosomes than the normal number.

Thalassemia. An inherited abnormality of the blood. Homozygotes for the gene concerned have severe anemia, called *thalassemia major.* Heterozygotes have a much milder condition called *thalassemia minor.*

Thermal agitation. Heat-induced movement of molecules.

Thymidine kinase (TK). An enzyme involved in the synthesis of nucleotides from which DNA is built.

Thymine. One of the organic bases (pyrimidines) found in DNA.

Trait association. Occurrence together in an individual of two or more traits, for reasons other than genetic linkage.

Transfer RNA (soluble RNA). RNA molecules attached to amino acids, one type of molecule for each of the 20 amino acids. Each molecule "recognizes" its corresponding code word in a messenger-RNA molecule, and brings its amino acid into alignment so that it may enter into formation of a polypeptide chain (see Fig. 4.5).

Transfusion syndrome. Passage of blood from one twin fetus to the other, so that one is anemic while the other has too much blood. Occurs if the twins have a common placenta, and an artery from one twin connects to a vein from the other.

Translocation. Chromosomal aberration in which a portion of one chromosome becomes attached to a nonhomologous chromosome (of a different pair) (see Fig. 20.2).

Reciprocal translocation. Mutual exchange of material between two nonhomologous chromosomes (see Fig. 20.2).

trans phase of linkage. *see* Repulsion linkage.

Triploid. Having three haploid sets of chromosomes, three of each kind of chromosome.

Trisomy. Having one more chromosome than the normal number; one kind of chromosome is present in triplicate.

Trisomy 21. *see* Down's syndrome.

Tritiated thymidine. Thymine compound rendered radioactive by incorporating radioactive hydrogen, ^3H (tritium).

Truncate distribution. An atypical sample of a population, certain types of individuals or matings being omitted.

Trypsin. A digestive enzyme that acts on proteins.

Turner's syndrome. Abnormal condition caused by presence of one X chromosome only, and no Y chromosome (XO). Characteristics are those of a juvenile female, but the ovaries are absent or vestigial.

Tyrosinase. An enzyme that acts on tyrosine; important in the synthesis of melanin.

Tyrosine. An amino acid; precursor of melanin.

Umbilical cord. An embryonic structure containing blood vessels and extending from the embryo to the placenta (see Figure 23.1).

Uracil. One of the organic bases (pyrimidines) found in RNA.

Valine. One of the amino acids entering into the sturcture of proteins, e.g., of hemoglobin.

Variable expressivity. *see* Expressivity.

Variance, mean intrapair. A statistic indicating the average amount of difference between members of pairs of twins.

Vasectomy. Closure by surgical means of the duct leading from the testis.

von Dungern-Hirszfeld hypothesis. The hypothesis that red blood cell antigens A and B are inherited on the basis of two pairs of genes which undergo independent assortment.

X chromosome. The sexes differ in possession of this chromosome: females have two X chromosomes, males only one.

Xeroderma pigmentosum. A skin disease characterized by development of numerous pigmented spots, which frequently become cancerous.

Xg blood groups. Blood groups based on the presence or absence of the antigen named Xg^a. Presence of the antigen is determined by a dominant, X-linked gene.

X-linked genes. Genes located in the X chromosome.

XO. *see* Turner's syndrome.

XXY. *see* Klinefelter's syndrome.

Y chromosome. A sex chromosome normally possessed by males only.

Y-linked genes. Genes located in the Y chromosome.

Zygote. A fertilized ovum, formed by the union of two gametes (ovum and sperm).

Index